C0-ASI-568

Particle Physics

Cargèse 1989

NATO ASI Series

Advanced Science Institutes Series

A series presenting the results of activities sponsored by the NATO Science Committee, which aims at the dissemination of advanced scientific and technological knowledge, with a view to strengthening links between scientific communities.

The series is published by an international board of publishers in conjunction with the NATO Scientific Affairs Division

A	**Life Sciences**	Plenum Publishing Corporation
B	**Physics**	New York and London
C	**Mathematical and Physical Sciences**	Kluwer Academic Publishers Dordrecht, Boston, and London
D	**Behavioral and Social Sciences**	
E	**Applied Sciences**	
F	**Computer and Systems Sciences**	Springer-Verlag
G	**Ecological Sciences**	Berlin, Heidelberg, New York, London,
H	**Cell Biology**	Paris, and Tokyo

Recent Volumes in this Series

Volume 222—Relaxation in Complex Systems and Related Topics
edited by Ian A. Campbell and Carlo Giovannella

Volume 223—Particle Physics: *Cargèse 1989*
edited by Maurice Lévy, Jean-Louis Basdevant, Maurice Jacob, David Speiser, Jacques Weyers, and Raymond Gastmans

Volume 224—Probabilistic Methods in Quantum Field Theory and Quantum Gravity
edited by P. H. Damgaard, H. Hüffel, and A. Rosenblum

Volume 225—Nonlinear Evolution of Spatio-Temporal Structures in Dissipative Continuous Systems
edited by F. H. Busse and L. Kramer

Volume 226—Sixty-Two Years of Uncertainty: Historical, Philosophical, and Physical Inquiries into the Foundations of Quantum Mechanics
edited by Arthur I. Miller

Volume 227—Dynamics of Polyatomic Van der Waals Complexes
edited by Nadine Halberstadt and Kenneth C. Janda

Volume 228—Hadrons and Hadronic Matter
edited by Dominique Vautherin, F. Lenz, and J. W. Negele

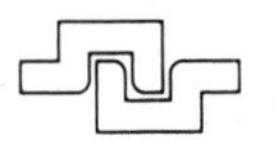

Series B: Physics

Particle Physics

Cargèse 1989

Edited by

Maurice Lévy and Jean-Louis Basdevant

Laboratory of Theoretical Physics and High Energies
Université Pierre et Marie Curie
Paris, France

Maurice Jacob

Theory Division
CERN
Geneva, Switzerland

David Speiser and Jacques Weyers

Institute of Theoretical Physics
Université Catholique de Louvain
Louvain-la-Neuve, Belgium

and

Raymond Gastmans

Institute of Theoretical Physics
Katholieke Universiteit Leuven
Leuven, Belgium

Plenum Press
New York and London
Published in cooperation with NATO Scientific Affairs Division

Proceedings of a NATO Advanced Study Institute on
Particle Physics,
held July 18–August 4, 1989,
in Cargèse, France

ISBN 0-306-43601-9

A Division of Plenum Publishing Corporation
233 Spring Street, New York, N.Y. 10013

Printed in the United States of America

PREVIOUS CARGÈSE SYMPOSIA PUBLISHED IN THE NATO ASI SERIES B: PHYSICS

Volume 173 PARTICLE PHYSICS: *Cargèse 1987*
edited by Maurice Lévy, Jean-Louis Basdevant, Maurice Jacob, David Speiser, Jacques Weyers, and Raymond Gastmans

Volume 156 GRAVITATION IN ASTROPHYSICS: *Cargèse 1986*
edited by B. Carter and J. B. Hartle

Volume 150 PARTICLE PHYSICS: *Cargèse 1985*
edited by Maurice Lévy, Jean-Louis Basdevant, Maurice Jacob, David Speiser, Jacques Weyers, and Raymond Gastmans

Volume 130 HEAVY ION COLLISIONS: *Cargèse 1984*
edited by P. Bonche, Maurice Lévy, Phillippe Quentin, and Dominique Vautherin

Volume 126 PERSPECTIVES IN PARTICLES AND FIELDS: *Cargèse 1983*
edited by Maurice Lévy, Jean-Louis Basdevant, David Speiser, Jacques Weyers, Maurice Jacob, and Raymond Gastmans

Volume 85 FUNDAMENTAL INTERACTIONS: *Cargèse 1981*
edited by Maurice Lévy, Jean-Louis Basdevant, David Speiser, Jacques Weyers, Maurice Jacob, and Raymond Gastmans

Volume 72 PHASE TRANSITIONS: *Cargèse 1980*
edited by Maurice Lévy, Jean-Claude Le Guillou, and Jean Zinn-Justin

Volume 61 QUARKS AND LEPTONS: *Cargèse 1979*
edited by Maurice Lévy, Jean-Louis Basdevant, David Speiser, Jacques Weyers, Raymond Gastmans, and Maurice Jacob

Volume 44 RECENT DEVELOPMENTS IN GRAVITATIONS: *Cargèse 1978*
edited by Maurice Lévy and S. Deser

Volume 39 HADRON STRUCTURE AND LEPTON–HADRON INTERACTIONS: *Cargèse 1977*
edited by Maurice Lévy, Jean-Louis Basdevant, David Speiser, Jacques Weyers, Raymond Gastmans, and Jean Zinn-Justin

Volume 26 NEW DEVELOPMENTS IN QUANTUM FIELD THEORY AND STATISTICAL MECHANICS: *Cargèse 1976*
edited by Maurice Lévy and Pronob Mitter

Volume 13 WEAK AND ELECTROMAGNETIC INTERACTIONS AT HIGH ENERGIES: *Cargèse 1975* (Parts A and B)
edited by Maurice Lévy, Jean-Louis Basdevant, David Speiser, and Raymond Gastmans

PREFACE

The 1989 Cargèse Summer Institute on Particle Physics was organized by the Université Pierre et Marie Curie, Paris (M. Lévy and J.-L. Basdevant), CERN (M. Jacob), the Université Catholique de Louvain (D. Speiser and J. Weyers) and the Katholieke Universiteit te Leuven (R. Gastmans), which, since 1975, have joined their efforts and worked in common. It was the twenty-sixth Summer Institute held at Cargèse and the tenth organized by the two Institutes of Theoretical Physics at Leuven and Louvain-la-Neuve.

The 1989 school centered on the following topics

- new experimental results
- strings, superstrings and conformal field theory
- lattice approximations.

Of the many new experimental results, we would like to mention especially those from SLAC presented by Professor G. Feldman. On the other hand, we had the tantalizing knowledge that LEP would begin to operate only right after the end of the school!

For this we received ample replacement : Professor J. Steinberger summed up all major CP violation experiments done to date and commented upon them. The reader will find also various other most interesting contributions, for instance on high energy ion beams. Once more theoreticians and experimentalists (this time more than usual) came together to discuss high energy particle physics.

We owe many thanks to all those who have made this Summer Institute possible!

Special thanks are due to the Scientific Committee of NATO and its President for a generous grant. We are also very grateful for the financial contributions given by the French Ministry of Foreign Affairs and the Institut National de Physique Nucléaire et de Physique des Particules.

We also want to thank Ms. M.-F. Hanseler for her efficient organizational help, Mr. and Ms. Ariano for their kind assistance in all material matters of the school, and, last but not least, the people from Cargèse for their hospitality.

Most of all, however, we would like to thank all the lecturers and participants : their commitment to the school was the real basis for its success.

M. Lévy
J.-L. Basdevant
M. Jacob
R. Gastmans
D. Speiser
J. Weyers

CONTENTS

Physics at LEP 1
J.-J. Aubert

Physics at SLC 25
G. Feldman

Physics in $p\bar{p}$ collisions 47
F. Pauss

Physics with High Energy Ion Beams 79
P. Sonderegger

Electroweak Interactions and LEP/SLC Experiments 97
G. Altarelli

The Geometrical Principles of String Theory Constructions 141
J. Govaerts

Two Topics in Quantum Chromodynamics 217
J.D. Bjorken

Hadronic Matrix Elements and Weak Decays in Lattice QCD 239
G. Martinelli

Experimental Status of CP Violation 297
J. Steinberger

Lee-Wick Model and Soliton Stars 335
R. Vinh Mau

Gluon Confinement in Chromoelectric Vacuum 343
R. Basu

Mean Field Theory and Beyond for Gauge-Higgs-Fermion Theories 353
S.W. de Souza

Index 357

PHYSICS AT LEP

Jean-Jacques Aubert

Centre de Physique des Particules de Marseille,
I N2 P3-CNRS/Université d'Aix-Marseille II, 70 route
Léon-Lachamp, Case 907, F-13288 Marseille Cedex 9, France

I - INTRODUCTION

By the time of the school the commissioning of LEP was just starting and obviously there were no physics results, not even a single Z° event. The first events have been observed since then in August.

So "physics at LEP" was still "Monte Carlo physics". As Altarelli covered also part of the topics, I limit myself to specific points of physics as the Z width measurements, the higgs detection, QCD and B physics.

II - DETECTORS AT LEP

There are four detectors : Aleph, Delphi, L3 and Opal installed respectively on pit 4,8,2,6. They are in place, more or less ready to take data.

The properties of the detectors have been described extensively in many places[1], Aleph has a good space resolution for the electromagnetic calorimeter, Delphi has Rich counters for particle identification, L3 has good energy resolution on leptons and Opal is a rather more conventional detector. Anyhow, being so close from the point where physics results will come out, the readiness and the skill of physicist will probably make more difference than detector design.

III - THE WIDTH OF THE Z

- In the standard model

$$\Gamma_Z = N_\nu \Gamma_\nu + \Gamma_e + \Gamma_\mu + \Gamma_\tau + \Gamma_{had}$$

Particle Physics: Cargèse 1989
Edited by M. Lévy *et al.*
Plenum Press, New York, 1990

$\Gamma_Z = 2570$ MeV with $\Gamma_u = 102 \times 3$ MeV (quark with charge 2/3)

$\Gamma_d = 131 \times 3$ MeV (quark with charge 1/3)

$\Gamma_e = \Gamma_\mu = \Gamma_\tau = 86$ MeV

$\Gamma_\nu = 170$ MeV

So a new quark, a new charged lepton or a new neutrino will increase Γ_Z in a known way (provided than $m_{new} < m_Z/2$)

- In supersymmetric models, the Z° can decay in a pair of charged higgsinos or winos

$$e^+e^- \to Z^0 \to \tilde{W}^+\tilde{W}^-$$

$$Z^0 \to \tilde{H}^+\tilde{H}^-$$

The contribution to an increase of the Z width can be large 800 MeV for winos, around 100 MeV for higgsinos or smaller depending on the precise model.

- In composite models as shown by F.M.Renard and F.Boudjema[2] W and Z are no more true gauge bosons. There may exist a composite boson sector with a rich spectrum of spin, flavour, color, states, scalars and vectors (isoscalar Y, excited Z*).

Even with very high mass (much larger than m_Z), mixing effects with the Z° modify its mass and coupling, it also affects the Z° width and can give a decrease of Γ_Z. This effect may contribute to a $\delta\Gamma_Z$ of the order of plus or minus 70 MeV.

So any deviation from the width of the standard model will be an indication for new physics and I am sure that it will then be easy to understand its origin.

As shown by Altarelli in the school, the theoretical calculation of the line shape is well under control in the standard model ; the radiative correction and the electroweak correction are computable.

IV - SENSITIVITY ON NEUTRINO COUNTING

They are many ways of computing the neutrinos species available at the Z° peak[3], I will restrict myself to two approaches

- Total width measurement : as we have seen

$$\Gamma_Z = N_\nu \Gamma_\nu + 3\,\Gamma_l + \Gamma_{had}$$

so $\Delta N_\nu = (\Gamma_Z^{meas} - \Gamma_Z^{SM}) / \Gamma_\nu^{SM}$,

ΔN_ν : number of neutrinos species

Γ_Z^{meas} : width of the Z measured

Γ_Z^{SM} : width of the Z calculated in the standard model

Γ_ν^{SM} : width induced by one neutrino species computed in the standard model

Since only Γ_Z is measured one needs only a pulse to pulse luminosity determination (not an absolute one). It is expected to be as good as 1%.

With 3pb^{-1} (10^5Z), the error on the total width will be around 30 MeV, the statistical error 20 MeV being smaller than the systematic one : 25 MeV.

With Γ_ν = 170 MeV, ΔN_ν will be around 0.2.

- Invisible width as shown by Gary Feldman[4]

$$\Gamma_{invisible} = \Gamma_Z - (\Gamma_e + \Gamma_\mu + \Gamma_\tau + \Gamma_{had})$$

$$\Gamma_e = \Gamma_\mu = \Gamma_\tau$$

and $\sigma^{peak}_{\mu\mu} = (12\pi/M^2_Z)\,(\Gamma_e\Gamma_\mu/\Gamma^2_Z)$

so $\Gamma_Z = (12\pi/\sigma^{peak}_{\mu\mu})^{1/2}\,(\Gamma_\mu/M_Z)$

and

$$\Gamma_{invisible} = \Gamma_\mu \left((12\pi/\sigma^{peak}_{\mu\mu})^{1/2} M_Z - 3 - \epsilon_{\mu\mu}N_{had}/\epsilon_{had}N_{\mu\mu}\right)$$

One measures $\sigma^{peak}_{\mu\mu}$. This requires an absolute knowledge of the luminosity. Γ_μ is computed in the standard model, $N_{had}/N_{\mu\mu}$ is measured, does not need any luminosity determination but the knowledge of efficiency, acceptance and background.

. How well N_{had} can be measured

$$N_{had} = (1/\epsilon_{had}\epsilon_{trig})\,(n_{obs} - L\sigma_{back1}\,\epsilon_{back1} - \ldots)$$

where ϵ are efficiencies, σ_{back1} are cross section for background 1, L the luminosity.

$$(\Delta N_{had}/N_{had})^2 = (1/\sqrt{n}_{obs})^2 + (\Delta\epsilon_{had}/\epsilon_{had})^2 + (\Delta\epsilon_{trig}/\epsilon_{trig})^2$$

$$+ (n_{back1}/n_{obs})^2 \left((\Delta L/L)^2 + (\Delta\sigma_{back1}/\sigma_{back1})^2 + (\Delta\epsilon_{back1}/\epsilon_{back1})^2 \right)$$
$+ \ldots$

-Potential background to hadronic events are 2γ events, $\tau^+\tau^-$ pairs and beam gas interactions. Only 2γ events may contribute to a sizeable error at LEP, if one considers that 2γ cross section is known with an uncertainty of the same magnitude $\Delta\sigma\gamma\gamma/\sigma\gamma\gamma \approx 0(1)$, the 2γ events will then contribute to $\Delta N_{had}/N_{had}$ with an error of 0.3%.

- Selection of hadronic events can be done by requesting some visible energy and more than few charged tracks :

with $E_{vis} \geq 0.1\, M_{Z^o}$ and $N_{ch} \geq 3$ $\epsilon_{had} = 99.5\%$. The estimated error using different generators is 0.2%.

With a more strict cut $E_{vis} > 0.2\, M_{Z^o}$ and $N_{ch} > 5$ $\epsilon_{had} = 98\%$ and the estimated error is 0.9%.

So $\Delta\epsilon_{had}/\epsilon_{had}$ can be kept below 0.5%.

- The trigger efficiency should not be a problem and $\Delta\epsilon_{trig}/\epsilon_{trig} = .3\%$

Overall $(\Delta N_{had}/N_{had}) = 0.7\%$ systematic + error on statistic

namely 0.7% with $1pb^{-1}$ (30 000Z°) or 0.2% with $10pb^{-1}$.

So with a small statistic the error on the number of hadronic events can be kept at the 1% level.

Delphi[5] has estimated in detail the accuracy which can be achieved as a function of different errors. Results are shown in fig 1a, b and c. Typical values are :

$\Delta L/L = 2\%$, $\Delta\epsilon/\epsilon = 1\%$, $\Delta N_\nu = 0.2$ for $1pb^{-1}$

$\Delta N_\nu = 0.17$ for $100pb^{-1}$

Mark II at SLC[6] has already given results with a small number of Z using different fits to the data.

- One can also estimate the number of neutrinos by measuring the reaction $e^+e^- \to \nu\bar{\nu}\gamma$. Extensive studies have been done running on the Z° peak or just above.

A rather good sensivity with a good background rejection can be better achieved running 4 GeV over the Z° peak : with $1pb^{-1}$ one can get $\Delta N\nu^{statist} = 0.35$. This is less powerful than measuring the invisible width.

V-DETECTION OF HIGGS

Standard model higgs

An up to date theoretical review has been done by Franzini, Lee and Taxil[7]. They critically review the different limits which

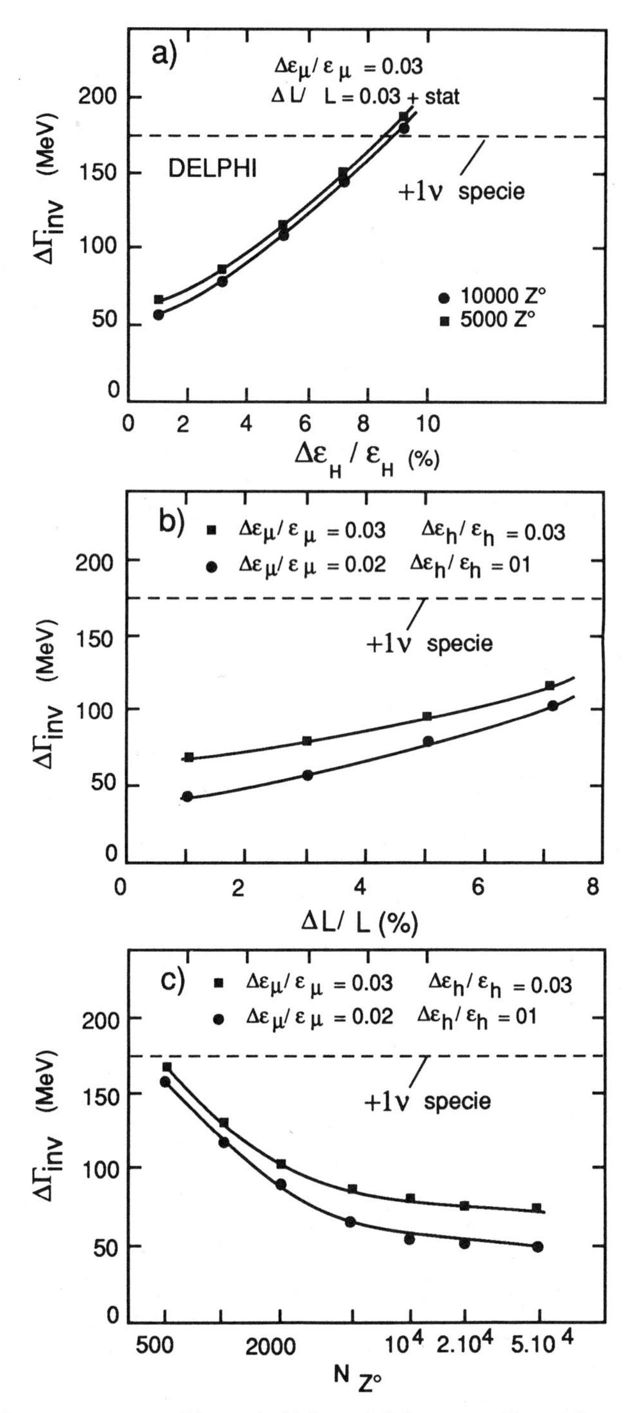

Fig 1 . error on the visible width as a function of : a) $\Delta\epsilon_H/\epsilon_H$ - b) $\Delta L/L$ - c) $N_{Z°}$

one gets from the experiments : muonic Xray, $\pi^+ \rightarrow e\nu H^\circ$, $K^\circ_L \rightarrow \pi^\circ H^\circ$, B decays, $Y \rightarrow H\gamma$... The conclusion is that $m_{H^\circ} < 4.5$ GeV is very unlikely but cannot be ruled out. (only 10 to 110 MeV and 210 - 280 MeV are formally excluded). At LEP 1, toponium is likely to be out of range so the best approach is to use the production $Z^\circ \rightarrow H^\circ f\bar{f}$.

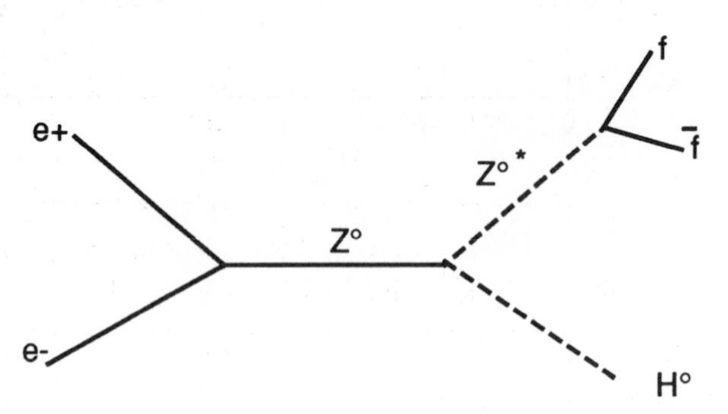

The relative cross section is shown on fig 2, the decay $H^\circ \rightarrow b\bar{b}$ is dominant above 2_{mb} and between 2_{mc} and 2_{mb} $H^\circ \rightarrow c\bar{c}$ dominates.

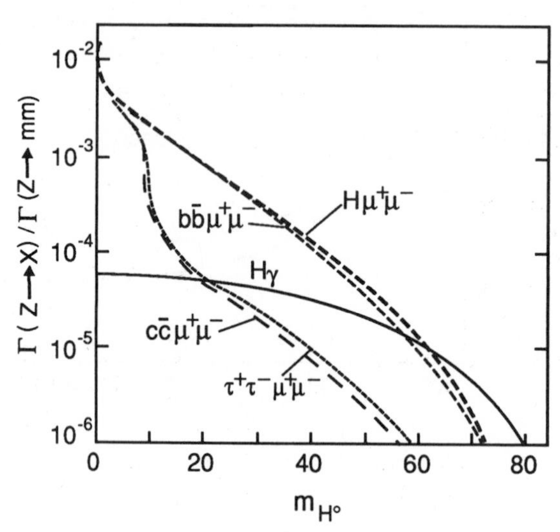

Fig 2 Relative higgs cross section as a function of m_{H° on the Zpeak

Assuming a detection efficiency of 50% for the process $Z^\circ \to H^\circ l^+l^-$, and a minimum of 10 observed events, fig 3 shows the highest detectable higgs mass as a function of the integrated luminosity.

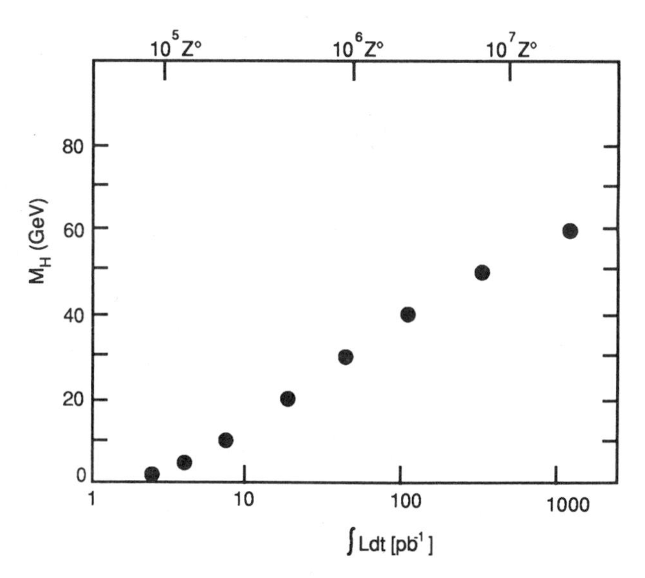

Fig 3 Highest detectable higgs mass versus integrated luminosity

The initial state radiative corrections decrease the higgs cross section by 30% and final state radiative corrections dilute the invariant mass reconstructed for H°.

<u>Detection of H°</u> : $e^+e^- \to e^+e^- + (H^\circ \to b\bar{b}$ jets)

One detects l^+l^- and computes $M^2{}_{recoil}$ =

$$(E_{cm} - E_{l^+} - E_{l^-})^2 - (\vec{p}_l{}^+ + \vec{p}_l{}^-)^2 .$$

There is a background from semileptonic decay of b and c and from misidentified leptons, by defining the sphericity axis of all particles but leptons, kinematics of H° events gives leptons with large p_t versus this sphericity axis. A cut on the arithmetic sum of the p_t of the two leptons eliminates this background.

Fig 4a and b shows the missing mass distribution of dilepton for 10 GeV (20) higgs final states.

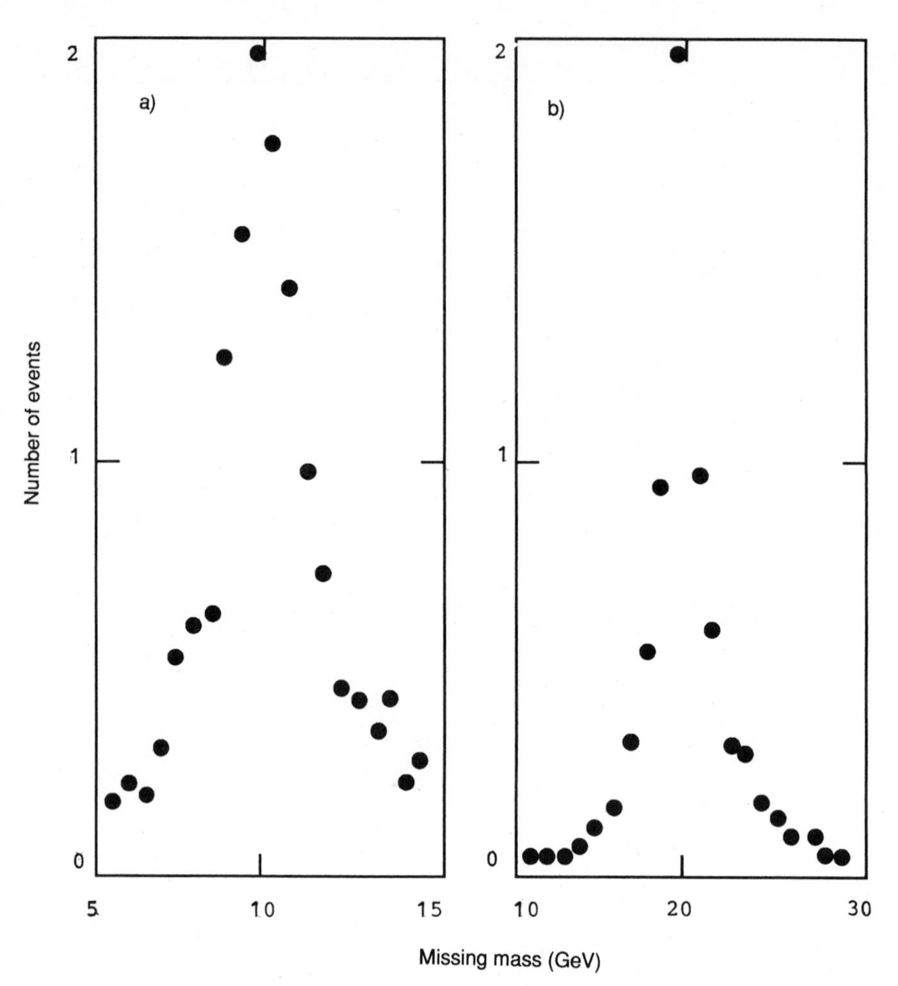

Fig 4 Missing mass distribution of dilepton and
a) 10 GeV higgs final state - b) 20 GeV higgs final state from L3 collaboration.

H° will be easy to detect up to 40GeV and more difficult above 50GeV.

<u>Remark</u> : $Z^\circ \to H^\circ \nu\bar{\nu}$ is 6 times larger than $Z^\circ \to H^\circ l^+l^-$ but it is obviously not as clean. Mickenberg et al[8].

<u>Peculiarities of low mass region</u>

For a very light higgs $100 < m_H < 200$ MeV the resolution on the recoil mass is not adequate so one should look at the H° decay

$e^+e^- \to l^+l^-$ $(H^\circ \to e^+e^-)$ but $c\,\tau_{H^\circ}$ is sizeable $\approx$ 10cm.

This gives a mean decay length decreasing from 50cm for 100 MeV mass to 13 cm for 200 MeV. Electrons from H° decay have around 3 to 4 GeV, they should be easily detectable.

For a higgs mass above 200MeV there are theoretical uncertainties in the branching ratio $H^\circ \to \mu^+\mu^-$. If it is 5% then $e^+e^- \to q\bar{q}$ $(H^\circ \to \mu^+\mu^-)$ provides 20-40 events signal for a million Z°. Table 1 gives the estimate of signal and background in Aleph:

Table 1 Signal and background for H° detection
10^7 Z°

$p_T^{\,l1}+p_T^{\,l2}$ cut	signal m_H=40GeV	Backgr	Signal m_H=50GeV	Backgr
15GeV	51.5	2.0	16.8	5.5
20GeV	50.0	1.2	15.6	3.7

<u>Beyond the minimal standard model Higgs</u>

There are many scenarios, let's examine the experimental consequences of a two Higgs doublet model ie a model with

2 charged higgs $H^{+/-}$
3 neutral higgs H° being the highest by convention
h° being the lighest
A° pseudoscalar

- $e^+e^- \to l^+l^-h^\circ$ is to be looked at in the same way as for the S.M., the sensitivity only will be different.

- $e^+e^- \to h^\circ A^\circ$ ($e^+e^- \to l^+l^-A^\circ$ is forbiden) no clear signature, look for $b\bar{b}\,b\bar{b}$ or $\tau^+\tau^-$ + jets.

Results from L3 are shown on fig 5 for QCD and higgs events.

VI - B PHYSICS AT LEP

1) LEP versus Argus/cleo and B factories

Table 2 gives some numbers of B events for LEP, Argus, cleo and a possible B factory

Table 2

	Lep $10^6 Z^\circ$	$10^7 Z^\circ$	Argus	Cleo	B factory
numb. of B	3.10^5	3.10^6	3.10^5	10^6	$10^8/10^9$
Bs	4.10^4	4.10^5			
$b\bar{b}$/had	22%	22%	20%	20%	20%
$b\bar{b}/c\bar{c}$	1.2	1.2			

At Lep energy, lepton identification is improved, B and $\bar{B}$ jets are separated (due to the Lorentz boost), secondary vertices are more easily measured but b quark does not give only a b meson but also a few more hadrons.

At Lep one can measure B lifetime, $b\bar{b}$ asymmetries and B°_s.

2) B tagging

The fragmentation of the b quark is harder than that of a light quark (adding a light q to b slows it down less than in the case $\bar{q}q'$ where q' is a light quark)

f(z) should be harder for b(c) than for light quarks.

$b\bar{b}$ tagging by semileptonic decays

Apart from photon conversion, pion, kaon decays and misidentification, electron (muon) comes mainly through semileptonic decay of B and C

(1) $e^+e^- \to B\bar{B} + \ldots$ $\bar{B} \to \ldots$
$B \to cl\nu$

(2) $e^+e^- \to B\bar{B} + \ldots$ $\bar{B} \to \ldots$
$B \to se\nu$

(3) $e^+e^- \to C\bar{C} + \ldots$ $\bar{C} \to \ldots$
$C \to se\nu$

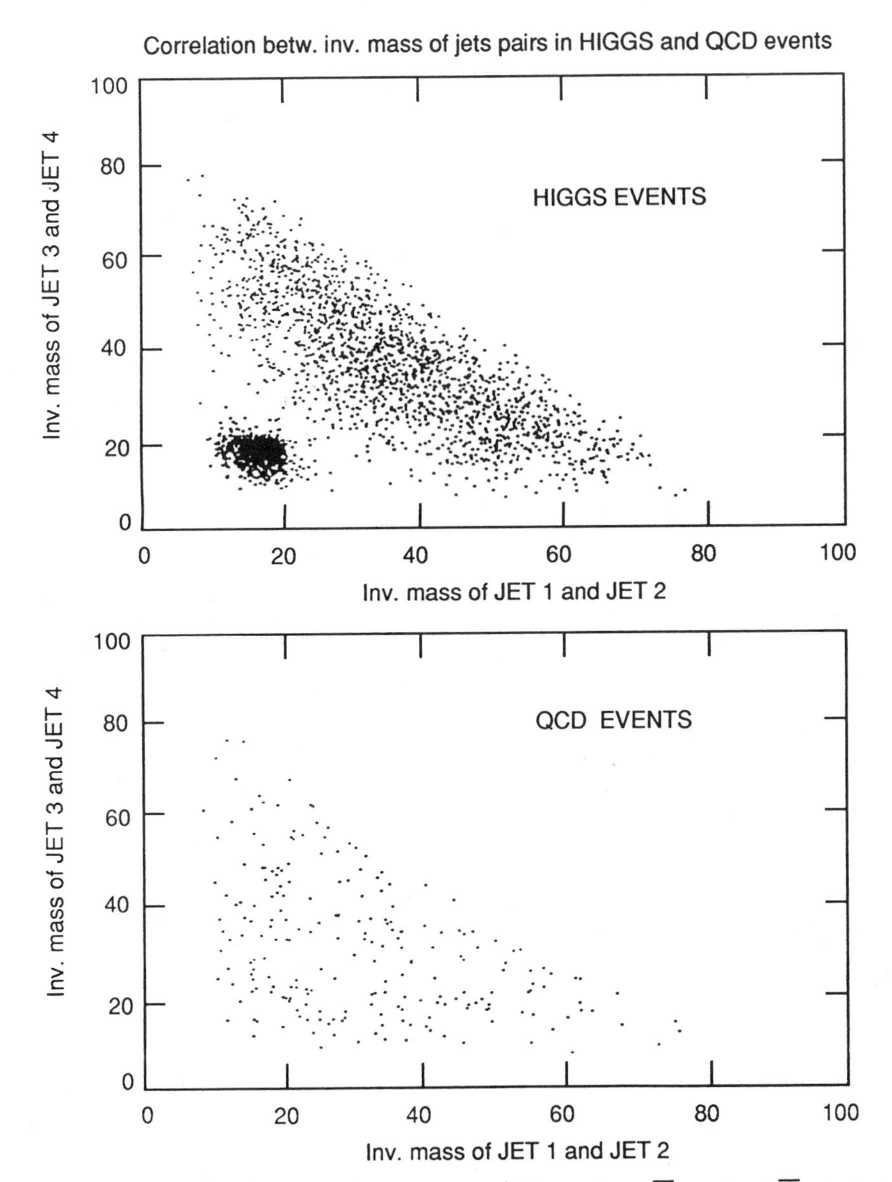

Fig 5 detection feasibility for $e^+e^- \to (H^\circ \to b\bar{b}) + (A^\circ \to b\bar{b})$ 4 jets
a) higgs events - b) QCD background from L3 collaboration.

The largest lepton p_T is produced by channel 1, in practice one looks for jets then computes $p_T^{\,l}$ inside the jet (this gives a mean value of p_T for b,c jets which is independent on the number of jets in the event).

Selecting events with large p_T will enlarge the b sample and decrease the c contamination.
ex) p_e > 3 GeV and p^T_e > 1 GeV gives a b tagging efficiency of 50% for the channel (1), with a purity of 80%. If one adds channel 2 and b -> τ -> l, the b tagging efficiency is increased to 60% and the purity is 90%.

Similar results are obtained with muons.

So with $10^5 Z^\circ$ one gets 16 000 $B\overline{B}$ with 3200 $B\overline{B}$ tagged with 1 lepton and 160 $B\overline{B}$ tagged with 2 leptons.

If needed, the purity can be improved by increasing the $p_T^{\,l}$ cut, by using the sphericity product or by using impact parameter cuts.

$b\overline{b}$ tagging by sphericity product[9]

Since m_b is large, b jets have a larger sphericity, so for $b\overline{b}$ events $S_1 S_2$ (sphericity product) will be larger than for $q\overline{q}$. Since there is not only two jets events and since the B do not decay in their center of mass one will find B jets, try an empirical Lorentz boost β on these two jets to transform B in its cm. In practice β=0.95 makes a good estimate.

This sphericity product can be applied on all B decays and can be mixed with other cuts.

ex : S1 x S2 > .2 leaves 40% of $B\overline{B}$ with a purity of 30%.

$b\overline{b}$ tagging by specific decays

- B -> J/ϕ K -> $l^+ l^-$ is a good way to tag B but BR = 10^{-3} only and at the end one will be left with only a few events. It may still be the good way to measure B°_s, B°_d, $B^{+/-}_u$ lifetimes separately.

- B -> $D^{*\,+}$ ... tags B and C quarks.

D* -> D° $\pi^{+/-}$ (57%))
) overall 2.4%
Kπ^+ (4.2%))

with $10^5 Z^\circ$ one ends up with 170 D^+ coming from $c\overline{c}$ events
216 D^+ coming from $b\overline{b}$ events

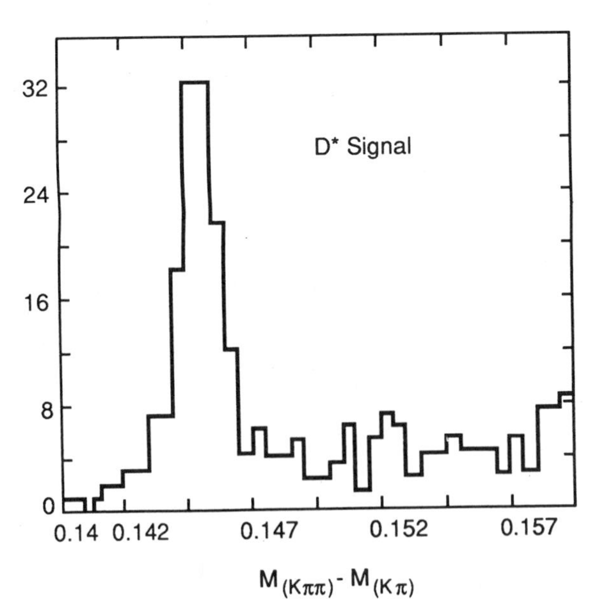

Fig 6 D* signal from Z decays (Aleph)

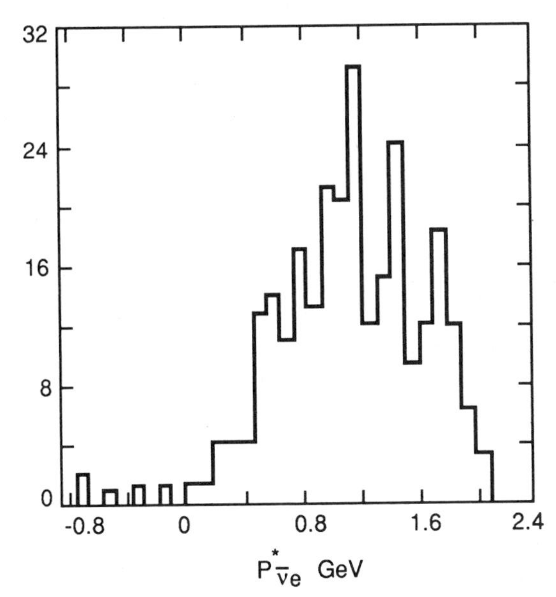

Fig 7 $P^{em}_{\nu e}$ distribution from semileptonic $B \to D^{*+} e^- \overline{\nu}e$ decays from Aleph.

to see the D* signal from the combinatorial background one benefits from kinematic properties

- $p_T^{\pi/D^\circ} < 40$ MeV/c in the lab

- since $p\pi/p_D = m_\pi/m_D$ $p\pi < 3$ GeV/c

- $m_{D^*} - m_D = m_\pi$, errors on $m_{D^*} - m_D$ are strongly correlated and the resolution on the mass difference is 2 MeV/c^2.

D* signal is shown on fig 6.

This D* identification can also be used in addition to lepton tagging and D*, e tagging will give access to cm kinematic of semileptonic $B^\circ_d \to l^- \overline{\nu}_e D^{*+}$

$p^{*cm}_\nu = (m_B^2 - m^2_{eD^{*+}})/2m_B$ is shown on fig 7

3) $\Gamma b\bar{b}$

$$\Gamma(Z \to q\bar{q}) = (G\mu\, m^3_Z/8\sqrt{2}\pi)\ (\beta\ (3-\beta^2/2)\ v_q^2 + \beta^2 a^2_q)$$

with $v_q = 2I_{q3} - 4\ Qq\ \sin^2\theta$
$Aq = 2I_{q3}$

On top of the born term there are QCD and electroweak corrections. A peculiarity of the $b\bar{b}$ width is a relatively larger sensitivity to the m_t mass through the following graph :

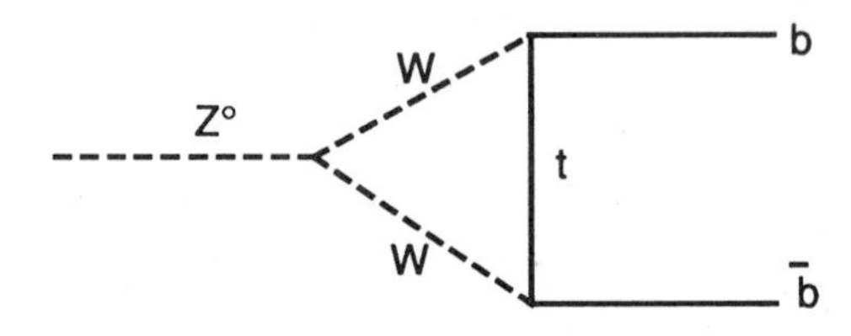

The ratio Γ/Γ_{born} as a function of m_t is shown on fig 8. Up to 2% effects are predicted for $m_t = 200$ GeV. The necessary experimental accuracy will certainly not be reached at the start of Lep.

4) $B^oB̄^o$ mixing

$B^o\overline{B}^o$ transition can occur through different graphs (fig 9) but the transition rate is dominated by t exchange and its probability is $\propto (Vtq)^2 \, m_t^2 \, f_B$.

One defines

$rq = \mathrm{Prob}\left((B^oq) \to (\overline{B}^oq)\right) / \mathrm{Prob}\left((B^oq) \to (B^oq)\right)$ $rq = \overline{rq}$ if CP invariance holds.

$\chi q = \mathrm{Prob}(B^oq) \to (\overline{B}^oq) / \left(\mathrm{Prob}(B^oq) \to (\overline{B}^oq) + \mathrm{Prob}(B^oq) + (B^oq)\right) = rq/(1+rq)$

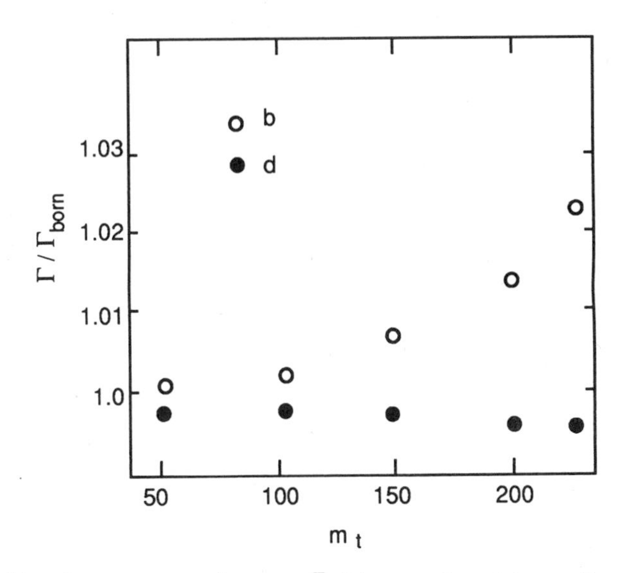

Fig 8 correction to $\Gamma_{b\overline{b}}$ as a function of m_t

Time evolution of $B\overline{B}$ mixing (assuming CP invariance)

$$H \begin{bmatrix} B^oq \\ \overline{B}^oq \end{bmatrix} = \begin{bmatrix} M - i\Gamma & M_{12} - i\Gamma_{12}/2 \\ M^*_{12} - i\Gamma^*_{12}/2 & M - i\Gamma/2 \end{bmatrix} \begin{bmatrix} B^oq \\ \overline{B}^oq \end{bmatrix}$$

with the eigenstates $|B^oq_{H,L}> = 1/\sqrt{2}\,(|B^oq> \pm |\overline{B}^oq>)$ and the eigenvalues $V_{H,L} = M_{H,L} - 1/2\Gamma_{H,L}$,
the mixing parameters are defined by :

$\Delta M = M_H - M_L \quad \Delta\Gamma = \Gamma_H - \Gamma_L$

$x = \Delta M/\Gamma \qquad y = \Delta\Gamma/2\,\Gamma$

then $\left(|B^\circ_H(t)\right> = e^{-i(M_H-1/2\,\Gamma_H)t}\,|B^\circ_H(o)>$

$\left(|B^\circ_L(t)\right> = e^{-i(M_L-1/2\,\Gamma_L)t}\,|B^\circ_L(o)>$

-> $\left(|B^\circ_q(t)\right> = a_+(t)\,|B^\circ_q(o)> + a_-(t)\,|\overline{B^\circ}_q(o)>$

$\left(|\overline{B^\circ}_q(t)\right> = a_-(t)\,|B^\circ_q(o)> + a_+(t)\,|\overline{B^\circ}_q(o)>$

with $a\pm(t)$ $= 1/2e^{-\Gamma t}(\mathrm{ch}\Delta\Gamma/2\; t \pm \cos\Delta Mt)$

$= 1/2e^{-\Gamma t}(\mathrm{ch}y\Gamma t \pm \cos x\Gamma t)$

->Prob $\left(B^\circ q(o)\text{->}B^\circ q(t)\right)$ + Prob $\left(B^\circ q(o)\text{->}\overline{B^\circ}q(t)\right) = e^{-\Gamma t}\mathrm{ch}y\Gamma t$

y small makes the expression very similar to the case without mixing.

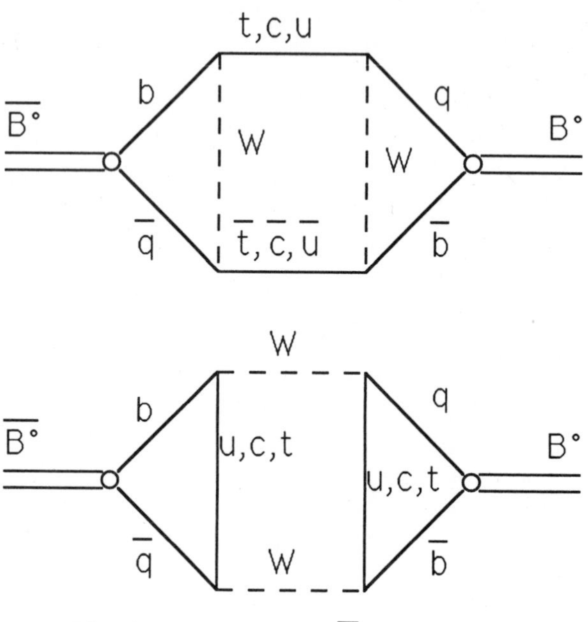

Fig 9 Graph for $b\overline{b}$ transition

The integrated probabilities are :

Prob $(B^\circ q\text{->}B^\circ q) = N\int_0^{+\infty} P\left(B^\circ q\text{->}B^\circ q(t)\right)dt = C(2+x^2_q - y^2_q)$

Prob $(B^\circ q\text{->}\overline{B^\circ}q) = N\int_0^{+\infty} P\left(B^\circ q\text{->}\overline{B^\circ}q(t)\right)dt = C(x^2_q + y^2_q)$

so $rq = (x^2_q + y^2_q)/(2+x^2_q + y^2_q)$ and $\chi q = (x^2_q + y^2_q)/2(1+x^2_q)$

The isocontour of rq are shown on fig 10.

Existing measurements

Argus and cleo measure $\chi d = 0.17 \pm 0.05$ and 0.15 ± 0.05 respectively. UA1 measures a mixing of $B^o{}_s$ and $B^o d$
$<\chi> = p_d <\chi_d> + p_s <\chi_s> = 0.12 \pm 0.05$

Calculation of the box diagram give $(\Delta\Gamma/\Delta M)$ small (0.05 for $m_t = 50$ GeV) so $y << x$ (this is also true for x_s)

From χ_d measurement one derives $x_d = 0.70 \pm 0.13$

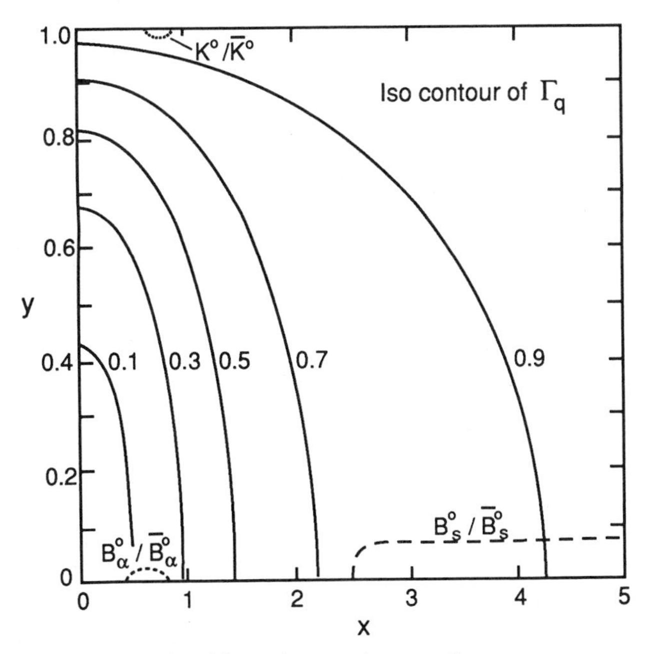

Fig 10 isocontour of rq

Theoretical expectation for $B^o s$

In the standard model $x_s/x_d = (f_{Bs}/f_{Bd})^2 \ (V_{ts}/V_{td})^2$

$(V_{ts}/V_{td})^2 > 6$ from τ_B and limits on V_{bu}

$f_{Bs}/f_{Bd} > 2$ from theoretical calculation

this constraints x_s to : $3 < x_s < 30$ and χ_s

$0.45 < \chi_s = (x_s{}^2 + y_s{}^2)/2(1+x_s{}^2) < 0.499$

Outside the standard model $\Delta\Gamma_s(y_s)$ is not necessarily small B_s and $\overline{B_s}$ may even decay to identical final states.

$(b\overline{s}) \to (c\overline{c}\ s\overline{s})$

$1 < (\tau_L/\tau_S) < 2$

and χ_s can be considerably lowered.

Experimental measurement at Lep

One should in principle be able to measure the time evolution of $B\overline{B}$ events. In practice one should get an accuracy on $\sigma\tau/\tau$ much better than 10% if one does not want to dilute B°s oscillation. At least for the begining only the time integrated distribution will be measured.

Using di-lepton events

The ratio of same charge leptons to all leptons is related to χ

$$R_{ll} = (N_{++} + N_{--}) / (N_{++} + N_{--} + N_{+-} + N_{-+}) = 2\chi(1-\chi)$$

with $\chi = p_d\ \chi_d + p_s\ \chi_s$

p_d fraction of b quark giving B_d = 1./2.3
p_s fraction of b quark giving B_s = 0.3/2.3
$\chi_d = 0.17 \quad \chi_s = 0.45$
$\chi = 0.129$

but at Lep there will be some mistagging due to b -> c -> e so $R_{obs} = 2(1-x)(x+\delta)/(1+\delta)$ where δ is the contamination (10 to 20%).

For 10^6Z one will get 2000 b -> l dileptons events with 600 background events (b->c->l), the systematic error on χ will be equal to the statistical error of 10%.

Fig 11 shows the expected measurement and resolution, one can see that no real constraint on x_s will come out of that measurement in the standard model.

Using identified B_s

One should find an other way to tag B°_s and look directly at B_s oscillation, for example tag lepton on one side and identify B on the other side may be a possibility.

5) Foward backward asymmetry, angular distribution

The forward backward asymmetry A_{FB} is related to vector and axial coupling

$A_{FB} = 3/4 \quad (2v_e a_e/(v_e^2+a_e^2)) \quad (2v_b a_b/(v_b^2+a_b^2))$

A_{FB} is sensitive to small changes in the electron coupling

$a_e = -1 \quad v_e = (-1 + 4\sin^2\theta w)$

$a_b = -1 \quad v_b = (-1 + 4/3\sin^2\theta w)$

with $\sin^2\theta w = 0.25$ $\partial A_{FB}/\partial \sin^2\theta w = 5.5$ so $A_{FB}B\overline{B}$ is very sensitive to $\sin^2\theta w$.

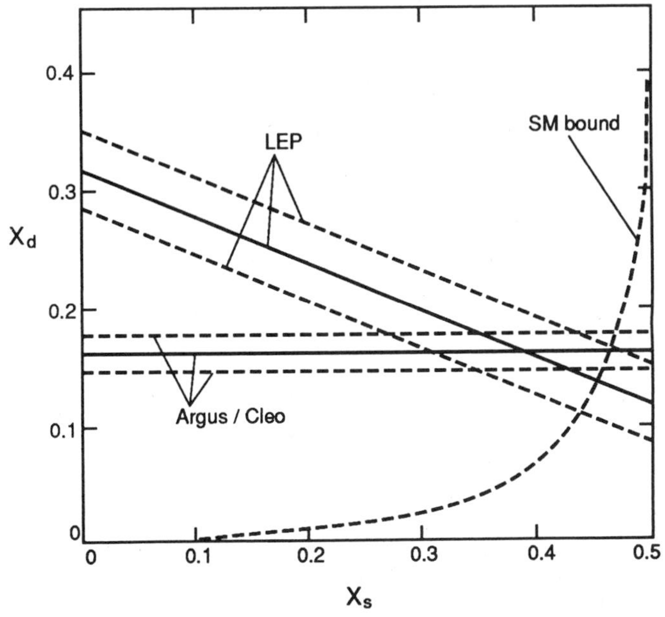

Fig 11 measurement of χ_d, χ_s, at Lep.

On the theoretical point of view, corrections have been computed. The main uncertainty concerns the top mass.

A_{FB} rises from 0.12 for m_t = 50 GeV to 0.14 for m_t = 200 GeV. The statistical error is shown on table 3

Systematic errors come from mixing, background and B direction estimation : the mixing decreases the observed forward backward

asymmetry $A_{FB}^{meas/mixing} = A_{FB}^{B\bar{B}} (1-2<\chi>)$

The actual knowledge on $<\chi>$ induces a systematic error $\Delta \sin^2\theta_w = 0.002$. Obviously it will be improved.

Background is mainly from b ->c->leptons. With $p^l > 5$ GeV and $p_t^l > 1$ GeV, the residual contamination δ is 20% of true b->e

$A_{FB}^{Meas} = A_{FB} \times (1-\delta)/(1+\delta)$,

Assuming $\Delta\delta/\delta = 10\%$ (fragmentation) $\Delta A_{FB}^{meas} = 0.005$ and the systematic error on $\sin^2\theta_w = 0.001$. Again this can be decreased within the time.

Overall the asymmetry measured with $B\bar{B}$ may be as sensitive as other mean at the beginning of Lep (before polarisation can be used).

Table 3

number of Z°	$\Delta A_{FB}^{B\bar{B}}$	$\Delta \sin^2\theta_w$
10^4	.07	0.01
10^5	.02	0.004
10^6	.007	0.001

VII - SOME REMARKS ON QCD AND α_s DETERMINATION

We shall review the state of the art in the determination of α_s measurement at Lep 1.

1) Status of α_s and Λ

- deep inelastic scattering

From muon proton deep inelastic scattering one gets at second order $\Lambda_{MS} = 200$ MeV ± 20MeV ± 60MeV the last error is the systematic one

$\alpha_s = .185 \pm .002 \pm .01$

It is rather difficult to do much better, because of experimental difficulties and because of the so call higher twist contribution.

- energy energy correlations in e^+e^- annihilation

One defines $EEC(\chi) = 1/N_{events} \; \Sigma_{events} \; \Sigma_{i,j} \; E_iE_j/s \;\; \delta(\chi_i-\chi_j)$

where χ_i is the angle of particle i

$\chi_i-\chi_j$ is the angle between i and j.

For 2 jets events $EEC(\chi)$ peaks at $\chi = 0°$ and 180°. For 3 jets events (all at 120° with equal sharing of energy) $EEC(\chi)$ populates the intermediate region, contributing to antisymmetric term in EEC.

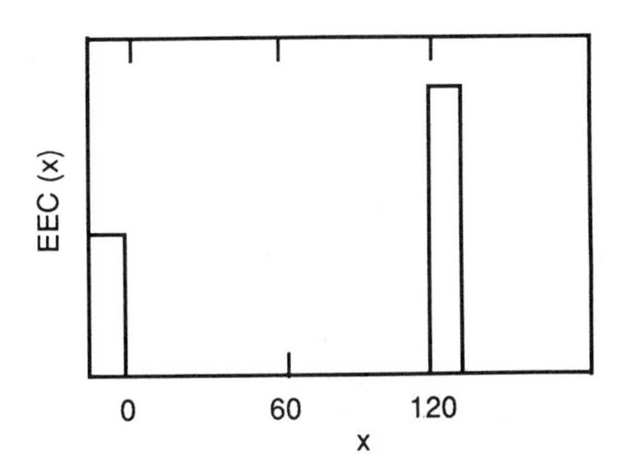

One defines $A(\chi) = EEC(\pi-\chi) - EEC(\chi)$. $A(\chi)$ vanishes for 2 jets events, is infrared finite (no cuts), is less sensitive to fragmentation. See Cello results on fig 12. Mark II quoted at $\sqrt{s} = 34$ GeV $\alpha s(34GeV) = 0.153 \pm 0.003 \pm 0.005$

The systematic error is computed assuming that the Lund model is a correct representation of the fragmentation, Cello comparing Lund and Hoyer fragmentation model, find larger systematic errors

$0.11 < \alpha s(44GeV) < 0.16$

So at Pep, Petra energy, energy energy correlation is still strongly influenced by fragmentation.

- Measurement of R below the Z peak

$R = \sigma_{had} / \sigma_{\mu\mu} = \omega e^2 \; q \; \tilde{R}$

$\tilde{R}$ is computed in QCD

$\tilde{R}=1+(\alpha_{\overline{MS}}(s)/\pi)+C_2(\alpha_{\overline{MS}}(s)/\pi)^2+C_3((\alpha_{\overline{MS}}(s)/\pi)^3)+...?$

$C_2 = 1.986 - 0.115 - n_f = 1.411 \; (n_f=5)$

$C_3 = (70.986 - 1.2n_f - 0.005 \; n_f^2) = 64.9$

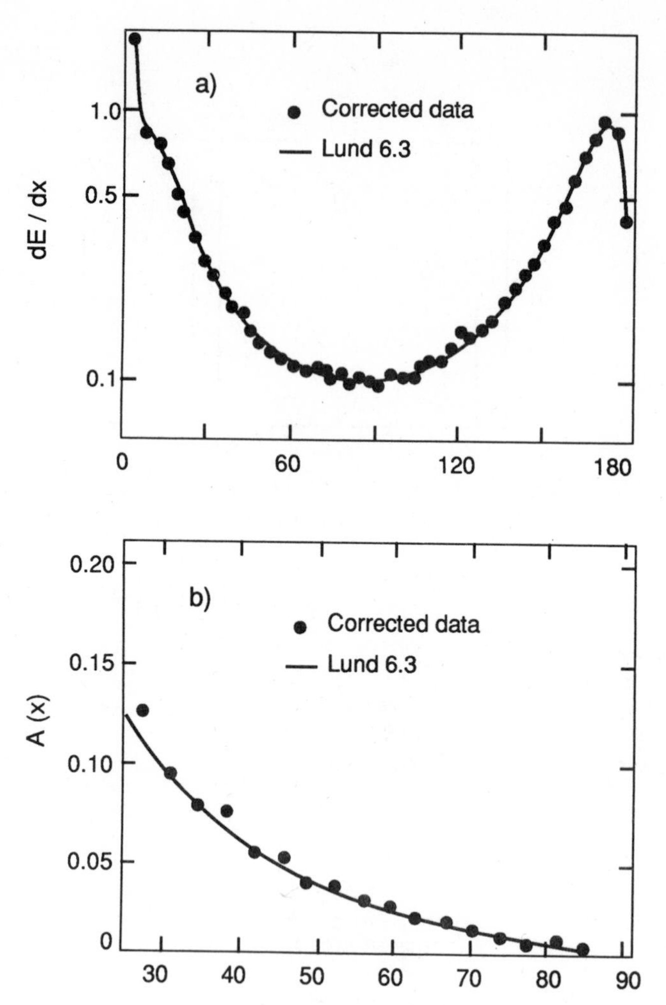

Fig 12 Cello on results on EEC(χ) and asymmetry A(χ).

A relative error on R of 0.005 translates into an error on α of 0.003. Note that at relatively low energy the expansion of $\tilde{R}$ may be divergent very soon[10]

$\tilde{R}$ = 1+0.0461 + 0.0030(second order) + 0.0063 (third order)
at S = 34 GeV and α_s = 0.145.
At Lep energy
$\tilde{R}$ = 1. + 0.0391 + 0.0022 + 0.0038 (S = 90GeV, αs = 0.123)

2) Measurement of R at Lep

One measures $R' = \sigma_{had}/\sigma_{\mu\mu}$ (with $1/R' \; \partial R'/\partial(\alpha_{s/\pi}) = 1$)

$$R' = \sigma_{had}/\sigma_{\mu\mu} = \left((N_H - N_{BH})/\Sigma_H\right) \; . \; \left(\epsilon_\mu/(N_\mu - N_{B\mu})\right)$$

Efficiencies and background have already been discussed
$\Delta R'/R'$ syst = 0.7%
$\Delta R'/R'$ statistical is dominated by $\Delta N_\mu/N_\mu = 1/\sqrt{N_\mu}$
ie 1% for $10 pb^{-1}$ or 0.3% for $100 pb^{-1}$
So R at lep provides a α_s determination with
$\Delta \alpha_s/\alpha_s$= (30% with $\int \mathcal{L} dt = 10\ pb^{-1}$
(20% with $\int \mathcal{L}\ dt$ = 100 pb-1

This determination does not depend on fragmentation but the accuracy is not good enough to see the effect of the running coupling constant.

Table 4

$\sqrt{s}$	$\Delta \alpha_s/\alpha_s$ Lund/Ali	$\Delta \alpha_s/\alpha_s$ Lund/Hoyer
30 GeV	.12	.22
90 GeV	.04	.12

3) Energy energy correlation at lep

Since the main error at PEP-PETRA is coming from fragmentation, results should improve at lep energy. Estimate of the error can be done by comparing the error induced on α_s using different models.

Using LUND, ALI, HOYER models and without any attempt to make them give the same "data" at lep energy one should be able to make an upper bound of the error. Results are shown on table 4.

The statistical error is already very small for 50k events and systematic errors due to uncertainty inside a given model is also small. Energy energy correlation seems to be the best way to measure α_s within a relative uncertainty of 10% at Lep energy.

It has been a pleasure for me to be in a very well organised school located in such a nice environment.

REFERENCES

1. Aleph : LEPC 83-2 p1 technical proposal
   The Aleph hanbook 1989, Aleph 89-77, W.Blum editor
   Delphi : LEPC 83-3-P2 technical proposal
   Opal : LEPC 83-4-P3 technical proposal
   L3 : LEPC 83-5-P4 technical proposal
2. F.Boudjema and F.M.Renard, Compositeness at Lep 100, June 1989
   Proceeding of the workshop on Z physics at Lep
   CERN yellow report, to appear
3. G.Barbiellini et al, Neutrino counting
   Proceeding of the workshop on Z physics at Lep
   CERN yellow report, to appear
4. G.Feldman, On the possibility of measuring the number of neutrino species to a precision of 1/2 species with only 2000 Z events
   Proceeding of the third Mark II workshop on SLC physics
   Slac report 315, July 1987
5. C.Matteuzzi, Delphi note 88-14, unpublished 1988
6. G.Feldman, Mass and width of the Z°, Results from Mark II
   Presented at this school
7. M.Drees et al, Higgs search at Lep, Proceeding of the workshop on Z physics at Lep, Preprint cern TH 5487/89
   CERN yellow report, to appear
8. E.Duchovni et al, Weizmann preprint Wis-88/39 (1988)
9. See for example R.Marshall e^+e^- annihilation at high energies, RAL 89-021, May 1989
10. J.D.Bjorken in proceeding of this school.

PRODUCTION AND DECAY OF Z BOSONS AT THE SLC*

G. J. Feldman

Stanford Linear Accelerator Center
Stanford University, Stanford, California 94309, USA

INTRODUCTION

My lectures at Cargèse covered the very first physics results from the SLAC Linear Collider (SLC). At the time of this writing (December 1989), it seems most sensible to present a review of the results that were presented at the school in an updated form. The organization of this report will be to give a brief introduction to linear colliders and the SLC, then to describe the MARK II detector, and finally to review the current status of the three major physics topics discussed at Cargèse:

1. the Z line shape, from which we deduce the Z mass and width, and the number of neutrino species,
2. the partonic structure of hadronic decays and a measurement of α_s, and
3. searches for new quarks and leptons.

LINEAR COLLIDERS AND THE SLC

The SLC is the first operating single-pass e^+e^- collider. We built it for two reasons:

1. to develop the technology which will be used for all future e^+e^- colliders with energies higher than 200 GeV, and
2. to make the first study at the Z mass.

Both reasons were essential, but the former will provide the lasting contribution of the SLC.

Why will linear colliders rather than storage rings provide the technology needed to explore higher energies in e^+e^- annihilation? Burton Richter studied the scaling laws for storage rings in 1976.[1] There are two factors in the cost of a high-energy storage ring. Most of the costs scale as the size of the ring—tunnels, magnets, vacuum systems, etc. The one cost that does not scale with the size of the ring is the RF system, which is required to make up the energy lost to synchrotron radiation. The voltage required to restore the lost energy is proportional to the fourth power of the energy and

⋆ Work supported by Department of Energy contract DE–AC03–76SF00515.

Particle Physics: Cargèse 1989
Edited by M. Lévy *et al.*
Plenum Press, New York, 1990

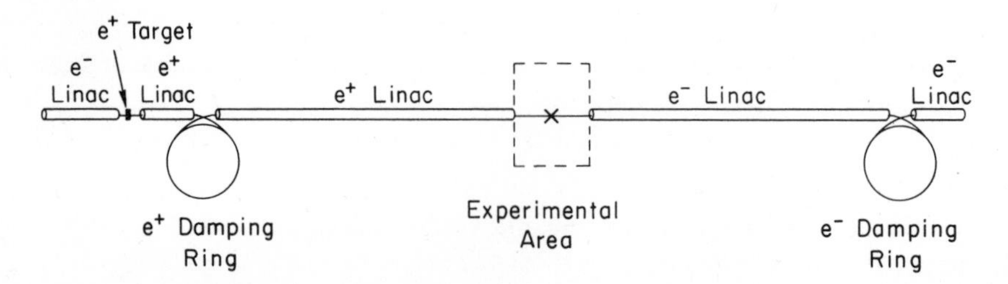

Fig. 1. Schematic of a generic linear collider.

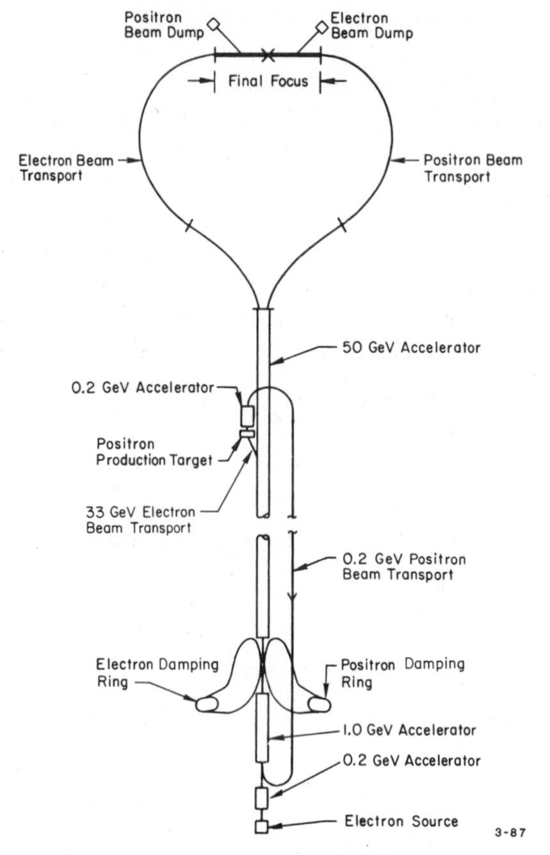

Fig. 2. Schematic of the SLC.

inversely proportional to the radius of curvature. Thus, simplifying Richter's argument considerably, we can write

$$C = \alpha R + \beta \frac{E^4}{R} \tag{1}$$

where C is the cost, R is the radius, E is the energy, and α and β are constants. Optimizing the cost by setting the derivative of Eq. (1) with respect to R to zero yields the result that both the cost and size of a storage ring scale with E^2.

We can thus estimate the cost of a 1 TeV storage ring by assuming that LEP II is an optimized 200 GeV storage ring and using this scaling law. The result is that such a ring would be 675 km in circumference and cost 17.5 billion dollars. Even by our new sense of reasonableness set by the SSC scale, this seems unreasonable and suggests that we should pursue an alternate technology. Both the cost and size of a linear collider, of course, scale with energy, making it appear to be a more promising approach.

Figure 1 shows a generic linear collider. It has three main accelerators: an electron linac to produce positrons, and positron and electron linacs to accelerate the beams to high energy. It also has two damping rings to reduce the emittance of the beams.

Figure 2 shows the only present example of a linear collider, the SLC. Please note that this design is topologically equivalent to the generic linear collider with the present SLAC linac serving as all three required linacs. A positron return line and two arcs have been added to transport the particles to the required locations; in principle, these transport lines do not affect the basic functioning of the collider.

The SLC was originally scheduled to begin taking physics data in January 1987. However, since it represented a new and difficult technology, we obtained the first reasonable luminosity in late March 1989, and observed the first Z boson decay on April 11, 1989. Since that time, we have collected 19 nb^{-1} of integrated luminosity and have observed about 500 Z decays.

THE MARK II DETECTOR

The MARK II detector began life as the second general purpose detector at the 7 GeV storage ring SPEAR. Later it was moved to the 29 GeV storage ring PEP. After it was selected to be the first SLC detector, it was upgraded, tested at PEP, and finally moved to the SLC. The MARK II Collaboration presently consists of approximately 130 physicists from nine institutions.[2]

A drawing of the MARK II detector[3] is shown in Fig. 3. The principal components that we will be interested in here will be the drift chamber, the calorimeters, and the luminosity monitors.

The drift chamber is a 72-layer, minijet cell, cylindrical chamber[4] immersed in 4.75 kG solenoidal magnetic field. It tracks charged particles in the region $|\cos\theta| < 0.92$, but the efficiency and momentum resolution begin to deteriorate at $|\cos\theta| = 0.82$. Without a vertex constraint, the momentum resolution is about 0.5% p (p in GeV/c).

There are two sets of electromagnetic calorimeters, which, together, detect photons in the region $|\cos\theta| < 0.96$. The central calorimeters are lead-liquid argon sandwich ionization chambers[5] with an energy resolution of about $14\%/\sqrt{E}$ (E in GeV). The forward and backward calorimeters are composed of lead-gas proportional tube sandwiches with energy resolution of about $22\%/\sqrt{E}$. Both calorimeters have a strip geometry with three or four strip directions for stereographic reconstruction.

Figure 4 shows a close-up of the region around the beam-line. Note the two luminosity monitors at small angles, the Small Angle Monitor (SAM), followed at smaller angles by the MiniSAM.

A typical hadronic Z decay is shown in Fig. 5(a). The two jet structure, shown graphically in the Lego plot of Fig. 5(b), and the charged multiplicity of about 20 tracks are typical of these events. About 7/8 of visible Z decays are into hadronic modes. The remainder are split among e, μ, and τ pairs. A τ pair decay is shown in Fig. 6, in which one of the τ's decays into a 16 GeV/c muon and the other decays into a 17 GeV/c electron.

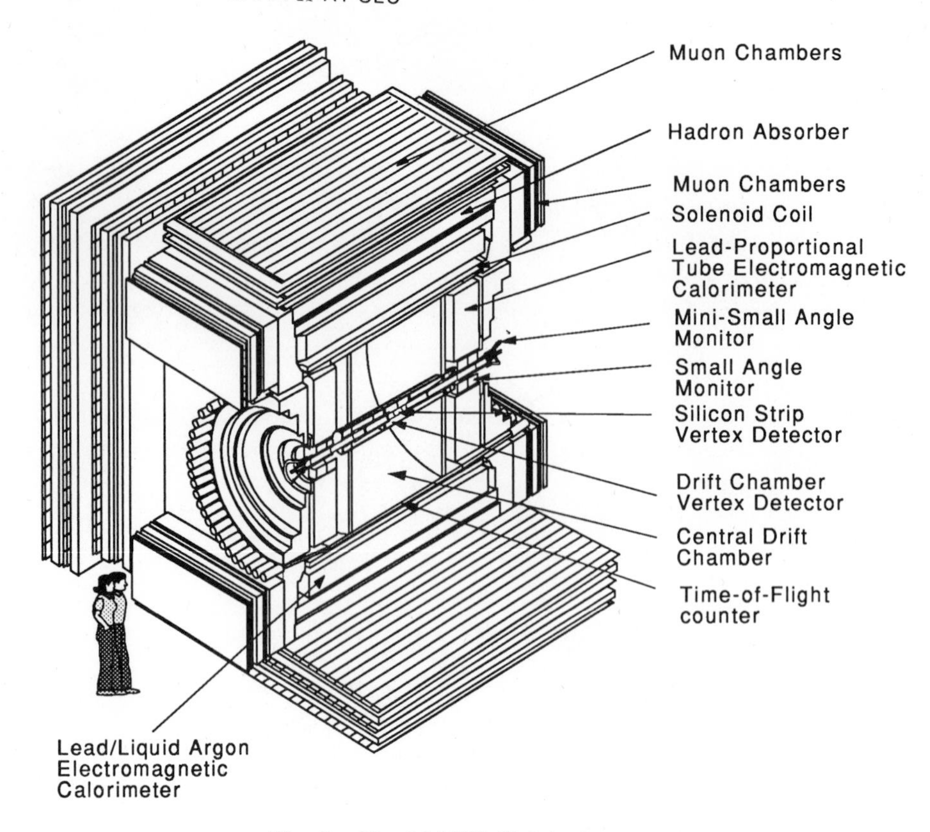

Fig. 3. The MARK II detector.

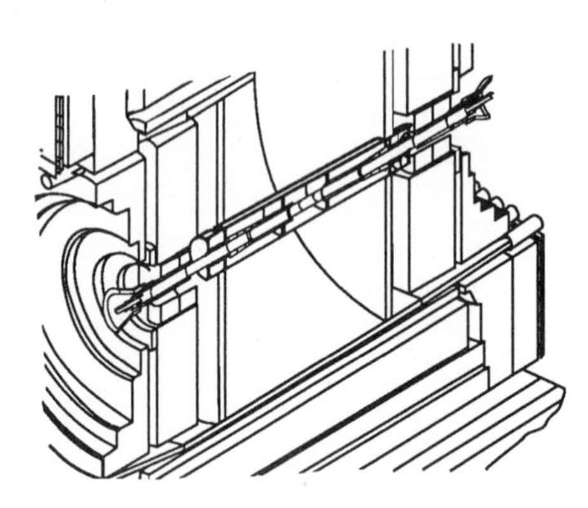

Fig. 4. Detail around the beam line of the MARK II detector.

The six main MARK II triggers are listed in Table 1. Monte Carlo simulations indicate that the charged and neutral triggers are 97% and 95% efficient for Z hadronic decays, respectively. In addition to being highly redundant, they are complementary in that the charged trigger is more efficient in the central region, while the neutral trigger is more efficient in the forward and backward regions. Together, they are calculated to be 99.8% efficient.

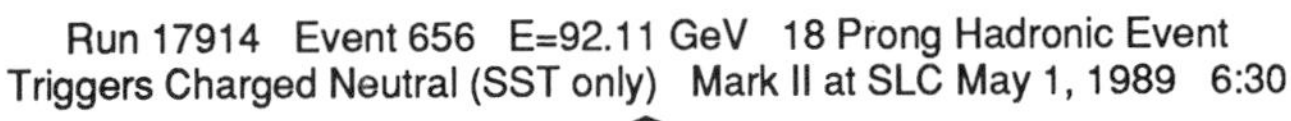

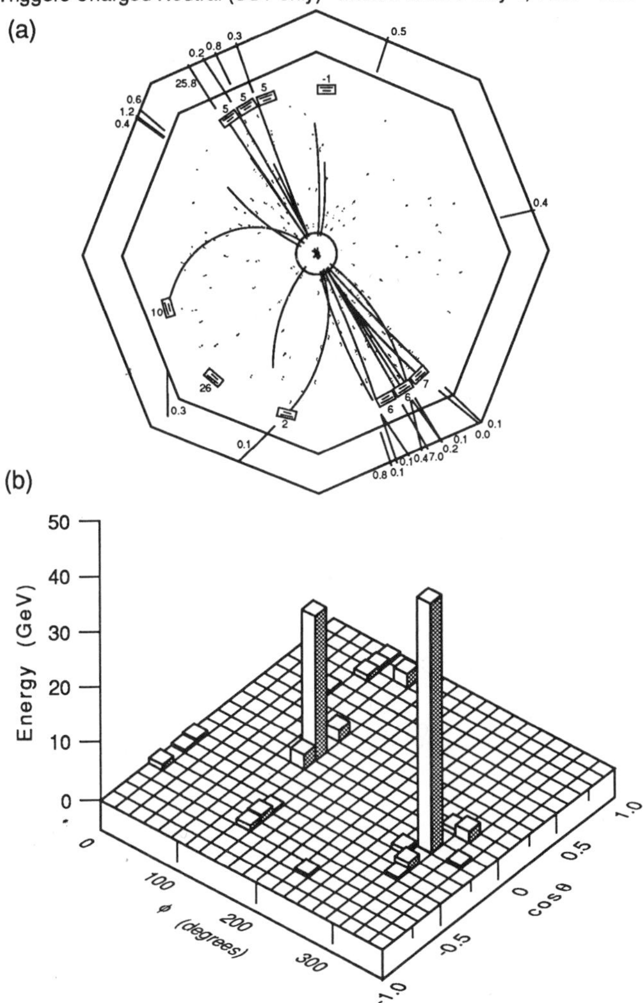

Fig. 5. (a) A computer reconstruction of a typical hadronic Z decay viewed along the beam axis. (b) A plot of the detected energy for this event as a function of the azimuthal angle and the cosine of the polar angle.

The random trigger is used to correct for beam-induced backgrounds. For example, in all MARK II SLC analyses, randomly triggered events are combined with Monte Carlo simulations of physical processes to give a complete simulation of both the physics and the backgrounds.

Z RESONANCE PARAMETERS[6]

We want to determine the Z boson resonance parameters by comparing the rate of Z formation in e^+e^- annihilation (Fig. 7) as a function of the center-of-mass energy, E, with that for a process with a known cross section, Bhabha (e^+e^-) scattering (Fig. 8). To accomplish this, we have to do four things:

1. measure E,

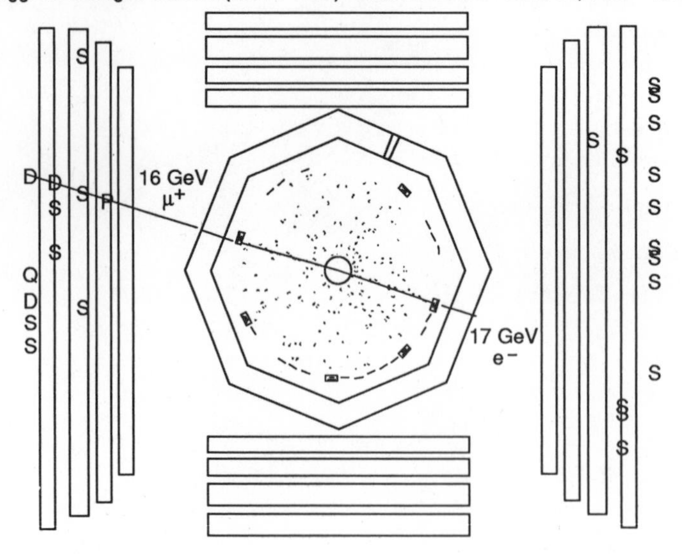

Fig. 6. A computer reconstruction of a Z decay into τ pairs.

Table 1. MARK II triggers.

Purpose	Trigger	Requirements
Z decays	charged	≥ 2 charged tracks with $p_t > 150$ MeV/c and $\lvert\cos\theta\rvert < 0.75$.
	neutral	A single deposition of ≥ 2.2 GeV in the endcap calorimeter or ≥ 3.3 GeV in the barrel calorimeter.
Luminosity	SAM	≥ 6 GeV in both detectors.
	MiniSAM	≥ 15 GeV in both detectors.
Diagnostic	random	Random beam crossings.
	cosmic	Taken between beam crossings.

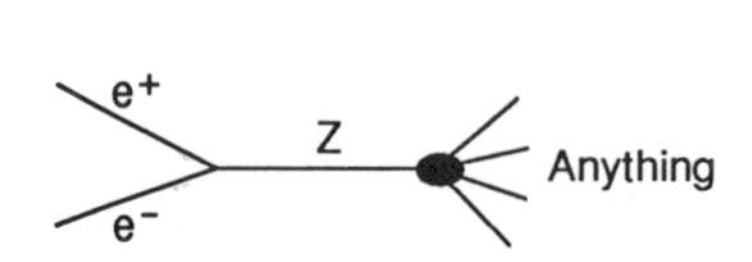

Fig. 7. Z formation in e^+e^- annihilation.

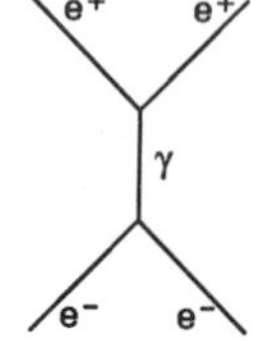

Fig. 8. Small-angle Bhabha scattering.

2. count Z's,
3. count Bhabha scatters, and
4. fit the ratios to obtain the Z parameters.

Absolute Energy Measurement

We have built spectrometers of a novel design to measure the absolute energies of both beams to high accuracy.[7] Figure 9 shows a schematic drawing of one of the spectrometers. The electron beam first passes through a horizontal bend and emits a horizontal swath of synchrotron radiation in the initial electron beam direction. It then passes through an accurately-measured spectrometer magnet which bends it down. Finally, it traverses a second horizontal bend to give another swath of synchrotron radiation in the direction of the outgoing beam. The two swaths of synchrotron radiation are intercepted by a phosphorescent screen. It is clear that the mean energy of beam can be measured from the knowledge of three quantities:

1. the magnetic-field integral of the spectrometer magnet,
2. the distance between the center of the spectrometer magnet and the screen, and
3. the distance between the two synchrotron radiation swaths on the screen.

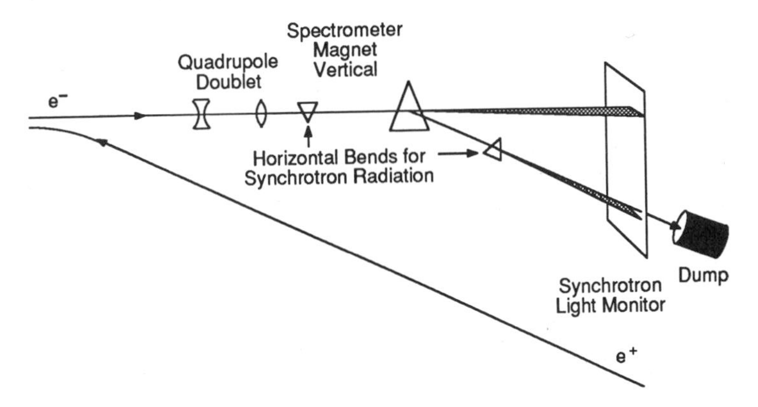

Fig. 9. Schematic of one of the energy spectrometers.

The spectrometer magnet field integral has been calibrated to a few parts in 10^5 by two independent techniques and is constantly measured by a rotating coil. The distance between the magnet center and the screen is determined to high precision by surveying techniques, and the distances on the screen are calibrated by accurately placed fiducial wires.

The systematic uncertainties in the measurement of each beam (itemized in Table 2) total to 20 MeV.

The energy spread of each beam is also measured to about 30% accuracy by the increased dispersion caused by the spectrometer magnet. The mean energy and energy spread are measured on every SLC pulse and are read by the MARK II on every trigger.

Table 2. Systematic uncertainties in the energy measurement of each beam.

Item	Uncertainty (MeV)
Magnetic measurement	5
Detector resolution	10
Magnet rotation	16
Survey	5
Total	20

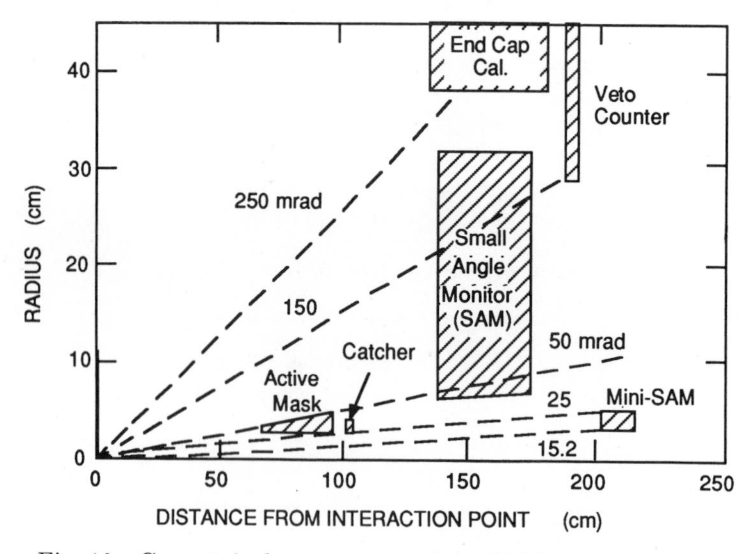

Fig. 10. Geometrical acceptances of the SAM and MiniSAM.

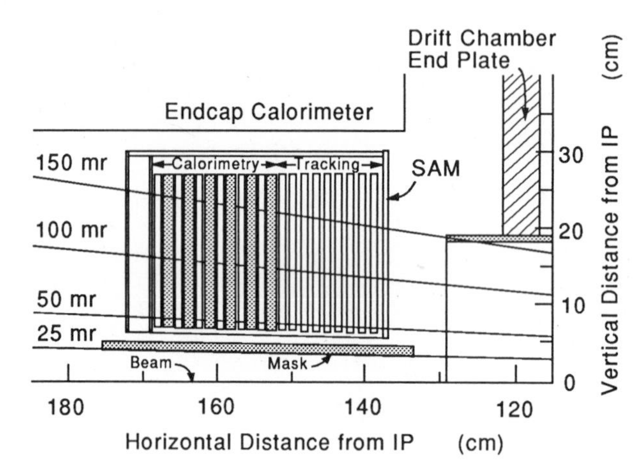

Fig. 11. The Small Angle Monitor (SAM).

Luminosity Measurements

To obtain the optimum absolute and relative luminosity measurements we use a well-defined fiducial region of the SAM to measure the absolute luminosity, while we use the total SAM and the MiniSAM to determine the relative, or point-to-point, luminosity. The geometrical acceptance of these detectors is illustrated in Fig. 10.

A drawing of the SAM is shown in Fig. 11. Each SAM consists of nine layers of drift tubes for tracking and a six-layer lead-proportional-tube sandwich for measuring the electron energy and position. A typical event is shown in Fig. 12. The tracking information is not always available due to backgrounds, but the calorimetric reconstruction of the electron pulse is unmistakable and background free. The angular resolution from the shower reconstruction is about a milliradian.

The technique for determining the absolute luminosity was to count events with both the electron and positron tracks in the angular region $65 < \theta < 160$ mrad with

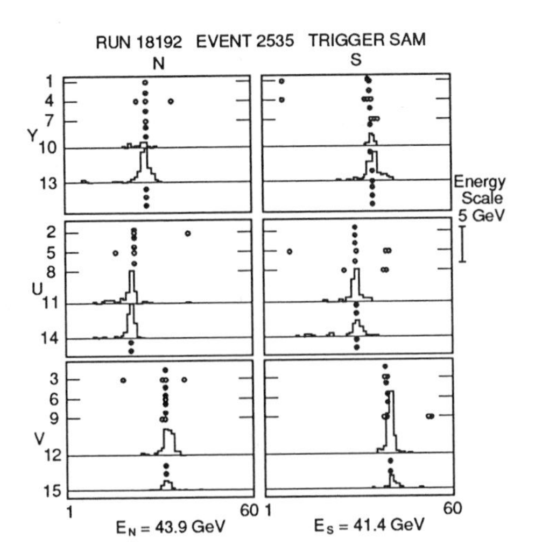

Fig. 12. A typical SAM event.

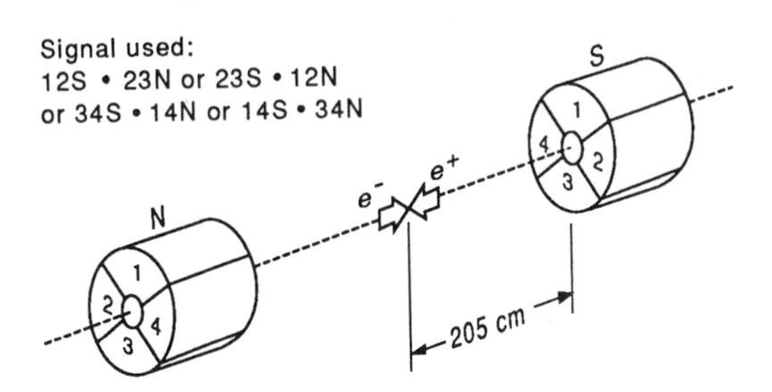

Fig. 13. The MiniSAMs.

unit weight and events with only one track in this with region with half weight. This is a standard technique to reduce the sensitivity of the measurement to possible misalignments and detector resolution.

The cross section corresponding to the precise region was calculated[8] to be 25.2 nb at 91.1 GeV.

The systematic uncertainties in the absolute luminosity measurement total 3.0% and are equally divided between unknown higher-order radiative corrections and the effect of detector resolution on the SAM precise region acceptance.

The most important part of the point-to-point luminosity measurement are the MiniSAMs, a drawing of which is shown in Fig. 13. These are simple tungsten-scintillator sandwiches divided into four quadrants which are separately read out.

The requirement for detecting a Bhabha scattering event in the MiniSAMs is:

1. a back-to-back pair of adjacent quadrants on each side, each of which has 25 GeV more energy than the other pair of quadrants on that side,
2. time-of-flight measurement in all quadrants with more than 18 GeV consistent with the Bhabha scattering.

The efficiency of the MiniSAMs varies from 91% to $>$ 99%, depending on scan point; and the backgrounds, measured from non-back-to-back quadrants, vary from 0 to 3.5%. We estimate the point-point systematic error to be the larger of 1% or the background subtraction.

Z Decay Event Selection

Z production is the dominant annihilation process, so the selection criteria can be quite loose. The only possible backgrounds come from beam-gas interactions and γ–γ interactions. Both of these process leave a large amount of energy in at most one of the forward-backward hemispheres.

Accordingly, the criteria for hadronic Z decays are:

1. ≥ 3 charged tracks from a cylindrical volume around the interaction point with a radius of 1 cm and a half-length of 3 cm, and
2. at least 0.05 E visible in both the forward and backward hemispheres.

These criteria give an efficiency of $94.5 \pm 0.5\%$.

The only major sources of backgrounds are from beam-gas and two photon interactions. With the above selection criteria, we have determined that both are negligible, much less than one event in the entire data sample.

To increase our statistical precision slightly, we also accept those leptonic Z decays for which the efficiency is high and the identification and interpretation is clear, namely, μ and τ pairs in the angular region $|\cos\theta| < 0.65$. To avoid backgrounds from γ–γ interactions, we require a minimum of 0.10 E visible energy for τ pairs. The efficiencies for detecting μ and τ pairs within the fiducial angular region are $99 \pm 1\%$ and $96 \pm 1\%$, respectively.

The data are shown in Fig. 14. Note that we plot an unusual quantity, but one that is closely related to what we actually measure, the cross sections for all hadronic decays and $\mu^+\mu^-$ and $\tau^+\tau^-$ with $|\cos\theta| < 0.65$.

Fits to the Data

We perform maximum-likelihood fits using Poisson statistics to a relativistic Breit–Wigner line shape

$$\sigma(E) = \frac{12\pi}{m^2}\frac{s\Gamma_{ee}(\Gamma - \Gamma_{inv})}{(s-m^2)^2 + s^2\Gamma^2/m^2}\left[1+\delta(E)\right] \quad , \tag{2}$$

where Γ is the total width and Γ_{inv} is the partial width into invisible decay modes, *i.e.*, into neutrinos and neutrino-like particles. Large effects due to initial state radiation, represented in Eq. (2) by $[1+\delta(E)]$, are calculated by an analytic form due to Cahn.[9] Alexander *et al.* have shown that this form has more than sufficient accuracy for our purposes.[10]

A Breit–Wigner shape has three parameters, a position, a width, and a height. We can fit for these three parameters as m, Γ, and Γ_{inv}, or equivalently, the number of neutrino species, N_ν.

The mass and width clearly determine the position and width of the resonance. The height is most sensitive to the third parameter, Γ_{inv}. This comes about because a Breit–Wigner is proportional to the partial width to the initial state times the partial width to the final state. The partial width to the initial state, e^+e^-, is well determined in the Standard Model. The final state can be taken to be all of the final states that we can see, in principle, in our detector, *i.e.*, all states except those into neutrino pairs (or pairs of neutrino-like objects).

Another way of viewing this is the following: If we could detect all of the final states, and if we integrated the resonance over energy, then we would find that the

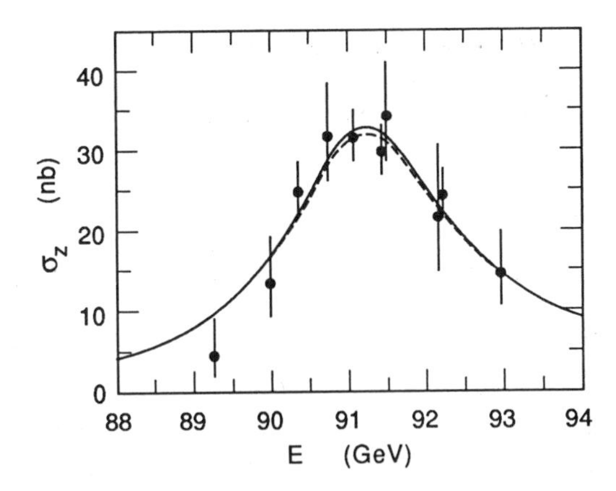

Fig. 14. The cross sections for all hadronic decays and $\mu^+\mu^-$ and $\tau^+\tau^-$ with $|\cos\theta| < 0.65$. The dashed curve represents the result of the Standard Model fit. The solid curve represents the free ν fit and the unconstrained fit, which are indistinguishable.

integral only depends on the width to the initial state. This is a statement that we produce a Z and that it subsequently decays with unit probability. What we do not detect, then, must be those decays into neutrino pairs.

We perform three fits which differ in their reliance on the Standard Model, and thus address different questions that one may wish to ask.

1. In the "Standard Model fit," m is the only parameter that is varied. The widths are taken to have their Standard Model values corresponding to the decays into five quarks and three charged and neutral leptons.
2. In the "free ν fit," both m and Γ_{inv} are allowed to vary. The visible width is constrained to its Standard Model value. The rationale for this twofold:
   (a) New particle production in the quark-lepton sector might be expected to show up first with the lightest of particles, which, from the three examples we have seen so far, are the neutrinos.
   (b) Visible new particle production would probably show up first in the observation of distinctive decays.
3. Finally, the "unconstrained fit" allows all three parameters to be varied.

These fits are displayed in Fig. 14. The mass values from the three fits are identical at

$$m = 91.14 \pm 0.12 \text{ GeV/c}^2 \quad . \tag{3}$$

All of the systematic errors in the mass determination are small, with the largest source of systematic uncertainty, included in the quoted errors, being 35 MeV for the absolute energy determination.

The total width Γ is only determined by the unconstrained fit. The value of

$$\Gamma = 2.42^{+0.45}_{-0.35} \text{ GeV} \tag{4}$$

is almost exactly the same as the Standard Model value of 2.45 GeV.[11] The errors on the width are large because a good measurement of the width requires substantial data at ± 2 GeV from the peak. We did not take very much data that far from the peak because the rate was just too low with our luminosity. Again all the systematic errors are small compared to the statistical error. The most significant contribution is 50 MeV from the uncertainty in the MiniSAM efficiency and background corrections.

The value

$$N_\nu = 2.8 \pm 0.6 \tag{5}$$

is taken from the free ν fit. This translates into the upper limit

$$N_\nu < 3.9 \text{ at } 95\% \text{ C.L.} \quad , \tag{6}$$

which provides strong evidence that the number of light neutrino species is limited to the three that we have already discovered. The quoted errors include a contribution of 0.45 from the uncertainty in the absolute luminosity measurement.

Relationship between the Mass and $\sin^2\theta_W$

The electroweak mixing angle, θ_W, can be expressed in terms of the Z mass by the relation

$$\sin 2\theta_W = \left(\frac{4\pi\alpha}{\sqrt{2}G_F m_Z^2(1-\Delta r)}\right)^{1/2} \quad , \tag{7}$$

where Δr represents weak radiative corrections. These corrections arise from loops and are sensitive to the masses of high mass particles.

The most common definition of $\sin^2\theta_W$ is the Sirlin form,[12] which is defined as

$$\sin^2\theta_W \equiv 1 - \frac{m_W^2}{m_Z^2} \tag{8}$$

For specific values of the two unknown masses in the Standard Model,

$$m_t = m_H = 100 \text{ GeV}/c^2 \quad , \tag{9}$$

our measured m_Z of 91.14 ± 0.12 implies

$$\sin^2\theta_W = 0.2304 \pm 0.0009 \quad . \tag{10}$$

The dependence of $\sin^2\theta_W$ on these two masses is shown in Fig. 15.

PARTONIC STRUCTURE OF HADRONIC DECAYS[13]

In addition to the underlying quark-antiquark structure of hadronic decays, QCD predicts the existence of multijet events due to gluon bremsstrahlung. By studying event shape parameters and counting jets, we can test the predictions of QCD and look for the existence of other processes.

In order to ensure well-measured momenta and a high tracking efficiency, the event selection is more restrictive for this study than for the Z line shape. Events were required to have at least seven charged tracks in the region $|\cos\theta| < 0.82$ and have at least 0.5 E visible in charged and neutral energy.

Shape Parameters

Figure 16 shows the distribution in thrust, T, which is defined

$$T = \left(\frac{\sum_i |p_i \cdot \hat{t}|}{\sum_i |p_i|}\right)_{max} \quad , \tag{11}$$

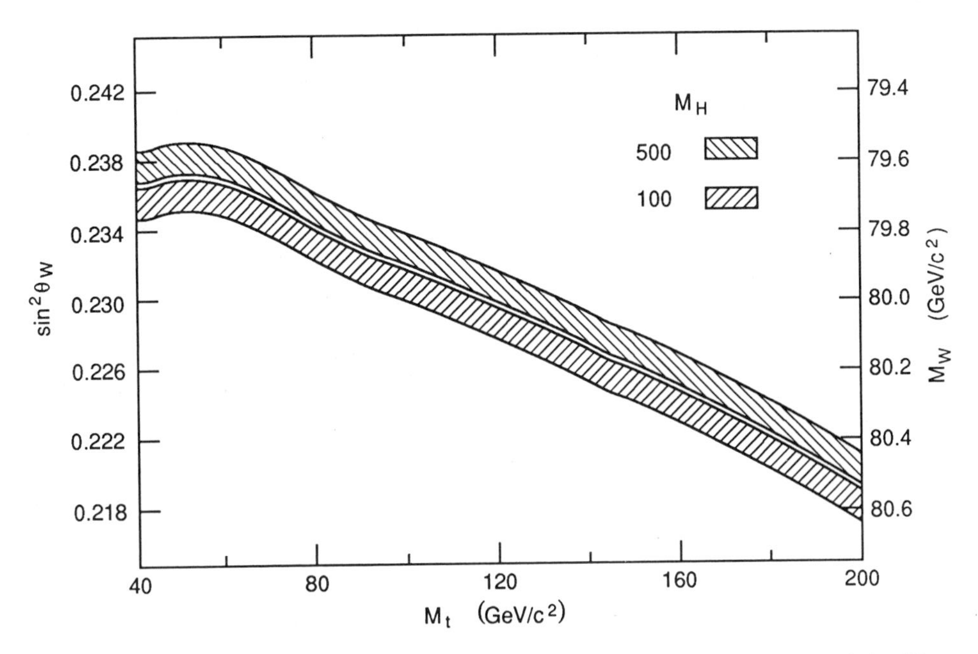

Fig. 15. $\sin^2\theta_W$ as a function of the mass of the top quark for two values of the Higgs boson mass. The bands represent ± 1 standard deviation about the values derived from Eq. (7).

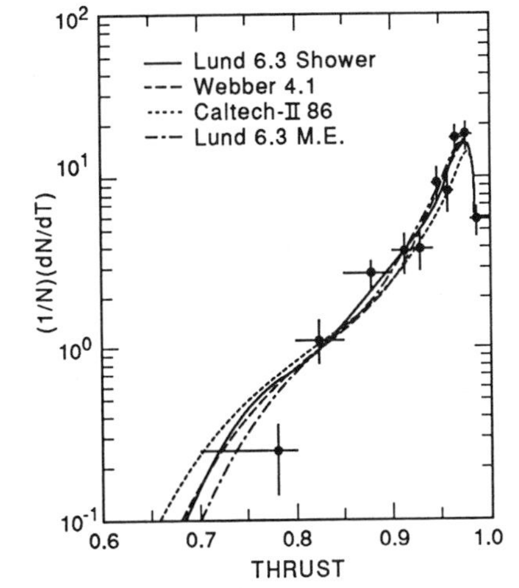

Fig. 16. Thrust values shown with four Monte Carlo simulations.

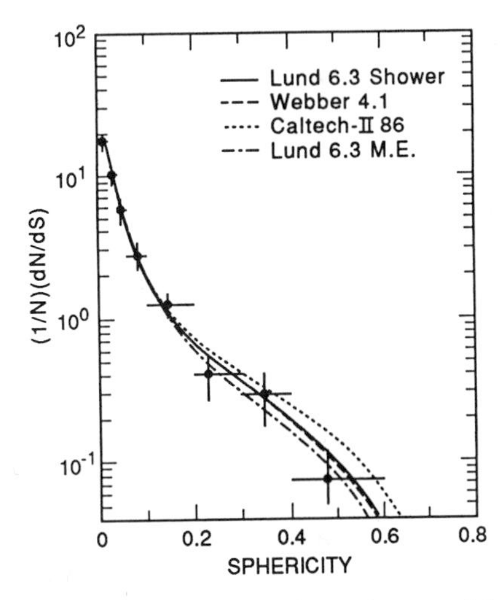

Fig. 17. Sphericity values shown with four Monte Carlo simulations.

where $\hat{t}$ is defined as the thrust axis. Thrust is a linear measure of the departure from a two-jet shape. The data are compared with four Monte Carlo simulations: the Lund shower parton model (Lund 6.3 shower),[14] the Webber–Marchesini parton shower model (Webber 4.1),[15] the Gottschalk–Morris parton shower model (Caltech–II 86),[16] and the Lund model based on the second-order QCD matrix element calculated by Gottschalk and Shatz (Lund 6.3 M.E.).[17] In general, the last of these models is not expected to agree well with the data, since it is explicitly incapable of producing more than four jets. For the thrust distribution, all of the models describe the data well.

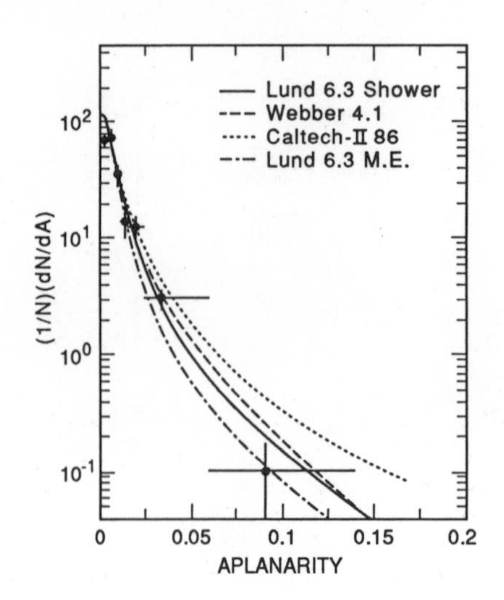

Fig. 18. Aplanarity values shown with four Monte Carlo simulations.

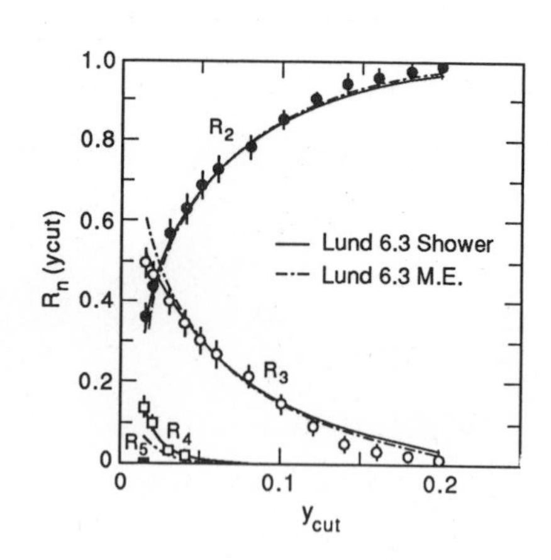

Fig. 19. The observed fraction of events with n jets as a function of y_{cut}. The curves show two Monte Carlo simulations.

Figure 17 shows the distribution in sphericity, S. This is also a measure of deviation from a two-jet shape, but using p_t^2 rather than p_t. It can be expressed as

$$S = \left(\frac{3}{2}\frac{\sum_i p_{t_i}^2}{\sum_i p_i^2}\right)_{min} . \tag{12}$$

Again the models describe the data well.

Figure 18 shows the distribution of aplanarity, A. It is a measure of amount of momentum out of an optimum plane. This is a measure of four or more jets, since conservation of momentum requires three jets to lie in a plane. Aplanarity can be written as

$$A = \left(\frac{3}{2}\frac{\sum_i p_{out_i}^2}{\sum_i p_i^2}\right)_{min} . \tag{13}$$

The parton shower models describe the data well, while the Lund matrix element model underestimates the amount of aplanarity, as expected.

Jet Counting

A cluster algorithm developed by the JADE group is used to count jets.[18] In each event, the quantity $y_{ij} = m_{ij}^2/E_{vis}^2$ is calculated for all pairs of particles i and j. The pair with the smallest invariant mass, m_{ij} is combined to form a pseudoparticle with four-momentum $p_i + p_j$. This procedure is repeated until the smallest y_{ij} exceeds an adjustable threshold value y_{cut}. The hadronic jets defined in this way have the property that they are very similar to partonic jets, as created in a QCD shower simulation.

Figure 19 shows the fraction of events with n jets as a function of y_{cut}. There is good agreement between the data and the Lund parton shower model.

Energy Dependence and Measurement of α_s

Figure 20 shows the mean values of three shape parameters and the fraction of three jet events for $y_{cut} = 0.08$ as a function of center-of-mass energy. The lower energy data come from the MARK II experiment[19] at 29 GeV and other experiments.[20–27]

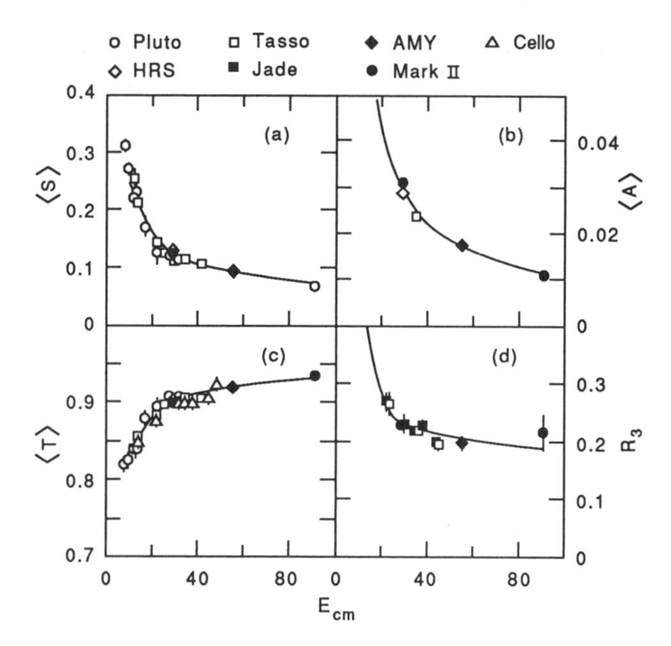

Fig. 20. The mean values of (a) sphericity, (b) aplanarity, (c) thrust, and (d) the three-jet fraction for a y_{cut} of 0.08 versus center-of-mass energy.

The curves show a strong dependence with energy at low energy and a more gradual dependence at high energy. The former is due to fragmentation effects, while the latter is due to the running of the strong coupling constant, α_s.

We have made a determination of α_s at 29 and 91 GeV with the MARK II[28] by counting the fraction of three jet events with y_{cut} between 0.04 and 0.14. Using the second-order calculation of Kramer and Lampe,[29] with $Q^2 = E^2$ in the $\overline{MS}$ renormalization scheme, we obtain

$$\alpha_s = 0.149 \pm 0.002 \pm 0.007 \text{ at } 29 \text{ GeV} \quad , \tag{14}$$

and

$$\alpha_s = 0.123 \pm 0.009 \pm 0.005 \text{ at } 91 \text{ GeV} \quad , \tag{15}$$

where the first error is statistical and the second systematic. These results, shown in Fig. 21, are consistent with the QCD prediction for the running of α_s.

SEARCHES FOR NEW QUARKS AND LEPTONS[30]

The great power of e^+e^- annihilation is that all pairs of fundamental particles with masses less than half the center-of-mass energy are copiously produced. We have started our searches for new particles with quarks and leptons, but the techniques are quite general and would have uncovered other types of new particles if they were present in sufficient numbers. Specifically, we have searched for the top quark, a fourth-generation charge $-1/3$ quark (b'), and heavy, unstable, neutral leptons. (We have, of course, also searched for new charged leptons, but due to limited statistics, our limits are not higher than those already obtained at TRISTAN.)

We expect new quarks and leptons to decay through virtual W decay. However, there are several other possibilities that we have explored. A b' quark could decay

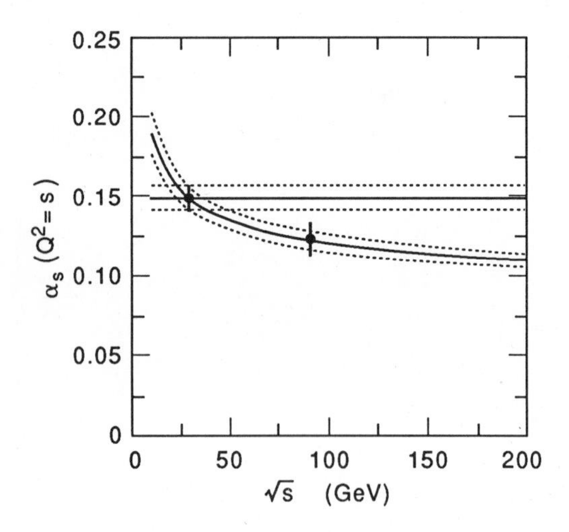

Fig. 21. The values of α_s determined from the three-jet fraction versus center-of-mass energy. The solid lines show a constant value of α_s and the values expected from QCD normalized to the data at 29 GeV. The dotted lines show the 1 σ errors on the values extrapolated from 29 GeV.

through penguin diagrams into b + gluon or $b + \gamma$ if there is sufficient mixing angle suppression of the charged current decays.[31] If charged Higgs bosons exist with masses less than those of new quarks or leptons, they, rather than virtual W's will mediate the decays of these fermions.

We use two complementary techniques to search for new particles: a search for isolated particles and a search for nonplanar events.

For both types of searches, the event selection is intermediate between that for the study of resonance parameters and that for the study of partonic properties. Explicitly, we require events to have at least six charged tracks in the region $|\cos\theta| < 0.85$ and have at least 0.1 E visible in charged and neutral energy. In addition, in order to insure that the events are well contained within the detector, the thrust axis of each event be in the region $|\cos\theta| < 0.80$.

In the search for isolated tracks, we define an isolation parameter, ρ, as follows: Excluding the candidate track, we use the jet-finding algorithm, with effectively a low value of y_{cut}, to form a number of jets. We then define

$$\rho \equiv \min_j[(2E(1 - \cos\theta_j))^{1/2}] \quad , \tag{16}$$

where E is the track energy in GeV and θ_j is the angle between the track and each jet axis. We define an isolated track to be one with $\rho > 1.8$.

Figure 22 shows the maximum ρ for each event along with the results of Monte Carlo simulations for the five known quarks, and, as an example, for a 35 GeV/c^2 top quark. The data agree well with the five-quark Monte Carlo, and only one event has an isolated track. The lower limits on masses of top and b' quarks decaying through virtual W bosons can be read off Fig. 23, which shows the expected number of events for these particles as a function of their masses.

Heavy neutral leptons will decay by mixing with light neutrinos, in analogy to the mixing which occurs among quarks. However, the mixing angles are completely

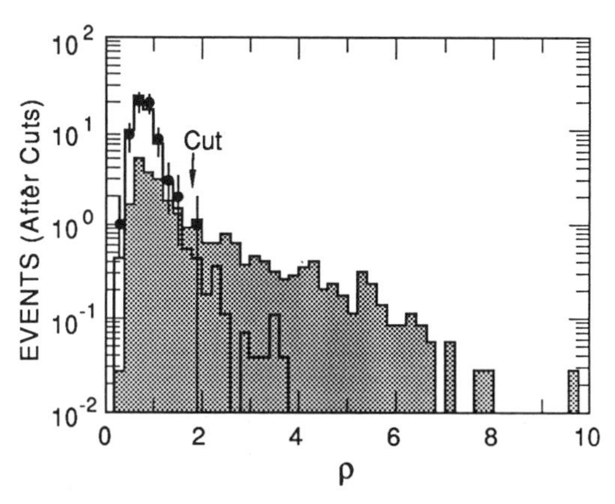

Fig. 22. The maximum isolation parameter ρ in each event. The solid line represents the result of a Monte Carlo simulation using the five known quarks. The shaded area shows the additional events that would be expected for a 35 GeV/c^2 top quark.

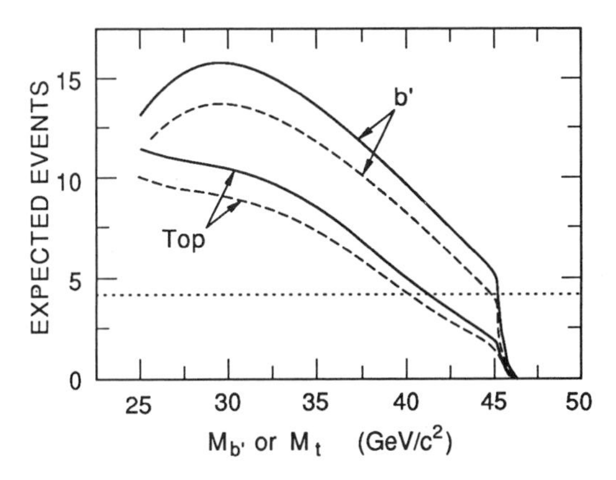

Fig. 23. The expected number of top and b' quark events (in which the quarks decay through virtual W's) with at least one isolated track are shown by the solid lines. The dashed curves indicate the central value minus the uncertainty from statistical and systematic errors. The dotted line represents the upper bound at 95% C.L. for one observed event with background subtracted.

unknown and could be quite small. We thus display results as a function of the mixing matrix element squared, $|U_{L^o\ell}|^2$. Figure 24 shows the results of the isolated track search proper.

For very small values of $\sum |U_{L^o\ell}|^2$, the lepton will live long enough to fail our normal vertex requirements. We have explicitly searched for such decays by searching for events with vertices away from the interaction point.[32] Figures 25 and 26 show the additional regions excluded, along with the results from previous experiments at lower energy.[33–37] In general, smaller values of $|U_{L^o\ell}|^2$ than excluded in Figs. 25 and 26, will be excluded by limits on Γ_{inv}.

A search for isolated photons, which are defined to be photons with $\rho > 3.0$, found no events and set a limit of $m_{b'} > 45.4$ GeV/c^2 at 95% C.L. for $B(b' \to b\gamma) \geq 25\%$.

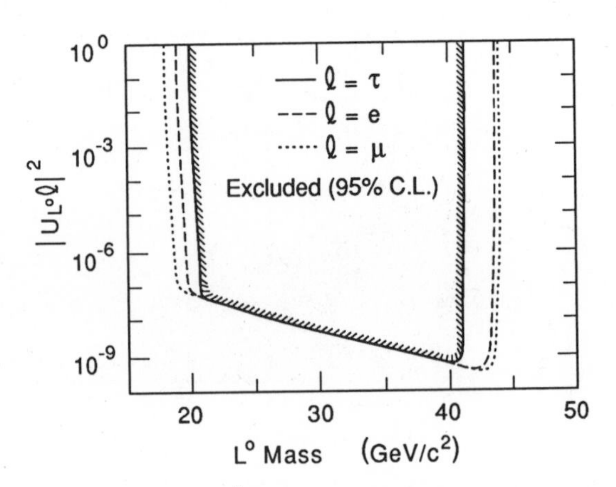

Fig. 24. Mass limits at the 95% C.L. for an unstable heavy neutral lepton L^0 as a function of mass and mixing matrix element squared for cases in which only one matrix element is important.

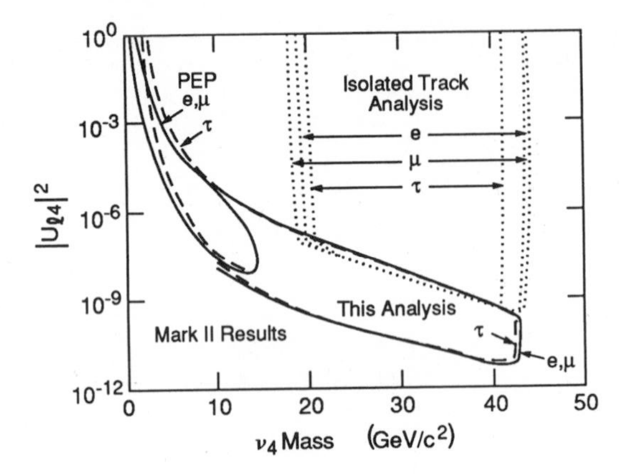

Fig. 25. Mass limits at the 95% C.L. for an unstable heavy neutral lepton ν_4 as a function of mass and mixing matrix element squared for cases in which only one matrix element is important. Also shown are the data from Fig. 24 and a MARK II search for detached vertices at PEP (Ref. 33).

The second type of search is for nonplanar events. Since three jets must lie in a plane, this, in effect, is a search for events with four or more jets. This is sensitive to new particle production since heavy new particles will decay into two or three jets. Since they are produced in pairs, they will yield events with four to six jets.

The variable which is used for this search is m_{out} defined as

$$m_{out} \equiv \frac{E_{cm}}{E_{vis}} \frac{1}{c} \sum |p_{out}| \quad , \tag{17}$$

where p_{out} is the momentum component out of the event plane as determined by the sphericity tensor, and the sum is taken over all charged and neutral particles. A nonplanar event is defined to be one with $m_{out} > 18$ GeV.

Figure 27 shows the distribution of m_{out} for the data along with the Monte Carlo simulation predictions for the five known quarks, and, as an example, a 35 GeV/c^2 b' quark decaying into a charged Higgs boson and a c quark.

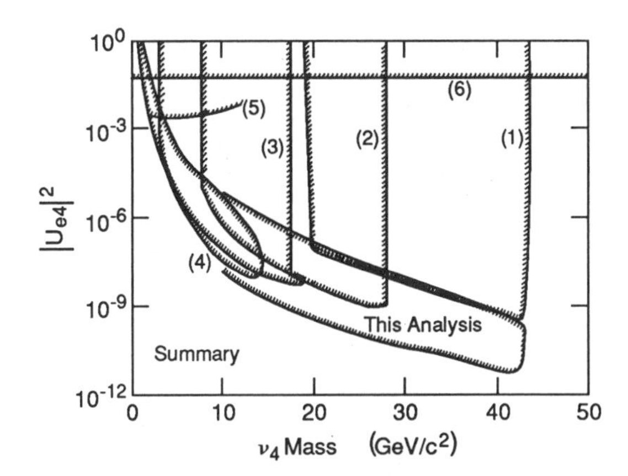

Fig. 26. Mass limits at the 95% C.L. for an unstable heavy neutral lepton ν_4 as a function of mass and mixing matrix element squared $|U_{L^o e}|^2$. Also shown are the data from (1) Fig. 24, (2) AMY (Ref. 35), (3) CELLO (Ref. 36), (4) MARK II at PEP (Ref. 33), (5) monojet searches at PEP (Ref. 37), and (6) universality (Ref. 34).

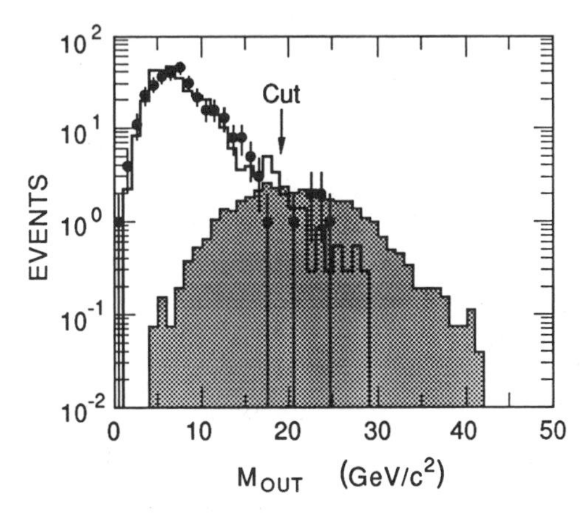

Fig. 27. The distribution of M_{out}. The solid line represents the result of a Monte Carlo simulation using the five known quarks. The shaded area shows the additional events that would be expected for a 35 GeV/c^2 b quark decaying into a charged Higgs boson and a c quark.

Six nonplanar events are found in the data, compared to five to twelve events expected from different Monte Carlo models of the known process. All of the limits are summarized in Table 3. The lower limits on the masses of new quarks and neutral leptons range from 40 GeV/c^2 to the kinematic limit of around 45 GeV/c^2.

New quarks that decay through virtual W's are ruled out by experiments at hadron colliders up to a mass of 77 GeV/c^2.[38,39] The new aspect of the limits in Table 3 are the ones which rule out all hadronic decay modes, which are difficult to detect with hadron colliders.

Although these limits are evaluated explicitly for the cases of new quarks and leptons, the techniques are fairly general and indicate that there is no new heavy particle production at the level of about 3% of the Z cross section.

Table 3. Summary of mass limits.

Particle	Decay Products (B.R. 100%)	Topology	Mass Limit (95% C.L.) (GeV/c^2)
top	bW^*	isolated track	40.0
	bW^*	m_{out}	40.7
	bH^+	m_{out}	42.5
b'	cW^*	isolated track	44.7
	cW^*	m_{out}	44.2
	cH^-	m_{out}	45.2
	b + gluon	m_{out}	42.7
	$b\gamma$, B.R.$\geq$ 25%	isolated photon	45.4
L^0	eW^*	isolated track	43.7
	μW^*	isolated track	44.0
	τW^*	isolated track	41.3

SUMMARY

The major physics results from the MARK II are:

- a precise measurement of the Z mass, $m = 91.14 \pm 0.12$ GeV/c^2,
- a measurement of the number of light neutrino species, $N_\nu = 2.8 \pm 0.6$, which corresponds to $N_\nu < 3.9$ at 95% C.L.,
- a study of the partonic structure of hadronic decays, which shows good agreement with the expectations from QCD,
- a measurement of the strong coupling constant, $\alpha_s = 0.123 \pm 0.009 \pm 0.005$ at 91 GeV, and
- a search for new quarks and leptons which sets lower limits on their masses in the range 40 to 45 GeV/c^2.

REFERENCES

1. B. Richter, *Nucl. Instrum. Meth.* **136**:47 (1976).
2. The nine MARK II institutions are: California Institute of Technology, University of California at Santa Cruz, University of Colorado, University of Hawaii, Indiana University, Johns Hopkins University, Lawrence Berkeley Laboratory, University of Michigan, and Stanford Linear Accelerator Center. The present members of the collaboration are: G. S. Abrams, C. E. Adolphsen, R. Aleksan, J. P. Alexander, D. Averill, J. Ballam, B. C. Barish, T. Barklow, B. A. Barnett, J. Bartelt, S. Bethke, D. Blockus, W. de Boer, G. Bonvicini, A. Boyarski, B. Brabson, A. Breakstone, F. Bulos, P. R. Burchat, D. L. Burke, R. J. Cence, J. Chapman, M. Chmeissani, D. Cords, D. P. Coupal, P. Dauncey, H. C. DeStaebler, D. E. Dorfan, J. M. Dorfan, D. C. Drewer, R. Elia, G. J. Feldman, D. Fernandes, R. C. Field, W. T. Ford, C. Fordham, R. Frey, D. Fujino, K. K. Gan, C. Gatto, E. Gero, G. Gidal, T. Glanzman, G. Goldhaber, J. J. Gomez Cadenas, G. Gratta, G. Grindhammer, P. Grosse-Wiesmann, G. Hanson, R. Harr, B. Harral, F. A. Harris, C. M. Hawkes, K. Hayes, C. Hearty, C. A. Heusch, M. D. Hildreth, T. Himel, D. A. Hinshaw, S. J. Hong,

D. Hutchinson, J. Hylen, W. R. Innes, R. G. Jacobsen, J. A. Jaros, C. K. Jung, J. A. Kadyk, J. Kent, M. King, S. R. Klein, D. S. Koetke, S. Komamiya, W. Koska, L. A. Kowalski, W. Kozanecki, J. F. Kral, M. Kuhlen, L. Labarga, A. J. Lankford, R. R. Larsen, F. Le Diberder, M. E. Levi, A. M. Litke, X. C. Lou, V. Lüth, G. R. Lynch, J. A. McKenna, J. A. J. Matthews, T. Mattison, B. D. Milliken, K. C. Moffeit, C. T. Munger, W. N. Murray, J. Nash, H. Ogren, K. F. O'Shaughnessy, S. I. Parker, C. Peck, M. L. Perl, F. Perrier, M. Petradza, R. Pitthan, F. C. Porter, P. Rankin, K. Riles, F. R. Rouse, D. R. Rust, H. F. W. Sadrozinski, M. W. Schaad, B. A. Schumm, A. Seiden, J. G. Smith, A. Snyder, E. Soderstrom, D. P. Stoker, R. Stroynowski, M. Swartz, R. Thun, G. H. Trilling, R. Van Kooten, P. Voruganti, S. R. Wagner, S. Watson, P. Weber, A. Weigend, A. J. Weinstein, A. J. Weir, E. Wicklund, M. Woods, G. Wormser, D. Y. Wu, M. Yurko, C. Zaccardelli, and C. von Zanthier.

3. G. S. Abrams *et al., Nucl. Instrum. Meth.* **A281**:55 (1989).
4. G. G. Hanson, *Nucl. Instrum. Meth.* **A252**:343 (1986).
5. G. S. Abrams *et al., IEEE Trans. Nucl. Sci.* **NS–25**:309 (1978) and **NS–27**:59 (1980).
6. G. S. Abrams *et al., Phys. Rev. Lett.* **63**:2173 (1989).
7. J. Kent *et al.,* SLAC–PUB–4922 (1989); M. Levi, J. Nash, and S. Watson, SLAC–PUB–4654 (1989); and M. Levi *et al.,* SLAC–PUB–4921 (1989).
8. F. A. Berends, R. Kleiss, and W. Hollik, *Nucl. Phys.* **B304**:712 (1988); S. Jadach and B. F. L. Ward, University of Tennessee report UTHEP–88–11–01 (1988).
9. R. N. Cahn, *Phys. Rev.* **D36**:2666 (1987), Eqs. (4.4) and (3.1).
10. J. Alexander *et al., Phys. Rev.* **D37**:56 (1988).
11. Calculated using the program EXPOSTAR, assuming $m_t = m_H = 100$ GeV/c^2. D. C. Kennedy *et al., Nucl. Phys.* **B321**:83 (1989).
12. A. Sirlin, *Phys. Rev.* **D22**:2695 (1980).
13. G. S. Abrams *et al., Phys. Rev. Lett.* **63**:1558 (1989).
14. T. Sjöstrand, *Comput. Phys. Commun.* **39**:347 (1986); T. Sjöstrand and M. Bengtsson, *Comput. Phys. Commun.* **43**:367 (1987); M. Bengtsson and T. Sjöstrand, *Nucl. Phys.* **B289**:810 (1987).
15. G. Marchesini and B. R. Webber, *Nucl. Phys.* **B238**:1 (1984); B. R. Webber, *Nucl. Phys.* **B238**:492 (1984).
16. T. D. Gottschalk and D. Morris, *Nucl. Phys.* **B288**:729 (1987).
17. T. D. Gottschalk and M. P. Shatz, *Phys. Lett.* **150B**:451 (1985); Caltech reports CALT–68–1172, –1173, –1199 (1985).
18. W. Bartel *et al., Z. Phys.* **C43**:325 (1986).
19. A. Petersen *et al., Phys. Rev.* **D37**:1 (1988); S. Bethke *et al., Z. Phys.* **C43**:325 (1989).
20. C. Berger *et al., Z. Phys.* **C12**:297 (1982).
21. M. Althoff *et al., Z. Phys.* **C22**:307 (1984).
22. D. Bender *et al., Phys. Rev.* **D31**:31 (1985).
23. S. Bethke *et al., Phys. Lett.* **B213**:235 (1988).

24. W. Braunschweig *et al., Z. Phys.* **C41**:359 (1988); W. Braunschweig *et al., Phys. Lett.* **B214**:286 (1988).

25. I .H. Park *et al., Phys. Rev. Lett.* **62**:1713 (1989).

26. H .J. Behrend *et al.,* DESY Preprint 89–019 (1989).

27. Y. K. Li *et al.,* KEK Preprint 89–34 (1989); Y. K. Li, S. Olsen (private communication).

28. S. Komamiya *et al.,* SLAC–PUB–5137 (1989).

29. G. Kramer and B. Lampe, *Fortschr. Phys.* **37**:161 (1989)

30. G. S. Abrams *et al., Phys. Rev. Lett.* **63**:2447 (1989).

31. V. Barger *et al., Phys. Rev.* **D30**:947 (1984); *Phys. Rev. Lett.* **57**:1518 (1986); W. Hou and R. G. Stuart, *Phys. Rev. Lett.* **62**:617 (1989).

32. C. K. Jung *et al.,* SLAC–PUB–5136 (1989).

33. C. Wendt *et al., Phys. Rev. Lett.* **58**:1810 (1987).

34. M. Gronau, C. N. Leung, and J. L. Rosner, *Phys. Rev.* **D29**:2359 (1984); V. Barger, W. Y. Keung, and R. J. Phillips, *Phys. Lett.* **B141**:126 (1984).

35. N. M. Shaw *et al., Phys. Rev. Lett.* **63**:1342 (1989).

36. H.-J. Behrend *et al., Z. Phys.* **C41**:7 (1988).

37. F. J. Gilman and S. H. Rhie, *Phys. Rev.* **D32**:324 (1985).

38. C. Albajar *et al., Z. Phys.* **C37**:505 (1988).

39. F. Abe *et al.,* University of Pennsylvania Report UPR–0172E.

PHYSICS IN $p\bar{p}$ COLLISIONS

Felicitas Pauss

CERN, Geneva, Switzerland

ABSTRACT

A review of experimental results obtained by UA1 and UA2 before the upgrading of the Proton–Antiproton Collider at CERN is presented. Topics covered in these lectures are the physics of the Intermediate Vector Bosons, heavy-flavour production, B^0–$\bar{B}^0$ mixing, the search for the top quark, and searches for physics beyond the Standard Model. We give preliminary results from the 1988/89 Collider run at CERN [UA1, upgraded UA2 and improved CERN $p\bar{p}$ Collider (ACOL)]. The physics potential of the full upgrading is discussed as well. Also, where available, results from the Fermilab Tevatron Collider are quoted. And finally, from the results obtained we infer the possible features of physics at future hadron colliders (LHC and SSC).

1. INTRODUCTION

In a series of data-taking runs between 1982 and 1985, an integrated luminosity of close to 1 pb^{-1} was accumulated at the CERN $p\bar{p}$ Collider by each of the two major experiments, UA1 and UA2. The study of general event characteristics of hadron–hadron collisions at $\sqrt{s}$ = 630 GeV has provided revealing information about the physics of the Standard Model (SM). A long list of important results covering a wide range of physics topics has come from the analyses of these data, e.g. the detailed measurements of jet production and fragmentation properties, the production and decay characteristics of the charged and neutral Intermediate Vector Bosons (IVBs), heavy-flavour production, B^0–$\bar{B}^0$ mixing, etc. The main objective of all these studies is to know if the SM of electroweak and strong interactions is verified by the data, or if deviations are found which would hint at new physics.

Tests of the electroweak sector have been performed in measuring the SM parameters (Section 2). The strong sector (QCD) has been explored by studying jets, photons, the production of IVBs (Section 2), and the production of heavy flavour (Section 3). Within the minimal version of the SM there are three more elementary particles still waiting to be discovered, namely the ν_τ, the top (t) quark, and the neutral Higgs boson. Even though the ν_τ has not yet been discovered directly, there are indirect arguments for its existence, e.g. observation of the $W \to \tau\nu$ decay at the rate consistent with e–μ–τ universality of the weak charged-current coupling at $Q^2 = m_W^2$. The upper limit of 1.2×10^{-3} (90% CL) on $BR(B^0 \to \mu^+\mu^- X)$ is strong indirect evidence that the t-quark exists. At the CERN $p\bar{p}$ Collider, the main sources of t-quarks are the decay $W \to t\bar{b}$, if kinematically allowed, and QCD production of $t\bar{t}$ pairs by gluon–gluon fusion or $q\bar{q}$ annihilation. At Fermilab, QCD $t\bar{t}$ production is the dominant production mechanism for all m_t. The search for the t-quark at CERN and Fermilab will be discussed in Section 4.

Particle Physics: Cargèse 1989
Edited by M. Lévy *et al.*
Plenum Press, New York, 1990

One of the most important discoveries still to be pursued is the experimental observation of the Higgs boson—which is believed to be the source of particle masses in the SM. The Higgs mass itself is, unfortunately, not predicted ($m_H \leqslant 1$ TeV/c^2), and the only 'safe' experimental limit on the Higgs mass is $m_H > 15$ MeV/c^2. There are two ways of searching for the Higgs at existing Colliders [1]: i) $p\bar{p} \to W^* \to WH + X$ (or $W \to W^*H$), i.e. the production of the Higgs with either a real or a virtual IVB; and ii) $p\bar{p} \to H + X$. In the first case the number of events expected is small, even for a very light Higgs. For example, for 1 pb^{-1}, one expects about four events (H + IVB) produced for $m_H < 1$ GeV/c^2 at $\sqrt{s} = 630$ GeV. A decay signature of $H \to e^+e^-$ ($\mu^+\mu^-$, $\pi^+\pi^-$, or K^+K^-) could lead to possible background-free detection of the Higgs, but, owing to the small branching ratios, a very high statistics W/Z data sample is required. At present $p\bar{p}$ Collider energies the dominant mechanism for Higgs production is through gluon–gluon fusion via a heavy quark loop. The Higgs decays into the heaviest fermion pair, i.e. into $b\bar{b}$, if $m_H > 10$ GeV/c^2. Even though in this case the rates are large enough to observe the Higgs up to masses of about 50 GeV/c^2 (about 100 GeV/c^2) at $\sqrt{s} = 630$ GeV (1.8 TeV), the background from QCD jets is about a factor of 10^5 above a possible $H \to b\bar{b}$ signal. It therefore seems very unlikely that the Higgs will be discovered at existing hadron colliders.

There is a general consensus, however, that the SM is not expected to be our ultimate theory. Theorists have proposed numerous possible solutions to the well-known problems of the SM. For example, compositeness [2] could solve the Flavour problem, i.e. the understanding of the number of matter species. The Hierarchy problem—the understanding of the origin of the different particle masses—could be solved by supersymmetry [3]. It is expected that new physics should show up for $m \leqslant O(\text{TeV}/c^2)$. Unfortunately, the precise mass scale of this new physics is, as in the case of the Higgs, not known. Searches for physics beyond the SM can therefore provide limits on possible new mass scales of new physics (Section 5).

This review includes mostly data from the pre-ACOL era. These results have already been summarized and published in various conference proceedings. Therefore, instead of repeating details for the selected physics topics, the relevant results are stated and the interested reader is referred to the original publications.

The CERN $p\bar{p}$ Collider has been undergoing an important upgrading programme (addition of a new separate antiproton collector ring and six-bunch operation of the SPS). The performance reached during the first run of ACOL at the end of 1987 was far below design value ($L \approx 50$ nb^{-1} only). However, the following two runs (autumn 1988 and spring 1989) have been very successful, reaching an order of magnitude increase in luminosity. Since the end of June 1989, the total integrated luminosity delivered by the machine is 8.4 pb^{-1}, out of which 4.7 pb^{-1} (6.7 pb^{-1}) are recorded on tape by UA1 (UA2). Preliminary results and expectations from the final analysis of the present data samples will be discussed.

The electromagnetic (e.m.) calorimeters of the UA1 detector have been removed for the 1988 and 1989 runs in order to prepare for the installation of the new uranium–TMP calorimeter [4]. The muon detection capability has been improved by adding additional iron shielding in the forward region. The calorimetric measurements rely on the hadron calorimeters alone.

UA2 has been running since 1987 with the upgraded detector [5]. Full e.m. and hadronic calorimeter coverage has been achieved with the addition of new end-cap modules covering the angular regions $6° < \theta < 40°$ with respect to the beam directions. The electron identification is improved by a new central detector assembly consisting of a jet chamber, a silicon-pad detector, two layers of transition radiation detectors, and a scintillating-fibre detector.

The first physics run at $\sqrt{s} = 1.8$ TeV $p\bar{p}$ collisions at the Fermilab Tevatron Collider took place in 1987, resulting in a delivered luminosity of ~ 50 nb^{-1}, out of which ~ 30 nb^{-1} have been collected by the Collider Detector at Fermilab (CDF) [6]. Published results from these data will be discussed in these lectures. The Tevatron started a long collider run in June 1988, which ended in June 1989, with excellent performance of the machine. The luminosity delivered is 9.5 pb^{-1}, with about 50% data-taking efficiency of the CDF detector (i.e. 4.7 pb^{-1}). Preliminary results and expectations from the final analysis of the present data sample will be discussed. The next Collider run at Fermilab is scheduled for the beginning of 1991.

Finally, some examples will be given of the possible features of physics at future hadron colliders (LHC and SSC), inferred from our present understanding of physics at hadron colliders.

2. PRODUCTION AND DECAY OF IVBs

2.1 IVB masses and Standard Model parameters

The pre-ACOL data samples used in the UA1 analysis correspond to integrated luminosities of 136 nb^{-1} and 568 nb^{-1} at the Collider energies of $\sqrt{s}$ = 546 GeV and 630 GeV, respectively. The UA2 results are based on 142 nb^{-1} and 768 nb^{-1} at the two energies. The full pre-ACOL statistics of UA1 [7] and UA2 [8] result in a total of about 650 $W \rightarrow (e, \mu, \tau)\nu$ decays and about 100 $Z \rightarrow e^+e^-$ or $\mu^+\mu^-$ decays, observed in both experiments. The major fraction of the statistics is dominated by the $W \rightarrow e\nu$ and $Z \rightarrow e^+e^-$ channels.

The W and Z masses are the two parameters that can be directly measured by the Collider experiments. The most precise values for m_W and m_Z obtained from the electron channel are [9]

$$m_W = 82.7 \pm 1.0 \pm 2.7 \text{ GeV}/c^2 \quad \text{(UA1)},$$

$$m_W = 80.2 \pm 0.6 \pm 0.5 \pm 1.3 \text{ GeV}/c^2 \quad \text{(UA2)},$$

$$m_Z = 93.1 \pm 1.0 \pm 3.1 \text{ GeV}/c^2 \quad \text{(UA1)},$$

$$m_Z = 91.5 \pm 1.2 \pm 1.7 \text{ GeV}/c^2 \quad \text{(UA2)}.$$

The first errors quoted are statistical, followed by the systematic uncertainties. In the case of m_W, UA2 quotes the systematic uncertainties separately for transverse-mass determination and energy scale. The errors are already dominated by the energy-scale uncertainties of the e.m. calorimeters (about $\pm 3\%$ for UA1 and about $\pm 1.5\%$ for UA2).

From a data sample corresponding to 25.3 nb^{-1} integrated luminosity at $\sqrt{s}$ = 1.8 TeV, CDF obtained 22 $W \rightarrow e\nu$ events from an analysis with missing transverse energy $\not{E}_T$ larger than 25 GeV and with electron transverse energy larger than 15 GeV. A fit of the expected spectrum to the measured transverse mass distribution of the 22 events leads to [10]

$$m_W = 80.0 \pm 3.3 \pm 2.4 \text{ GeV}/c^2 \quad \text{(CDF)},$$

where the first error is statistical, and the second error, which is systematic, is dominated by the uncertainty in the absolute energy scale.

Preliminary results on m_W and m_Z from the new data samples have been presented by UA2 and CDF. The improved mass values quoted by UA2 obtained from $W \rightarrow e\nu$ and $Z \rightarrow e^+e^-$ decays are [11]:

$$m_W = 80.0 \pm 0.4 \text{ (stat.)} \pm 0.4 \text{ (syst.)} \pm 1.2 \text{ (scale) GeV}/c^2 \quad \text{(UA2)}$$

$$m_Z = 90.2 \pm 0.6 \text{ (stat.)} \pm 1.4 \text{ (scale) GeV}/c^2 \quad \text{(UA2)}.$$

CDF obtained new values for m_W and m_Z by combining both the e- and μ-channel. The resulting preliminary values quoted are [12]:

$$m_W = 80.0 \pm 0.6 \text{ (stat. + syst.)} \pm 0.2 \text{ (scale) GeV}/c^2 \quad \text{(CDF)}$$

$$m_Z = 90.9 \pm 0.3 \text{ (stat. + syst.)} \pm 0.2 \text{ (scale) GeV}/c^2 \quad \text{(CDF)}.$$

These numbers have to be compared with the first m_Z measurement by the Mark II detector at the SLAC Linear Collider [13]

$$m_Z = 91.11 \pm 0.23 \text{ GeV}/c^2 \qquad \text{(Mark II)} .$$

One of the important tests of the electroweak model is the need for radiative corrections. The relation between the radiative correction (Δr) and m_W and m_Z is given by

$$(1 - \Delta r) = (A^2/m_W^2)/[1 - (m_W^2/m_Z^2)] ,$$

where $A = (\pi\alpha/\sqrt{2}G_F)^{1/2} = 37.2810 \pm 0.0003$ GeV. With this relation, we obtain [9]:

$$\Delta r = 0.036 \pm 0.100 \pm 0.067 \qquad \text{(UA1)} ,$$

$$\Delta r = 0.068 \pm 0.087 \pm 0.030 \qquad \text{(UA2)} .$$

A more precise value of Δr can be arrived at by using $\sin^2 \theta_w = 0.233 \pm 0.003 \pm 0.005$ from a world average of neutral-current neutrino experiments as additional input [14]:

$$\Delta r = 0.127 \pm 0.023 \pm 0.060 \qquad \text{(UA1)} ,$$

$$\Delta r = 0.068 \pm 0.022 \pm 0.032 \qquad \text{(UA2)} .$$

The experimental results are summarized in Fig. 1, where correlations between the uncertainties of the m_Z and m_Z–m_W measurements are shown. As can be seen from the error ellipse, the measurements of the IVB masses are not yet sensitive to the radiative corrections.

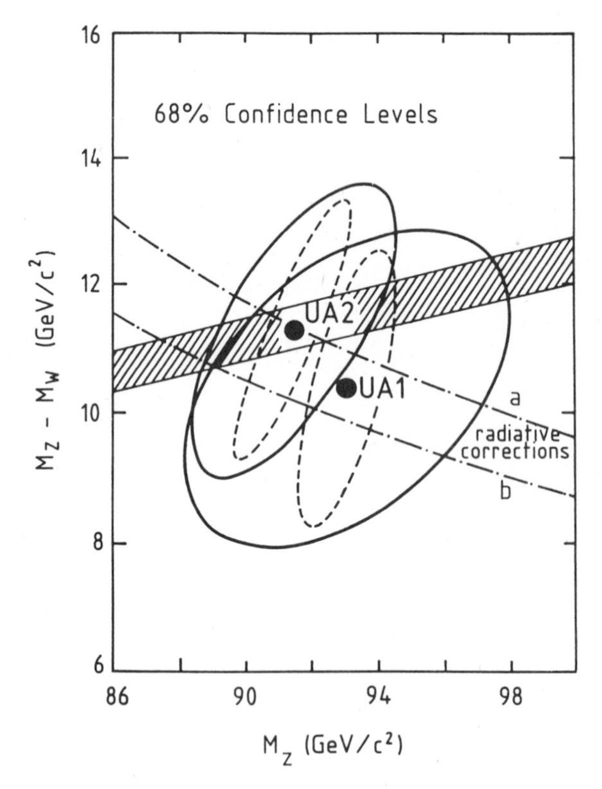

Fig. 1 Confidence contours (68%) in the (m_Z–m_W) versus m_Z plane, taking into account the statistical error (dashed line) and with statistical and systematic errors combined in quadrature (solid line). Only published pre-ACOL results are shown. The shaded region is allowed by the average of recent low-energy measurements of $\sin^2 \theta_w$. The upper dashed-dotted curve (a) includes radiative corrections, the lower one (b) is without radiative corrections.

The one-loop radiative correction on the IVB masses has been calculated [15] to be

$$\Delta r = 0.0713 \pm 0.0013\,,$$

assuming that $m_t = 45\ \mathrm{GeV}/c^2$ and $m_H = 100\ \mathrm{GeV}/c^2$. The value of Δr decreases with increasing t-quark mass ($m_t \approx 250\ \mathrm{GeV}/c^2$ leads to $\Delta r = 0$). Hence a precise measurement of Δr provides information on m_t. A global fit to all neutral-current data and to the W and the Z mass would correspond to $m_t < 180\ \mathrm{GeV}/c^2$ (90% CL) for $m_H = 100\ \mathrm{GeV}/c^2$ and assuming that only three fermion families exist [16].

Recently a new global fit to electroweak data has been performed, including the recent SLC, CDF, and UA2 measurements, and new CHARM II data, which results in [17]

$$m_t = 132^{+31}_{-37}\ \mathrm{GeV}/c^2\,.$$

2.2 IVB production cross-sections and e–μ–τ universality

Figures 2a,b show the UA1 and UA2 values for $\sigma\cdot\mathrm{BR}(W\to\ell\nu)$ and $\sigma\cdot\mathrm{BR}(Z\to\ell^+\ell^-)$ at $\sqrt{s} = 546$ GeV and at $\sqrt{s} = 630$ GeV. The experimental errors are dominated by systematic uncertainties in the luminosity in the case of $W \to e\nu$ and by statistics in the case of $Z \to e^+e^-$ [9]. Also shown are the theoretical predictions and their error bands on the total W and Z production cross-sections taken from Ref. [18]. The leptonic branching ratios for $W \to \ell\nu$ and $Z \to \ell^+\ell^-$ have been calculated for $m_t > (m_W - m_b)$. The theoretical predictions are affected by several uncertainties, e.g. i) various choices of structure functions and Q^2-scale, ii) uncertainty in $\mathrm{BR}(W \to \ell\nu)$ and $\mathrm{BR}(Z \to \ell^+\ell^-)$ due to the as yet unknown t-quark mass. The sensitivity of the experimental results to the t-quark mass and to the choice of structure functions [19] is illustrated in Figs. 3a,b.

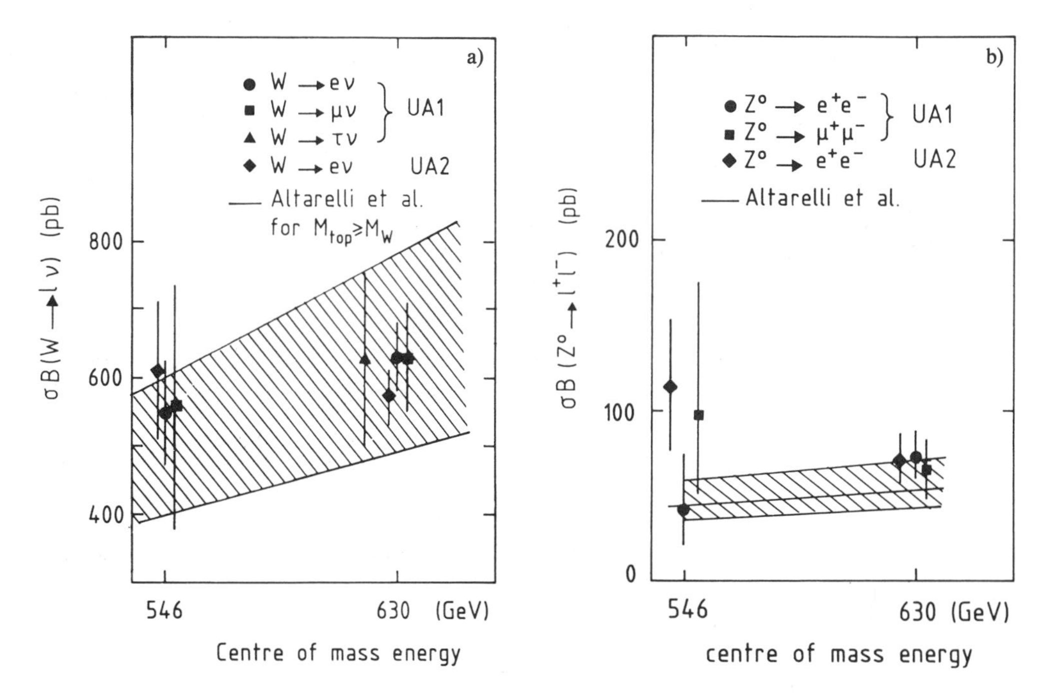

Fig. 2 Partial cross-sections versus c.m. energy: a) for $W \to \ell\nu$ production and b) for $Z \to \ell\ell$ production. The theoretical predictions and their error bands on the total W and Z production cross-sections are taken from Ref. [18], and the leptonic branching ratios for $W \to \ell\nu$ and $Z \to \ell\ell$ have been calculated under the assumption that $m_t > (m_W - m_b)$.

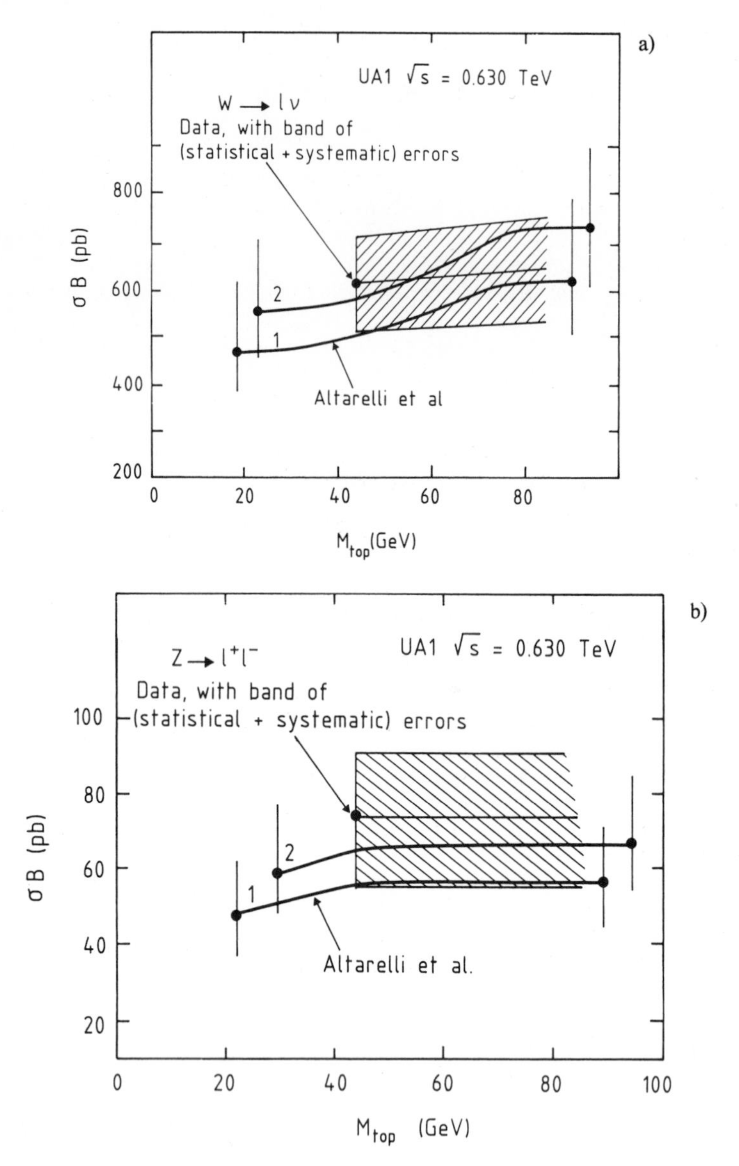

Fig. 3 Measurements and theoretical predictions for a) $W \rightarrow \ell\nu$ and b) $Z \rightarrow \ell\ell$ partial cross-sections as a function of m_t. The line labelled '1' represents the predictions [18] used in Figs. 2a,b. The line labelled '2' includes a 20% increase due to the BCDMS results on structure function parametrization [19]. The uncertainties are indicated by the error bars. The hatched band represents the measurements.

An update on the cross-section measurement has been presented by UA2 using the new data samples. The preliminary results quoted are [20]

$$\sigma \cdot BR(W \rightarrow e\nu) = 630 \pm 20\,(stat.) \pm 50\,(syst.) \quad pb$$

$$\sigma \cdot BR(Z \rightarrow e^+e^-) = 61 \pm 7\,(stat.) \pm 5\,(syst.) \quad pb\,.$$

The W production cross-section at the Tevatron energy of 1.8 TeV is expected to be about a factor of 3 larger compared with the CERN Collider energy of 0.63 TeV. CDF has published the measurement of $\sigma \cdot BR(W \rightarrow e\nu)$ obtained from a data sample of 25 nb^{-1} integrated

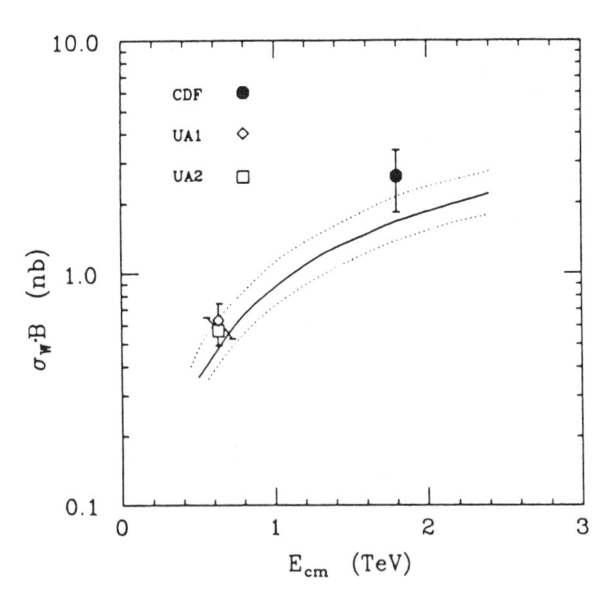

Fig. 4 Cross-section times branching ratio for $W \to e\nu$ versus E_{cm} measured by UA1, UA2 and CDF [10]. The prediction is from Ref. 18, adjusted for a W mass of 80 GeV/c^2. The dotted lines indicate the 1σ error limits for the theoretical curve.

luminosity. Using the 22 observed $W \to e\nu$ events and correcting for background and efficiency, CDF obtains [10]

$$\sigma \cdot BR = 2.6 \pm 0.6 \pm 0.5 \text{ nb} .$$

This result is shown together with the UA measurements in Fig. 4. The expected increase in W production is observed.

The UA1 data provide the first experimental verification of the universality of weak charged- and neutral-current coupling at $Q^2 = m_W^2$ (m_Z^2). Defining the weak charged-coupling constants by $(g_i/g_j)^2 = \Gamma(W \to \ell_i\nu_i)/\Gamma(W \to \ell_j\nu_j)$, and similarly for the weak neutral-coupling constants k, UA1 obtains [9]

$$g_\mu/g_e = 1.00 \pm 0.07 \text{ (stat.)} \pm 0.04 \text{ (syst.)} ,$$

$$g_\tau/g_e = 1.01 \pm 0.10 \text{ (stat.)} \pm 0.06 \text{ (syst.)} ,$$

$$k_\mu/k_e = 1.02 \pm 0.15 \text{ (stat.)} \pm 0.04 \text{ (syst.)} .$$

2.3 IVB transverse-momentum distribution

In the framework of the QCD improved Drell-Yan model for quark–antiquark annihilation, the radiation of gluons from the incoming partons gives rise to a non-zero IVB transverse momentum and leads to the associated emission of high transverse-momentum jets. The normalized W transverse-momentum distributions obtained by UA1 and UA2 are shown in Fig. 5 [21]. The UA2 results are shown only for $p_T^W > 15$ GeV/c. The experimental data are compared with the theoretical predictions of Ref. [18]; also shown are the theoretical uncertainties indicated by the hatched band. The measurements agree well with the QCD prediction, except possibly at the very highest values of p_T^W. For $p_T^W > 60$ GeV/c, the theory predicts (0.8 ± 0.3) events for a total of three events observed (one event by UA2 and two events by UA1). The UA1 events have been discussed in detail in Ref. [22]. These events contain two high-p_T jets with an invariant mass compatible with the W mass, which makes their interpretation in terms of QCD correction to the W production somewhat unlikely. From the preliminary p_T^W and p_T^Z distributions shown by CDF and UA2, no striking deviations from

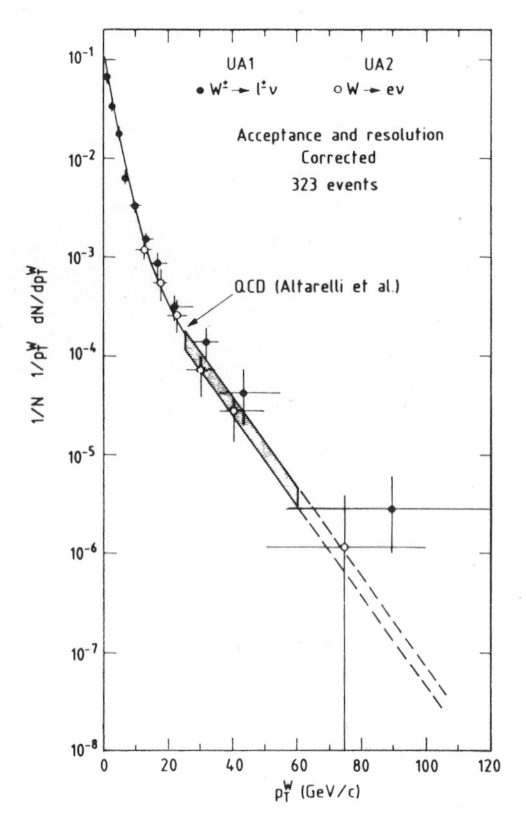

Fig. 5 W transverse-momentum distributions from UA1 (full points) and UA2 (open points). Only data with $p_T^W >$ 15 GeV/c are shown for UA2. The curves are QCD predictions as reported in Ref. [22], based on calculations of Ref. [18] and extrapolating for high p_T^W.

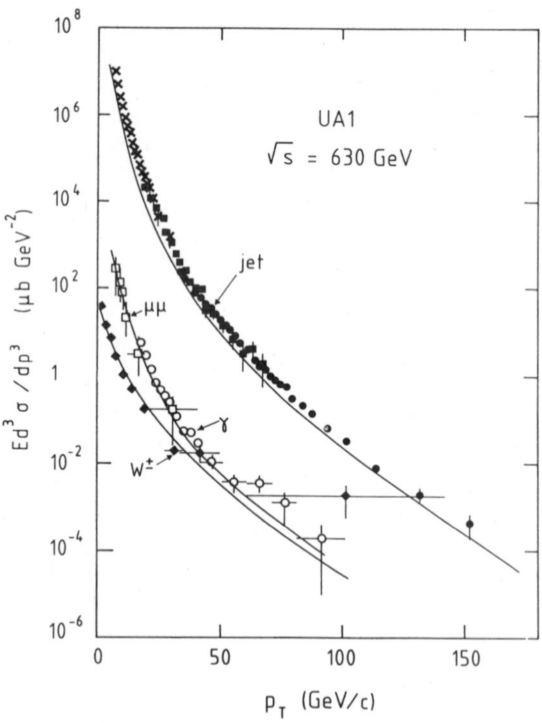

Fig. 6 Inclusive transverse-momentum distributions for W's, jets, direct photons, and nearly on-shell photons from low-mass dimuons as measured in the UA1 experiment [9]. The curves are QCD predictions as described in the text.

the QCD predictions are observed. UA1 has reported two new high-p_T^W events in the μ-channel. The final analyses of the large new data samples at CERN and Fermilab are expected to clarify the question of a possible excess of high-p_T W's.

Figure 6 displays the inclusive p_T cross-sections for jets [23], direct photons [24], low-mass Drell–Yan $\mu^+\mu^-$ pairs [25], and W production [22], as measured by UA1. The QCD predictions are from Stirling [26], scaled by 1.5 for the jet cross-section, from Aurenche et al. [27] for direct photons, and the line describing the W cross-section is from Altarelli et al. [18]. For p_T^W, $p_T^\gamma \geqslant 50$ GeV/c, the effect of finite W and Z masses are reduced relative to photon production, and the absolute production cross-sections for W and γ become comparable. This is a direct manifestation of electroweak unification, the ratio of W and γ coupling to quarks being $\sin\theta_w$.

2.4 Measurement of α_s using W + jet data

Jets produced in association with the W provide a means of measuring α_s. To first approximation, the yield of W + 1-jet events is proportional to α_s. UA2 used the measured one-jet to zero-jet ratio by comparing this experimental value with the QCD-predicted value obtained from Monte Carlo simulation of W production [28]. Requiring that the experimental ratio R_{exp} be reproduced by the Monte Carlo ratio $R_{MC}(\alpha_s)$, the following value for α_s was obtained:

$$\alpha_s(m_W^2) = 0.13 \pm 0.03\,(\text{stat.}) \pm 0.03\,(\text{syst. 1}) \pm 0.02\,(\text{syst. 2})\,,$$

where syst.1 represents the experimental systematic error and syst.2 is an estimate of the theoretical uncertainty. This measurement represents the first α_s determination that is (almost) free of K-factor effects in hadronic collisions. The result of this analysis is compared with other experimental determinations of α_s in Fig. 7 [28].

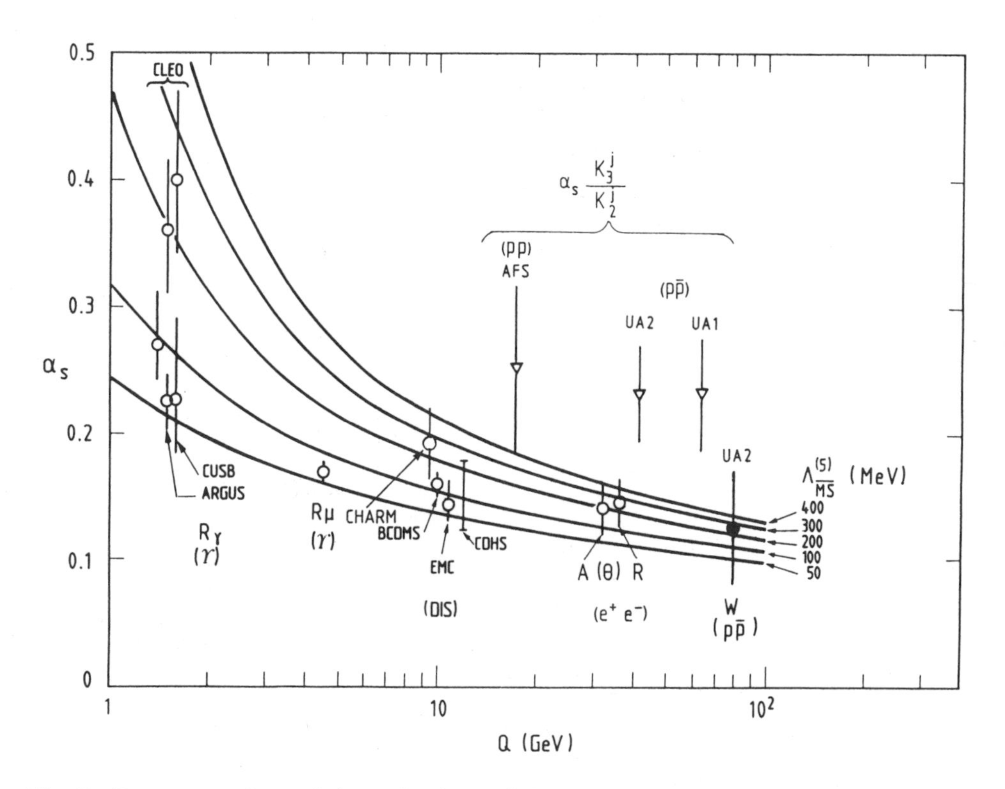

Fig. 7 Recent experimental determinations of the strong coupling constant compared with second-order QCD predictions [28].

2.5 Number of light neutrino species

The measured ratio $R = \sigma_W \cdot BR(W \to \ell\nu)/\sigma_Z \cdot BR(Z \to \ell^+\ell^-)$ depends on the t-quark mass and on the number of neutrino species via the branching ratios. The systematic measurement error largely cancels in the ratio R, and hence the available Z statistics limits the current precision on R. Combining all the available data of UA1 and UA2 (assuming lepton universality) results in [8, 29]

$$R = 8.4^{+1.2}_{-0.9} ,$$

$$R < 10.1 \qquad \text{at } 90\% \text{ CL} .$$

The theoretical prediction for R depends on: i) the ratio of the total production cross-section $R_0 = \sigma_W/\sigma_Z$ (an important source of uncertainty comes from the choice of structure functions; recent calculations of R_0 give values between 3.0 and 3.4 with a typical error of 0.15); ii) the partial width of the IVBs, which are precisely determined by the SM; iii) the total decay widths of W and Z, which depend on the t-quark mass and on the number of light neutrino species (N_ν).

The combined (UA1 and UA2) limit on N_ν as a function of m_t is shown in Fig. 8 [29], using the conservative theoretical input for $\sigma_W/\sigma_Z = 3.13$. Assuming that there is no W decay into fourth-generation leptons and that $m_\nu < m_Z/2$, one obtains

$$N_\nu < 5.7 \qquad (90\% \text{ CL}) .$$

If a new heavy charged lepton exists with a mass corresponding to the UA1 lower limit of 41 GeV/c^2, the limit obtained becomes less restrictive by one neutrino type. The data are not yet accurate enough to place an upper limit on the t-quark mass.

Preliminary results on R have been presented by UA2 [20] and CDF [30]:

$$R = 10.35^{+1.5}_{-1.0} (\text{stat.}) \pm 0.3 (\text{syst.}) \qquad (\text{UA2})$$

$$9.1 < R < 12.3 \qquad \text{at } 90\% \text{ CL}$$

$$R = 10.28 \pm 0.76 (\text{stat.}) \pm 0.45 (\text{syst.}) \qquad \text{CDF}$$

$$9.1 \leq R \leq 11.4 \qquad \text{at } 90\% \text{ CL} .$$

These preliminary values of R agree well with the SM, $N_\nu = 3$, and a heavy t-quark.

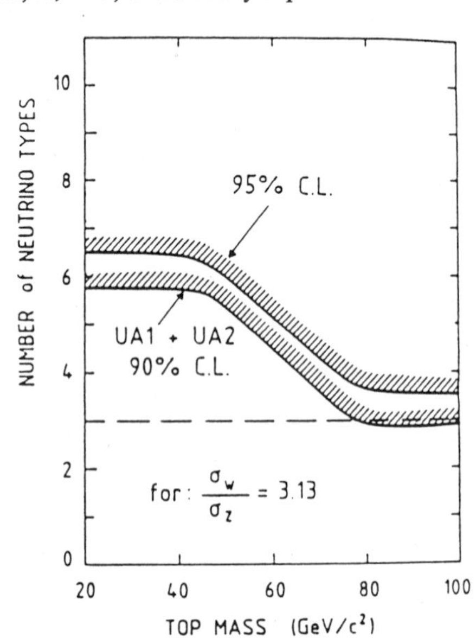

Fig. 8 The total number of light-neutrino species as a function of the t-quark mass, deduced from the combined UA1 and UA2 data.

2.6 Outlook

Although the production cross-sections for W and Z increase with increasing $\sqrt{s}$, one expects that precision measurements in the W,Z sector will become more difficult at higher $\sqrt{s}$. For example, the factor of ~ 3 increase of σ_W and σ_Z at the Tevatron compared with that at CERN has to be compared with about a factor of 10 increase in QCD two-jet production ($p_T^{jet} \approx 40$ GeV/c), thus resulting in a worse signal-to-background ratio at the Tevatron. In addition, the mean p_T^W at the Tevatron is expected to be almost twice that at CERN, thus significantly broadening the Jacobian peak observed in $W \rightarrow e\nu$ decay, and affecting the precision measurement of m_W.

On the other hand, CDF has demonstrated already that systematic errors from the p_T^W contribution on the m_W determination can be kept small. Clearly it is necessary to wait for the final result on m_W from UA2 and CDF to study the limiting factors for the systematic errors in each case.

From an integrated luminosity collected during the 1988/89 runs of 4.7 pb^{-1} for UA1, 6.7 pb^{-1} for UA2, and 4.7 pb^{-1} for CDF, a total of 1600 (5000) reconstructed $W \rightarrow e\nu$ decays and 200 (500) reconstructed $Z \rightarrow e^+e^-$ in UA2 (CDF) can be expected. UA1 (CDF) will have collected about 500 (2500) $W \rightarrow \mu\nu$ and about 100 (200) $Z \rightarrow \mu^+\mu^-$ decays.

With the existing data sample and the expected future Collider runs, it should be possible to perform detailed tests of the SM, e.g. make a precise measurement of the W mass.

Using 10 pb^{-1} integrated luminosity UA2 estimated the expected error on the W and Z mass determination to be [31]

$$\delta m_W \approx \pm 0.22\,(\text{stat.}) \pm 0.20\,(\text{syst. 1}) \pm 0.80\,(\text{syst. 2})\ \text{GeV}/c^2$$

$$\delta m_Z \approx \pm 0.25\,(\text{stat.}) \pm 0.92\,(\text{syst. 2}) \qquad \text{GeV}/c^2$$

where the errors are due to statistics (stat.), method (syst. 1) and calibration (syst. 2), assuming an uncertainty of $\pm 1\%$ on the absolute energy scale. This last error cancels in the measurement of $R = m_W/m_Z$; the expected precision on this ratio is

$$\delta(m_W/m_Z) \approx \pm 0.004\,.$$

The precise measurement of m_Z (to ± 50 MeV/c^2) expected from LEP allows to calibrate the mass scale of the hadron collider experiments. For UA2 this leads to an overall error on m_W (adding all contributions in quadrature) of

$$\delta m_W \approx \pm 0.37\ \text{GeV}/c^2\,.$$

UA1 estimated the error on m_W, assuming 50 pb^{-1} and an absolute energy scale uncertainty of ~ 0.5% obtainable with the uranium–TMP calorimeter, to be

$$\delta m_W \approx \pm 0.10\,(\text{stat.}) \pm 0.15\,(\text{syst.})\ \text{GeV}/c^2\,,$$

again assuming a precise Z-mass value from LEP. This accuracy on m_W is unlikely to improve significantly until LEP operates above the threshold for W pair production.

At LHC and SSC energies, about 10^9 W's and Z^0's will be produced per year at a luminosity of 10^{33} cm^{-2} s^{-1}. This has to be compared with ~ 10^7 Z's produced in three years of running at LEP with $L = 1.7 \times 10^{31}$ cm^{-2} s^{-1}, and ~ 10^4 W^+W^- at LEP200. Is it possible to use the LHC/SSC to search for rare W and Z decays, either to probe new physics or to perform precision measurements of the SM? For the Z, the very unfavourable signal-to-background ratio at hadron colliders is to be contrasted with the complete purity of the signal at e^+e^- Z factories; thus, looking for rare decays will very likely be the domain of LEP. However, observing the conventional decays $Z \rightarrow e^+e^-$ and $\mu^+\mu^-$ at hadron colliders is of great importance for testing perturbative QCD.

The same argument holds for the standard W-decays. Note that the calculations depend on parton distributions at small $x \approx 0.002$ [compared with $x \approx 0.15$ (0.04) at CERN (Fermilab) energies].

With ~ 10^9 W events produced per year, a branching ratio of BR $\geq 10^{-6}$–10^{-7} can in principle be reached. Looking for hadronic decay modes seems rather discouraging owing to the overwelming QCD-jet background. Leptonic decay modes are detectable, but it should be remembered that the W rapidity distribution is flat for $|y| \leq 4$ at SSC energies. For more details on possible detection of rare W-decays see Ref. 32.

It is clear that the sensitivity for observing interesting purely leptonic decay modes will be limited by large background—therefore rather challenging requirements for the detector performances are necessary.

3. HEAVY-FLAVOUR PRODUCTION

3.1 Inclusive b-quark production

At the CERN Collider, the predicted cross-sections for charm and beauty production are about 100 μb and 10 μb, respectively, according to QCD estimates. These cross-sections need to be extracted from a large background of hadronic production due to gg → gg. Furthermore, it is necessary to distinguish between charm and beauty production, which are expected to have comparable cross-sections at large p_T. Charm and beauty quarks can be tagged by the presence of one or more muons coming from the semileptonic decay of the heavy quarks. These muons are expected to be embedded in hadronic jets.

Inclusive muon production has been studied in UA1 for $p_T^\mu > 6$ GeV/c and within a pseudorapidity range $|\eta| < 1.5$ [33]. Figure 9 shows the p_T spectrum for muons after the decay background has been subtracted. The decays of charged π and K mesons amount to about 70% at $p_T^\mu = 6$ GeV/c and decrease to about 20% at $p_T^\mu = 20$ GeV/c. The solid curve is the prediction for the sum of all expected contributions listed in Fig. 9. The agreement with the data shown in this figure is very good. In the interval of $10 < p_T^\mu < 15$ GeV/c, heavy-flavour production is expected to be the dominant contribution. The harder fragmentation of the b-quark, compared with charm, leads on the average to a harder p_T spectrum of muons; thus

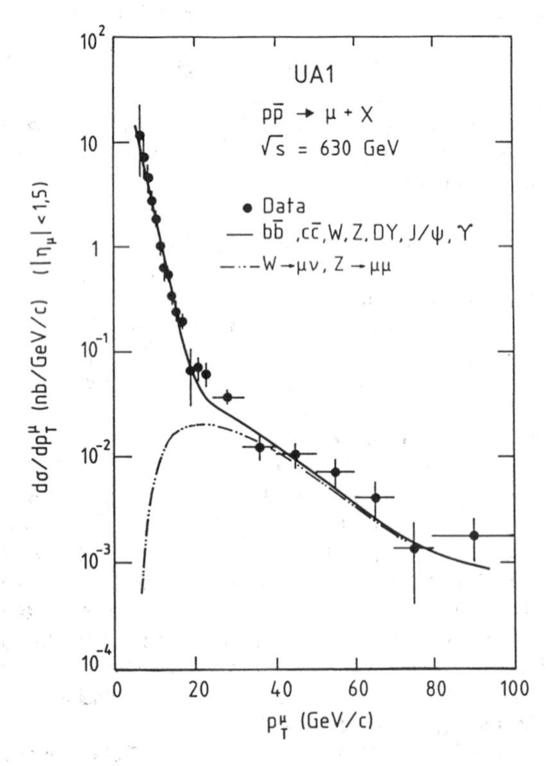

Fig. 9 Inclusive transverse-momentum distribution for muons, after decay background subtraction. Predictions made with ISAJET are shown for comparison [33].

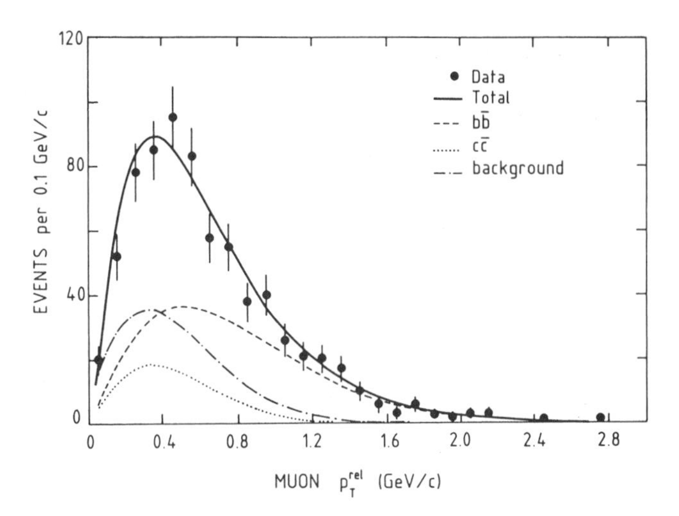

Fig. 10 Distribution of the transverse momentum relative to the nearby jet, for muons of $p_T > 10$ GeV/c in a sample of events with at least one jet of $E_T > 10$ GeV [33].

$b\bar{b}$ is favoured over $c\bar{c}$ once a p_T cut is applied to the muons. An independent estimate of the beauty fraction is made by fitting the distribution of p_T^{rel}, the transverse momentum of the muons with respect to the nearest jet axis [33]. Figure 10 shows the p_T^{rel} distribution together with the fitted contributions from beauty, charm, and decay background, which for single-muon events results in

$$b\bar{b}/(b\bar{b} + c\bar{c}) = (76 \pm 12)\% \ .$$

The b-quark production cross-section has been evaluated using four different data sets [33]:
- J/ψ events: $p_T(J/\psi) > 5$ GeV/c;
- low-mass dimuons: $2m_\mu < m_{\mu\mu} < 6$ GeV/c^2; $p_T^{\mu 1} > 3$ GeV/c; $p_T^{\mu 2} > 3$ GeV/c;
- high-mass dimuons: $m_{\mu\mu} > 6$ GeV/c^2; $p_T^{\mu 1} > 3$ GeV/c; $p_T^{\mu 2} > 3$ GeV/c;
- single muons: $p_T^\mu > 6$ GeV/c.

The measured lepton cross-sections $\sigma(\mu)$ are related to the cross-sections for producing b-quarks, with quark transverse momentum above p_T^{min}, by the following expression:

$$\sigma(p_T^b > p_T^{min}) = \sigma(\mu)[\sigma_{MC}(p_T^b > p_T^{min})/\sigma_{MC}(\mu)] \ ,$$

where the cross-sections with subscript 'MC' are evaluated with the ISAJET Monte Carlo program [34]. The b-quark cross-section is then quoted for $p_T^b > p_T^{min}$, where p_T^{min} is defined in such a way that 90% of the muon events have $p_T^b > p_T^{min}$. Figure 11 shows the resulting inclusive b-quark cross-sections for six values of p_T^{min}, and for $|y^b| < 1.5$. The solid curve is the next-to-leading-order calculation of Nason et al. [35], and the dashed line corresponds to the full ISAJET prediction. The level of agreement between the cross-section measurements made with quite different techniques (using all available μ data samples) and the theoretical predictions are very satisfactory.

The total cross-section for beauty production is obtained by normalizing the $O(\alpha_s^3)$ prediction to the data and extrapolating to $p_T^{min} = 0$. Excluding the high-mass dimuon measurement as well as the measurement for $p_T^{min} > 15$ GeV/c results in [33]

$$\sigma(p\bar{p} \to b \text{ or } \bar{b}, |y| < 1.5) = 14.7 \pm 4.7\ \mu b \ .$$

Finally, extrapolating to all rapidities gives

$$\sigma(p\bar{p} \to b\bar{b}) = 10.2 \pm 3.3\ \mu b \ .$$

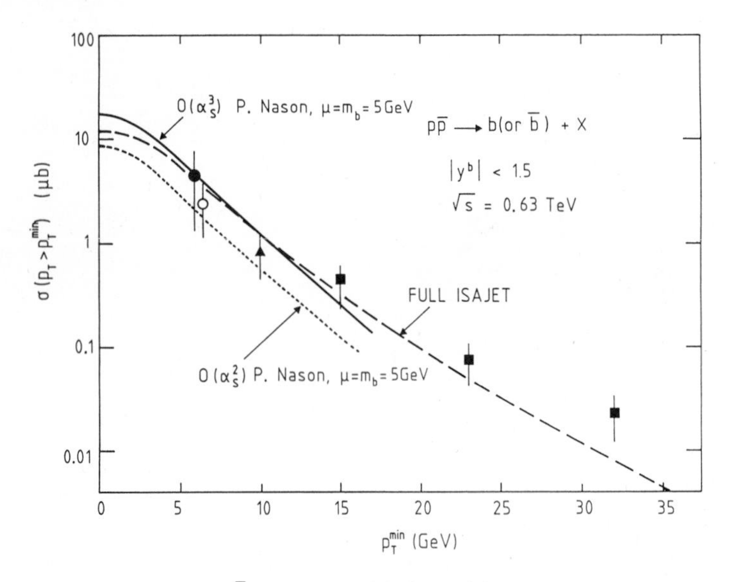

Fig. 11 Cross-section for b (or $\bar{b}$) quark production with transverse momentum above p_T^{min} and $|y| < 1.5$ [33]. The interpretation of the symbols is as follows: (■) single muons, (▲) low-mass dimuons, (○) high-mass dimuons, and (•) J/ψ. Also shown are predictions from ISAJET, as well as from second- and third-order QCD calculations [35].

This can be compared with a recent theoretical calculation [36], which gives

$$\sigma(p\bar{p} \to b\bar{b}) = (12 \,^{+\,7}_{-\,4})\ \mu b \qquad \text{for} \qquad m_b = 5\ \text{GeV}/c^2\ ,$$

$$\sigma(p\bar{p} \to b\bar{b}) = (19 \,^{+\,10}_{-\,8})\ \mu b \qquad \text{for} \qquad m_b = 4.5\ \text{GeV}/c^2\ .$$

Within the large experimental and theoretical errors, there is good agreement between the QCD prediction and the UA1 data.

3.2 B^0–$\bar{B}^0$ oscillations

Evidence for B^0–$\bar{B}^0$ oscillations has been reported by UA1 from the observation of an apparent excess of like-sign dimuon events [37]. Oscillation in the B_d^0–$\bar{B}_d^0$ system has been observed by ARGUS and CLEO [38].

An experimentally accessible quantity that measures the degree of oscillations at the Collider is the fraction of wrong-sign decays:

$$\chi = P(B^0 \to \bar{B}^0 \to \mu^-)/P(B^0 \to \mu^\pm)\ .$$

The experimental result obtained by UA1 yields

$$\chi = 0.121 \pm 0.047\ ,$$

which is an average over all beauty states. Oscillations can occur for the neutral-meson states B_d^0 and B^0_s. The corresponding mixing parameters χ_d and χ_s are related to χ via the expression

$$\chi = f_d\chi_d + f_s\chi_s\ ,$$

where $f_{d(s)}$ is the fraction of beauty quarks hadronizing into $B_{d(s)}$ mesons. To obtain a limit on χ_d and χ_s from UA1 data, the values $f_d = 0.36$ and $f_s = 0.18$ have been used, based on the ratio of K^+/π^+ measured at the CERN Intersecting Storage Rings (ISR) [39]. Figure 12a shows the resulting ± 1 standard-deviation band obtained by UA1, compared with the

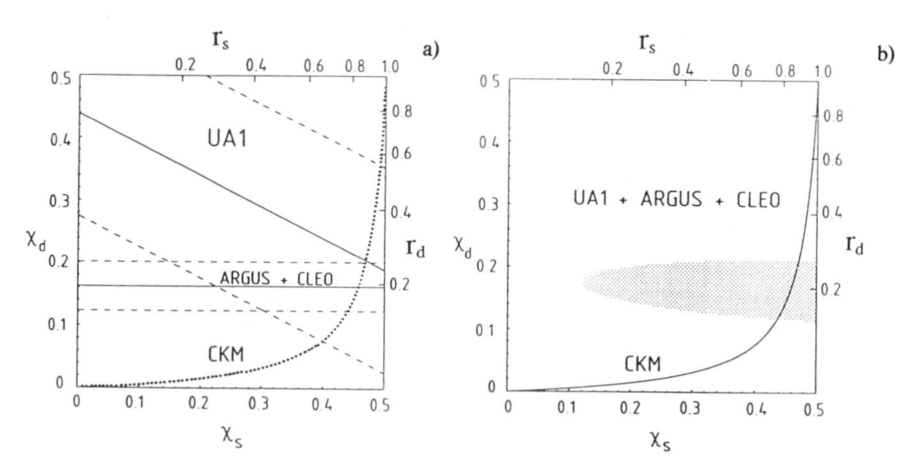

Fig. 12 Beauty-oscillation results in the χ_s–χ_d plane: a) one-standard-deviation bands, and b) combined 90% CL limit. The region below the line is allowed by the present constraints from the KM matrix, assuming three generations [40].

weighted average of the ARGUS and CLEO results ($\chi_d = 0.162 \pm 0.039$). Figure 12b shows the corresponding combined 90% CL limit. The line in Fig. 12b indicates the upper limit on χ_d as a function of χ_s, which can be derived from existing limits on the Kobayashi–Maskawa (KM) matrix elements, assuming that we have a SM with three families. Combining this constraint with the experimental limits, only a small allowed region in the χ_s–χ_d plane remains, indicating almost maximal mixing in the B_s^0 channel [40].

4. SEARCH FOR HEAVY QUARKS (t, b′)

In the framework of the SM, the t-quark should exist and its mass should be smaller than about 200 GeV/c^2 in order to be consistent with low-energy data. Thus, taking into account the lower limit on m_t from Mark II [41], the t-quark mass is expected to be in the range

$$38.5 < m_t \leqslant 180 \text{ GeV}/c^2 .$$

An indirect lower mass limit of about 50 GeV/c^2 has also been obtained from analyses of B^0–$\bar{B}^0$ mixing [42]. From a new global fit to electroweak data a value of $m_t = 132^{+31}_{-37}$ GeV/c^2 is obtained [17].

The most promising channel for finding the t-quark at the CERN Collider is the decay $W \to t\bar{b}$, followed by the semileptonic decay of the t-quark. This channel dominates t-quark production at $\sqrt{s} = 630$ GeV if m_t is in the range 40–70 GeV/c^2. For m_t outside this range, direct $t\bar{t}$ production is the dominant process. The t-quark production cross-section for $t\bar{b}$ can be derived from the measurement $\sigma \cdot \mathrm{BR}(W \to \ell\nu)$, taking into account the m_t dependence of the branching ratio for $W \to t\bar{b}$ and QCD corrections to the partial decay widths. The $t\bar{t}$ production cross-section can be estimated, using perturbative QCD, at the next-to-leading order and taking into account theoretical uncertainties. This leads, for example — assuming that $m_t = 40$ GeV/c^2 — to $\sigma(t\bar{b}) = 1.41$ nb and $\sigma(t\bar{t}) = 0.64$ nb at $\sqrt{s} = 0.630$ TeV. The production through the decay $Z \to t\bar{t}$ has a comparatively negligible cross-section and is not considered here.

UA1 has searched for heavy quarks using the distinctive signature of the semileptonic decay modes in the e and μ channels [43]. The experimental statistics available (~ 0.7 pb^{-1}) allowed a systematic investigation of the region $m_t \leqslant 45$ GeV/c^2. A signal for the t-quark was searched for in the isolated e/μ + $\geq$1-jet channel. The characteristic topology of a t-quark decaying semileptonically allows other sources of charged leptons to be discriminated, such as W, Z, Drell–Yan, J/ψ, Υ, $b\bar{b}$, and $c\bar{c}$ — the main physics backgrounds to a new heavy-quark signal in the UA1 analysis.

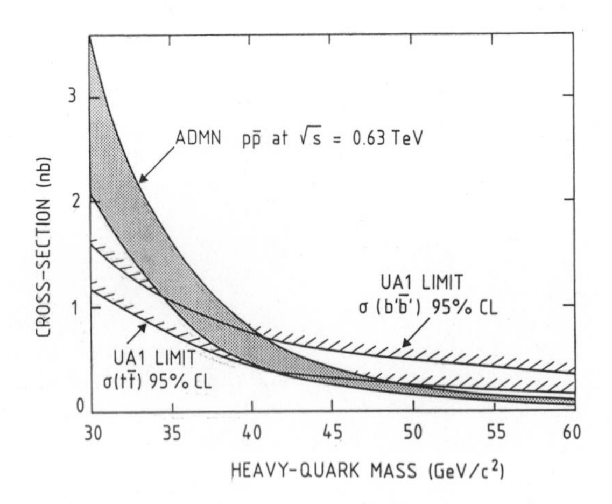

Fig. 13 The UA1 experimental upper limits (95% CL) on $t\bar{t}$ and $b'\bar{b}'$ production cross-sections in $p\bar{p}$ collisions at $\sqrt{s} = 0.63$ TeV. The shaded area is the predicted cross-section for heavy quarks, from Ref. [36].

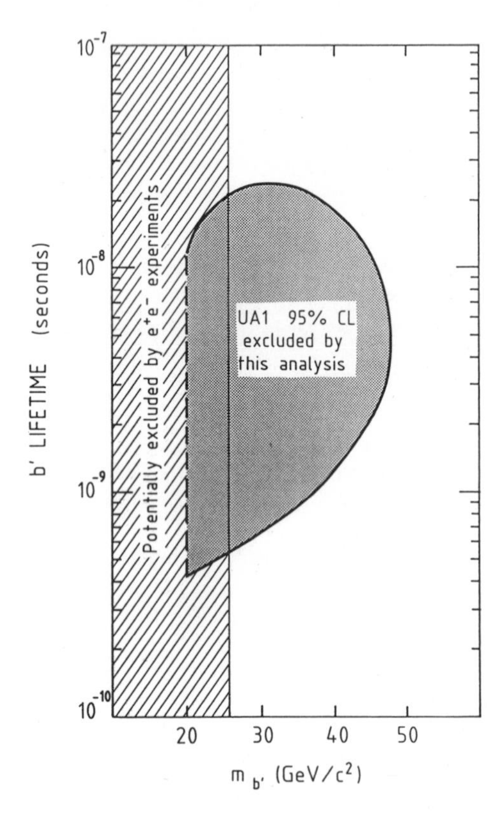

Fig. 14 The region excluded at 95% CL by UA1 in the $m_{b'}$–$\tau_{b'}$ plane, for a long-lived b′-quark [44].

Since no evidence for a t-quark signal was found in the data (~ 0.7 pb^{-1}), limits have been placed on the t-quark production cross-section in $p\bar{p}$ collisions. A lower limit on m_t is obtained by comparing this experimental limit with the theoretical prediction of the cross-section, which is a function of m_t. The contribution from $W \to t\bar{b}$ decays is fixed by the measured $\sigma \cdot BR(W \to e\nu)$, therefore the result is a limit on $\sigma(t\bar{t})$. UA1 has published lower limits of 44 GeV/c^2 and 32 GeV/c^2 on the masses of the t and b′ (fourth-generation) quarks, respectively, at the 95% CL [43]. The results depend on the $O(\alpha_s)$ QCD prediction for the heavy-quark production cross-section. Using the $O(\alpha_s^3)$ calculation of Nason et al. [35] and the DFLM structure functions, Altarelli et al. [36] have revised these limits to

$$m_t > 41\ \mathrm{GeV/c^2} \qquad (95\%\ \mathrm{CL})\,,$$

$$m_{b'} > 34\ \mathrm{GeV/c^2} \qquad (95\%\ \mathrm{CL})\,,$$

as shown in Fig. 13. Since no contribution from $W \to t'\bar{b}'$ is included, the 95% CL upper limit on $\sigma(b'\bar{b}')$ is larger than the $\sigma(t\bar{t})$ limit.

To derive the b′-quark limit, it was assumed that the b′ is lighter than the t- and t′-quarks, so the b′ decays to a c-quark or a u-quark. Since there exists no lower bound on $V_{b'u}$ and $V_{b'c}$, these couplings could be very small, thus leading to long lifetimes of the b′. If the b′-quark has such a long lifetime, the quoted UA1 limit on $\sigma(b'\bar{b}')$ is no longer valid, because the selection of lepton candidates requires that the lepton track points back to the primary vertex of the $p\bar{p}$ interaction. However, such a long-lived b′ provides an alternative signature: the b′, after travelling some distance, decays and produces a secondary vertex of charged particles in the UA1 central drift chamber and a high-momentum jet in the calorimeter. This event topology has been used to search for long-lived heavy quarks. The result of this search is plotted in Fig. 14 and details of the analysis can be found in Ref. [44].

4.1 Outlook

One of the most important analysis topics of the new data of UA1, UA2, and CDF ($\sim 5\ \mathrm{pb}^{-1}$, $\sim 7\ \mathrm{pb}^{-1}$, and $\sim 5\ \mathrm{pb}^{-1}$, respectively), is the search for the t-quark.

The $p\bar{p} \to$ t-quark production cross-section (from both $t\bar{b}$ and $t\bar{t}$) as a function of m_t at $\sqrt{s} = 0.63$ and 1.8 TeV is shown in Fig. 15. At CERN energies the $W \to t\bar{b}$ production dominates for $m_t \lesssim 70\ \mathrm{GeV/c^2}$, which has the advantage of a simpler topology than for $t\bar{t}$ and

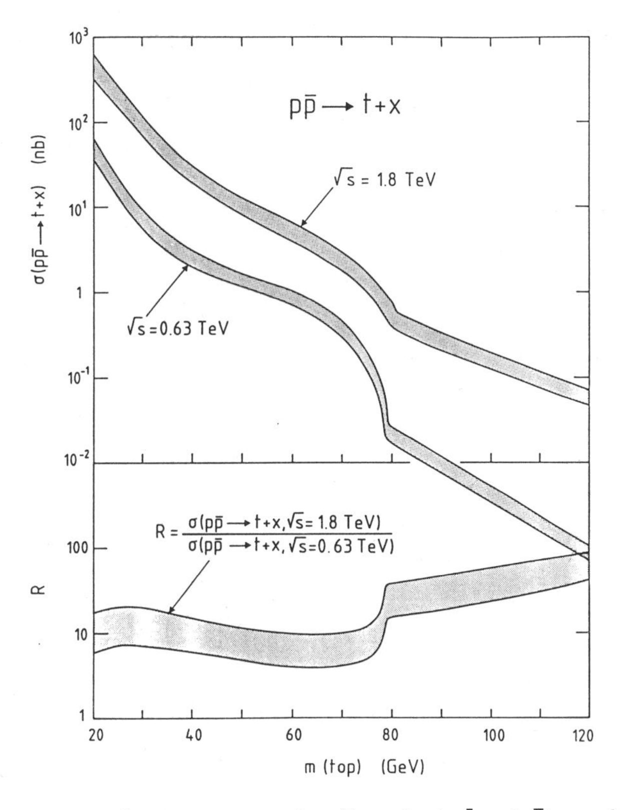

Fig. 15 The $p\bar{p} \to$ top production cross-section (from both $t\bar{t}$ and $t\bar{b}$) as a function of m_t at $\sqrt{s} = 0.63$ and 1.8 TeV.

a more precisely known production rate, since normalization to the measured $W \rightarrow e\nu$ cross-section is possible. At Fermilab, QCD $t\bar{t}$ production is the dominant production mechanism for all m_t. From Fig. 15 it is evident that the Fermilab Collider has a substantial advantage over the CERN Collider in t-production rate. Especially at large m_t (e.g. $m_t = 100$ GeV/c^2), the $t\bar{t}$ cross-section increases by a factor of ~ 40 between $\sqrt{s} = 0.63$ and 1.8 TeV. If $m_t > m_W$, the interesting situation occurs that the t-quark decays into a real W + b and the process $p\bar{p} \rightarrow t\bar{t} + X$ leads to final states with two real W's and two b-quarks. The average p_T of each t-quark is $\sim m_t/2$. The W—because of its large mass—will carry most of the transverse momentum of the parent t-quark, and will therefore be produced at large p_T ($p_T^W \approx 45$ GeV/c). The $p\bar{p} \rightarrow t\bar{t} + X \rightarrow W^+W^-b\bar{b} + X$ final state has therefore a very distinctive experimental signature: large transverse momentum p_T^t, highly unbalanced transverse momenta of the two t-quark decay products ($p_T^W \gg p_T^b$), and unbalanced W decay products, characteristic of large p_T^W production.

Preliminary results on the search for the t-quark using the new data samples are available from all three Collider experiments.

The new data of UA1 allow the t-quark to be searched for in the muon channel only, since the e.m. calorimeters have been removed during the 1988/89 data-taking period. Applying analysis techniques similar to those discussed in Section 4 to the new data, UA1 does not observe a signal for t-quark production in the single-muon and dimuon data samples. Combining the additional data from 1988 and 1989 (μ + jet, $\mu\mu$) and data collected before 1988 (μ + jet, $\mu\mu$, e + jet, and e + μ), UA1 obtained a new, still preliminary, limit on m_t [45]:

$$m_t > 61 \text{ GeV}/c^2 \qquad (95\% \text{ CL}),$$

$$m_t > 66 \text{ GeV}/c^2 \qquad (90\% \text{ CL}),$$

and for the b′-quark, the new UA1 lower limit on the b′ mass is

$$m_{b'} > 41 \text{ GeV}/c^2 \qquad (95\% \text{ CL}).$$

UA2 also presented preliminary results from the 1988 and 1989 data on the t-quark search in the e-channel [46]. The selection requires missing transverse momentum $\not{p}_T > 15$ GeV, $p_T^e > 11.5$ GeV and $15 < m_T(e\nu) < 50$ GeV/c^2 for the t signal region. An increase in acceptance is expected if the $\not{p}_T$ is reduced to 10 GeV; however, this requires more rejection power against background due to conversions and overlaps, i.e. tighter quality cuts on the electron candidate. The data are consistent with the expected background for both selections, where the dominant physics contribution comes from W + jet production. The preliminary UA2 limit on m_t using 7.1 pb^{-1} yields [46]

$$m_t > 67 \text{ GeV}/c^2 \qquad (95\% \text{ CL})$$

$$m_t > 70 \text{ GeV}/c^2 \qquad (90\%) \quad ,$$

and similarly for the b′ quark, the UA2 lower limit on the b′ mass is

$$m_{b'} > 53 \text{ GeV}/c^2 \qquad (95\% \text{ CL}).$$

The search strategy at the Fermilab Tevatron is different from that at the CERN $p\bar{p}$ Collider; this is because the $W \rightarrow t\bar{b}$ contribution is small and $t\bar{t}$ can be observed in many channels. Owing to the large cross-section and large integrated luminosity, CDF can afford to look for $e^{\pm}\mu^{\mp}$ events from $t\bar{t}$ decay. This channel provides a very clear signature, no background from W + jets, and negligible background from $b\bar{b}$ due to p_T and isolation cuts. The preliminary result from this channel leads to a t mass limit of [47]

$$30 < m_t < 72 \text{ GeV}/c^2 \qquad (95\% \text{ CL}).$$

CDF has also presented preliminary results from the search for the t-quark in the e + jet channel. Again, the data are consistent with the expected contributions and no t signal has been observed. This translates to a limit of [47]

$$40 < m_t < 77 \text{ GeV/c} \qquad (95\% \text{ CL}) .$$

Clearly, the final CDF limit on m_t from all existing data samples will exceed m_W. The ultimate mass-reach for a heavy t-quark from CDF assuming 200 pb^{-1} is estimated to be ~ 180 GeV/c^2 from the dilepton channel and ~ 250 GeV/c^2 from the single-lepton channel [48].

Heavy-quark production at the LHC/SSC will be dominated by pair-production through gg fusion [49]. The total production cross-section for heavy quark pairs as a function of m_Q is plotted in Fig. 16a at $\sqrt{s}$ = 20 TeV. The available centre-of-mass energy is an important parameter for heavy-quark production, e.g. for m_Q = 0.5 TeV/c^2, the cross-section decreases by a factor of ~ 10 if $\sqrt{s}$ decreases from 20 to 10 TeV. Assuming that $Q \rightarrow qW$, the most promising decay mode to be detected is $Q\bar{Q} \rightarrow (W \rightarrow \ell\nu) + 4$ jets. For one year of running at $L = 10^{33}\ \text{cm}^{-2}\ \text{s}^{-1}$, a mass reach of [49]

$$m_Q \approx 0.8 \text{ TeV/c}^2 \qquad \text{at the LHC} ,$$

$$m_Q \approx 1.0 \text{ TeV/c}^2 \qquad \text{at the SSC} ,$$

can be expected provided that the main background from W + jet production is kept small. This can be achieved by cutting in a range of p_T^e values, for which the signal exceeds the inclusive $W \rightarrow e\nu$ background, as can be seen in Fig. 16b.

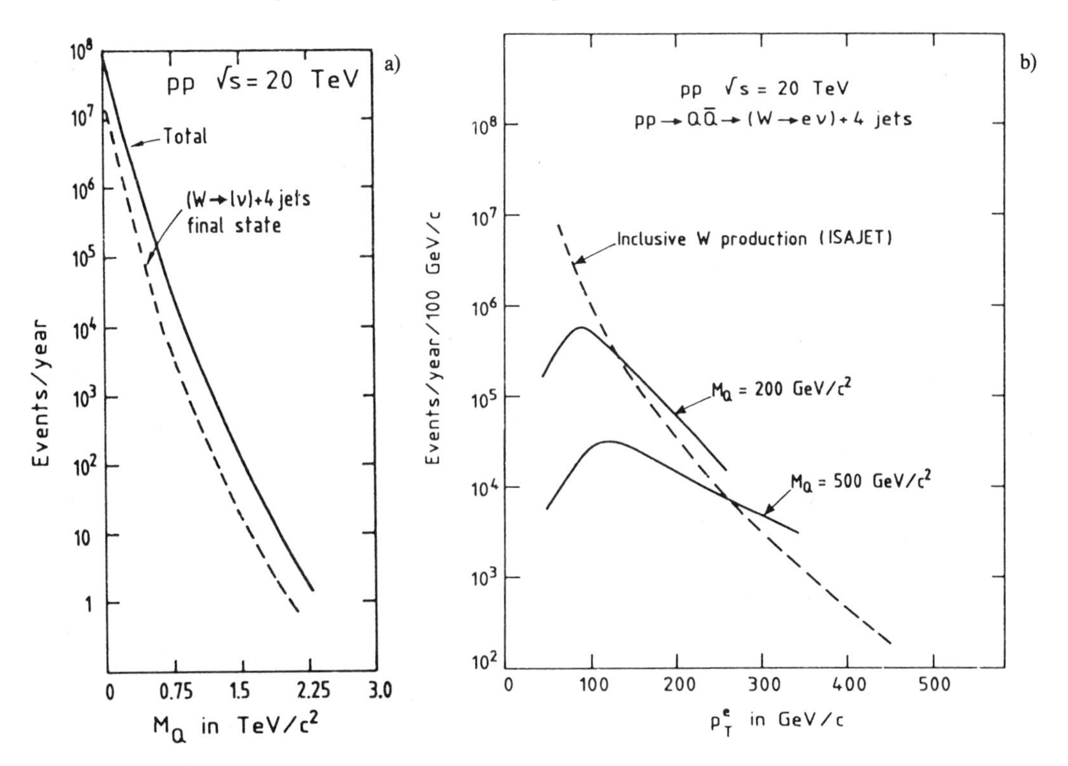

Fig. 16 a) The total production cross-section for heavy-quark pairs as a function of the heavy-quark mass m_Q at $\sqrt{s}$ = 20 TeV. Also shown is the cross-section for $pp \rightarrow Q\bar{Q} \rightarrow (W \rightarrow \ell\nu) + 4$ jets, assuming that $Q \rightarrow Wq$ [49].

b) The inclusive electron transverse momentum p_T^e spectrum from $W \rightarrow e\nu$ decays (dashed line) and for the process $pp \rightarrow Q\bar{Q} \rightarrow (W \rightarrow e\nu) + 4$ jets, for two different mass values of the heavy-quark Q (solid lines) [49].

5. SEARCHES FOR NEW PHYSICS

5.1 Search for heavy leptons and the number of neutrino species

An analysis of events with large missing transverse energy ($\not{E}_T$) due to the production of one or more energetic neutrinos or weakly interacting particles can extend our understanding of the SM, and is a sensitive way to search for new physics. A fourth generation of quarks and leptons would contain a charged heavy lepton L, together with its associated ν_L. Assuming that there is a universal strength of coupling to W and Z particles, and that $m_L < m_W$, the heavy lepton can be produced in W decays. One then looks for a process $p\bar{p} \to W + X$ (with $W \to L\nu$, and $L \to q\bar{q} + \nu$). The experimental signature would be events with one or more jets plus large $\not{E}_T$. This decay pattern gives events that are qualitatively similar to $W \to \tau\nu$ ($\tau \to$ hadrons $+ \nu$), except for the phase-space factor and the typically larger invariant mass of the $q\bar{q}$ system. UA1 has used the $\not{E}_T$ + jet data sample—which has led to the identification of $W \to \tau\nu$ decays—to place a limit on m_L [50]. Using the calculated heavy-lepton contribution and taking into account the existence of the additional neutrino (ν_L) coupling to the Z, it was found that

$$m_L > 41 \text{ GeV}/c^2 \qquad (90\% \text{ CL}) .$$

Similarly, the rate of events from the process $p\bar{p} \to Z + X$ ($Z \to \nu\nu$) is proportional to the number of light-neutrino species coupling to the Z. If the Z is produced in association with an energetic jet (Z production at high p_T), such events are expected to appear as hadronic events with large values of $\not{E}_T$. The same data sample and method as were used for the heavy-lepton search have been employed by UA1 [50] to derive a limit on the number of light-neutrino species:

$$N_\nu \leqslant 10 \qquad (90\% \text{ CL}) .$$

5.2 Searches for supersymmetric particles

As already mentioned, supersymmetric theories are attractive because they provide solutions for some of the fundamental problems of the SM. In supersymmetric theories, fermions and bosons are linked by a common symmetry, every particle having a supersymmetric partner, a 'sparticle'. The coupling of these new particles is well defined and therefore production cross-sections can be calculated. Supersymmetric particles are pair-produced (e.g. $p\bar{p} \to \tilde{g}\tilde{g}$), they decay into lighter SUSY particles (e.g. $\tilde{g} \to q\bar{q}\tilde{\gamma}$), and the lightest SUSY particle (LSP) is expected to be stable and non-interacting, and thus will escape detection. In many models, the LSP is assumed to be the photino ($\tilde{\gamma}$).

At hadron colliders the dominant final states containing SUSY particles are $\tilde{g}\tilde{g}$, $\tilde{g}\tilde{q}$, and $\tilde{q}\bar{\tilde{q}}$ [51]. In the case where $m_{\tilde{q}} > m_{\tilde{g}}$, the squark decays into $q\tilde{g}$ and the gluino decays into $q\bar{q}\tilde{\gamma}$; in the case where $m_{\tilde{q}} < m_{\tilde{g}}$, the decay modes are $\tilde{q} \to q\tilde{\gamma}$ and $\tilde{g} \to \bar{q}\tilde{q}$ or $(q\bar{\tilde{q}})$. In both cases, the event topology consists of hadronic jets (from the fragmentation of outgoing q and $\bar{q}$) and $\not{E}_T$ (from the outgoing $\tilde{\gamma}$'s). The number of jets depends upon the production sub-process and the $\tilde{q}$ and $\tilde{g}$ masses. When $m_{\tilde{q}} \ll m_{\tilde{g}}$, $\tilde{q}\bar{\tilde{q}}$ pair-production dominates the production of strongly interacting sparticles, whereas $\tilde{g}\tilde{g}$ dominates if $m_{\tilde{q}} \gg m_{\tilde{g}}$.

UA1 has performed a search for gluinos and squarks [52] using the jet(s) + $\not{E}_T$ signature (assuming a stable $\tilde{\gamma}$, which escapes detection). The domain of the ($m_{\tilde{g}}$, $m_{\tilde{q}}$) plane excluded by the UA1 analysis is shown in Fig. 17. In the case where $m_{\tilde{g}} \gg m_{\tilde{q}}$, the lower bound $m_{\tilde{q}} > 45$ GeV/c^2 is obtained. Similarly, if $m_{\tilde{q}} \gg m_{\tilde{g}}$, the lower bound $m_{\tilde{g}} > 53$ GeV/c^2 is found. For $m_{\tilde{g}} \approx m_{\tilde{q}}$, a more stringent limit of $\tilde{m} > 75$ GeV/c^2 is obtained owing to the increase in the number of events expected from $\tilde{g}\tilde{q}$ production. The results are valid for photino masses up to ~ 20 GeV/c^2. For $m_{\tilde{\gamma}} > 20$ GeV/c^2, however, the squark and gluino mass limits decrease rapidly.

In the search for $\tilde{g}$ and $\tilde{q}$ by UA2 [53] two cases were considered: i) the photino is the LSP and is therefore stable, and ii) the photino decays into $\tilde{H}\,\gamma$, where $\tilde{H}$ escapes detection. In the first case, a somewhat lower mass limit for $\tilde{q}$ and $\tilde{g}$ has been obtained, compared with that from UA1. In the case of an unstable photino, the better granularity of the UA2 calorimeter allows an analysis of events containing two photons + jets in the final state. In this analysis,

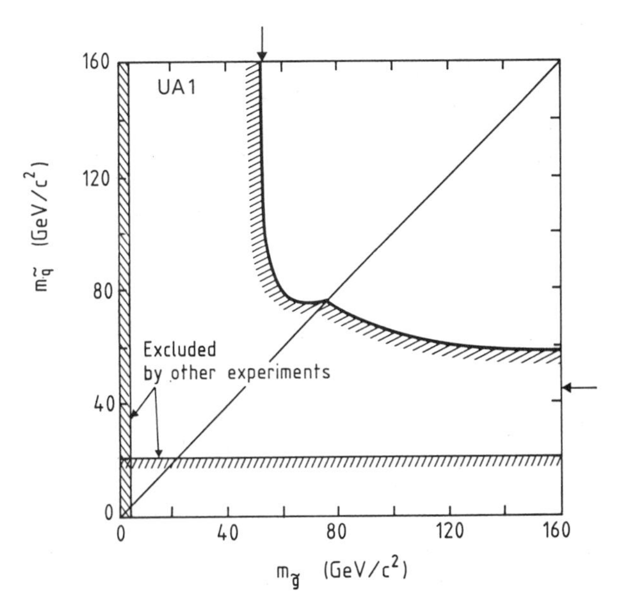

Fig. 17 Limits on squark and gluino masses (90% CL) obtained by UA1. The arrows indicate the asymptotic values of the 90% CL contours as the squark or gluino mass becomes infinitely large [52].

UA2 excludes gluino masses between 15 and 20 GeV/c^2 and squark masses between 9 and 46 GeV/c^2. Again, assuming $m_{\tilde{g}} \approx m_{\tilde{q}}$, a limit of $\tilde{m} > 60$ GeV/c^2 is obtained.

UA2 has presented preliminary results on squark and gluino searches using the new data samples [54]. The improved values quoted are $m_{\tilde{g}} > 84.5$ GeV/c^2, $m_{\tilde{q}} > 76$ GeV/c^2, and assuming $m_{\tilde{g}} \approx m_{\tilde{q}}$, a limit of $\tilde{m} > 107$ GeV/c^2 is obtained (all at 90% CL).

CDF has performed a search for gluinos and squarks using the jet(s) + $\not{E}_T$ signature (assuming a stable $\tilde{\gamma}$, which escapes detection). The data sample used corresponds to 25.3 nb^{-1} of integrated luminosity [55]. The domain in the ($m_{\tilde{g}}$, $m_{\tilde{q}}$) plane excluded by the CDF analysis is shown in Fig. 18. The asymptotic mass limits are $m_{\tilde{g}} > 73$ GeV/c^2 and $m_{\tilde{q}} > 74$ GeV/c^2 at

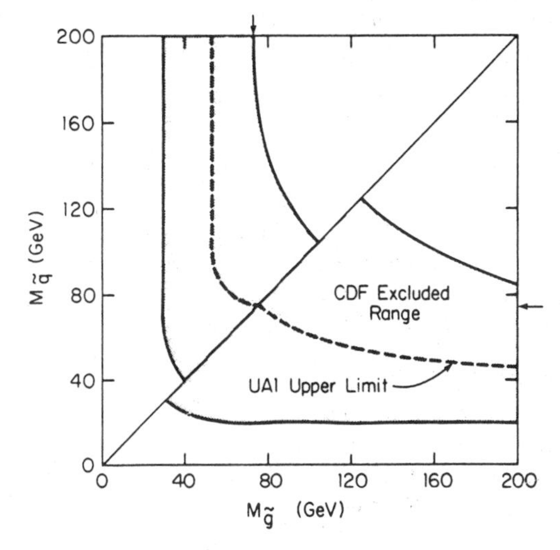

Fig. 18 Limits on squark and gluino masses (90% CL) obtained by CDF. The arrows indicate the asymptotic values of the 90% CL contours, where the $\tilde{g}$ or $\tilde{q}$ mass is very large. The lower-limit curve corresponds to a CDF acceptance of 0.1% [55].

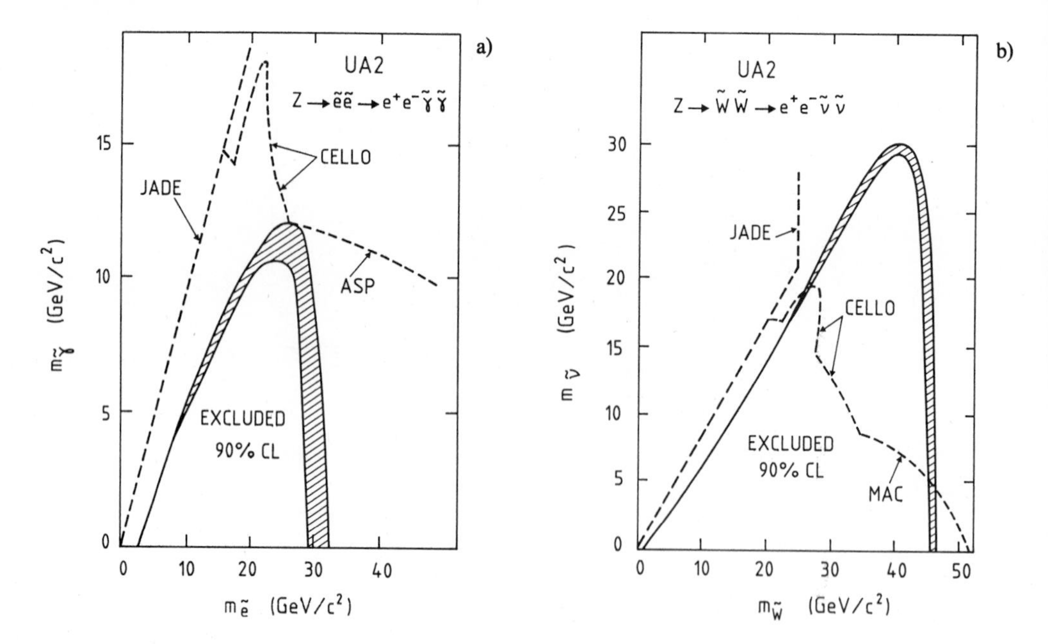

Fig. 19 Limits (90% CL) on a) $\tilde{e}$ and $\tilde{\gamma}$ masses and b) $\tilde{W}$ and $\tilde{\nu}$ masses obtained by UA2 [53]. The hatched area illustrates the systematic uncertainty. Also shown are limits from e^+e^- experiments.

90% CL. The lower-limit curve in Fig. 18 has been chosen where the acceptance dropped to 0.1%. The quoted results are valid for photino masses up to 30 GeV/c^2 in the high-mass region. From the present data sample CDF expects to reach gluino and squark mass limits of about 150 GeV/c^2.

Supersymmetric leptons can be searched for in $W \rightarrow \tilde{\ell}\tilde{\nu}$ and $Z \rightarrow \tilde{\ell}^+ \tilde{\ell}^-$ decays. In each case $\tilde{\ell} \rightarrow \ell\tilde{\gamma}$ is assumed to dominate, and one expects $\ell + \not{E}_T$ final states from W decays and $\ell^+\ell^- + \not{E}_T$ final states from Z decays. The former may be distinguished from the conventional $W \rightarrow \ell\nu$ decays by its softer lepton spectrum and different angular distribution. The $Z \rightarrow \tilde{\ell}^+\tilde{\ell}^-$ decay can be distinguished from conventional Drell–Yan background by the additional $\not{E}_T$ signature.

The Z data sample can also be used to extract a limit for the mass of the wino under the assumption that the decay $Z \rightarrow \tilde{W}\tilde{W}$ is kinematically possible. In this case one expects to observe again $\ell^+\ell^-$ pairs in association with large $\not{E}_T$, under the assumption that $\tilde{W} \rightarrow e\tilde{\nu}$, where $\tilde{\nu}$ escapes detection (stable $\tilde{\nu}$ or $\tilde{\nu} \rightarrow \nu\tilde{\gamma}$).

Figures 19a,b show the 90% CL for $m_{\tilde{e}}$ and $m_{\tilde{W}}$ as a function of $m_{\tilde{\gamma}}$ and $m_{\tilde{\nu}}$, respectively, as obtained by UA2 [53]. In the $\tilde{W}$ case the UA2 result significantly improves the values reached in e^+e^- experiments so far. Using the $W \rightarrow e\nu$ data samples, UA1 [9] obtained a selectron mass limit of $\tilde{m} > 32$ GeV/c^2, for $m_{\tilde{W}}, m_{\tilde{Z}} \gg m_{\tilde{e}}$, and $m_{\tilde{e}} = m_{\tilde{\nu}}$.

5.3 Searches for additional vector bosons

Additional vector bosons—presumably heavy—are predicted in many possible extensions of the SM [56]. These heavy particles (W′,Z′) would be detected in UA1 and UA2, depending on their mass, their coupling to quarks, and their branching ratio to leptons.

In the UA1 data sample, no $W \rightarrow e\nu$ and no $Z \rightarrow e^+e^-$ candidates have been observed with electron–neutrino transverse mass or e^+e^- mass in excess of the expected distribution for Standard Model IVB decays [9]. These results have been used to set limits on the production and decay of more massive W(Z)-like objects. Normalizing the cross-section times branching ratio $(\sigma'\cdot BR')_{IVB'}$ to those of the SM vector bosons results in limits on $m_{W'}$ and $m_{Z'}$ as a

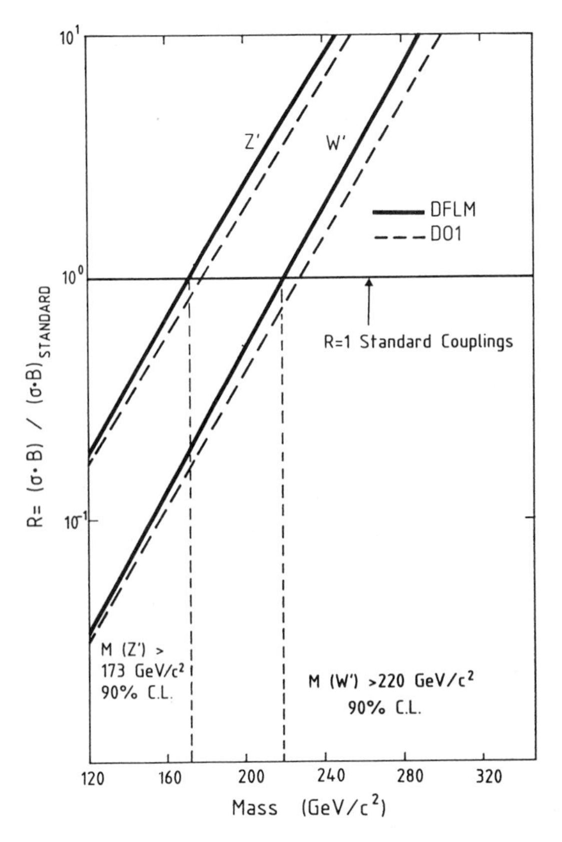

Fig. 20 The W′ (Z′) lower mass limits (90% CL) shown as a function of the W′ (Z′) coupling strength obtained by UA1 [9].

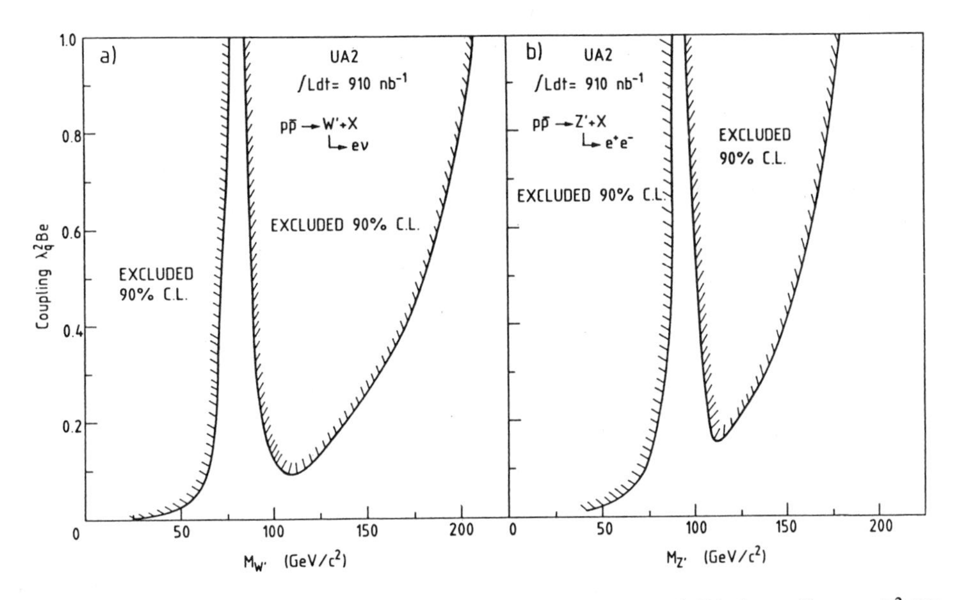

Fig. 21 Limits (90% CL) on additional vector bosons W′ and Z′ depending on $\lambda_q^2 \cdot BR_e$ obtained by UA2 [53]: a) for the W′ obtained from an analysis of single-electron candidates and b) for the Z′ obtained from an analysis of electron pairs.

function of this ratio, as shown in Fig. 20. For standard coupling (R = 1), UA1 obtained the limits [9]

$$m_{W'} > 220\ \mathrm{GeV}/c^2\,, \qquad m_{Z'} > 173\ \mathrm{GeV}/c^2 \qquad (90\%\ \mathrm{CL})\,.$$

From the analysis of the final UA2 data sample, the general limits for $m_{W'}$ and $m_{Z'}$ as a function of $\lambda_q^2 \cdot BR_e$ (coupling times branching ratio) are derived, as shown in Fig. 21 [53]. For standard couplings, i.e. $\lambda_q^2 \cdot BR_e = 1$, UA2 obtains

$$m_{W'} > 209\ \mathrm{GeV}/c^2\,, \qquad m_{Z'} > 180\ \mathrm{GeV}/c^2 \qquad (90\%\ \mathrm{CL})\,.$$

Most models derived from superstring theories predict at least one extra Z′ with $\lambda_q^2 \approx 0.2$ and BR_e varying from 0.3 to 1.0. For the resulting values of $\lambda_q^2 \cdot BR_e$ between 0.06 and 0.2, UA2 does not exclude any significant mass values.

5.4 Axigluon search

Axigluons are predicted by gauge theories based on the chiral colour group $SU(3)_L \times SU(3)_R$ [57]. In these theories, the $SU(3)_L \times SU(3)_R$ symmetry is broken down into the QCD $SU(3)_c$ symmetry at some scale that is comparable to the weak scale. A model-independent prediction of all chiral colour theories is the existence of a new massive colour octet of gauge bosons, the axigluons. The dominant decay channel of the axigluon is expected to be into two jets. The width of the axigluon depends on the number of particles into which it can decay. Within the model, the width is expected to be in the range $0.1\ m_A < \Gamma < 0.4\ m_A$.

A search for axigluons in the two-jet channel has been performed by UA1 [58]. Assuming that there is no interference between the normal QCD jet-production mechanism and the axigluon, the QCD and axigluon contributions have been added to obtain the total theoretical prediction in the presence of the axigluon. The data are fitted to the prediction using the maximum likelihood method. Figure 22 displays the limit, at 95% CL, on the production rate

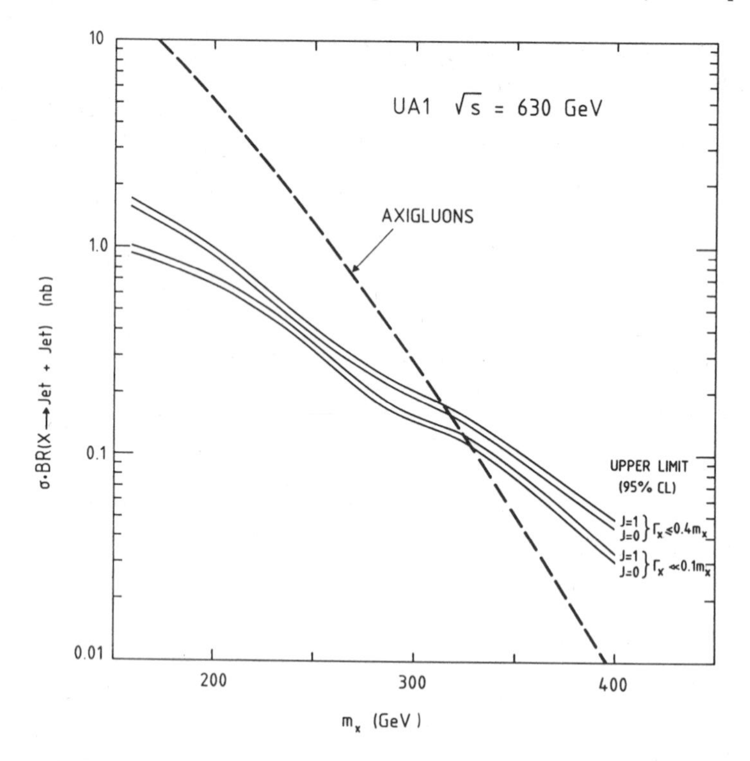

Fig. 22 Cross-section times branching ratio limits (95% CL) for the spin-0 and spin-1 particle X as a function of the mass m_X [58].

of axigluons in $p\bar{p}$ collisions at $\sqrt{s}$ = 630 GeV. It can be seen that the axigluon is excluded at the 95% CL for

$$150 \text{ GeV}/c^2 < m_A < 315 \text{ GeV}/c^2 .$$

This result has been generalized to obtain limits on the production rate of a heavy-state X decaying into two jets, for very small and very large widths, and also for the two spin cases (0 and 1). These limits are also shown in Fig. 22.

5.5 Quark substructure

As already mentioned, there are several theoretical motivations for compositeness. The proliferation of flavour may be a hint that quarks and leptons are composite. A possible manifestation of quark and lepton compositeness could be a new contact interaction having the new strong interaction scale Λ_c [59]. At the CERN Collider, the effect of a four-quark interaction term ($\sim 1/\Lambda_c$) would manifest itself in observed deviations from the QCD jet-production cross-section. Such deviations would be observable at scattering energies below Λ_c, and have been searched for by UA1, UA2 and CDF.

UA2 has concentrated on the overall normalization of the large-p_T jet cross-section [60]. Finite values of Λ_c would produce an excess of events at large p_T compared with the standard QCD prediction ($\Lambda_c = \infty$). Taking into account the theoretical and experimental uncertainties, UA2 obtained the limit

$$\Lambda_c > 370 \text{ GeV} \qquad (95\% \text{ CL}) .$$

UA1 has used the angular distribution of high-mass jet pairs to set a limit on Λ_c [61]. Figure 23 shows the measured angular distribution, where χ is defined as $(1 + \cos\theta)$ over $(1 - \cos\theta)$, θ being the angle between the axis of the jet pair and the beam direction in the jet–jet rest frame [62]. A finite quark size would modify this angular distribution in such a way that more events are expected at small χ values. By varying the value of the parameter Λ_c in the fit to the measured angular distribution, and taking into account the systematic uncertainty on the jet-energy scale, UA1 obtained

$$\Lambda_c > 415 \text{ GeV} \qquad (95\% \text{ CL}) .$$

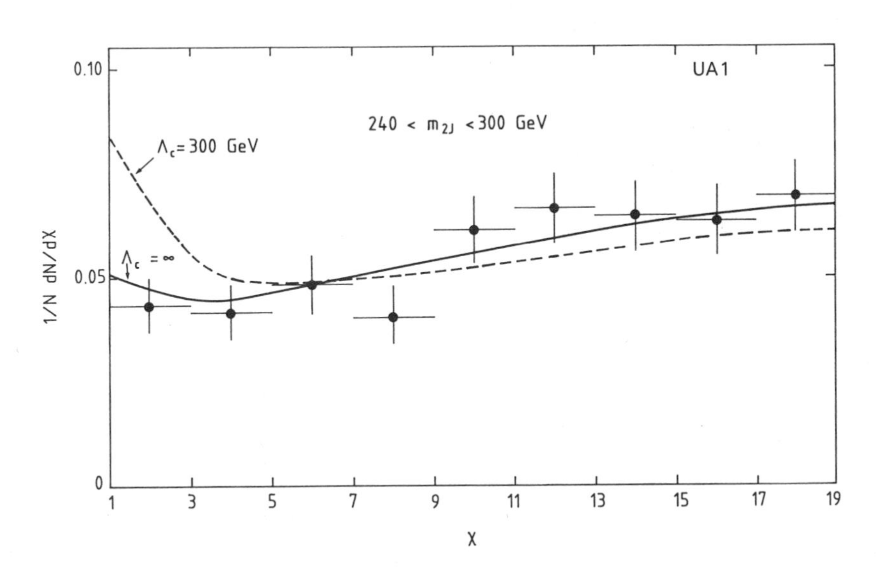

Fig. 23 Normalized angular distribution for very high mass jet-pairs as a function of χ. The solid curve is a QCD prediction ($\Lambda_c = \infty$); the dotted curve corresponds to $\Lambda_c = 300$ GeV, which is clearly excluded by the UA1 data [61].

To search for possible quark substructure, CDF has compared the measured inclusive jet cross-section at $\sqrt{s} = 1.8$ TeV with the predictions of leading-order QCD modified by the addition of a contact interaction with scale Λ_c [63]. Jet transverse energies up to 250 GeV have been observed, and the data with $E_T^{jet} < 130$ GeV were used to normalize the prediction. Taking into account Poisson fluctuations in the highest E_T-bins, the absence of jets with $E_T >$ 255 GeV, and uncertainties in structure functions and process-scales, CDF obtained the limit

$$\Lambda_c > 700 \text{ GeV} \qquad (95\% \text{ CL}) .$$

5.6 Outlook

Searches for new physics have so far been fruitless at both CERN and Fermilab. However, in several cases, significant lower mass limits could be placed on hypothetical new particles.

What is the discovery potential of hadron colliders in the near and the distant future?

Searches for new particles using the new data samples of UA1, UA2, and CDF can be divided into two classes: a) where $\sqrt{s}$ does not play an important role, and b) where the production cross-section is strongly dependent on $\sqrt{s}$.

Examples for class (a) are
- fourth-generation of charged leptons searched for in W decays,
- scalar leptons ($\tilde{e}$, $\tilde{\mu}$, $\tilde{\nu}$) and $\widetilde{W}$ searched for in W and Z decays.

One expects sensitivity up to the kinematic limit using present data. Figure 24 shows, as an example, the expected limits obtainable by UA2 for 10 pb^{-1} from the decay $Z \to \tilde{e}\tilde{e} \to ee\tilde{\gamma}\tilde{\gamma}$ [64].

Examples for class (b) are
- gluino and squark,
- additional vector bosons,
- compositeness scale Λ_c.

We have already seen the obvious benefit of having higher $\sqrt{s}$ in the search for gluinos and squarks. The mass reach at the Tevatron was about 20 GeV higher for about a factor of 30 less integrated luminosity. From the present data samples, one expects limits of the order of

$$\widetilde{m} \approx 120 \text{ GeV/c}^2 \qquad (\text{UA2}) ,$$

$$\widetilde{m} \approx 200 \text{ GeV/c}^2 \qquad (\text{CDF}) .$$

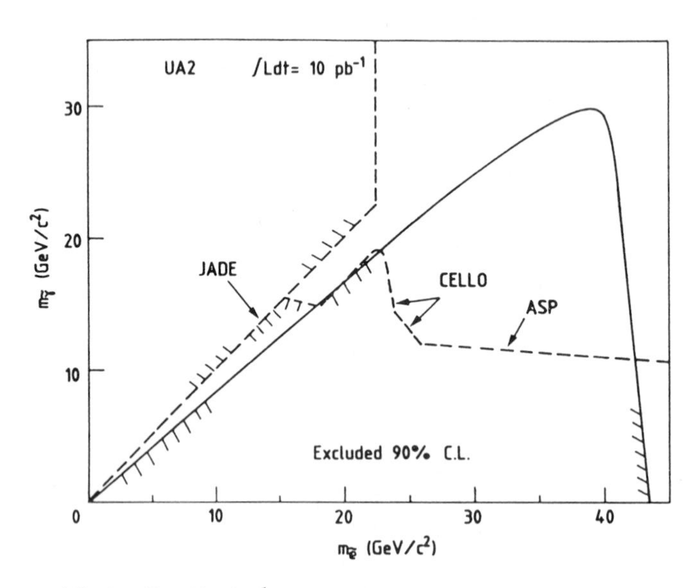

Fig. 24 Expected limits (for 10 pb^{-1}) on $\tilde{e}$ and $\tilde{\gamma}$ at 90% CL from a search for $Z \to \tilde{e}\tilde{e} \to ee\tilde{\gamma}\tilde{\gamma}$ decays in UA2 [64].

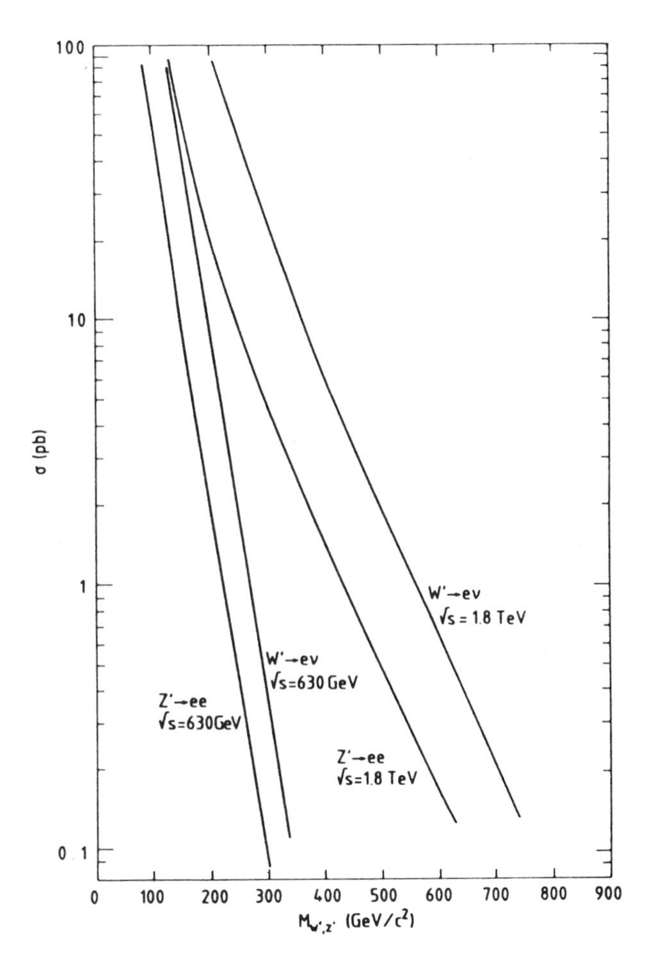

Fig. 25 Production of W′ and Z ′ at ACOL and Tevatron as a function of their mass, assuming standard couplings to leptons and quarks [64].

Because of the absence of background in the decay of additional vector bosons (W′, Z′) into leptons, hadron colliders make a large mass range accessible to experiments. Figure 25 shows the production cross-section for W′ and Z ′ at $\sqrt{s}$ = 0.63 and 1.8 TeV, assuming standard couplings to leptons and quarks, and $W' \rightarrow e\nu$ and $Z' \rightarrow ee$ decays. The corresponding sensitivity from the present data samples is expected to be

for $m_{W'}$: ~ 300 GeV (UA2) and ~ 600 GeV (CDF),

for $m_{Z'}$: ~ 250 GeV (UA2) and ~ 520 GeV (CDF).

An improvement in the limit on Λ_c of ≥ 650 GeV from CERN data will be achievable. CDF expects to reach Λ_c > 1.6 TeV from present data samples, again demonstrating the clear benefit of higher $\sqrt{s}$.

Finally, we discuss the possible discovery potential of LHC/SSC with respect to two topics: supersymmetry and Z′.

The detection of supersymmetric particles has received a lot of attention in various physics studies dealing with future colliders [65]. Present limits on sparticles are not yet problematic for low-energy supersymmetry. Sparticle masses may still be too high to be produced at the CERN $p\bar{p}$ Collider, at the Tevatron, at LEP, and at the SLC. If this is the case, it will be necessary to await the next generation of high-energy colliders, probing parton–parton collisions at energies of the order of 1 TeV. However, we have to keep in mind that even if the sparticle masses are larger than a few hundred GeV, nearly all low-energy SUSY models imply the existence of at least a light Higgs scalar with mass less than $O(m_Z)$. Therefore, the non-observation of a neutral Higgs boson up to LEP200 energies would put rather tight constraints on possible supersymmetric parameters [66].

Squarks and gluinos should be copiously produced at the LHC/SSC. For example, for a total integrated luminosity of 10^{40} cm^{-2} = 10 fb^{-1} (corresponding to one year of running at L = 10^{33} cm^{-2} s^{-1}), we expect ~ 10^4 gluino pairs to be produced at the SSC for $m_{\tilde{g}}$ = 1.5 TeV, or at the LHC for $m_{\tilde{g}}$ = 1 TeV. The sparticle signatures depend on the details of the assumed masses and decay patterns. The 'classical' signature of large $\not{E}_T$ plus two or more high-E_T jets results from the assumption that $\tilde{q}$ and $\tilde{g}$ will decay directly to the LSP. This assumption has been used in the analysis of UA1, UA2, and CDF data. Although these signatures should remain valid at LHC and SSC energies, it has been emphasized that for large squark and gluino masses, direct decays to the LSP becomes less important [67]. Instead, the decays to heavier neutralinos and charginos become dominant. These cascade decays generally occur for both the produced gluinos and squarks, and result in a large variety of possible final states, often containing W and Z bosons in a complex event topology.

Detailed studies of signal and backgrounds have been performed for the direct decays of sparticles into the LSP [68]. Figure 26 shows an example of the $\not{E}_T$ spectrum using one of the possible event selections at LHC energies. The resulting signal-to-background ratios are typically ~ 10:1 at large $\not{E}_T$. Taking into account uncertainties involved in these calculations, it was concluded that squark and gluino masses, up to ~ 1 TeV/c^2 are potentially accessible for 10 fb^{-1} integrated luminosity. Extrapolating the LHC study to SSC energies, a discovery limit of ~ 1.5–2 TeV/c^2 can be expected for the gluino and squark mass, because the signal cross-section increases much faster than the background cross-sections for higher $\sqrt{s}$. The analysis discussed above has been extended to study the feasibility of searching for $\tilde{g}$ and $\tilde{q}$ if the LHC runs at a luminosity of ~ 5 × 10^{34} cm^{-2} s^{-1} [69]. It was concluded that, provided the experimental problems associated with running at high luminosity can be overcome, an extension of the discovery limit for $\tilde{g}$ and $\tilde{q}$ of masses ~ 2 TeV/c^2 should be feasible at the LHC.

A study of gluino final states containing Z's from cascade decays was carried out for SSC energies [67]. It was concluded that it is possible to find event signatures that give promising results for signal-to-background ratios even for these complex decay modes.

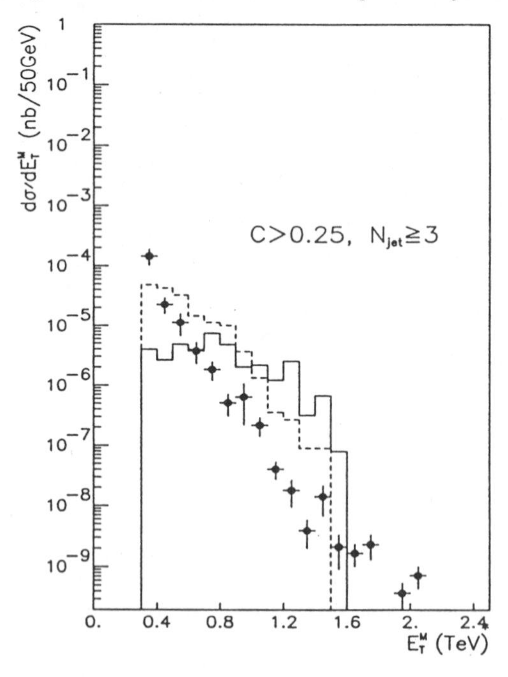

Fig. 26 Missing transverse energy distribution at $\sqrt{s}$ = 17 TeV after possible selection cuts for squark (solid histogram) and gluino (dashed histogram) production and for the total background (points with error bars). A squark and gluino mass of 1 TeV/c^2 is assumed [68].

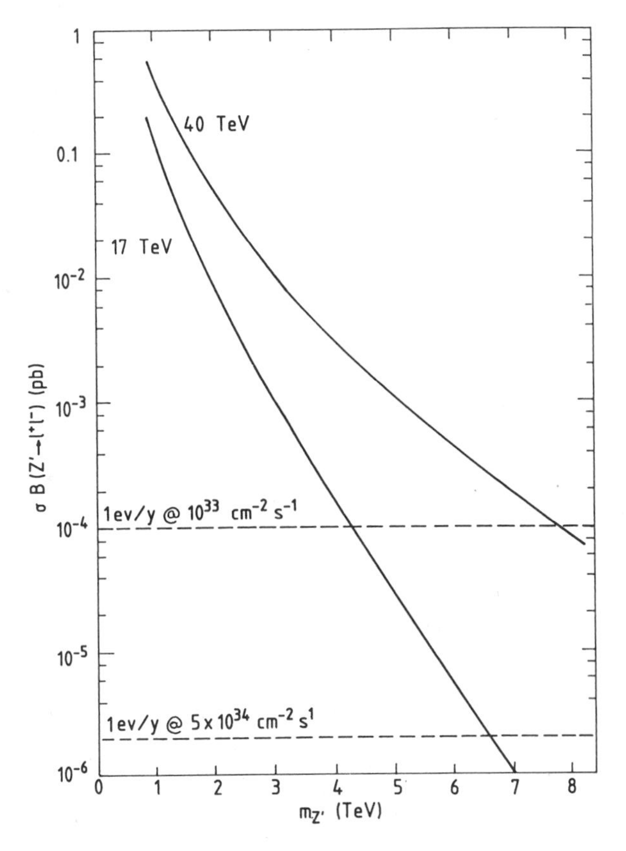

Fig. 27 Cross-section times maximal $\ell^+\ell^-$ branching ratio for Z′ in a minimal rank-5 superstring-inspired model at $\sqrt{s}$ = 17 and 40 TeV [70].

Additional vector bosons are predicted in many possible extensions of the Standard Model. In order to study the discovery potential for an additional neutral gauge boson Z′ at the LHC/SSC, a superstring-inspired model was used [70]. This model predicts one (or two) new Z′ with masses in the range of ≤ 1 TeV/c^2. Figure 27 shows the predicted $\sigma \cdot BR(Z' \to \ell^+\ell^-)$ for the minimal superstring model at LHC and SSC energies. Since no significant background is expected for the leptonic decay modes, it was concluded that a search can be made for Z′ with masses up to 4 TeV (6 TeV) using the $Z' \to \ell^+\ell^-$ decay channel at the LHC for 10 fb^{-1} (500 fb^{-1}) integrated luminosity. The corresponding discovery limit at the SSC is 7 TeV for 10 fb^{-1} integrated luminosity. This shows that the discovery potential for additional vector bosons is very high in hadron–hadron machines, and also demonstrates the benefit of having higher available centre-of-mass energies.

6. CONCLUSIONS

The data collected between 1982 and 1985 by the UA1 and UA2 experiments at the CERN $p\bar{p}$ Collider, corresponding to integrated luminosities of about 1 pb^{-1}, have provided many important physics results. Excellent agreement with the Standard Model has been observed. Searches for new physics have so far been fruitless; however, in several cases, significant lower mass limits could be placed on hypothetical new particles.

First results from CDF at the Fermilab Collider are in agreement with the expectations owing to the increase in $\sqrt{s}$, and no deviations from the Standard Model have been observed. The benefit of the larger $\sqrt{s}$ has been best demonstrated in searches for new physics ($m_{\tilde{g}}$, $m_{\tilde{q}}$, Λ_c).

From the data taken with ACOL and at Fermilab, it is expected that the Standard Model predictions can be tested more precisely, and that hopefully at least one of its missing elements—the top quark—will be found. The preliminary results on searches for the t-quark lead to the conclusion that the t-quark must be very heavy compared with other known quarks and leptons, provided the Standard Model is corrrect and that no new physics influences the production and decay properties of the t-quark. The goal of discovering new physics still remains, but will in some cases be more easily accessible at the Tevatron and at LEP.

Acknowledgement

I wish to thank the Scientific Reports Editing and Text Processing Section at CERN for their competence and patience in preparing the final version of this paper.

REFERENCES

[1] S.L. Glashow et al., Phys. Rev. **D18** (1978) 1724.
E. Eichten et al., Rev. Mod. Phys. **56** (1984) 579.
[2] M.E. Peskin, Proc. Int. Symp. on Lepton and Photon Interactions at High Energies, Kyoto, 1985, eds. M. Konuma and C. Takahashi (Kyoto Univ., 1985), p. 714.
H. Harari, Proc. 5th Topical Workshop on Proton–Antiproton Collider Physics, Saint-Vincent, 1985, ed. M. Greco (World Scientific, Singapore, 1985), p. 429.
[3] P. Fayet and S. Ferrara, Phys. Rep. **32** (1977) 249.
H.P. Nilles, Phys. Rep. **110** (1984) 1.
J. Ellis et al., Nucl. Phys. **B238** (1984) 453.
H.E. Haber and G.L. Kane, Phys. Rep. **117** (1985) 75.
[4] J. Dowell (UA1 Collab.), Proc. 6th Topical Workshop on Proton–Antiproton Collider Physics, Aachen, 1986, eds. K. Eggert et al. (World Scientific, Singapore, 1987), p. 419.
[5] C. Booth (UA2 Collab.), ibid., p. 381.
[6] F. Abe et al., Nucl. Instrum. Methods Phys. Res., **A271** (1988) 387.
[7] UA1 Collaboration papers:
G. Arnison et al., Phys. Lett. **122B** (1983) 103; **126B** (1983) 398; **129B** (1983) 273; **134B** (1984) 469; **147B** (1984) 241; **166B** (1986) 484; Europhys. Lett. **1** (1986) 327.
C. Albajar et al., Phys. Lett. **185B** (1987) 233.
[8] UA2 Collaboration papers:
M. Banner et al., Phys. Lett. **122B** (1983) 476.
P. Bagnaia et al., Phys. Lett. **129B** (1983) 130; Z. Phys. **C24** (1984) 1.
J.A. Appel et al., Z. Phys. **C30** (1986) 1.
R. Ansari et al., Phys. Lett. **186B** (1987) 440.
[9] C. Albajar et al. (UA1 Collab.), preprint CERN-EP/88-168 (1988), submitted to Z. Physik.
[10] F. Abe et al. (CDF Collab.), Phys. Rev. Lett. **62** (1989) 1005.
[11] K. Einsweiler, talk presented at the 8th Topical Workshop on $p\bar{p}$ Collider Physics, Castiglione, 1989.
[12] T. Phillips, talk presented at the same Workshop as Ref. [11].
[13] G.S. Abrams et al. (Mark II Collab.), Phys. Rev. Lett. **63** (1989) 724.
[14] H. Abramowicz et al., Phys. Rev. Lett. **57** (1986) 298.
J.V. Allaby et al., Z. Phys. **C36** (1987) 611.
D. Bogert et al., Phys. Rev. Lett. **55** (1985) 1969.
P. Reutens et al., Phys. Lett. **152B** (1985) 404.

[15] F. Jegerlehner, Z. Phys. **C32** (1986) 425.
W.J. Marciano and A. Sirlin, Phys. Rev. **D22** (1980) 2695 and **D29** (1984) 945.
[16] U. Amaldi et al., Phys. Rev. **D36** (1987) 1385.
J.V. Allaby et al., preprint CERN-EP/87-140 (1987).
G. Costa et al., Nucl. Phys. **B297** (1988) 244.
G. Gounaris and D. Schildknecht, preprint CERN-TH.4940 (1987).
[17] J. Ellis and G.L. Fogli, preprint CERN-TH.5511/89 (1989).
[18] G. Altarelli et al., Nucl. Phys. **B157** (1979) 461 and **B246** (1984) 12.
G. Altarelli et al., Z. Phys. **C27** (1985) 617.
[19] R. Voss, Proc. Int. Symp. on Lepton and Photon Interactions at High Energies, Hamburg, 1987, eds. W. Bartel and R. Rückl (North-Holland Publ. Co., Amsterdam, 1988), p. 525.
[20] M. Lefebvre, talk presented at the same Workshop as Ref. [11].
[21] P. Jenni, same Proceedings as Ref. [19], p. 341.
[22] C. Albajar et al. (UA1 Collab.), Phys. Lett. **193B** (1987) 389.
[23] G. Arnison et al. (UA1 Collab.), Phys. Lett. **172B** (1986) 461.
[24] C. Albajar et al. (UA1 Collab.), Phys. Lett. **209B** (1988) 385.
[25] C. Albajar et al. (UA1 Collab.), Phys. Lett. **209B** (1988) 397.
[26] W.J. Stirling, Proc. 6th Topical Workshop on Proton–Antiproton Collider Physics, Aachen, 1986, eds. K. Eggert, H. Faissner and E. Radermacher (World Scientific, Singapore, 1987), p. 301.
[27] P. Aurenche et al., Nucl. Phys. **B297** (1988) 661; Phys. Lett. **140B** (1984) 87.
[28] R. Ansari et al. (UA2 Collab.), Phys. Lett. **215B** (1988) 175.
[29] C. Albajar et al. (UA1 Collab.), Phys. Lett. **198B** (1987) 271.
[30] T. Kamon, talk presented at the same Workshop as Ref. [11].
[31] L. Di Lella, Proc. 7th Topical Workshop on Proton–Antiproton Collider Physics, Fermilab, 1988, eds. R. Raja et al. (World Scientific, Singapore, 1989), p. 849.
[32] F.E. Paige, Proc. Workshop on Experiments, Detectors, and Experimental Areas for the Supercollider, Berkeley, USA, 1987, eds. R. Donaldson and M. Gilchriese (World Scientific, Singapore, 1989), p. 750.
[33] C. Albajar et al. (UA1 Collab.), Z. Phys. **C37** (1988) 489; Phys. Lett. **186B** (1987) 237; **200B** (1988) 380; **213B** (1988) 405.
[34] F.E. Paige and S.D. Protopopescu, Brookhaven National Laboratory report, BNL-38034 (1986).
[35] P. Nason, S. Dawson and R.K. Ellis, Nucl. Phys. **B303** (1988) 607.
[36] G. Altarelli et al., preprint CERN-TH.4978/88 (1988).
[37] C. Albajar et al. (UA1 Collab.), Phys. Lett. **186B** (1987) 247.
[38] H. Albrecht et al., Phys. Lett. **192B** (1987) 246.
H. Schröder, Proc. 24th Int. Conf. on High-Energy Physics, Munich, 1988, eds. R. Kotthaus and J.H. Kühn (Springer-Verlag, Berlin, Heidelberg, 1989), p. 73.
[39] A. Breakstone et al., Phys. Lett. **135B** (1984) 510.
[40] K. Eggert and H.G. Moser, Proc. 7th Topical Workshop on Proton–Antiproton Collider Physics, Fermilab, 1988, eds. R. Raja et al. (World Scientific, Singapore, 1989), p. 599.
[41] A. Weinstein, talk presented at the 14th Int. Symp. on Lepton and Photon Interactions, Stanford, 1989.
[42] J. Ellis, J.S. Hagelin and S. Rudaz, Phys. Lett. **192B** (1987) 201.
V. Barger, T. Hau and D.V. Nanopoulos, Phys. Lett. **194B** (1987) 312.
I.I. Bigi and A.I. Sanda, Phys. Lett. **194B** (1987) 307.
L.L. Chau and W.Y. Keung, UC Davis preprint, UCD-87-02 (1987).
H. Harari and Y. Nir, Stanford preprint SLAC-PUB-4341 (1987).
G. Altarelli and P.J. Franzini, preprint CERN-TH.4745/87 (1987).
H. Albrecht et al. (ARGUS Collab.), Phys. Lett. **192B** (1987) 245.
[43] C. Albajar et al. (UA1 Collab.), Z. Phys. **C37** (1988) 504.
[44] J. Kroll, Proc. 24th Int. Conf. on High-Energy Physics, Munich, 1988, eds. R. Kotthaus and J.H. Kühn (Springer-Verlag, Berlin, Heidelberg, 1989), p. 1416.

[45] A. Nisati, talk presented at the same Workshop as Ref. [11].
S. Lammel, talk presented at the same Workshop as Ref. [11].
[46] D. Buskulic, talk presented at the same Workshop as Ref. [11].
[47] L. Galtieri, talk presented at the same Workshop as Ref. [11].
H.H. Williams, talk presented at the same Workshop as Ref. [11].
[48] M. Shochet, talk presented at the same Workshop as Ref. [11].
[49] D. Froidevaux, Proc. Workshop on Physics at Future Accelerators, La Thuile, Italy, 1987, and CERN, Geneva, Switzerland, ed. J. Mulvey, CERN 87-07 (1987), Vol. I, p. 61.
[50] C. Albajar et al. (UA1 Collab.), Phys. Lett. **185B** (1987) 241.
[51] S. Dawson et al., Phys. Rev. **D31** (1985) 1581.
[52] C. Albajar et al. (UA1 Collab.), Phys. Lett. **198B** (1987) 261.
[53] R. Ansari et al. (UA2 Collab.), Phys. Lett. **195B** (1987) 613.
[54] V. Vercesi, talk presented at the same Workshop as Ref. [11].
[55] F. Abe et al. (CDF Collab.), Phys. Rev. Lett. **62** (1989) 1825.
[56] P. Langacker et al., Phys. Rev. **D30** (1984) 1470.
E. Cohen et al., Phys. Lett. **165B** (1985) 76.
M.B. Green, J.H. Schwarz and E. Witten, Superstring theory (Cambridge Univ. Press, Cambridge, 1987).
F. del Aguila et al., Nucl. Phys. **B284** (1987) 530 and **B287** (1987) 419.
[57] P.H. Frampton and S.L. Glashow, Phys. Lett. **190B** (1987) 157.
[58] P.A. Sphicas, Proc. 7th Topical Workshop on Proton–Antiproton Collider Physics, Fermilab, 1988, eds. R. Raja et al. (World Scientific, Singapore, 1989), p. 239.
[59] E. Eichten et al., Phys. Rev. Lett. **50** (1983) 811.
[60] J.A. Appel et al. (UA2 Collab.), Phys. Lett. **160B** (1985) 349.
[61] G. Arnison et al. (UA1 Collab.), Phys. Lett. **177B** (1986) 244.
[62] J.C. Collins and D.E. Soper, Phys. Rev. **D16** (1977) 2219.
[63] F. Abe et al. (CDF Collab.), Phys. Rev. Lett. **62** (1989) 613.
[64] D. Froidevaux and P. Jenni, Proton–Antiproton Collider Physics, eds. G. Altarelli and L. Di Lella (World Scientific, Singapore, 1989), p. 323.
[65] For a recent review, see
F. Pauss, Proc. 24th Int. Conf. on High-Energy Physics, Munich 1988, eds. R. Kotthaus and J.H. Kühn (Springer-Verlag, Berlin, Heidelberg, 1989), p. 1275, and references therein.
[66] R. Barbieri, Proc. Rencontres de Physique de la Vallée d'Aoste, La Thuile, Italy, 1987, ed. M. Greco (Editions Frontières, Gif-sur-Yvette, 1987), p. 523.
J. Gunion et al., Univ. Calif. Davis report UCD-88-11 and LBL-25033 (1988).
G.F. Giudice, Phys. Lett. **B208** (1988) 315.
[67] R.M. Barnett et al., Proc. Workshop on Experiments, Detectors, and Experimental Areas for the Supercollider, Berkeley, 1987, eds. R. Donaldson and H. Gilchriese (World Scientific, Singapore, 1989), p. 178.
H. Baer and E. Berger, Phys. Rev. **D34** (1986) 1361.
H. Baer et al., Phys. Rev. **D36** (1987) 96.
[68] R. Batley, Proc. Workshop on Physics at Future Accelerators, La Thuile, Italy, 1987, and CERN, Geneva, Switzerland, ed. J. Mulvey, CERN 87-07 (1987), Vol. II, p. 109.
[69] F. Pauss et al., The Feasibility of Experiments at High Luminosity at the Large Hadron Collider, ed. J. Mulvey, CERN 88-02 (1988), p. 79.
[70] J. Ellis and F. Pauss, Proc. Workshop on Physics at Future Accelerators, La Thuile, Italy, 1987, and CERN, Geneva, Switzerland, ed. J. Mulvey, CERN 87-07 (1987), Vol. I, p. 80.
J. Ellis, Proc. 1987 ICFA Seminar on Future Perspectives in High-Energy Physics, BNL, Upton, NY, USA, October 1987, ed. P.F. Dahl, BNL 52114, p. 117.

PHYSICS WITH HIGH ENERGY ION BEAMS

Peter Sonderegger

CERN, Geneva, Switzerland

INTRODUCTION

With the recent advent of ultrarelativistic ion beams, a new frontier has become open to scrutiny: the physics at high energy density. It is at an energy density high enough above nuclear energy density (0.15 GeV/fm^3) that matter is expected, by QCD, to exist in a quark-gluon plasma (QGP) phase. If the phase transition is of first order, a mixed phase would exist, with QGP bubbles in a hadron gas. The quest of the quark-gluon plasma is the main goal of experiments with high energy ion beams.

The introduction of thermodynamics into particle physics, advocated by Hagedorn since 1965 [1], has become irreversible with the advent of the ion beams.

The present paper is intended to be, at best, a progress report. After two runs of 17 days each (at CERN), and a few years of work of dedicated but not numerous theorists, the greater part of the road towards a scientific standard comparable with LEP or collider physics is still ahead of us.

QGP AND PHASE TRANSITION

The idea that at high enough temperature and/or matter density the hadrons melt into a plasma of quarks and gluons has been first suggested and studied within the MIT bag model [2], and then taken up by lattice QCD [3] (Fig. 1 shows a schematic phase diagram). The Monte Carlo technique on a 4−dimensional lattice (with the time taken imaginary so that it plays the role of temperature) was first applied on gluons only (quarks were static, i.e. heavy and not affected); then dynamical (light) quarks were also incorporated [4] (Fig. 2; for a pedagogical introduction, see [5]). Results show a transition temperature T_c near 160 − 200 MeV, a transition baryon density ρ_c many times nuclear density, and a strong indication that chiral symmetry restoration (which makes a.o. quark masses small) happens also during the deconfinement transition. The transition has been mostly assumed to be of first order (i.e. like for boiling or freezing water), but recent calculations (gluons only) on much bigger lattices [6] as made possible by dedicated computers, shed some doubt on this important point (calculations with dynamical quarks are under way).

The first order transition implies a mixed phase extending in energy density from ε_H to ε_Q (Fig. 3), with the ratio between ε_Q and ε_H given by the ratio between the respective degrees of freedom, 37 for gluons and u and d quarks, and a number of hadrons ranging from 3 (pions only) to considerably more if mesons, baryons and resonances are taken into account. Even in the latter case, one expects a fairly extended region in ε where temperature and pressure stay constant (or vary little, in the case of a second order transition).

Particle Physics: Cargèse 1989
Edited by M. Lévy *et al.*
Plenum Press, New York, 1990

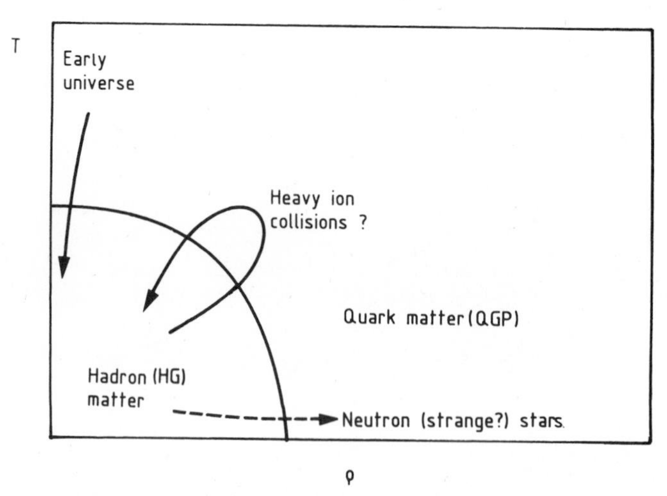

Fig. 1 Schematic temperature-baryon density phase diagram showing the deconfinement transition. The quark-gluon plasma phase (quark matter) prevailed in the first microseconds after the Big Bang; it may also be the end state of certain heavy stars which are commonly thought to be neutron stars. High energy collisions of heavy ions *may* lead to very short lived and very tiny quark matter states.

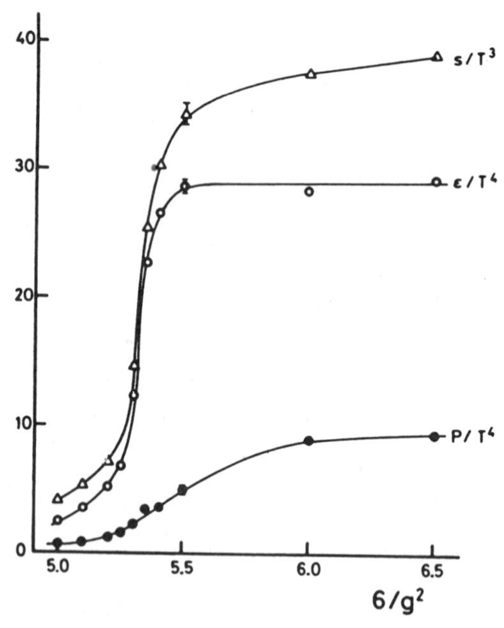

Fig. 2 Lattice calculations (here with interacting quarks [4]) show that at a certain temperature T_c (T is related to $6/g^2$) the energy density jumps, indicating a phase transition. The pressure is continuous across the transition. Entropy density (which is the sum of the two, divided by the temperature) also jumps; it is measured by the detected multiplicity, while the temperature reflects in the average transverse momentum of the produced particles.

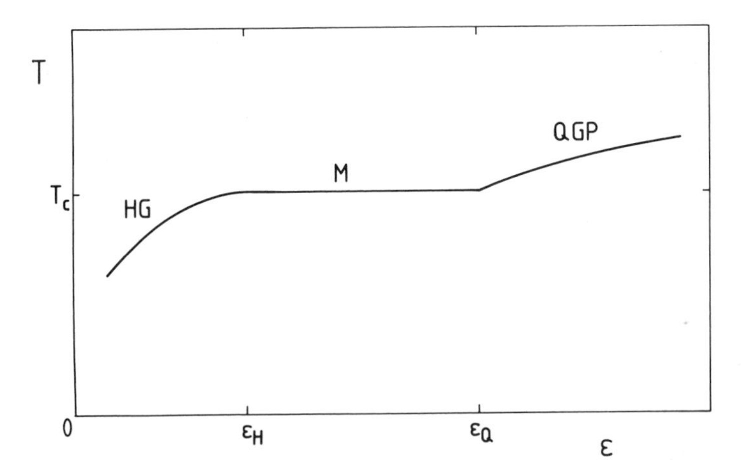

Fig. 3 The temperature increases, both for a hadron gas (HG) and a quark-gluon plasma (QGP), with the fourth root of energy density. In case of a first order phase transition (as suggested by Fig. 2) there is an extended mixed phase (M) of constant temperature and pressure, between two limiting energy densities ε_H and ε_Q, propertional to the number of quantum carriers in the two phases. ε_Q is a multiple of ε_H.

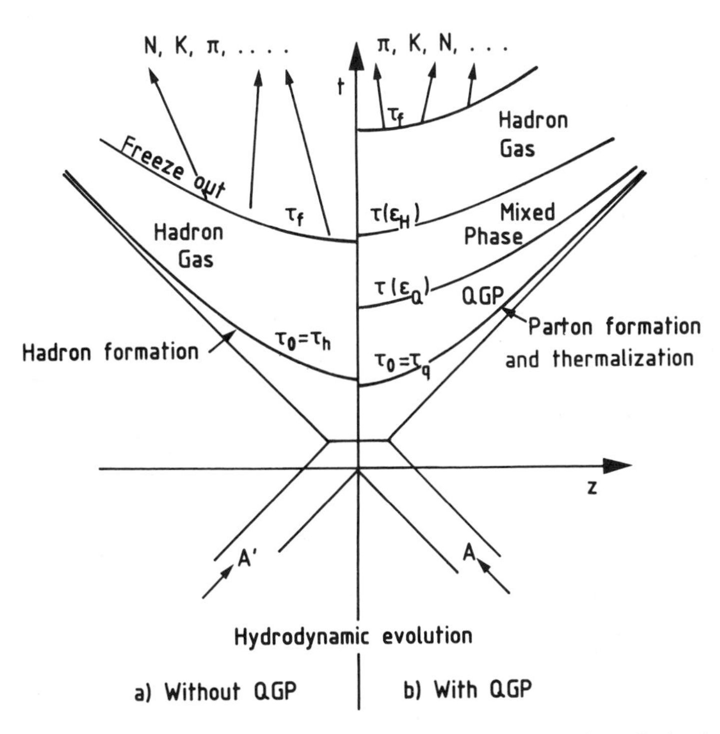

Fig. 4 An ion–ion (A–A′) collision viewed in a z–t plane (z is the beam direction). The schematic but conceivable scenario assumed here has a QGP formation stage (followed by a mixed phase) in the forward but not in the backward hemisphere (proper times $\tau=\sqrt{t^2-z^2}$). The developing art of experimental ion physics is aimed at deducing from the hadrons emerging after freeze-out whether the reaction has gone through a QGP state or not, and which were its properties.

Where would QGP occur in nature? The Universe was in a hot and dilute state of QGP for the first 10 microseconds or so (Fig. 1). The question whether neutron stars contain in fact a form of cold and very dense QGP (maybe strange matter; see below) is being debated. We address here the possibility that an extremely short-lived and tiny form of QGP can a) be produced and b) if produced, detected, in the laboratory, using presently available or future ion beams and conventional detectors. Both these issues were entirely open when the exploitation of ultrarelativistic ion beams was decided; and in truth, they still are.

ION COLLISIONS, SIGNATURES, ENERGY DENSITY

Fig. 4 shows a space-time picture of an ion-ion collision for two different scenarios, assumed to take place in the forward and backward hemisphere respectively. In one case, energy is dissipated and hadrons are formed after a formation time τ_o; these may interact until they are far enough apart so that freeze-out occurs, at time τ_f. In the other case, it is the quarks and gluons which form and thermalize within a time τ_o which may be of the same order of magnitude as in the previous case (~ 1 fm/c which would correspond to a transverse mass of ~ 200 MeV). A plasma phase may set in, followed by a mixed phase during which hadronization takes place, and ending up in a hadron gas phase, possibly with hadronic final state interactions till freeze-out time τ_f. Plasma lifetimes of many, maybe tens of fm are sometimes anticipated, so τ_f could be substantially longer in the case involving a QGP phase.

The successive expansion of the system is described by hydrodynamics; the expansion is inversely proportional to the initial size and is therefore much faster along the beam than in the transverse directions.

The second law of thermodynamics has tricky consequences for the hadronization process which does not simply consist in recombining pairs of quarks and antiquarks into mesons: that case would imply a large loss of entropy since both relativistic quarks and mesons have some 4 units of entropy each, as well as the gluons which have also to be disposed of. Hadronization is pictured as a subtle combination of cooling and soft quark production by gluons [7].

Fig. 4 is meant to show the difficulty of knowing what happened in the early stages of an interaction when one observes only, or mostly, the pions produced during its very latest stage. The proposed signatures for QGP include some characteristics of these mesons formed or decoupled long after a possible QGP phase has ceased to exist: the dimension of the interaction volume at freeze-out as measured via pion interferometry; the average p_T of the produced pions which signals whether temperature and/or pressure have increased or not with energy density; rapidity fluctuations signalling the coexistence of QGP bubbles and hadrons; the strangeness content which is expected to increase towards thermal equilibrium through multiple interactions (of any kind). Other signatures depend directly on what happened during the initial phases of the interaction: direct emission of real and virtual photons; and Debye colour screening which prevents heavy mesons of large radius (e.g. the J/Ψ) from forming inside QGP. Many other possible signatures have been suggested; the above are the ones which have presently some theoretical and experimental significance.

Which energy density will result for a given beam, target, and energy in case of a central collision? Bjorken [8] has suggested a flexible and relativistically sound form of Landau's celebrated energy density [9] (= c.m. energy / Lorentz contracted overlap volume of the colliding nuclei): The volume element is transformed $dx \cdot dy \cdot dz = dx \cdot dy \cdot \tau \cdot dY$ so that its length, instead of being allowed to shrink indefinitely at higher and higher energy, is now linked to the formation time τ of the produced partons (Y is the rapidity). In the absence of a credible theoretical estimate of the dissipated energy, the present practice uses the measured transverse energy E_T instead (which would equal $\pi/4 \cdot E_{cm}$ for maximum energy dissipation). The Bjorken energy density then reads

$$\epsilon_{Bj} = dE_T/dY \cdot 1/(S_T \cdot \tau_o) \qquad (1)$$

with τ_0 of order 1 fm/c, and the transverse area $S_T = \pi \cdot r_0^2 \cdot Ap^{2/3}$ in the case of a central collision of a projectile of atomic number Ap on a heavier target, where $r_0 = 1.15$ fm. For peripheral collisions, it is customary to divide the (smaller) E_T by a (smaller) effective overlap area S_T computed by some geometrical model.

The energy density variable ε is of central importance and governs in particular the probability of forming a QGP. Hence all experiments measure E_T and its angular dependence, or at least, the charged multiplicity, from which E_T obtains by taking into account the neutral/charged ratio and by multiplying by the estimated average transverse mass.

THE EXPERIMENTAL PROGRAMS AND THEIR HISTORY

Early explorations of ion reactions were made during the last two years of ISR operation with $d-d$ and $\alpha-\alpha$ reactions, which showed collective effects [10] and with cosmic rays beyond 1 TeV by the JACEE collaboration (Japanese-American Collaborative Emulsion Experiment) [11], which gave the first indication of unexpected behaviour at high energy density (see later).

On the nuclear physics side, ion collisions had become long ago a privileged technique towards exploration of highly deformed high spin states, of hot nuclei, of hard photon emission, and of the nuclear equation of state [12]. The latter was in particular studied at the Bevalac. Recently, the flow angle phenomenon was discovered there in medium ion collisions near 1 GeV/nucleon by the GSI-LBL group [13]. Secondary particles are found to peak at a non-zero angle which depends on the size of colliding nuclei and on the impact parameter. This phenomenon has not yet been tackled by the high energy ion experiments.

The GSI-LBL group (H. Gutbrod, R. Stock et al.) decided to extend this line of research to higher energies and submitted in 1980 a proposal to expose both the Plastic Ball and a streamer chamber to a 10 GeV/nucleon ion beam at the PS. An ion source was to be built in Grenoble with German funds. The proposal was accepted and the two effectively separated experiments were named WA80 (Plastic Ball, to be enriched with multiplicity and photon detectors) and NA35 (Streamer chamber). After CERN had imposed a major reorientation towards top SPS energies, and after the source had been successfully coupled to the refurbished Linac I, two ion runs of 17 days each took place, the first with a ^{16}O beam at the end of 1986, briefly at 60 and mainly at 200 GeV/nucleon, and the second with a ^{32}S beam in 1987, at 200 GeV/nucleon. It had also been decided to carry out a wider exploratory programme, using the best available apparatus but little fresh money, which was to include four experiments born out of the particle physics tradition. NA34 (HELIOS) was based on the celebrated R807 U calorimeters and on the NA3 dimuon spectrometer, and completed by an external magnetic spectrometer, a multiplicity detector, and the team and experience of WA75, another former dimuon experiment. The most modern dimuon experiment, NA10, was complemented by a specially fast and radiation resistent optical fibre calorimeter to become NA38. By 1987, two more experiments started, NA36 which was based on the EHS spectrometer and a TPC, and WA85, based on the Omega spectrometer, whose detectors were desensitized in the low p_T, i.e. very high multiplicity regions. Both experiments aimed at measuring strange particle and antiparticle production. The programme also included four small and about ten emulsion experiments.

At the same time, the AGS at Brookhaven had been equipped with an ion source and ran with ^{16}O and ^{28}Si beams at up to 14.5 GeV/nucleon, from late 1986 onwards, for three major experiments: E802 (particle identification), E814 (forward particle production), and E810 (a TPC in the MPS spectrometer magnet).

What follows is based on the success of this materially modest effort, which involved however the energies of some 500 physicists.

EXPERIMENTAL RESULTS

We will work our way through the main results obtained so far, starting with general characteristics, attained transverse energies and estimated energy densities, discussing then the freeze-out volume measurements which may give indications on plasma lifetimes, and reviewing finally the signatures proper, of which few have been adequately studied so far.

Multiplicity distribution

The two relevant variables for particle production are the transverse momentum p_T, and either the rapidity Y,

$$Y = 1/2 \ln ((E+p_L)/(E-p_L)) = \ln ((E+p_L)/m_T) \quad (2)$$

(where the transverse mass is given by $m_T^2 = m^2 + p_T^2$) which is usually taken in the nucleon-nucleon centre of mass, and which is shifted by a constant by a Lorentz boost; or the pseudorapidity η,

$$\eta = 1/2 \ln ((1+\cos\Theta)/(1-\cos\Theta)) = -\ln \mathrm{tg}\,(\Theta/2) \quad (2')$$

usually given in the lab, and identical to Y for m = 0; this is the old variable used in cosmic ray events. The rapidity shift from c.m. to lab is 3.03 at 200 GeV/nucleon.

The charged pion multiplicity is an approximate gaussian in Y or η centered on the overall c.m. system (i.e., for a central O−Au collision, the c.m. of the O nucleus and a cylinder of the same section as the O projectile cut into the Au target). Its r.m.s. width (in Y or η) increases with energy (in the N−N c.m.) somewhat slower than the domain in Y, but not with ion size, from 1 at 14.5 GeV/nucleon [14] to 1.5 at 200 GeV/nucleon for p−p, p−Xe [15] and O−Au [14]. It would be 0.88 (in η) in case of an isotropic fireball.

Transverse Energy, Stopping Power, and Energy Density

The operational definition of the transverse energy E_T is

$$E_T = \Sigma \quad E_i \cdot \sin\Theta_i \quad (3)$$

where the sum runs over all secondaries, and E_i is the total energy for mesons, and the kinetic energy for nucleons. This is roughly (and neglecting antinucleon production) what is measured in calorimeters, in the experimentalist's way of assessing energy dissipation. The ratio E_T/multiplicity is close to 0.6 GeV (= $n_{total}/n_{charged} \cdot \langle m_T \rangle$) and rather constant at CERN energies. The same proportionality factor links the η distributions of E_T and n_{ch} provided the experimental resolutions in η are adequate.

Fig. 5 shows certain E_T distributions measured by NA34 [16] for collisions of a sulphur beam with various targets, for an almost complete angular coverage. There is a direct relation between impact parameter and E_T: peripheral interactions produce small E_T, while central interactions correspond to the end of the plateau and mean E_T values proportional to the target length. The naive expectation

$$E_T\,(\max) \simeq A_{projectile}^{2/3} \cdot A_{target}^{1/3} \quad (4)$$

is rather well verified except maybe at small rapidities where sidewards intranuclear cascading contributes to E_T.

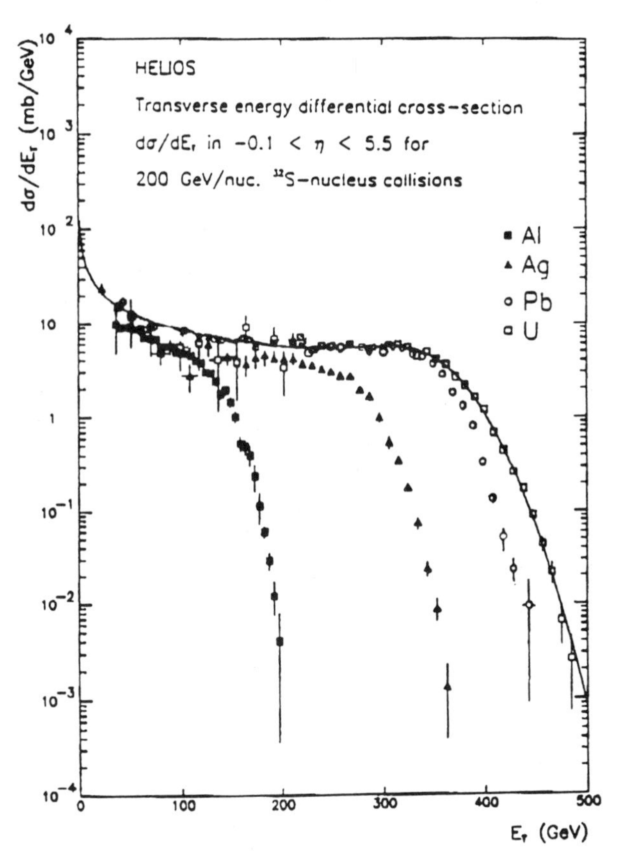

Fig. 5 E_T distributions from NA34, with almost full angular coverage ($-0.1 < \varepsilon < 5.5$), for collisions of a ^{32}S beam with various targets: ^{27}Al, almost a collision among equal nuclei, without a plateau, and the heavier Ag, Pb and U targets for which the probability of central and almost central collisions with maximal E_T is large and leads to a plateau up to a maximal E_T value which depends on the target size. The slope of the drop beyond the plateau is roughly conditioned by the square root of the number of detected particles for a spherical target nucleus such as Pb. With ^{238}U, a strongly deformed nucleus, the slope is much gentler, allowing to reach exceptionally high transverse energies.

The S−Al data show a smoothly falling behaviour also typical of A−A collisions for which central collisions have vanishing probability; a plateau is characteristic for collisions with targets much larger than the projectile. The tail is thought to reflect multiplicity fluctuations; its width is therefore proportional to $\sqrt{E_T(\max)}$ for collisions with spherical targets such as Pb; and considerably wider for the strongly deformed ^{238}U nucleus which, if centrally hit along its major axis, produces as much transverse energy as a nucleus of A ~ 420.

Stopping power is defined as the ratio between the maximal $E_T(\max)$ and the kinematical limit $E_{T,\lim}$ corresponding in particular to full thermalization and given by

$$E_{T,\lim} = \pi/4 \cdot (E_{asym.c.m.} - N_{part} \cdot M_p) \tag{5}$$

where the c.m. kinetic energy is calculated for the asymmetric sphere-cylinder system described above and $\pi/4$ is the spherical average of $\sin\Theta$ over a sphere. N_{part} is the sum of participant (i.e. colliding, non spectator) nucleons in the projectile and in the target ion. The stopping power reaches, on a heavy target and for the few events at the end of the tail, 90 % for ^{16}O at 60 GeV/c, and lower values of 70 and 65 % for ^{16}O and ^{32}S at 200 GeV/c [16]. The corresponding energy density ϵ_{Bj} is 2.6, 3.3 and 3.3 GeV/fm^3 for the three cases: the maximal energy density increases somewhat with energy, but is independent of the projectile mass (as expected when inserting (4) in (1)). Even the projected Pb injector is not expected to drastically enlarge the energy density domain.

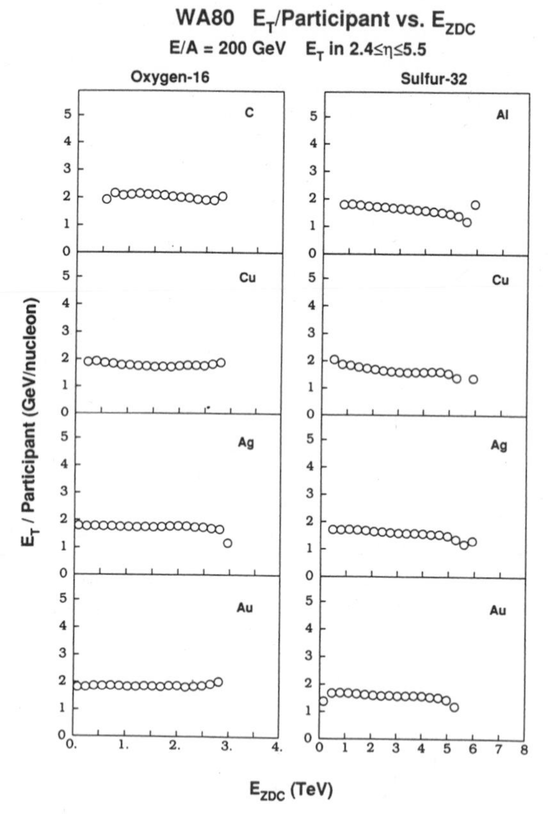

Fig. 6 The "dream rule", E_T=const. · N_{part}, illustrated by WA80 data [17] which cover a wide range of projectile/target combinations and number of participants. Each of the non-participating, spectator nucleons of the projectile deposits 200 GeV in the Zero Degree Calorimeter (ZDC) which acts therefore virtually as a counter of projectile participants. The target participants are then computed from the geometry of colliding spheres.

The determination of ϵ_{Bj} actually implies a calculation of the needed transverse area S_T from E_T using a nuclear geometry and some particle production model. The task is not straightforward, especially for peripheral reactions, and therefore the "dream rule" discovered by WA80 [17]

$$E_T = K \cdot N_{part} \tag{6}$$

is welcome, even if its basis is not understood, its striking simplicity not withstanding. As Fig. 6 shows, K is virtually constant = 1.8 − 2 GeV.

HBT Interferometry and Freeze-out Volume

The Hanbury − Brown − Twiss two-boson interferometry [18] has been used for some 30 years first to measure star sizes using photons, and then also sizes of high energy reaction volumes [19] using mesons. The basic principle is the following: If two sources at a distance 2R emit bosons in a totally uncorrelated way, then a detector will see an interference pattern in the sense that two identical bosons with zero relative three-momentum will be found in a given event $1+\lambda$ times more often than boson pairs with very different momenta. The width of the peak, in momentum difference squared, equals the inverse radius squared of the emitting system in the direction parallel to the momentum difference. The chaoticity coefficient λ reaches its maximum possible value of 1 if there is no phase correlation between emitted particles − the natural situation both for a star and for a collision of medium or heavy ions.

The beauty of the streamer chamber experiment NA35 was to obtain correlation functions between pairs of negative particles (mainly π^-) within less than a year after taking the 200 GeV/nucleon O − Au data, in spite of the very time consuming analysis of heavily loaded photographs, with rather singular results [20]: The transverse radius was found, for central collisions, to be of 4 fm for a $1 < y < 2$ rapidity interval, not far from the 3 fm radius of the oxygen ion, but of 8 fm for the $2 < y < 3$ interval where particle production is maximal − an intriguing feature since it may indicate the formation of a plasma of a lifetime of several fm/c. Data with more statistics are expected (from S − S collisions) and needed: in the published data, the difference in particle density is not large between the two intervals (38 vs. 43 negative particles per unit of rapidity); and the chaoticity parameter is uncomfortably small, namely 0.3, for the first rapidity interval.

Fig. 7 shows a spectacular achievement: a measurement of an interference effect from a single high multiplicity event. We follow the responsible team in advocating single event interferometry as a basic

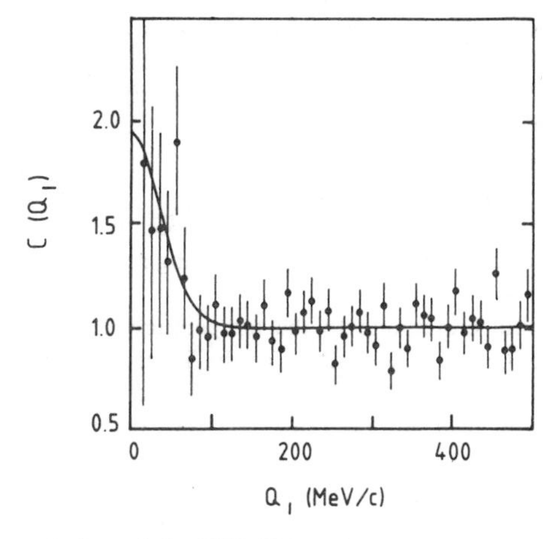

Fig. 7 The most daring of the H-B-T interferometry plots (Correlation coefficient for pairs of negative tracks vs. their squared momentum difference) concerns a single O − Au event (200 GeV/nucleon) with 104 π^- and 3581 useable pairs; a signal can be seen, and anticipates the feasibility, with the huge multiplicities expected from future Pb ion beams, of freezeout volume measurements on single events.

tool for future Pb beam interactions, where some 30 times more pairs per event are expected. The measured radii can then be correlated to a potential signal on an event by event basis. Clearly, the streamer chamber would be replaced by other, faster detectors, maybe TPC's.

Signature: Strangeness

There is a broad consensus that p−p collisions, because the initial state is non-strange, do not produce enough strange particles to saturate chemical equilibrium, so that further final state interactions of any kind will enhance the strange to non-strange ratio. The idea that QGP should cause an exceptional strangeness enhancement comes from the Boltzmann factor $\exp(-2 \cdot m_T/T)$ being so much larger for a pair of s quarks of mass 160 MeV created in a QGP than for a pair of kaons created by a final state hadronic interaction [7]. But it has been argued that entropy conservation during hadronization of the QGP would produce so many more non-strange hadrons that little would be left of that factor [21]. The biggest effect is still expected, needless to say, for the least accessible channels: strange antibaryons.

The situation on the experimental front is also far from clear. A strong enhancement above earlier p−p and p−Pb data is seen by E802 in ^{28}Si−Au reactions at 14.5 GeV/nucleon in the K^+/π^+ ratio, but not in K^-/π^-, especially at larger p_T of order 1 GeV/c [22]. The underlying experimental achievement is an unprecedented ± 75 psec time-of-flight resolution. NA35 analyze the V^0's seen in the streamer chamber and find, for central S−S reactions, a Λ/charged ratio 2.5 times higher than both for peripheral S−S and for p−p (earlier data) reactions [23]. On the other hand, the Λ/negatives ratio hardly increases from p−Au to central O−Au reactions [24].

NA38 has a statistically powerful strangeness indicator which is N_{++} / N_{--}, the ratio of positive to negative like-sign decay muon pairs. Since there are two times more K^+ than K^- at $p_T \sim .6 - 1$ GeV/c, and since kaons decay ≃ 4 times more often to muons than pions, many more $\mu^+\mu^+$ than $\mu^-\mu^-$ are expected. The ratio is given, for isoscalar projectiles ($\pi^+/\pi^- \simeq 1$), by

$$N_{++} / N_{--} = (1 + 4 \cdot K^+ / \pi^+)^2 / (1 + 4 \cdot K^- / \pi^-)^2 \qquad (7)$$

and is found to be ≃ = 1.7, and almost independent of transverse energy [25]. This does not allow us to conclude that the strangeness content does not increase with E_T (any increase in K^+/π^+ would leave the ratio invariant if it were 1.3 times larger than a corresponding increase in K^-/π^-); this data can however be used as a significant test for any quantitative model or parametrisation of K/π ratios.

Data from the more specialized experiments NA36 (TPC in EHS) and WA85 (Omega) on (anti−)hyperon and kaon production are eagerly expected; both experiments have already shown that the challenge of automatic V^0 reconstruction in a high multiplicity environment can be met.

Signature: Average Transverse Momentum

The four experiments operational since 1986 (NA35, WA80, NA34 and NA38) measure p_T distributions in one way or another, and find the following main features:

a. the figure which would correspond to the temperature in a Boltzmann approximation, and which is extracted either as $\langle p_T \rangle /2$ or as the inverse of the exponential slope (fitted for intermediate p_T, $.5 < p_T < 2$ GeV/c or so), is found to reach rapidly (as a function of E_T or, better, of ε_{Bj}) a plateau around 180 − 210 MeV for p, O and S projectiles on heavy targets;

b. when normalized to proton data, the ion data exhibit a relative excess at small p_T which is not understood; a possible contribution of an uncontrolled electron production is not always ruled out;

c. another excess is found at large p_T, and may be related to the "Cronin effect" [26]: an increase faster than linear with A found in particle production in p−A reactions at intermediate p_T, and attributed to collective (multiple scattering) effects.

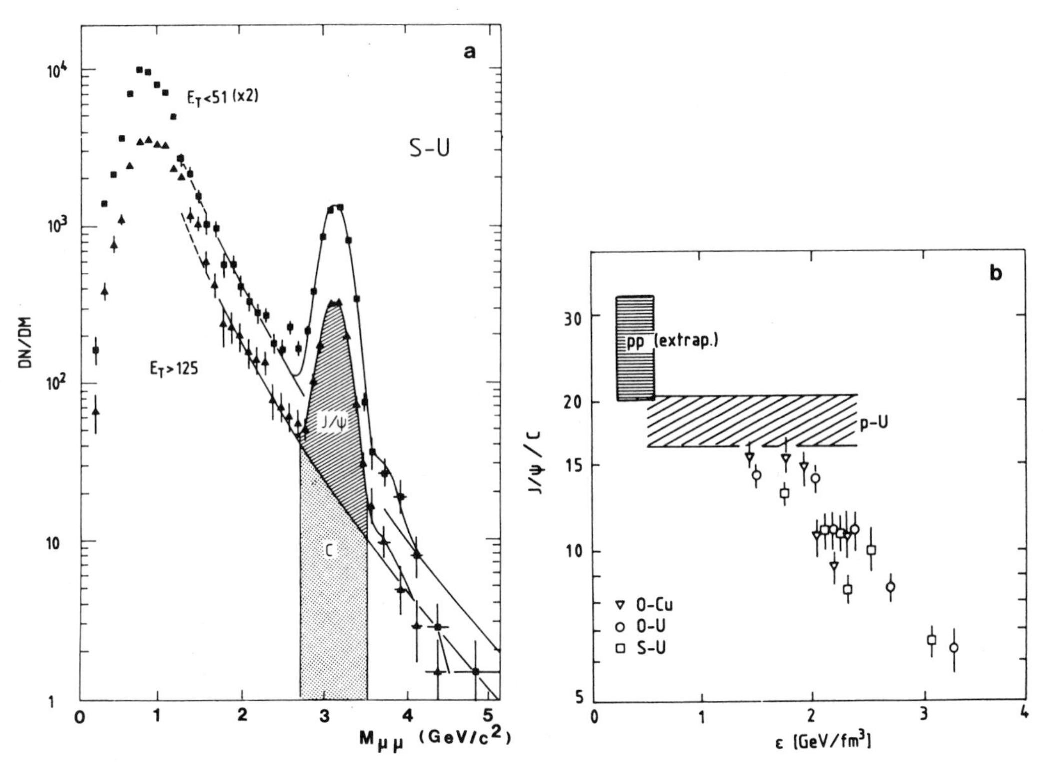

Fig. 8 The J/Ψ suppression measured by NA38 [39] is exemplified (a) by comparing dimuon mass spectra from $^{32}S-^{238}U$ collisions in the lowest and the highest of 6 E_T intervals; a variation (by a factor of 2) of the J/Ψ signal as compared to the (Drell – Yan like) continuum is clearly seen. The J/Ψ to Continuum ratios are then plotted (b) versus Bjorken's energy density ε, for the studied O – Cu, O – U and S – U systems. All data points fall on a common pattern. However, even the low E_T points do not yield low ε points (since the overlap area S_T is also small). The measured p – U data did not, for various reasons, yield reliable ε values; they were however used to obtain extrapolated p – p data for which ε is indeed low. The problem of assessing full scale ε dependences will generalize with the appearence of more potential signals. Here, a kink seen in a more complete and continuous curve would certainly favour the QGP hypothesis, with its plausible threshold in ε, over alternative models which invoke hadron-$J/\Psi \rightarrow \bar{D}-D$ reactions in an extraordinarily dense hadron gas.

While experimentalists were somehow frustrated in their hope for high $<p_T>$ at the very highest E_T attainable, others advocated the very constancy of $<p_T>$ as being *the* experimental proof of the existence of QGP and even of a first-order transition, since any reasonable hydrodynamic model would predict the gradual increase of $<p_T>$ at low ε to continue indefinitely, unless a mixed phase (constant temperature and pressure) should arise [27].

Attempts are under way [28] to analyze the p_T, or, rather, m_T distributions of identified particles which are becoming available (V^o's from NA35, π^o's from WA80, and p, π^- and possibly soon K's from the NA34 external spectrometer) in view of signs of a collective flow. A common slope would confirm the thermodynamical expectations, while thermal motion superimposed on a common flow velocity would result in mass-dependent slopes.

Signature: Direct Photons (Real and Virtual)

The emission of real or virtual photons by quarks and gluons in prolonged interaction has been thought of as a privileged signature for QGP for some time [29], [30]. Both direct photons of 200 MeV temperature and also very soft photons of bremsstrahlung type (very low p_T and x_F) like those found in much greater abundance than expected in K^+-p from hadron bremsstrahlung [31] and very recently (by the NA34−1 proton beam experiment) in p−A [32], are deemed of interest.

Unfortunately, there is at present no experimental evidence in ion reactions in spite of a series of attempts. Direct photons must be disentangled from π^o decay photons which are orders of magnitude more numerous; neither NA34 [33] nor WA80 [34] went beyond limits of ~ 10 % direct photons after accounting for decay photons from π^o's (and other parent mesons) obtained from π^- (NA34) or directly measured π^o (WA80) spectra. Recent refined calculations [35] may explain while NA38 failed to see a virtual photon signal in the 1.2 − 2 GeV mass range where they are not as prominent as was originally hoped. The low mass dimuons will be amenable to closer scrutiny with the improved NA34 setup planned for 1990.

Signature: Heavy Meson Suppression by Debye Colour Screening

The so far most promising signature was the last to be proposed, at a time when the NA38 experiment was already built [36]: Debye colour screening will operate in a quark-gluon plasma, provided the temperature is well above the critical temperature T_c, over distances larger than e.g. the J/Ψ radius, so that the c and $\bar{c}$ will combine with ordinary (anti−)quarks and J/Ψ production will be suppressed, with respect to e.g. the not affected Drell−Yan muon pairs. (The corresponding enhancement in $D-\bar{D}$ production is thought to be undetectable).

A factor of 2 in J/Y suppression relative to Drell−Yan type continuum pairs was found both in O−U [37] and in S−U [38] reactions (Fig. 8a). In Fig. 8b data from all studied systems (O−Cu, O−U, S−U and tentatively p−U) are plotted against a common variable, ε_{Bj} [39]. One sees that the suppression is a rather smooth function of ε_{Bj}; that there is a certain lack of low ε_{Bj} points so that it is not easy to check for a threshold phenomenon (slope discontinuity) which would be a more clearcut indication of QGP; and, incidentally, that ε_{Bj} is indeed a meaningful variable which bears out a common behaviour for very different systems. The problem with obtaining low ε_{Bj} points arises because the low E_T values, for a given system, correspond to peripheral collisions and have to be divided by the overlap area which is the also small and, furthermore, subject to uncertainties such as nuclear geometry. A more precise phenomenology of the process (in particular regarding the found p_T dependence of the suppression) is needed since hadronic reinteraction mechanisms have been tailored to also be compatible with the data (albeit with somewhat far-fetched parameters) [40].

OUTLOOK, EXOTICS

A heavy ion source capable of injecting up to 10^8 Pb ions per burst into the CERN accelerators is planned, although not yet financed. It will allow to study the interaction properties and QGP signatures on a much broader basis, from 1993 onwards, provided the rather formidable experimental problems connected with particle multiplicity (several thousand for a central Pb – Pb collision) can be mastered. In the following, we wish to mention two rather exotic sidelines which are opened by the prospects of high energy interactions with truly large interaction volumes.

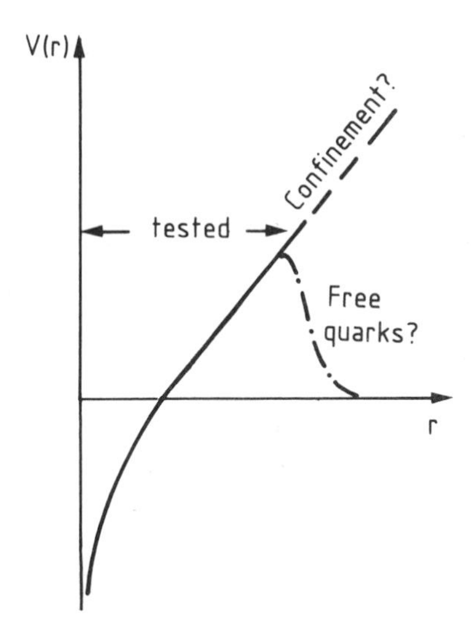

Fig. 9 The QCD potential $V(r) = \alpha_S / r + \sigma \cdot r$ has only been tested (e.g. by quarkonium spectroscopy) over a distance of a few fermi. If free quarks were discovered this would modify V(r) (e.g. lower dotted curve) but not jeopardize the standard model, provided the quarks had been released from reaction volumes larger than the tested range of V(r).

The call for higher energies per nucleon is motivated by intriguing results from JACEE and more from the Tevatron. It will have to wait for RHIC (a 100 GeV/nucleon heavy ion collider planned at BNL), since adapting HERA or the LEP – SPS (+ Bypass) complex for ion collisions is not a likely prospect. The extreme energies attainable with heavy ions at LHC will open the challenging possibility of producing the Higgs boson in photon-photon collisions. We will also touch these two fields.

Bigger Interaction Volumes for Free Quark Search

The idea of confinement stems entirely from the non-discovery of free quarks. The QCD potential (Fig. 10) has meanwhile been tested over several fermi (e.g. heavy quarkonium spectroscopy); it could however break down at larger distances. All high sensitivity quark searches done so far used reactions where both the projectile radius and the Lorentz contracted reaction volume thickness did not exceed one fermi. Field theoretical models of $SU(3)_c$ broken within a higher unified symmetry have been proposed by de Rujula et al. [41] and by Slansky et al. [42], which have free quarks and free diquarks respectively. Therefore, it appears that a search for (di)quarks of fractional charge would make sense in a situation where quark formation takes place over volumes larger than a few fermi (the interaction area of a central Pb−Pb collision has a radius of 6 fermi). During hadronization some quarks and antiquarks might loose track of their partners and emerge as such from the reaction.

Larger Interaction Volumes and the Search for Strange Matter

E. Witten has suggested [43] that strange matter, i.e. matter made of (many) up, down and strange quarks in similar proportions and contained in a single bag, could be absolutely stable, being energetically more favorable than nuclear matter above some large enough baryon number A_{min} (for $A=1$, the Λ is well known to be metastable). It has been shown [44], [52] that this is the case, within the bag model, for reasonable values of the bag constant and the strange quark mass; the latter must obviously be smaller than the Fermi level of up and down quarks. Five years of debate, recently reviewed by Alcock and Olinto [45], have left unscathed this extravagant hypothesis that strange matter, rather than the world we know, should be the true ground state of matter. Indeed, a decay of the universe to its (hypothetical) ground state would have to be seeded by an instantaneous transition of a nucleus of size A_{min} ($\sim$ 100) to the minimal stable lump of strange matter − a weak interaction of order Amin whose probability is zero for all practical purposes. The original suggestion [43] that dark matter might be strange matter produced in the early universe seems now very unlikely [45] [46]: conditions were such that any strange matter would have evaporated during hadronization. The possibility that neutron stars, either all or at least those originating from blue giants, are in fact strange matter stars has been revived [47] by the SN1987A pulsar with its extraordinary rotation speed and the suggestion that it has a Jupiter-like planet quietly revolving within the size of the parent star [48]. Strange matter debris of collisions of strange stars in cosmic rays and on earth have been discussed [49]. Centauro's might be among them. A search for backscattering of uranium ions (possible only from a target heavier than uranium) has yielded an upper limit of occurrence on earth [50].

A unique chance to attempt production and detection of tiny lumps of strange matter ("strangelets") [51] [52] [53] is given by the planned high energy Pb ion beam. If QGP is formed in the fragmentation regions, $s-\bar{s}$ production is favoured since the Fermi levels of u and d quarks are occupied; K^+ and K^0 ($u-\bar{s}$ and $d-\bar{s}$) emission is a favoured process which leaves behind the u,s,d raw material needed for strangelet formation. Unlike the situation in the early universe, there is no further energy input, so that smooth cooling (by expansion and meson and photon emission) down to temperatures $<$ 2 MeV where strange matter becomes stable is at least conceivable. Subsequent detection is a matter of standard mass spectroscopy.

Higher Energy for Reaching QGP

Have the CERN 200 GeV/nucleon ^{16}O and ^{32}S beams allowed us to explore a mixed quark-gluon and hadron phase? And, if so, how will we reach the pure QGP? A cosmic ray and a Tevatron experiment may help us to formulate a very tentative working hypothesis. The JACEE experiment [11] sent emulsion chambers completed by Pb plate − emulsion electromagnetic calorimeters in baloons to 40 km altitude and found a number of multiTeV events mostly produced by cosmic ions. Of these, the ones at the highest energy densities, between 3 and 8 GeV/fm^3 if computed according to formula (1), had values of $<p_T>$, as inferred from isolated photon showers at large angles, around 1 GeV/c, way above the standard 400 MeV/c. Similarly, the Tevatron collider experiment E−735 [54] finds rather high $<p_T>$ values for centrally produced hadrons when the accompanying multiplicity density exceeds 28. This would correspond to $\varepsilon_{Bj} > 5$ GeV/fm³ if we use (1) with $r_0 = 1.15$ fm. It seems then that it should be worthwhile to double the available energy density. If the latter is proportional to the p−p and p−$\bar{p}$ measured multiplicities, then N−N c.m. energies of 100 GeV or more instead of the present 20 GeV would be required. RHIC (200 GeV) would then be a good bet, but adapting an existing machine, HERA, or machine complex, LEP−SPS (140 GeV in both cases), might be a promising goal possibly within shorter delays.

Very High Energy: the Heavy Ion Higgs Window

Much higher N−N center-of-mass energies of 6 TeV for colliding Pb beams would be reached in LHC. Besides hadronic interactions, reactions between the Coulomb fields of the nuclei become important. The direct formation of the Higgs boson via $\gamma-\gamma \rightarrow H$ becomes possible, the mass window being $85 < M_H < 170$ GeV, and the dominant decay $H \rightarrow b-\bar{b}$ (if the Higgs is lighter than two top masses) [55]. If a luminosity of 10^{28} can be reached, well above the presently expected values for colliding Pb beams, and sustained for reasonable times in spite of beam loss by electromagnetic excitations [56] which are not fully known, then one $H \rightarrow b-\bar{b}$ event per month would result. It would have to be identified statistically, as a Jacobian peak, amidst 10^7 non resonant $\gamma-\gamma \rightarrow b-\bar{b}$ events; there would be a total number of 10^{11} hadronic reactions per month.

CONCLUSIONS

The challenge of studying experimentally ion−ion reactions with multiplicities up to 500 has been met. The energy and particle flows have been measured and understood in geometric terms; the stopping power at 200 GeV/nucleon is not maximal, but energy densities of 3 GeV/fm³ or slightly more have been reached. Pion interferometry is being used as a tool to measure reaction volumes and, indirectly, lifetimes. Among the proposed signatures, the average transverse momenta and the J/Ψ over Continuum ratio have yielded significant results for which mechanisms alternative to QGP are being investigated. The situation of the strangeness signature is not fully clarified, neither theoretically (is it a signature for QGP, or just for intense final state interactions?) nor experimentally, sustained efforts by several experiments notwithstanding. First searches for virtual and real direct photons have been carried out with limited means, and have not been successful. The field of correlations of signatures with rapidity fluctuations, with volumes or lifetimes, or with other signatures is still untouched.

It seems therefore premature to draw any conclusions regarding the QGP issue, after only 34 days of ion beams. There is however already a conspicuous harvest which includes intriguing experimental hints, considerable progress in understanding methodology (which may be going the hard way: the only probably unambiguous single signatures are the free quarks and the strangelet, and the prospect of finding these is not a likely one), and the progressive unification of particle physics and thermodynamics. Finally, the very fact that so many physicists work so hard in this difficult and not straightforward field, may be taken as a good indication that physics with ion beams has a future.

References

[1] R. Hagedorn, Suppl. Nuovo Cim. 2 (1965) 147; also preprint TH-3014-CERN (1980); Riv. Nuovo Cim. 6 (1983) 1.

[2] Reviewed in H. Satz, ed., Statistical mechanics of quarks and hadrons, Bielefeld Symposium, 1980 (North Holland, 1981).

[3] QCD on a lattice: K.G. Wilson, Phys. Rev. D10 (1974) 2445. Monte Carlo: M. Creutz, Phys. Rev. Lett. 45 (1980) 313. Phase transition: T. Celik et al., Phys. Lett. 125B (1983) 411, and J. Kogut et al., Phys. Rev. Lett. 50 (1983) 393.

[4] K. Redlich, H. Satz, Phys. Rev. D33 (1986) 3747.

[5] Th. DeGrand, Quark Matter '84 Proceedings (Springer 1985), p. 17

[6] P. Bacilieri et al., Phys. Rev. Lett. 14 (1988) 1545 and Phys. Lett. B224 B224 (1989) 333; N. Christ et al., Phys. Rev. Lett. 14 (1988) 2050; J.B. Kogut, D.K. Sinclair, Phys. Lett. B229 (1989) 107.
[7] P. Koch, B. Muller, J. Rafelski, Phys. Rep. 142 (1986) 167.
[8] J.D. Bjorken, Phys. Rev. D27 (1983) 140.
[9] L.D. Landau, Izv. Akad. Nauk. SSSR 17 (1953) 51.
[10] M. Faessler, Nucl. Phys. A434 (1985) 563c.
[11] T. Burnett et al., Phys. Rev. Lett. 57 (1986) 3249; also 50 (1983) 2062.
[12] Proc. Internat. Conf. on Nucleus-Nucleus Collisions, St.Malo, France, 6-11 June 1988, GANIL, Caen, France (1988); and Nucl. Phys. A488 (1988).
[13] H.A. Gustafsson et al., Phys. Rev. Lett. 52 (1984) 1590.
[14] H. Von Gersdorff et al. (KLM collab.), Phys. Rev. C39 (1989) 1385.
[15] C. DeMarzo et al., Phys. Rev. D26 (1982) 1019; 29 (1984) 363; 29 (1984) 2476.
[16] J. Schukraft et al., Nucl. Phys. A498 (1989) 79c; see also Proc. 24th Rencontres de Moriond (1989), to be published.
[17] G.R. Young et al., Nucl. Phys. A498 (1989) 53c.
[18] R. Hanbury-Twiss, R.Q. Brown, Phil. Mag. 45 (1954) 663.
[19] G. Goldhaber et al., Phys. Rev. 120 (1960) 300.
[20] T.J. Humanic et al., Z. Phys. C79 (1988) 79.
[21] K.S. Lee et al., Phys. Rev. C37 (1988) 1452.
[22] Y. Miake et al., Z. Phys. C138 (1988) 135; and T. Abbott et al., Nucl. Phys. A498 (1989) 67c.
[23] M. Gazdzicki et al., Nucl.Phys. A498 (1989) 375c.
[24] A. Bamberger et al., Z. Phys. C43 (1989) 25.
[25] P. Sonderegger et al., Z. Phys. C38 (1988) 129.
[26] D. Antreasyan et al., Phys.Rev. D19 (1979) 764.
[27] G. Bertsch et al., Phys. Rev. D37 (1988), 1202.
[28] K.S. Lee, U. Heinz, Z. Phys. C43 (1989) 425.
[29] E.L. Feinberg, Izv. Akad. Nauk Ser. Fiz. 26 (1962) 62; Nuovo Cim. 34A (1976) 391.
[30] E.V. Shuryak, Phys. Lett. 78B (1978) 150.
[31] P.V. Chliapnikov et al., Phys. Lett. 141B (1984) 276.
[32] U. Goerlach et al., in " XXIV Conference on High Energy Physics, Munich, August 4 − 10, 1988 ", Springer Verlag, 1989, p. 1412.
[33] T. Akesson et al., preprint CERN-EP/89-113; subm. to Z. Phys. C.
[34] H. Gutbrod, private communication (June 1989).
[35] K. Kajantie and P.V. Ruuskanen, Z. Phys. C44 (1989) 167.
[36] T. Matsui, H. Satz, Phys. Lett. 178B (1986) 416.
[37] C. Baglin et al., Phys. Lett. B220 (1989) 471.
[38] J.Y. Grossiord et al., Nucl. Phys. A498 (1989) 249c.
[39] P. Sonderegger, in " XXIV Conference on High Energy Physics, Munich, August 4-10, 1988 ", Springer Verlag, 1989, p. 1369.
[40] The subject is reviewed, and references given, in R.A. Salmeron, "The Suppression of the J/Ψ..", Ecole Polytechnique preprint (1989), to be published in Proc. 24th Rencontres de Moriond.
[41] A. de Rujula et al., Phys. Rev. D17 (1978) 285.
[42] R. Slansky et al., Phys. Rev. Lett. 47 (1981) 887.
[43] E. Witten, Phys. Rev. D30(1984) 272.
[44] E. Farhi and R.L. Jaffe, Phys. Rev. D30 (1984) 2379; see also ref. GRE88.
[45] C. Alcock, A. Olinto, Ann. Rev. Nucl. Part. Sci. 38 (1988) 161.
[46] C. Alcock et al., Phys. Rev. D39 (1989) 1233.
[47] J.A. Frieman, A. Olinto, preprint Fermilab Pub 88-129 A.
[48] J.A. Kristian et al., Nature 338 (1989) 234.
[49] A. de Rujula et al., Nature 312 (1984) 734.

[50] M. Brügger et al., Nature 337 (1989) 434.
[51] H.C. Liu et al., Phys. Rev. D30 (1984) 1137.
[52] C. Greiner et al., Z.Phys. C38 (1988) 283.
[53] G.L. Shaw et al., Nature 337 (1989) 463.
[54] T. Alexopoulos et al., Phys. Rev. Lett. 60 (1988) 1622.
[55] M. Drees, J.V. Ellis, D. Zeppenfeld, Phys. Lett. B223 (1989) 454, which references the original work of E. Papageorgiu and M. Grabiak et al..
[56] J.C. Hill et al., Phys. Rev. Lett. 60 (1988) 999.

ELECTROWEAK INTERACTIONS AND LEP PHYSICS

G. Altarelli

CERN

Geneva, Switzerland

1. INTRODUCTION

These lectures on electroweak interactions start with a summary of the Glashow–Weinberg–Salam theory [1] and then cover more advanced subjects of present interest in phenomenology: the Higgs sector and the open problem of the experimental investigation on the origin of the Fermi scale of mass $G_F^{-1/2}$; the experimental tests of the theory; and finally, an outlook on LEP physics.

The modern electroweak theory inherits the phenomenological successes of the $(V-A) \otimes (V-A)$ four-fermion low-energy description of weak interactions [2], and provides a well-defined and consistent theoretical framework including weak interactions and quantum electrodynamics in a unified picture.

As an introduction, in the following we recall some salient physical features of the weak interactions. The weak interactions derive their name from their intensity. At low energy the strength of the effective four-fermion interaction of charged currents is determined by the Fermi coupling constant G_F. For example, the effective interaction for muon decay is given by

$$\mathcal{L}_{eff} = (G_F/\sqrt{2}) \, [\bar{\nu}_\mu \gamma_\alpha (1-\gamma_5)\mu][\bar{e}\gamma^\alpha(1-\gamma_5)\nu_e] \,, \tag{1.1}$$

with [3]

$$G_F = 1.16637(2) \times 10^{-5}\,\mathrm{GeV}^{-2} \,. \tag{1.2}$$

In natural units $\hbar = c = 1$, G_F has dimensions of $(\mathrm{mass})^{-2}$. As a result, the intensity of weak interactions at low energy is characterized by $G_F E^2$, where E is the energy scale for a given process ($E \approx m_\mu$ for muon decay). Since

$$G_F E^2 = G_F m_p^2 \, (E/m_p)^2 \simeq 10^{-5} \, (E/m_p)^2 \,, \tag{1.3}$$

where m_p is the proton mass, the weak interactions are indeed weak at low energies (of order m_p). The quadratic increase with energy cannot continue for ever, because it would lead to a violation of unitarity. In fact, at large energies the propagator effects can no longer be neglected, and the current–current interaction is resolved into current–W gauge boson vertices connected by a W propagator. The strength of the weak interactions at high energies is then measured by g_W, the

Particle Physics: Cargèse 1989
Edited by M. Lévy *et al.*
Plenum Press, New York, 1990

W-μ-ν_μ coupling, or, even better, by $\alpha_W = g_W^2/4\pi$ analogous to the fine-structure constant α of QED. In the standard electroweak theory, we have

$$\alpha_W = \sqrt{2}\, G_F\, m_W^2/\pi \simeq \alpha/\sin^2\theta_w \cong 1/30\,. \tag{1.4}$$

That is, at high energies the weak interactions are no longer so weak.

The range r_w of weak interactions is very short: it is only with the experimental discovery of the W and Z gauge bosons that it could be demonstrated that r_w is non-vanishing. Now we know that

$$r_w = \hbar/m_W c \simeq 2 \times 10^{-16}\,\text{cm}\,, \tag{1.5}$$

corresponding to $m_W \approx 81$ GeV. This very large value for the W (or the Z) mass makes a drastic difference, compared with the massless photon and the infinite range of the QED force. The experimental limits on the photon mass [3] are listed in the following. From a laboratory experiment, one obtains $m_\gamma < 10^{-14}$ eV by a method based on the vanishing of the electric field inside a cavity with conducting walls, predicted by Gauss' law. In fact, the exact r^{-2} behaviour of the electric field corresponds to $m_\gamma = 0$. From the observed distribution of planetary magnetic fields (the field should be damped by an extra factor $e^{-m_\gamma r}$ if $m_\gamma \neq 0$) the Pioneer probe to Jupiter obtained $m_\gamma < 6 \times 10^{-16}$ eV. Finally, indirect evidence from galactic magnetic fields indicates that $m_\gamma < 3 \times 10^{-27}$ eV. Thus, on the one hand, there is very good evidence that the photon is massless. On the other hand, the weak bosons are very heavy. A unified theory of electroweak interactions has to face this striking difference.

Another apparent obstacle in the way of electroweak unification is the chiral structure of weak interactions: in the massless limit for fermions, only left-handed quarks and leptons (and right-handed antiquarks and antileptons) are coupled to W's. This clearly implies parity and charge-conjugation violation in weak interactions.

The universality of weak interactions and the algebraic properties of the electromagnetic and weak currents [the conservation of vector currents (CVC), the partial conservation of axial currents (PCAC), the algebra of currents, etc.] have been crucial in pointing to a symmetric role of electromagnetism and weak interactions at a more fundamental level. The old Cabibbo universality for the weak charged current [4]:

$$J_\alpha^{\text{weak}} = \bar{\nu}_\mu\gamma_\alpha(1-\gamma_5)\mu + \bar{\nu}_e\gamma_\alpha(1-\gamma_5)e + \cos\theta_C\,\bar{u}\gamma_\alpha(1-\gamma_5)d + \sin\theta_C\,\bar{u}\gamma_\alpha(1-\gamma_5)s + \ldots\,, \tag{1.6}$$

suitably extended, is naturally implied by the standard electroweak theory. In this theory the weak gauge bosons couple to all particles with couplings that are proportional to their weak charges, in the same way as the photon couples to all particles in proportion to their electric charges ($d' = \cos\theta_C\, d + \sin\theta_C\, s$ is the weak-isospin partner of u in a doublet).

Another crucial feature is that the charged weak interactions are the only known interactions that can change flavour: charged leptons into neutrinos or up-type quarks into down-type quarks. On the contrary, there are no flavour-changing neutral currents at tree level. This is a remarkable property of the weak neutral current, which is explained by the introduction of the GIM mechanism [5] and has led to the successful prediction of charm.

The natural suppression of flavour-changing neutral currents, the separate conservation of e, μ, and τ leptonic flavours, the mechanism of CP violation [6] through the phase in the quark-mixing matrix, are all crucial features of the Standard Model. Many examples of new physics tend to break the selection rules of the standard theory. Thus the experimental study of rare flavour-changing transitions is an important window on possible new physics.

In the following sections we shall see how these properties of weak interactions fit into the standard electroweak theory.

2. GAUGE THEORIES

In this section we summarize the definition and the structure of a gauge Yang–Mills theory [7, 8]. We will list here the general rules for constructing such a theory. Then in the next section these results will be applied to the electroweak theory.

Consider a Lagrangian density $\mathcal{L}[\phi, \partial_\mu\phi]$ which is invariant under a D dimensional continuous group of transformations:

$$\phi' = U(\theta^A)\phi \qquad (A = 1, 2, \ldots, D)\,. \tag{2.1}$$

For θ^A infinitesimal, $U(\theta^A) = 1 + ig\, \Sigma_A\, \theta^A T^A$, where T^A are the generators of the group Γ of transformations (2.1) in the (in general reducible) representation of the fields ϕ. Here we restrict ourselves to the case of internal symmetries, so that T^A are matrices that are independent of the space–time coordinates. The generators T^A are normalized in such a way that for the lowest dimensional non-trivial representation of the group Γ (we use t^A to denote the generators in this particular representation) we have

$$\mathrm{tr}(t^A t^B) = \tfrac{1}{2}\delta^{AB}\,. \tag{2.2}$$

The generators satisfy the commutation relations

$$[T^A, T^B] = iC_{ABC}T^C\,. \tag{2.3}$$

In the following, for each quantity V^A we define

$$\mathbf{V} = \sum_A T^A V^A\,. \tag{2.4}$$

If we now make the parameters θ^A depend on the space–time coordinates $\theta^A = \theta^A(x_\mu)$, $\mathcal{L}[\phi, \partial_\mu\phi]$ is in general no longer invariant under the gauge transformations $U[\theta^A(x_\mu)]$, because of the derivative terms. Gauge invariance is recovered if the ordinary derivative is replaced by the covariant derivative:

$$D_\mu = \partial_\mu + ig\mathbf{V}_\mu\,, \tag{2.5}$$

where V_μ^A are a set of D gauge fields (in one-to-one correspondence with the group generators) with the transformation law

$$\mathbf{V}'_\mu = U\mathbf{V}_\mu U^{-1} - (1/ig)(\partial_\mu U)U^{-1}\,. \tag{2.6}$$

For constant θ^A, $\mathbf{V}$ reduces to a tensor of the adjoint (or regular) representation of the group:

$$\mathbf{V}'_\mu = U\mathbf{V}_\mu U^{-1} \simeq \mathbf{V}_\mu + ig[\theta, \mathbf{V}_\mu]\,, \tag{2.7}$$

which implies that

$$V'^C_\mu = V^C_\mu - gC_{ABC}\theta^A V^B_\mu\,, \tag{2.8}$$

where repeated indices are summed up.

As a consequence of Eqs. (2.5) and (2.6), $D_\mu\phi$ has the same transformation properties as ϕ:

$$(D_\mu\phi)' = U(D_\mu\phi)\,. \tag{2.9}$$

Thus $\mathcal{L}[\phi, D_\mu\phi]$ is indeed invariant under gauge transformations. In order to construct a gauge-invariant kinetic energy term for the gauge fields V^A, we consider

$$[D_\mu, D_\nu]\phi = ig\{\partial_\mu \mathbf{V}_\nu - \partial_\nu \mathbf{V}_\mu + ig[\mathbf{V}_\mu, \mathbf{V}_\nu]\}\phi \equiv ig\mathbf{F}_{\mu\nu}\phi \,, \tag{2.10}$$

which is equivalent to

$$F^A_{\mu\nu} = \partial_\mu V^A_\nu - \partial_\nu V^A_\mu - gC_{ABC}V^B_\mu V^C_\nu \,. \tag{2.11}$$

From Eqs. (2.1), (2.9), and (2.10) it follows that the transformation properties of $F^A_{\mu\nu}$ are those of a tensor of the adjoint representation

$$\mathbf{F}'_{\mu\nu} = U\mathbf{F}_{\mu\nu}U^{-1} \,. \tag{2.12}$$

The complete Yang–Mills Lagrangian, which is invariant under gauge transformations, can be written in the form

$$\mathcal{L}_{YM} = -\frac{1}{4}\sum_A F^A_{\mu\nu}F^{A\mu\nu} + \mathcal{L}[\phi, D_\mu\phi] \,. \tag{2.13}$$

For an Abelian theory, as for example QED, the gauge transformation reduces to $U[\theta(x)] = \exp[ieQ\theta(x)]$, where Q is the charge generator. The associated gauge field (the photon), according to Eq. (2.6), transforms as

$$V'_\mu = V_\mu - \partial_\mu\theta(x) \,. \tag{2.14}$$

In this case, the $F_{\mu\nu}$ tensor is linear in the gauge field V_μ, so that in the absence of matter fields the theory is free. On the other hand, in the non-Abelian case the $F^A_{\mu\nu}$ tensor contains both linear and quadratic terms in V^A_μ, so that the theory is non-trivial even in the absence of matter fields.

3. THE STANDARD MODEL OF THE ELECTROWEAK INTERACTIONS

In this section, we summarize the structure of the standard electroweak Lagrangian and specify the couplings of $W^\pm$ and Z, the intermediate vector bosons (IVBs).

For this discussion we split the Lagrangian into two parts by separating the Higgs boson couplings:

$$\mathcal{L} = \mathcal{L}_{symm} + \mathcal{L}_{Higgs} \,. \tag{3.1}$$

We start by specifying $\mathcal{L}_{symm}$, which involves only gauge bosons and fermions:

$$\mathcal{L}_{symm} = -\tfrac{1}{4}\sum_{A=1}^{3} F^A_{\mu\nu}F^{A\mu\nu} - \tfrac{1}{4}B_{\mu\nu}B^{\mu\nu} + \bar{\psi}_L i\gamma^\mu D_\mu\psi_L + \bar{\psi}_R i\gamma^\mu D_\mu\psi_R \,. \tag{3.2}$$

This is the Yang–Mills Lagrangian for the gauge group SU(2) ⊗ U(1) with fermion matter fields. Here

$$B_{\mu\nu} = \partial_\mu B_\nu - \partial_\nu B_\mu \qquad \text{and} \qquad F^A_{\mu\nu} = \partial_\mu W^A_\nu - \partial_\nu W^A_\mu - g\,\epsilon_{ABC}\,W^B_\mu W^C_\nu \tag{3.3}$$

are the gauge antisymmetric tensors constructed out of the gauge field B_μ associated with U(1), and W^A_μ corresponding to the three SU(2) generators; ϵ_{ABC} are the group structure constants [see

Eqs. (2.3)] which, for SU(2), coincide with the totally antisymmetric Levi-Civita tensor (recall the familiar angular momentum commutators). The normalization of the SU(2) gauge coupling g is therefore specified by Eq. (3.3).

The fermion fields are described through their left-hand and right-hand components:

$$\psi_{L,R} = [(1 \mp \gamma_5)/2]\psi\ , \qquad \bar{\psi}_{L,R} = \bar{\psi}[(1 \pm \gamma_5)/2]\ , \tag{3.4}$$

with γ_5 and other Dirac matrices defined as in the book by Bjorken–Drell [9]. In particular, $\gamma_5^2 = 1$, $\gamma_5^\dagger = \gamma_5$. Note that, as given in Eq. (3.4),

$$\bar{\psi}_L = \psi_L^\dagger\gamma_0 = \psi^\dagger[(1-\gamma_5)/2]\gamma_0 = \bar{\psi}[\gamma_0(1-\gamma_5)/2]\gamma_0 = \bar{\psi}[(1+\gamma_5)/2]\ .$$

The matrices $P_\pm = (1 \pm \gamma_5)/2$ are projectors. They satisfy the relations $P_\pm P_\pm = P_\pm$, $P_\pm P_\mp = 0$, $P_+ + P_- = 1$.

The sixteen linearly independent Dirac matrices can be divided into γ_5-even and γ_5-odd according to whether they commute or anticommute with γ_5. For the γ_5-even, we have

$$\bar{\psi}\Gamma_E\psi = \bar{\psi}_L\Gamma_E\psi_R + \bar{\psi}_R\Gamma_E\psi_L \qquad (\Gamma_E \equiv 1, i\gamma_5, \sigma_{\mu\nu})\ , \tag{3.5}$$

whilst for the γ_5-odd,

$$\bar{\psi}\Gamma_O\psi = \bar{\psi}_L\Gamma_O\psi_L + \bar{\psi}_R\Gamma_O\psi_R \qquad (\Gamma_O = \gamma_\mu, \gamma_\mu\gamma_5)\ . \tag{3.6}$$

In the Standard Model the left and right fermions have different transformation properties under the gauge group. Thus, mass terms for fermions (of the form $\bar{\psi}_L\psi_R$ + h.c.) are forbidden in the symmetric limit. In particular, all ψ_R are singlets in the minimal Standard Model. But for the moment, by $\psi_{L,R}$ we mean a column vector, including all fermions in the theory that span a generic reducible representation of SU(2) ⊗ U(1). The standard electroweak theory is a chiral theory, in the sense that ψ_L and ψ_R behave differently under the gauge group. In the absence of mass terms, there are only vector and axial vector interactions in the Lagrangian that have the property of not mixing ψ_L and ψ_R. Fermion masses will be introduced, together with $W^\pm$ and Z masses, by the mechanism of symmetry breaking.

The covariant derivatives $D_\mu\psi_{L,R}$ are explicitly given by

$$D_\mu\psi_{L,R} = \left[\partial_\mu + ig\sum_{A=1}^{3} t^A_{L,R}W^A_\mu + ig'{}^1\!/_2 Y_{L,R}B_\mu\right]\psi_{L,R}\ , \tag{3.7}$$

where $t^A_{L,R}$ and $^1\!/_2 Y_{L,R}$ are the SU(2) and U(1) generators, respectively, in the reducible representations $\psi_{L,R}$. The commutation relations of the SU(2) generators are given by

$$[t^A_L, t^B_L] = i\,\epsilon_{ABC}\,t^C_L \qquad \text{and} \qquad [t^A_R, t^B_R] = i\,\epsilon_{ABC}\,t^C_R\ . \tag{3.8}$$

We use the normalization Eq. (2.2) [in the fundamental representation of SU(2)]. The electric charge generator Q (in units of e, the positron charge) is given by

$$Q = t^3_L + {}^1\!/_2 Y_L = t^3_R + {}^1\!/_2 Y_R\ . \tag{3.9}$$

Note that the normalization of the U(1) gauge coupling g′ in Eq. (3.7) is now specified.

All fermion couplings to the gauge bosons can be derived directly from Eqs. (3.2) and (3.7). The charged-current (CC) couplings are the simplest. From

$$g(t^1W^1_\mu + t^2W^2_\mu) = g\{[(t^1 + it^2)/\sqrt{2}]\,[(W^1_\mu - iW^2_\mu)/\sqrt{2}] + \text{h.c.}\} = g\{[(t^+W^-_\mu)/\sqrt{2}] + \text{h.c.}\}\ , \tag{3.10}$$

where $t^{\pm} = t^1 \pm it^2$ and $W^{\pm} = (W^1 \pm iW^2)/\sqrt{2}$, we obtain the vertex

$$V_{\bar{\psi}\psi W} = g\bar{\psi}\gamma_\mu\,[(t_L^+/\sqrt{2})\,(1-\gamma_5)/2 + (t_R^+/\sqrt{2})\,(1+\gamma_5)/2]\psi W_\mu^- + \text{h.c.} \tag{3.11}$$

In the neutral current (NC) sector, the photon A_μ and the mediator Z_μ of the weak NC are orthogonal and normalized linear combinations of B_μ and W_μ^3:

$$A_\mu = \cos\theta_w\, B_\mu + \sin\theta_w\, W_\mu^3 \qquad \text{and} \qquad Z_\mu = -\sin\theta_w\, B_\mu + \cos\theta_w\, W_\mu^3 . \tag{3.12}$$

Equations (3.12) define the weak mixing angle θ_w. The photon is characterized by equal couplings to left and right fermions with a strength equal to the electric charge. Recalling Eq. (3.9) for the charge matrix Q, we immediately obtain

$$g\sin\theta_w = g'\cos\theta_w = e , \tag{3.13}$$

or equivalently,

$$\begin{aligned} \text{tg}\,\theta_w &= g'/g , \\ e &= gg'/\sqrt{g^2 + g'^2} . \end{aligned} \tag{3.14}$$

Once θ_w has been fixed by the photon couplings, it is a simple matter of algebra to derive the Z couplings, with the result

$$\Gamma_{\bar{\psi}\psi Z} = (g/2\cos\theta_w)\bar{\psi}\gamma_\mu\,[t_L^3(1-\gamma_5) + t_R^3(1+\gamma_5) - 2Q\sin^2\theta_w]\psi Z^\mu , \tag{3.15}$$

where $\Gamma_{\bar{\psi}\psi Z}$ is a notation for the vertex. In the minimal Standard Model, $t_R^3 = 0$ and $t_L^3 = \pm 1/2$.

In order to derive the effective four-fermion interactions that are equivalent, at low energies, to the CC and NC couplings given in Eqs. (3.11) and (3.15), we anticipate that large masses, as experimentally observed, are provided for $W^{\pm}$ and Z by $\mathcal{L}_{\text{Higgs}}$. For left–left CC couplings, when the momentum transfer squared can be neglected with respect to m_W^2 in the propagator of Born diagrams with single W exchange, from Eq. (3.11) we can write

$$\mathcal{L}_{\text{eff}}^{\text{CC}} \simeq (g^2/8m_W^2)\,[\bar{\psi}\gamma_\mu(1-\gamma_5)t_L^+\psi][\bar{\psi}\gamma^\mu(1-\gamma_5)t_L^-\psi] . \tag{3.16}$$

By specializing further in the case of doublet fields such as ν_e–e^- or ν_μ–μ^-, we obtain the tree-level relation of g with the Fermi coupling constant G_F measured from μ decay [see Eq. (1.2)]:

$$G_F/\sqrt{2} = g^2/8m_W^2 . \tag{3.17}$$

By recalling that $g\sin\theta_w = e$, we can also cast this relation in the form

$$m_W = \mu_{\text{Born}}/\sin\theta_w , \tag{3.18}$$

with

$$\mu_{\text{Born}} = (\pi\alpha/\sqrt{2}\,G_F)^{1/2} \simeq 37.2802\ \text{GeV} , \tag{3.19}$$

where α is the fine-structure constant of QED ($\alpha \equiv e^2/4\pi = 1/137.036$).

In the same way, for neutral currents we obtain in Born approximation from Eq. (3.15) the effective four-fermion interaction given by

$$\mathcal{L}_{\rm eff}^{\rm NC} \simeq \sqrt{2}\, G_F \varrho_0 \bar{\psi}\gamma_\mu [\ldots] \psi \bar{\psi}\gamma^\mu [\ldots] \psi \,, \tag{3.20}$$

where

$$[\ldots] \equiv t_L^3 (1 - \gamma_5) + t_R^3 (1 + \gamma_5) - 2Q \sin^2\theta_w \tag{3.21}$$

and

$$\varrho_0 = m_W^2 / m_Z^2 \cos^2\theta_w \,. \tag{3.22}$$

All couplings given in this section are obtained at tree level and are modified in higher orders of perturbation theory. In particular, the relations between m_W and $\sin\theta_w$ [Eqs. (3.18) and (3.19)] and the observed values of ϱ ($\varrho = \varrho_0$ at tree level) in different NC processes, are altered by computable electroweak radiative corrections, as discussed in Section 5.

The gauge-boson self-interactions can be derived from the $F_{\mu\nu}$ term in $\mathcal{L}_{\rm symm}$, by using Eq. (3.12) and $W_\mu^\pm = (W^1 \pm iW^2)/\sqrt{2}$. Defining the three-gauge-boson vertex as in Fig. 1, we obtain ($V \equiv \gamma, Z$)

$$\Gamma_{W^-W^+V} = i g_{W^-W^+V} [g_{\mu\nu}(q - p)_\lambda + g_{\mu\lambda}(p - r)_\nu + g_{\nu\lambda}(r - q)_\mu] \,, \tag{3.23}$$

with

$$g_{W^-W^+\gamma} = g \sin\theta_w = e \qquad \text{and} \qquad g_{W^-W^+Z} = g \cos\theta_w \,. \tag{3.24}$$

The partial widths for the decay of W and Z into a massless fermion–antifermion pair, including first-order strong and electroweak radiative corrections, are given by

$$\Gamma(W \to f\bar{f}') = N(G_F m_W^3 / 6\pi\sqrt{2})(1 + \delta_f^W) \tag{3.25}$$

$$\Gamma(Z \to f\bar{f}) = N(G_F \varrho_0 m_Z^3 / 24\pi\sqrt{2})[1 + (1 - 4|Q_f| \sin^2\theta_w)^2](1 + \delta_f^Z) \,, \tag{3.26}$$

where $\varrho_0 = 1$ for doublet Higgses; δ_f^W and δ_f^Z are known electroweak radiative corrections, to be discussed in Sections 5 to 9 and are given explicitly, for example, in Ref. [10]; and

$$\begin{aligned} N &= 1 && \text{for leptons}\,, \\ N &= 3(1 + \alpha_s/\pi + \ldots) && \text{for quarks}\,. \end{aligned} \tag{3.27}$$

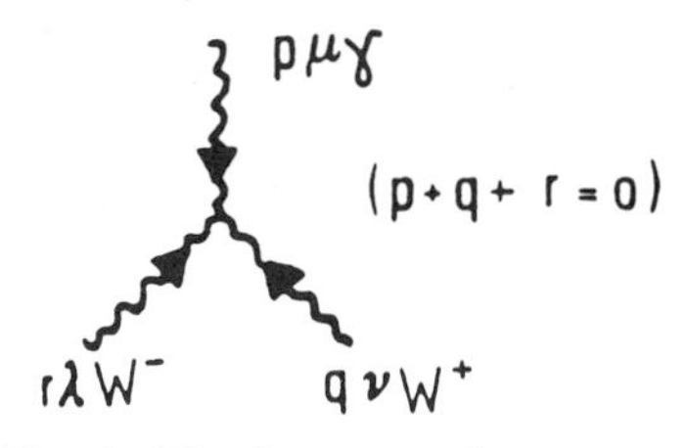

Fig. 1 The three-gauge boson vertex

Here α_s is the QCD coupling. We can define α_s according to the $\overline{MS}$ prescription. In this case the second-order correction is also known [11]. From experimental determinations of Λ_{QCD} [12], we obtain a value of $\alpha_s(m_Z)$ in the range

$$\alpha_s(m_Z) \simeq 0.11 \pm 0.01 \,. \tag{3.28}$$

In the case of W decays, formula (3.25) refers to the sum of all down-quarks associated with a given up-quark. For a particular down-quark, q', a factor $|V_{qq'}|^2$ would appear, where $V_{qq'}$ is the relevant term of the Cabibbo–Kobayashi–Maskawa matrix ($\Sigma_{q'}|V_{qq'}|^2 = 1$ by unitarity). Note that by writing $\Gamma_{Born} \approx G_F M^3$, instead of $\Gamma_{Born} \approx g^2 M \approx \alpha M/\sin^2\theta_w$, we make the Born approximation more precise (i.e. $\delta_f^{W,Z}$ smaller). In fact the radiative corrections are large on both α and $\sin^2\theta_w$. On the other hand, there are no leading logs in the scale dependence of the Fermi coupling G_F [13]; $\delta_f^{W,Z}$ are very small ($\delta_f \leqslant 0.1\%$) if $m_t \leqslant m_Z$ and there are no additional heavy-fermion generations [provided the physical values of $m_{W,Z}$ and $\sin^2\theta_w$ are inserted in Eqs. (3.25) and (3.26)]. The electroweak radiative corrections become considerably larger if m_t is large or if weak isospin multiplets with large mass splittings exist [10]. We shall come back to this case in Sections 5 to 9. With the exception of this case, for example, we obtain

$$\Gamma(W \to e\nu) \;= 224 \;\; (m_W/80\ \text{GeV})^3\ \text{MeV}\,, \tag{3.29}$$

$$\Gamma(Z \to e^+e^-) = \;83.5\ (m_Z/91\ \text{GeV})^3\ \text{MeV}\,, \tag{3.30}$$

$$\Gamma(Z \to \nu_e\bar{\nu}_e) \;= 166 \;\; (m_Z/91\ \text{GeV})^3\ \text{MeV}\,. \tag{3.31}$$

Note that the dependence of $\Gamma(Z \to ee)$ on $\sin^2\theta_w$ is very small for $\sin^2\theta_w \simeq 0.23$. If a careful determination of the partial widths is required, one should implement the dependence of $\sin^2\theta_w$ on m_Z given by Eq. (5.1) and take δ_f^W and δ_f^Z into account, which are large if $m_t > m_Z$. This will be done in Sections 5 to 9.

For $W \to t\bar{b}$ and $Z \to t\bar{t}$ (and perhaps $Z \to b\bar{b}$) the quark-mass corrections cannot be neglected. The widths in Born approximation are then modified as follows [14]:

$$\begin{aligned}\Gamma_{Born}(W \to t\bar{b}) &= (G_F m_W^3/2\pi\sqrt{2})(\{[(m_W^2 - m_t^2 - m_b^2)^2 - 4m_b^2 m_t^2]/m_W^4\}^{1/2} \\ &\quad \times \{1 - (m_t^2 + m_b^2)/2m_W^2 - \tfrac{1}{2}[(m_t^2 + m_b^2)/m_W^2]^2\}) \\ &\simeq (G_F m_W^3/2\pi\sqrt{2})\,(1 - \epsilon)^2\,(1 + \epsilon/2)\,,\end{aligned} \tag{3.32}$$

$$\Gamma_{Born}(Z \to Q\bar{Q}) = (G_F \varrho_0 m_Z^3/8\pi\sqrt{2})\,[\beta^3 + (1 - 4|Q_Q|\sin^2\theta_w)^2\,\tfrac{1}{2}\,\beta(3 - \beta^2)]\,, \tag{3.33}$$

with $\epsilon = m_t^2/m_W^2$ and $\beta = [1 - (4m_Q^2/m_Z^2)]^{1/2}$. The QCD corrections of order α_s are also modified with respect to the massless case and are given in Ref. [15].

The leptonic branching ratios are independent of m_W and m_Z (at fixed $\sin^2\theta_w$). We obtain

$$\begin{aligned}&BR(W \to e\nu) \simeq 0.089\,, \qquad m_t \simeq 40\ \text{GeV}\,, \\ &BR(W \to e\nu) \simeq 0.109\,, \qquad m_t > m_W - m_b\,,\end{aligned} \tag{3.34}$$

$$BR(Z \to e^+e^-) \simeq 0.034\,, \tag{3.35}$$

$$BR(Z \to \sum_{i=e,\mu,\tau} \nu_i\bar{\nu}_i) \simeq 0.20\,, \tag{3.36}$$

if $m_t > m_Z/2$ and $\sin^2\theta_w \simeq 0.23$.

The total widths Γ_W and Γ_Z are given to a good approximation by summing up the rates for $V \to f\bar{f}$ discussed above, because the rare decays of W/Z in the Standard Model are really rare and can be neglected. For example, for $m_t > m_W - m_b$, we obtain

$$\Gamma_W \equiv \Gamma(W \to e\nu)/BR(W \to e\nu) \simeq 2.06(m_W/80\ \text{GeV})^3\ \text{GeV}\,. \tag{3.37}$$

We now turn to the Higgs sector [16] of the electroweak Lagrangian. Here we simply review the formalism of the Higgs mechanism applied to the electroweak theory. In the next section we shall discuss in more detail the physics of the electroweak symmetry breaking. The Higgs Lagrangian is specified by the gauge principle and the requirement of renormalizability to be

$$\mathcal{L}_{Higgs} = (D_\mu\phi)^\dagger(D^\mu\phi) - V(\phi^\dagger\phi) - \bar{\psi}_L\Gamma\psi_R\phi - \bar{\psi}_R\Gamma^\dagger\psi_L\phi^\dagger\,, \tag{3.38}$$

where ϕ is a column vector including all Higgs fields; it transforms as a reducible representation of the gauge group. The quantities Γ (which include all coupling constants) are matrices that make the Yukawa couplings invariant under the Lorentz and gauge groups. The potential $V(\phi^\dagger\phi)$, symmetric under $SU(2) \otimes U(1)$, contains, at most, quartic terms in ϕ so that the theory is renormalizable. Spontaneous symmetry breaking is induced if the minimum of V—which is the classical analogue of the quantum mechanical vacuum state (both are the states of minimum energy)— is obtained for non-vanishing ϕ values. Precisely, we denote the vacuum expectation value (VEV) of ϕ, i.e. the position of the minimum, by v:

$$\langle 0|\phi(x)|0\rangle = v \neq 0\,. \tag{3.39}$$

The fermion mass matrix is obtained from the Yukawa couplings by replacing $\phi(x)$ by v:

$$M = \bar{\psi}_L\mathcal{M}\psi_R + \bar{\psi}_R\mathcal{M}^\dagger\psi_L\,, \tag{3.40}$$

with

$$\mathcal{M} = \Gamma\cdot v\,. \tag{3.41}$$

In the minimal Standard Model, where all left fermions ψ_L are doublets and all right fermions ψ_R are singlets, only Higgs doublets can contribute to fermion masses. There are enough free couplings in Γ, so that one single complex Higgs doublet is indeed sufficient to generate the most general fermion mass matrix. It is important to observe that by a suitable change of basis we can always make the matrix $\mathcal{M}$ Hermitian, γ_5-free, and diagonal. In fact, we can make separate unitary transformations on ψ_L and ψ_R according to

$$\psi_L' = U\psi_L\,, \qquad \psi_R' = V\psi_R \tag{3.42}$$

and consequently

$$\mathcal{M} \to \mathcal{M}' = U^\dagger\mathcal{M}V\,. \tag{3.43}$$

This transformation does not alter the general structure of the fermion couplings in $\mathcal{L}_{symm}$. For quarks, the Cabibbo–Kobayashi–Maskawa unitary transformation relates the mass eigenstates d, s, and b to the CC eigenstates d′, s′, b′, i.e. the states coupled by W emission to u, c, and t, respectively. The NC is then automatically diagonal in flavour at tree level (GIM mechanism [5]). In the case of leptons, if the neutrinos are massless then clearly there is no mixing.

If only one Higgs doublet is present, the change of basis that makes M diagonal will at the same time diagonalize also the fermion–Higgs Yukawa couplings. Thus, in this case, no flavour-changing neutral Higgs exchanges are present. This is not true, in general, when there are several Higgs doublets. But one Higgs doublet for each electric charge sector—i.e. one doublet coupled only to u-type quarks, one doublet to d-type quarks, one doublet to charged leptons—would also be all right [17], because the mass matrices of fermions with different charges are diagonalized separately. For several Higgs doublets it is also possible to generate CP violation by complex phases in the Higgs couplings [18]. In the presence of six quark flavours, this CP violation mechanism is not necessary. In fact, at the moment, the simplest model with only one Higgs doublet seems adequate for describing all observed phenomena.

We recall that the Standard Model, with N fermion families with the observed quantum numbers, is automatically free of γ_5 anomalies [19] owing to cancellation of quarks with lepton loops.

We now consider the gauge-boson masses and their couplings to the Higgs. These effects are induced by the $(D_\mu\phi)^\dagger(D^\mu\phi)$ term in $\mathcal{L}_{\text{Higgs}}$ [Eq. (3.38)], where

$$D_\mu\phi = \left[\partial_\mu + ig\sum_{A=1}^{3} t^A W_\mu^A + ig'(Y/2)\, B_\mu\right]\phi\,. \tag{3.44}$$

Here t^A and $\frac{1}{2}Y$ are the SU(2) ⊗ U(1) generators in the reducible representation spanned by ϕ. Not only doublets but all non-singlet Higgs representations can contribute to gauge-boson masses. The condition that the photon remains massless is equivalent to the condition that the vacuum is electrically neutral:

$$Q|v\rangle = (t^3 + \tfrac{1}{2}Y)|v\rangle = 0\,. \tag{3.45}$$

The charged W mass is given by the quadratic terms in the W field arising from $\mathcal{L}_{\text{Higgs}}$, when $\phi(x)$ is replaced by v. We obtain

$$m_W^2 W_\mu^+ W^{-\mu} = g^2\,|(t^+ v/\sqrt{2})|^2\, W_\mu^+ W^{-\mu}\,, \tag{3.46}$$

whilst for the Z mass we get [recalling Eq. (3.12)]

$$\tfrac{1}{2}m_Z^2 Z_\mu Z^\mu = |[g\cos\theta_w\, t^3 - g'\sin\theta_w\,(Y/2)]v|^2\, Z_\mu Z^\mu\,, \tag{3.47}$$

where the factor of ½ on the left-hand side is the correct normalization for the definition of the mass of a neutral field. By using Eq. (3.45), relating the action of t^3 and $\frac{1}{2}Y$ on the vacuum v, and Eqs. (3.14), we obtain

$$\tfrac{1}{2}m_Z^2 = (g\cos\theta_w + g'\sin\theta_w)^2\,|t^3 v|^2 = (g^2/\cos^2\theta_w)\,|t^3 v|^2\,. \tag{3.48}$$

For Higgs doublets

$$\phi = \begin{pmatrix}\phi^+\\ \phi^0\end{pmatrix}, \qquad v = \begin{pmatrix}0\\ v\end{pmatrix}, \tag{3.49}$$

we have

$$|t^+ v|^2 = v^2\,, \qquad |t^3 v|^2 = \tfrac{1}{4}v^2\,, \tag{3.50}$$

so that

$$m_W^2 = {}^1\!/\!_2 g^2 v^2 , \qquad m_Z^2 = {}^1\!/\!_2 g^2 v^2/\cos^2\theta_w . \tag{3.51}$$

Note that by using Eq. (3.17) we obtain

$$v = 2^{-3/4}\, G_F^{-1/2} = 174.1 \text{ GeV} . \tag{3.52}$$

It is also evident that for Higgs doublets

$$\varrho_0 = m_W^2/m_Z^2 \cos^2\theta_w = 1 . \tag{3.53}$$

This relation is typical of one or more Higgs doublets and would be spoiled by the existence of Higgs triplets etc. In general,

$$\varrho_0 = \sum_i \left[(t_i)^2 - (t_i^3)^2 + t_i\right] v_i^2 \Big/ \sum_i 2(t_i^3)^2 v_i^2 \tag{3.54}$$

for several Higgses with VEVs v_i, weak isospin t_i, and z-component t_i^3. These results are valid at the tree level and are modified by calculable electroweak radiative corrections, as discussed in Section 5.

If only one Higgs doublet is present, then the fermion–Higgs couplings are in proportion to the fermion masses. In fact, from the Yukawa couplings $g_{\phi\bar{f}f}$ $(\bar{f}_L \phi f_R$ + h.c.), the mass m_f is obtained by replacing ϕ by v, so that $g_{\phi\bar{f}f} = m_f v$.

With only one complex Higgs doublet, three out of the four Hermitian fields are removed from the physical spectrum by the Higgs mechanism and become the longitudinal modes of W^+, W^-, and Z. The fourth neutral Higgs is physical and should be found. If more doublets are present, two more charged and two more neutral Higgs scalars should be around for each additional doublet.

Finally, the couplings of the physical Higgs H to the gauge bosons can be simply obtained from $\mathcal{L}_{Higgs}$, by the replacement

$$\phi(x) = \begin{pmatrix} \phi^+(x) \\ \phi^0(x) \end{pmatrix} \rightarrow \begin{pmatrix} 0 \\ v + (H/\sqrt{2}) \end{pmatrix} , \tag{3.55}$$

[so that $(D_\mu\phi)^\dagger (D^\mu\phi) = {}^1\!/\!_2(\partial_\mu H)^2 + \ldots$], with the result

$$\begin{aligned} \mathcal{L}|H,W,Z| = {} & g^2 (v/\sqrt{2})\, W_\mu^+ W^{-\mu} H + (g^2/4)\, W_\mu^+ W^{-\mu} H^2 \\ & + [(g^2\, v\, Z_\mu Z^\mu)/(2\sqrt{2}\cos^2\theta_w)]H + [g^2/(8\cos^2\theta_w)]\, Z_\mu Z^\mu H^2 . \end{aligned} \tag{3.56}$$

We have thus completed our summary of the standard electroweak theory and of the $W^\pm$, Z couplings.

4. THE HIGGS AND BEYOND: THE PROBLEM OF THE FERMI SCALE

The gauge symmetry of the Standard Model was difficult to discover because it is well hidden in nature. The only observed gauge boson that is massless is the photon. The graviton is still unobserved even at the classical level of gravitational waves; the gluons are presumed massless but unobservable because of confinement, and the W and Z weak bosons carry a heavy mass. Actually the main difficulty in unifying weak and electromagnetic interactions was the fact that e.m. interactions have infinite range ($m_\gamma = 0$), whilst the weak forces have a very short range, owing to $m_{W,Z} \neq 0$.

The solution of this problem is in the concept of spontaneous symmetry breaking, which was borrowed from statistical mechanics.

Consider a ferromagnet at zero magnetic field in the Landau–Ginzburg approximation. The free energy in terms of the temperature T and the magnetization **M** can be written as

$$F(\mathbf{M}, T) \simeq F_0(T) + \tfrac{1}{2}\mu^2(T)\mathbf{M}^2 + \tfrac{1}{4}\lambda(T)(\mathbf{M}^2)^2 + \dots . \tag{4.1}$$

This is an expansion which is valid at small magnetization, and which is the analogue in this context of the renormalizability criterion; $\lambda(T) > 0$ is assumed for stability; F is invariant under rotations, i.e. all directions of **M** in space are equivalent. The minimum condition for F reads

$$\partial F/\partial M = 0 , \qquad [\mu^2(T) + \lambda(T)\mathbf{M}^2]\mathbf{M} = 0 . \tag{4.2}$$

There are two cases. If $\mu^2 > 0$, then the only solution is $\mathbf{M} = 0$, there is no magnetization, and the rotation symmetry is respected. If $\mu^2 < 0$, then another solution appears, which is

$$|\mathbf{M}_0|^2 = -\mu^2/\lambda . \tag{4.3}$$

The direction chosen by the vector $\mathbf{M}_0$ is a breaking of the rotation symmetry. The critical temperature T_{crit} is where $\mu^2(T)$ changes sign:

$$\mu^2(T_{crit}) = 0 . \tag{4.4}$$

It is simple to realize that the Goldstone theorem holds. It states that when spontaneous symmetry breaking takes place, there is always a zero-mass mode in the spectrum. In a classical context this can be proven as follows. Consider a Lagrangian

$$\mathcal{L} = |\partial_\mu \phi|^2 - V(\phi) \tag{4.5}$$

symmetric under the infinitesimal transformations

$$\phi \to \phi' = \phi + \delta\phi , \qquad \delta\phi_i = i\,\delta\theta\, t_{ij}\phi_j . \tag{4.6}$$

The minimum condition on V that identifies the equilibrium position (or the ground state in quantum language) is

$$(\partial V/\partial \phi_i)\,(\phi_i = \phi_i^0) = 0 . \tag{4.7}$$

The symmetry of V implies that

$$\delta V = (\partial V/\partial \phi_i)\,\delta\phi_i = i\,\delta\theta\,(\partial V/\partial\phi_i)t_{ij}\phi_j = 0 . \tag{4.8}$$

By taking a seconde derivative at the minimum $\phi_i = \phi_i^0$ of the previous equation, we obtain

$$\partial^2 V/\partial\phi_k\partial\phi_i\,(\phi_i = \phi_i^0)t_{ij}\phi_i^0 + (\partial V/\partial\phi_i)\,(\phi_i = \phi_i^0)t_{ik} = 0 . \tag{4.9}$$

The second term vanishes owing to the minimum condition, Eq. (4.7). We then find

$$\partial^2 V/\partial\phi_k\partial\phi_i\,(\phi_i = \phi_i^0)t_{ij}\phi_j^0 = 0 . \tag{4.10}$$

The second derivatives $M_{ki}^2 = (\partial^2 V/\partial\phi_k\partial\phi_i)(\phi_i = \phi_i^0)$ define the squared mass matrix. Thus the above equation in matrix notation can be read as

$$M^2 t\phi^0 = 0\,, \tag{4.11}$$

which shows that if the vector $(t\phi^0)$ is non-vanishing, i.e. there is some generator that shifts the ground state into some other state with the same energy, then $t\phi^0$ is an eigenstate of the squared mass matrix with zero eigenvalue. Therefore, a massless mode is associated with each broken generator.

When spontaneous symmetry breaking takes place in a gauge theory, the massless Goldstone mode exists, but it is unphysical and disappears from the spectrum. It becomes, in fact, the third helicity state of a gauge boson that takes mass. This is the Higgs mechanism. Consider, for example, the simplest Higgs model described by the Lagrangian

$$\mathcal{L} = -{}^1\!/_4 F_{\mu\nu}^2 + |(\partial_\mu - ieA_\mu)\phi|^2 + {}^1\!/_2\mu^2\phi^*\phi - (\lambda/4)(\phi^*\phi)^2\,. \tag{4.12}$$

Note the 'wrong' sign in front of the mass term for the scalar field ϕ, which is necessary for the spontaneous symmetry breaking to take place. The above Lagrangian in invariant under the U(1) gauge symmetry

$$A_\mu \rightarrow A'_\mu = A_\mu - (1/e)\partial_\mu\theta(x)\,, \qquad \phi \rightarrow \phi' = \phi \exp[i\theta(x)]\,. \tag{4.13}$$

Let $\phi^0 = v \neq 0$, with v real, be the ground state that minimizes the potential and induces the spontaneous symmetry breaking. Making use of gauge invariance, we can make the change of variables

$$\phi(x) \rightarrow (1/\sqrt{2})[\varrho(x) + v]\exp[i\zeta(x)/v]\,, \qquad A_\mu(x) \rightarrow A_\mu - (1/ev)(\partial_\mu\zeta(x)\,. \tag{4.14}$$

Then $\varrho = 0$ is the position of the minimum, and the Lagrangian becomes

$$\mathcal{L} = -{}^1\!/_4 F_{\mu\nu}^2 + {}^1\!/_2 e^2v^2A_\mu^2 + {}^1\!/_2 e^2\varrho^2A_\mu^2 + e^2\varrho v A_\mu + \mathcal{L}(\varrho)\,. \tag{4.15}$$

The field $\zeta(x)$, which corresponds to the would-be Goldstone boson, disappears, whilst the mass term ${}^1\!/_2 e^2v^2A_\mu^2$ for A_μ is now present; ϱ is the massive Higgs particle.

The Higgs mechanism is realized in well-known physical situations. For a superconductor in the Landau–Ginzburg approximation the free energy can be written as

$$F = F_0 + {}^1\!/_2\mathbf{B}^2 + |(\nabla - 2ie\mathbf{A})\phi|^2/4m - \alpha|\phi|^2 + \beta|\phi|^4\,. \tag{4.16}$$

Here **B** is the magnetic field, $|\phi|^2$ is the Cooper pair (e^-e^-) density, 2e and 2m are the charge and mass of the Cooper pair. The 'wrong' sign of α leads to $\phi \neq 0$ at the minimum. This is precisely the non-relativistic analogue of the Higgs model of the previous example. The Higgs mechanism implies the absence of propagation of massless phonons (states with dispersion relation $\omega = kv$ with constant v). Also the mass term for **A** is manifested by the exponential decrease of **B** inside the superconductor (Meissner effect).

Thus the Higgs effect is not an abstract device, but it is endowed with a precise physical reality. In the electroweak theory it is absolutely necessary in order to give masses to the W's and Z^0 and to the fermions, as well as to ensure the correct high-energy behaviour required by renormalizability.

However a more profound physical reality could be hidden behind or accompany the Higgs formalism. In fact, the Higgs mechanism is at present without experimental support. Actually the clarification of the physical origin of the electroweak symmetry breaking is one of the most important problems for experimental particle physics in the next decade. There are arguments indicating that the minimal Standard Model with fundamental Higgs fields cannot be the whole story and that some kind of new physics must necessarily appear near the Fermi scale. The most famous argument of this type is based on the so-called 'hierarchy problem', which we now describe.

There is no unification of the fundamental forces in the Standard Model, because a separate gauge group and coupling is introduced for each interaction. On the other hand, the structural unity implied by the common, restrictive property of gauge invariance strongly suggests the possibility that all the observed interactions actually stem from a unified theory at some more fundamental level. The idea of unification at energies of order $m_{GUT} \approx 10^{15}$ GeV, below the energy scale where quantum gravity becomes effective at masses of the order of the Planck mass, $m_P \approx 10^{19}$ GeV, has been much studied in recent years. However, the question remains whether unification without the inclusion of quantum gravity is really plausible. It is clear that the inclusion of gravity must induce major changes in the physics of the Standard Model at energies of order m_P and possibly even below.

Thus, at least because of the fact that gravity is not included in the Standard Model, new physics must necessarily emerge at some large energy scale Λ (equivalent to some small distance scale). Then the problem is to understand what order of magnitude can reasonably be expected for Λ. In particular, we can ask whether it is natural to expect that Λ may be as large as m_P or m_{GUT}. In other words, Is it possible that the Standard Model holds without any new physical input up to the energy scale of quantum gravity? The answer is probably negative, because then we could not naturally explain the enormous value of m_P/m_W, i.e. the ratio between the Planck and the Fermi scales of masses.

To develop this point further, we recall that in the Standard Model the fermion and vector-boson masses are all specified in terms of the VEV of the Higgs field v, according to Eqs. (3.41) and (3.51). The value of v is determined by the curvature scale of the Higgs potential V:

$$V(\phi) = -\tfrac{1}{2}\mu^2 \phi^\dagger\phi + (\lambda/4)(\phi^\dagger\phi)^2 , \tag{4.17}$$

according to

$$v = \mu/\sqrt{\lambda} . \tag{4.18}$$

The observed values of the masses require for v (and, therefore, roughly for μ as well) that $v \approx 10^2$ GeV [see Eq. (3.52)].

If $\Lambda \approx (10^{15}\text{–}10^{19})$ GeV, then we face the problem of justifying the presence of two so largely different mass scales in a single theory (the so-called hierarchy problem). In general, if Λ is very large in comparison with μ, then, even if we set by hand a small value for μ at the tree level, the radiative corrections would make μ increase up to nearly the order of Λ. This problem is particularly acute in theories with scalars, as in the Standard Model, because the degree of divergence of mass corrections is quadratic, whilst the same divergences are only logarithmic for spin-$\tfrac{1}{2}$ fermions.

One general way out would be that the limit $\mu \to 0$ corresponds to an increase of the symmetry of the theory. In fact, the observed value of μ^2 can be decomposed as

$$\mu^2 = \mu_0^2 + \delta\mu^2 , \tag{4.19}$$

where μ_0 is the tree-level value and $\delta\mu^2$ arises from the loop quantum corrections. If no new symmetry is induced when $\mu \to 0$ and no other non-renormalization theorem is operative, then a small value for μ^2/Λ^2 can only arise from an unbelievably precise cancellation between μ_0^2/Λ^2 and $\delta\mu^2/\Lambda^2$. If, however, $\mu = 0$ leads to an additional symmetry, then $\delta\mu^2$ must be proportional to μ_0^2, because for μ_0

$= 0$ both the tree diagrams and the loop corrections must respect the symmetry. Then, if one starts from a small value of μ_0^2/Λ^2, the radiative corrections would preserve the smallness of μ^2/Λ^2.

For fermions, chiral symmetry is added when $m_f \rightarrow 0$, because the axial currents are also conserved in this limit, as their divergence is proportional to the fermion mass. Chiral symmetry and the logarithmic degree of divergence for fermion masses considerably alleviate the hierarchy problem in theories with no fundamental scalars.

In the Standard Model no additional symmetry is gained for $\mu_0 = 0$. This is also seen from the explicit formula for $\delta\mu^2$ at the one-loop level, which shows that $\delta\mu^2$ is not proportional to μ_0^2:

$$\mu^2/\Lambda^2 = (\mu_0^2/\Lambda^2) + (1/128\pi^2)(d/dv)^2 \sum_J (2J + 1)(-1)^{2J} m_J^2(v) + \ldots, \tag{4.20}$$

where terms which vanish with $\Lambda \rightarrow \infty$ are indicated by the dots. The sum over J includes both particles and antiparticles (counted separately) of spin J and mass m_J (expressed as a function of v). Thus we are forced to the conclusion that in the Standard Model the natural value for $\delta\mu^2/\Lambda^2$ is of order one or so. Therefore, as Λ can be interpreted as the energy scale where some essentially new physical ingredient becomes important, we are led to expect that the validity of the present framework cannot be extended beyond $\Lambda \approx (1\text{–}10)$ TeV.

The problem of explaining the Fermi scale is seen to be closely connected to the Higgs mechanism and to the consequent presence of scalars, which makes the problem of testing the Higgs sector particularly crucial.

One possible solution is that the Higgses are really scalar fundamental fields, but naturality is restored by supersymmetry.

Supersymmetry (SUSY for short) [20] relates bosons and fermions, so that in a multiplet that forms one representation of supersymmetry there is an equal number of bosonic and fermionic degrees of freedom. This implies that SUSY generators are spin-½ charges Q_α. SUSY leads to an extension of the Poincaré algebra. Besides the obvious algebraic relations between Q_α and the Poincaré generators, which specify the spinorial transformations of Q_α under Lorentz transformations and its invariance under translations, the essentially new relation is the anticommutator

$$\{Q_\alpha, \bar{Q}_\beta\} = -2(\gamma_\mu)_{\alpha\beta}P^\mu, \tag{4.21}$$

where P^μ is the energy-momentum four-vector, which generates space–time translations.

If all fundamental symmetries are gauge symmetries, then also SUSY is presumably a local symmetry. This immediately leads to the realm of gravity. In fact, the product of two local SUSY transformations is a translation with space–time-dependent parameters, as follows from Eq. (4.21). But a translation with space–time-dependent parameters is a general coordinate transformation. As ordinary gravity can be seen to arise from gauging the Poincaré group, a similar gauging of the Poincaré algebra enlarged by SUSY generators leads to an extended version of gravity, called supergravity. In fact, supersymmetry and supergravity play a crucial role in most present attempts at constructing a sensible theory of quantum gravity, including the very interesting superstring theories [21] that at present represent the most advanced and promising project for a theory of gravity (and of all particle interactions).

Theorists like SUSY for several reasons. SUSY is the maximum symmetry compatible with a non-trivial S-matrix in a local relativistic field theory. The powerful constraints between couplings and masses implied by SUSY drastically reduce the degree of singularity of field theory as deduced from power counting. In some cases a finite field theory is even obtained. For example, $N = 4$ extended SUSY Yang–Mills theories are finite in four dimensions. In general, more powerful non-renormalization theorems are deduced for SUSY theories. This property may solve many naturalness problems of the Standard Model. In particular, the hierarchy problem can be solved in SUSY theories because the mass divergences of bosons and fermions become the same, and the quadratic divergences of scalars are reduced to logarithmic divergences as for spin-½ fermions. Thus in SUSY theories, we automatically obtain

$$(d/dv)^2 \sum_J [(2J + 1)(-1)^{2J} m_J^2(v)] = 0 , \tag{4.22}$$

because of a cancellation between bosons and fermions. Of course the cancellation is only exact in the limit of unbroken SUSY. But we know that SUSY must be broken because the SUSY partners of ordinary particles have not been observed. In broken SUSY, the Λ, appearing in Eq. (4.20), can be identified with the mass scale that is typical of SUSY partners of ordinary states. Thus if Λ is of order $G_F^{-1/2}$ or at most O(1 TeV), then the hierarchy problem would be solved in a natural way. This is the only known way out of the hierarchy problem compatible with fundamental scalar Higgs fields.

We have seen that most theorists working on quantum gravity and superstrings tend to consider SUSY as 'established' at m_P and beyond. For economy, we are naturally led to also try to use SUSY at low energy to solve some of the problems of the Standard Model, including the hierarchy problem. It is therefore very important that it was indeed shown [22] that models where SUSY is softly broken by gravity do offer a viable alternative. We stress again that the supersymmetric way is very appealing to theorists. In fact, it would represent the ultimate triumph of a continuous line of progress obtained by constructing field theories with an increasing degree of exact and/or broken symmetry and applying them to fundamental interactions. Also the value of the ratio of knowledge versus ignorance would be remarkably large in the case of SUSY: the correct degrees of freedom for a description of physics up to gravity would have been identified, the Hamiltonian would be essentially known, and the theory would, to a large extent, be computable up to m_P.

The alternative main avenue to physics beyond the Standard Model is compositeness or, more generally, the existence of new strong forces. For example, the electroweak symmetry could be broken by condensates of new fermions attracted by a new force with $\Lambda_{new} \approx m_F \approx G_F^{-1/2}$ (Λ_{new} being the analogous quantity of Λ_{QCD}), as in technicolour theories [23]. Or the Higgs scalar can be a composite of new fermions bound by a new force [24]. In the last two cases there are unsolved problems related to the fermion masses. Or the $SU(2) \otimes U(1)$ gauge symmetry can be a low-energy fake [25]. At high energies $\gtrsim m_F$, the W and Z^0 would be resolved into their constituents.

Another possibility is that the Higgs mass becomes very large [26] [O(1 TeV)]. Then, as we shall see, the weak interactions become strong and a spectroscopy of tightly bound weak interaction resonances appears (e.g. WW, WZ, or ZZ states).

However, it is fair to say that the compositeness alternative is not at all so neat and clearly formulated as the supersymmetric option. On the contrary, in many respects the compositeness way is not well defined at all and leads to many unsolved problems.

Of course, the two avenues are not necessarily mutually exclusive, and theoretical frameworks where both appear have been considered.

Searching for the standard Higgs particle appears to be a good way to organize the experimental solution of the symmetry-breaking problem. Note that the information we have regarding the mass of the standard Higgs is not very abundant. The experimental lower bound [27] on the Higgs mass from atomic and nuclear physics (absence of long-range forces),

$$m_H \geqslant 15 \text{ MeV} , \tag{4.23}$$

is rather solid, unless perhaps for $m_H < 2m_e$. The limit $m_H \gtrsim 325$ MeV, which was claimed in the past to derive from $K^+ \to \pi^+ + H$, was later criticized [28]. In a recent reanalysis [29] of the problem, the excluded domain was reported to be $50 < m_H < 140$ MeV. However, some model dependence is clearly unavoidable in this case (from hadronic matrix elements). The most important experimental input is the lower limit on m_H that can be obtained from the search for the decay $\Upsilon \to H + \gamma$ [30]. If we assume the validity of the theoretical prediction from the Wilczek [31] formula plus first-order QCD corrections [32], then CUSB data lead to $m_H > 5$ GeV (or $m_H < 2m_\mu$). However, the QCD corrections are large so that the higher orders can also be of importance. Also, corrections to the non-relativistic approximation could be sizeable. It is therefore difficult to fix a precise value for the lower bound. However, I think that the data are by now good enough to make a minimal standard Higgs of a few GeV unlikely. A much better limit will soon be available from LEP. Of course, it is sufficient to introduce two Higgs doublets in order to evade this limit (by taking suitable values for the ratio of VEVs).

Vacuum stability can be considered as a criterion for deriving theoretical lower limits on the Higgs mass [33]. At tree level the scalar potential is given by

$$V(\phi) = \mu^2|\phi|^2 - (\mu^2/2v^2)|\phi|^4 . \tag{4.24}$$

The quantum corrections can be computed by expanding in the number of loops. At one loop we obtain

$$V(\phi) = \mu^2|\phi|^2 - (\mu^2/2v^2)|\phi|^4 + \gamma|\phi|^4 [\ln(|\phi|^2/v^2) - (1/2)] , \tag{4.25}$$

with

$$\gamma = \left[3 \sum_{\text{vectors}} m_V^4 + \sum_{\text{scalars}} m_S^4 - 4 \sum_{\text{fermions}} m_f^4\right] \Big/ (64\pi^2 v^4) . \tag{4.26}$$

It is simple to check that, also in the corrected form, v is an extremum of $V(\phi)$. In the minimal Standard Model with one Higgs doublet and three fermion families, we obtain

$$\gamma = (3m_Z^4 + 6m_W^4 + m_H^4 - 12m_t^4) \,/\, (64\pi^2 v^4) . \tag{4.27}$$

(The extra factor of three in front of m_t^4 is, of course, due to colour). From the requirements that $|\phi| = v$ be a minimum and that $V(\phi) \to +\infty$ for $|\phi| \to \infty$, we obtain the lower bound shown in Fig. 2. In particular, for small m_t we obtain the Linde–Weinberg [34] bound $m_H \gtrsim 7$ GeV. For large m_t, large values of m_H are needed in order to stabilize the potential at $|\phi| \to \infty$ (the one-loop approximation has to be improved by renormalization group techniques in this region, as was done in the

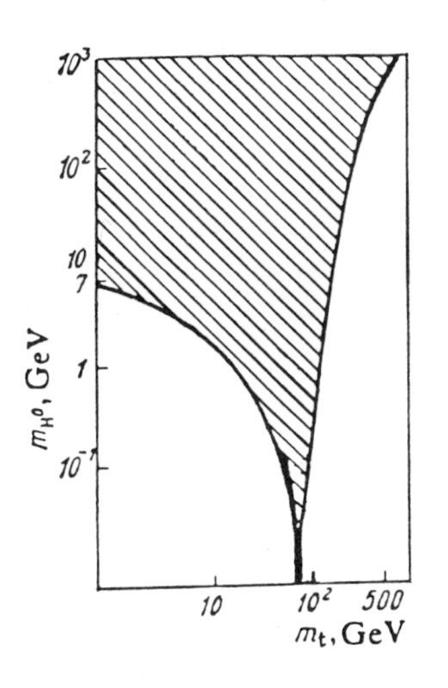

Fig. 2 The dashed region is the allowed region for the minimal Higgs mass from vacuum stability [33]

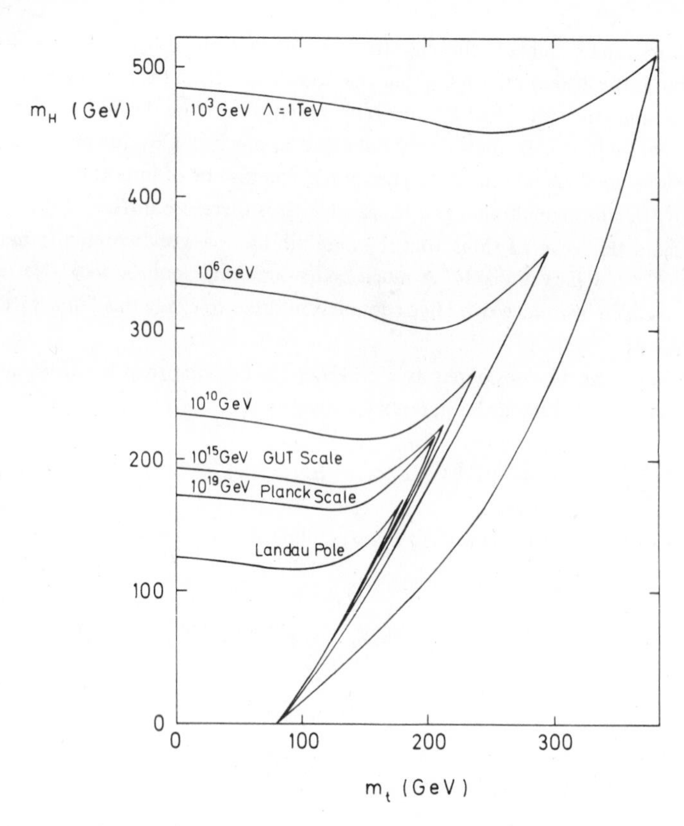

Fig. 3 The allowed parameter space of m_H and m_t for various embedding scales Λ (see Ref. [38]). The area around the origin bounded by the various curves is allowed. The curve labelled 'Landau pole' is obtained from Ref. [37]. The horizontal lines come from avoiding triviality, and the vertical lines are determined from $\lambda(t)$ becoming negative at scales lower than Λ.

computations of Fig. 3). These are certainly very interesting results. The problem is that the limit evaporates if a) $m_t \approx 80$ GeV, or b) there are more Higgs doublets, or c) we allow for a metastable vacuum with long enough lifetime. In these cases there is no limit for $80 \leqslant m_t \leqslant 200$ GeV [35].

On the other hand, the theoretical upper bounds on m_H are all based, in one form or another, on requiring perturbation theory to hold up to some large energy Λ. But this perhaps appealing requirement is in no way necessary. As is well known, the coupling of the quartic term $\lambda(\phi^+\phi)^2$ in the Higgs potential increases with m_H^2, because $m_H^2 \propto \lambda/G_F$. In addition, for a given m_H, λ increases logarithmically with energy since the theory is not asymptotically free in the Higgs sector. Then, requiring perturbation theory to be valid up to $\Lambda = m_P$ leads to [36]

$$m_H \leqslant 200 \text{ GeV} . \tag{4.28}$$

Similarly, from the requirement that problems due to the possible triviality of the $\lambda\phi^4$ theory be avoided, the limit

$$m_H \leqslant 125 \text{ GeV} \tag{4.29}$$

was obtained [37]. The upper curves of Fig. 3, taken from Ref. [38], are obtained merely by imposing that the theory remains non-trivial up to a given scale Λ. The same picture has been essentially confirmed by computer simulations of the electroweak theory on the lattice [39]. However, if m_H is made to increase, no physical contradiction is actually met. All that happens is that at $m_H \gtrsim 1$ TeV, the Born amplitudes for longitudinal gauge-boson scattering violate unitarity [26], manifesting the breakdown of perturbation theory. The helicity-zero state of gauge bosons is obviously connected to symmetry breaking, because it does not exist for massless vector bosons. For $m_H \sim 1$ TeV the weak interactions become strong. The Higgs boson becomes very broad,

$$\Gamma_H \approx \tfrac{1}{2}(m_H)^3 \qquad (\Gamma_H,\, m_H \text{ in TeV})\,. \tag{4.30}$$

We have already remarked that the search for the standard Higgs particle appears to be a good way to organize the experimental programme for a solution to the problem of the origin of the Fermi scale. On the basis of the previous arguments, it is believed [40] that if a set of experiments are sensitive enough to be able to detect the Higgs with mass up to O(1 TeV), then a great discovery will in any case be made. Either one finds the Higgs, or new physics, or both. At the very least, one would observe the onset of a new regime with strong weak-interactions.

The best opportunity to find the Higgs in the near future is offered by LEP experiments. At LEP1 or at the SLAC Linear Collider (SLC) we will look for the Higgs in Z decays. The most important modes are

$$Z \to H\ell\bar{\ell} \tag{4.31}$$

and (to a lesser extent)

$$Z \to H\gamma\,. \tag{4.32}$$

The corresponding branching ratios are shown in Fig. 4. With an integrated luminosity of around 500 pb^{-1}, the Higgs can be found if $m_H \lesssim (40\text{–}50)$ GeV [41].

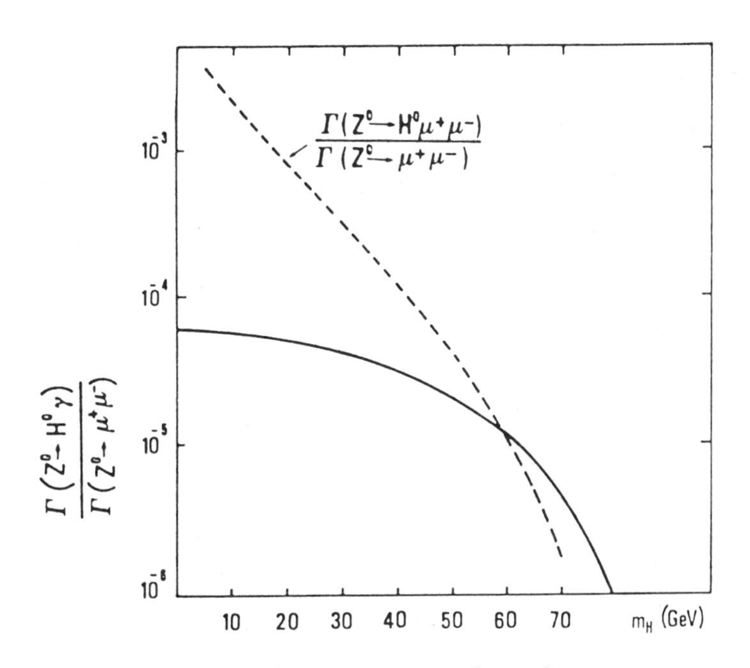

Fig. 4 The branching ratios $\Gamma(Z^0 \to H^0\mu^+\mu^-)/\Gamma_{\mu\mu}$ and $\Gamma(Z^0 \to H^0\gamma)/\Gamma_{\mu\mu}$ as functions of m_H, for $\sin^2\theta_w = 0.23$.

At LEP 2 [42], with $\sqrt{s}$ = 160–200 GeV, the important process is

$$e^+e^- \to ZH \quad \hookrightarrow b\bar{b} \tag{4.33}$$

The discovery range is easily extended up to $m_H \approx 70$ GeV. With more difficulty, one could possibly get up to $m_H \approx m_Z$. In this range of mass the additional problem is the confusion of the Higgs signal with the W and Z peaks in the relevant mass distributions.

Only supercolliders can continue the Higgs search beyond $m_H \simeq m_Z$: hadronic supercolliders such as the Superconducting Super Collider (SSC) ($\sqrt{s} \simeq 40$ TeV), or the Large Hadron Collider (LHC) ($\sqrt{s}$ = 16–17 TeV), or e^+e^- future linear colliders such as the CERN Linear Collider (CLIC) ($\sqrt{s} \simeq 2$ TeV). The problem of producing and detecting the Higgs particle at supercolliders has been much studied recently in connection with the SSC, the LHC, and CLIC. In particular, the problem of observing the Higgs at the LHC and CLIC was analysed in great detail at the La Thuile–CERN Workshop on Physics at Future Accelerators [43], and a comparison with the SSC was made. The luminosity for the SSC, the LHC, and CLIC was in most cases assumed to be $L = 10^{33}\ cm^{-2}\ s^{-1}$.

The results obtained at the La Thuile Workshop on the Higgs problem can be summarized as follows. The relevant range of Higgs masses must be separated into two intervals: the intermediate-mass Higgs: $m_{W/Z} < m_H \leqslant 200$ GeV, and the heavy Higgs: $m_H \geqslant 200$ GeV. The problem of observing the Higgs is completely different in the two cases. In fact, for an intermediate-mass Higgs, the main decay mode is into the heaviest pair of quarks allowed by phase space, whilst a heavy Higgs decays mainly into WW or ZZ pairs.

The observation of an intermediate-mass Higgs is certainly possible at e^+e^- linear colliders with $L = 10^{33}\ cm^{-2}\ s^{-1}$ and $\sqrt{s} \simeq 1$–2 TeV. On the contrary, at hadron–hadron colliders the observation of an intermediate-mass Higgs presents extremely difficult problems because of the QCD background. This is true for both the single-Higgs production ($pp \to HX$) and the associated Higgs–W production ($pp \to HWX$). Since the intermediate-Higgs production cross-section (mainly due to gluon fusion) is quite large (at $\sqrt{s}$ = 40 TeV, $\sigma \approx 10^2$ pb for $m_H \approx 150$ GeV and is 3–4 times smaller at $\sqrt{s} \approx 16$ TeV), one can hope to make use of the rare decay modes of the Higgs, e.g. $H \to \gamma\gamma$ and $\tau^+\tau^-$, or $\ell\ell\ell\ell$. The possibility of seeing the rare decay modes is for $m_H < 2m_t$. Then the main decay mode of the Higgs is $H \to b\bar{b}$, its total width is not too large, and the rare branching ratios are enhanced. Even in this favourable case the detection of rare modes is very difficult, the $\gamma\gamma$ mode being the most promising. Concerning the heavy Higgs, the most reliable strategy appears to be the search for the channel $H \to ZZ \to \ell^+\ell^-\nu\bar{\nu}$ (with $\ell = e, \mu$). This method has the advantage of avoiding all problems connected with QCD backgrounds, which are always present with quarks in the final states of WW or ZZ decays. Also, no additional problems are induced if $m_t > m_W$ and the main decay mode of the t-quark is $t \to Wb$. In this case, for each $t\bar{t}$ pair produced, there would be a WW pair and the background to $H \to WW$ would substantially increase. The disadvantage is that there is a reduced range of observable Higgs masses because of the small branching ratio $BR \approx 8 \times 10^{-3}$ for $H \to ZZ \to \ell^+\ell^-\nu\bar{\nu}$. The conclusion at La Thuile was that the Higgs can be discovered through this channel at the LHC for only $m_H \leqslant 0.6$ TeV (with $L = 10^{33}\ cm^{-2}\ s^{-1}$).

The discovery range guaranteed by the leptonic modes would be extended if the hadronic modes could be disentangled from the background. Whilst $H \to 4$ jets is hopeless, it was pointed out at La Thuile that $H \to WW \to \ell\nu + 2$ jets could be extracted from the background if an efficient system of quark tagging could be implemented. In fact, the dominant production mechanism for heavy Higgses is through WW fusion (Fig. 5). The two outgoing quarks that radiated the incoming W pair have a transverse momentum of order m_W and large longitudinal momentum. If both these quarks could be detected at very small angles (typically $\theta \approx 5°$ at the LHC), then the background processes could be sufficiently suppressed. Clearly, this poses formidable problems for the calorimetry, because of the

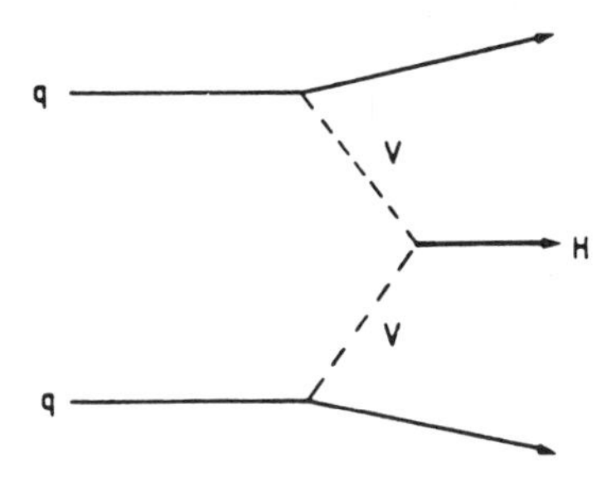

Fig. 5 The Feynman graph for WW or ZZ fusion into an on-shell Higgs boson

enormous level of radiation near the beam pipe. However, a quantitative analysis of the background rejection factor that can be obtained by tagging the quarks was done by Kleiss and Stirling [44]. Their conclusion was that the tagging method is a plausible possibility at the LHC. The discovery range for m_H could then be extended at the LHC to $m_H \simeq 0.8$ TeV with $L = 10^{33}\ cm^{-2}\ s^{-1}$ (at least for $m_t < m_W$ which, however, is by now unlikely).

With higher luminosity [45], from $H \to ZZ \to \ell\ell\ell\ell$, one could reach at the LHC:

$$m_H \leqslant 0.7\ \text{TeV} \qquad (L = 5 \times 10^{34}\ cm^{-2}\ s^{-1}, \ell = \mu),$$

$$m_H \leqslant 0.8\ \text{TeV} \qquad (L = 5 \times 10^{34}\ cm^{-2}\ s^{-1}, \ell = e, \mu),$$

or

$$m_H \leqslant 0.6\ \text{TeV} \qquad (L = 5 \times 10^{33}\ cm^{-2}\ s^{-1}, \ell = e, \mu).$$

A recent reappraisal [46] of the discovery potential of the heavy Higgs at the SSC, with $L = 10^{33}\ cm^{-2}\ s^{-1}$ and no quark tagging, reached the conclusions that a) the Higgs particle can be observed up to $m_H \leqslant 0.8$ TeV (at most) from $H \to ZZ \to \ell\ell\ell\ell$, $\ell = e, \mu$; b) the possibility of reaching larger Higgs masses by using $H \to ZZ \to \ell\ell\nu\bar{\nu}$ depends on detailed studies of the calorimeter hermeticity and response (owing to the large Higgs width and the small signal-to-background ratio for m_H large).

It is thus clear that the operation of the LHC at high luminosity, $L \approx 1\text{–}5 \times 10^{34}\ cm^{-2}\ s^{-1}$, and/or the possibility of realizing the quark tagging technique, would push the LHC discovery potential for the heavy Higgs substantially closer to that of the SSC in the basic configuration.

Finally, we recall that at CLIC the heavy Higgs can be detected in the hadronic mode $H \to 4$ jets up to a mass $m_H \leqslant 0.6\text{–}0.8$ TeV with $L = 10^{33}\ cm^{-2}\ s^{-1}$.

5. BASIC RELATIONS FOR PRECISION TESTS

In the standard electroweak theory, there are a number of basic relations that one wants to verify experimentally with the best possible precision. The same quantity $\sin^2\theta_w$ appears in all these relations.

First, $\sin^2\theta_w$ can be measured from the value of m_W. Starting from the tree-level relations [Eqs. (3.13) and (3.17)], $\sin^2\theta_w = e^2/g^2$ and $G_F/\sqrt{2} = g^2/8m_W^2$, we obtain

$$\sin^2\theta_w = (\pi\alpha/\sqrt{2}\,G_F)/[m_W^2(1 - \Delta r)] = (37.2802\ \text{GeV})^2/[m_W^2(1 - \Delta r)]\,, \tag{5.1}$$

where $\Delta r \neq 0$ owing to the effect of radiative corrections.

Then $\sin^2\theta_w$ is also related to the ratio of the vector boson masses. At tree level [see Eq. (3.22)],

$$\sin^2\theta_w = 1 - m_W^2/\varrho_0 m_Z^2 , \tag{5.2}$$

where $\varrho_0 = 1$ in the Standard Model with only doublets of Higgs bosons. In general, this relation is also modified by radiative corrections:

$$\sin^2\theta_w = 1 - m_W^2/\varrho_{mass} m_Z^2 , \tag{5.3}$$

with $\varrho_{mass} = \varrho_0(1 + \delta\varrho_{mass})$. However, as we shall discuss in detail later, Eq. (5.2) with $\varrho_0 = 1$ is often adopted as a definition of $\sin^2\theta_w$ at all orders. Clearly in this case, $\varrho_{mass} = 1$ by definition.

Finally, $\sin^2\theta_w$ can be obtained for neutral-current couplings. At tree level, the four-fermion interaction from Z exchange is given [see Eq. (3.20)] by

$$M_{if} = [\sqrt{2}\, G_F m_Z^2/D(s)]\, \varrho_0\, (J_3^i - 2\sin^2\theta_w J_{em}^i)(J_3^f - 2\sin^2\theta_w J_{em}^f) , \tag{5.4}$$

where D(s) is the Z propagator, and J_3^f and J_{em}^f are the weak isospin and electromagnetic currents for the fermion f. Excluding pure QED corrections, electroweak radiative corrections modify M_{if} according to

$$M_{if} = [\sqrt{2}\, G_F m_Z^2/D(s)]\, \varrho_{if}\, (J_3^i - 2k_i \sin^2\theta_w J_{em}^i)(J_3^f - 2k_f \sin^2\theta_w J_{em}^f) + \dots , \tag{5.5}$$

where $\varrho_{if} = \varrho_0(1 + \delta\varrho_{if})$; $k_a = 1 + \delta k_a$ (a = i,f) are in general different for different fermions and depend on the scheme adopted (for example, δk_a depend on the definition of $\sin^2\theta_w$). The ellipsis indicates possible additional non-factorizable terms.

6. INPUT PARAMETERS

For LEP 1/SLC physics*), a self-imposing set of input parameters is given by α, α_s, G_F, m_Z, m_f, and m_H. In fact, m_Z has now been (and will be even more) precisely measured. Clearly, $G_F = 1.166389(22) \times 10^{-5}\ \mathrm{GeV}^{-2}$ [48] is conceptually less simple than $\alpha_{weak} = g_2^2/4\pi$ (which would more naturally accompany $\alpha = 1/137.036$ and α_s) or $\sin^2\theta_w$ or m_W, but is known with almost absolute accuracy. Among the quark and lepton masses m_f, the main unknown is the t-quark mass m_t. Our ignorance of m_t is at present a serious limitation for precise tests of the electroweak theory because the radiative corrections depend strongly on m_t. The light quark masses are not very accurately known, but their effective values in all the relevant cases [for instance, in the evaluation of Δr in Eqs. (5.1)] can be obtained from low- and intermediate-energy data. The Higgs mass m_H is almost totally unknown. The sensitivity of the radiative corrections to m_H in the Standard Model with doublet Higgses is small. It fixes the required level of accuracy, as the experimental exploration of the symmetry-breaking sector of the theory is of the utmost importance. The best value of α_s at the Z mass is at present given by $\alpha_s(m_Z) = 0.11 \pm 0.01$ [12]. The QCD corrections to processes involving quarks are typically of order α_s/π, so that the present error on α_s leads to a few per mille relative uncertainty in the corresponding predictions.

*) A thorough and updated review of physics at the Z pole in e^+e^- annihilation can be found in the final report [47] of the Workshop on Z Physics at LEP 1 held at CERN during 1989. In the following, I will make large use of the material collected in the various chapters of this book, which is in three volumes.

7. LARGE CONTRIBUTIONS TO RADIATIVE CORRECTIONS

The most quantitative contributions to the radiative corrections arise from large logarithms [e.g. terms of the form $(\alpha/\pi \ln (m_Z/m_{f_\ell}))^n$, where f_ℓ is a light fermion], and from quadratic terms in m_t, i.e. terms proportional to $G_F m_t^2$.

The sequences of leading and close-to-leading logarithms are fixed [49] by well-known and consolidated techniques (β functions, anomalous dimensions, penguin-like diagrams, etc.). For example, large logarithms dominate the running of α from m_e, the electron mass, up to m_Z, with the result that [48]

$$\alpha(m_Z)/\alpha = 1/(1 - \delta\alpha) . \tag{7.1}$$

At present, the best value of $\delta\alpha$, obtained by extracting the relevant effective light-quark masses from the data on $e^+e^- \rightarrow$ hadrons, is given by [48]

$$\delta\alpha = 0.0601 + (40/9)(\alpha/\pi) \ln (m_Z/91 \text{ GeV}) \pm 0.0009 . \tag{7.2}$$

Large logarithms of the form $[\alpha/\pi \ln (m_Z/\mu)]^n$ also enter, for example, in the relation between $\sin^2 \theta_w$ at the scales m_Z (LEP, SLC) and μ (e.g. the scale of low-energy neutral-current experiments).

The quadratic dependence on m_t [50] (and on other possible widely broken isospin multiplets from new physics) arises because, in spontaneously broken gauge theories, heavy loops do not decouple. On the contrary, in QED or QCD the running of α and α_s at a scale Q is not affected by heavy quarks with mass $M \gg Q$. According to an intuitive decoupling theorem [51], diagrams with heavy virtual particles of mass M can be ignored at $Q \ll M$ provided that the couplings do not grow with M and that the theory with no heavy particles is still renormalizable. In spontaneously broken gauge theories, one important difference is in the longitudinal modes of weak gauge bosons. These modes are generated by the Higgs mechanism, and their couplings grow with masses (as is also the case for the physical Higgs couplings). The upper limit on m_t from radiative corrections arises from this phenomenon. Other subtler sources of non-decoupling are related, for example, to the presence of chiral anomalies [19] (which may not be completely cancelled if heavy particles are removed). Another very important consequence is that precision tests of the electroweak theory are sensitive to new physics even if the new particles are too heavy for their direct production. With the value of m_t being continuously pushed up by experiment, the quantitative importance of the terms of order $G_F m_t^2$ is increasingly large. Both the large lograithms and the $G_F m_t^2$ terms have a simple structure and are to a large extent universal, i.e. common to a wide class of processes. Their study is important for an understanding of the pattern of radiative corrections. One can also derive approximate formulae (e.g. improved Born approximations, see Section 9), which can be useful in cases where a limited precision may be adequate.

8. DETERMINATION OF $\sin^2 \theta_w$ FROM m_Z

Once α, α_s, G_F, m_Z, m_f, and m_H have been chosen as input parameters, $\sin^2 \theta_w$ is a derived quantity. We now consider the calculation of $\sin^2 \theta_w$ beyond the tree approximation. A precise definition of $\sin^2 \theta_w$ must be specified before its value can be computed. An infinite number of definitions are possible, and many have been actually proposed and studied. Physical results are of course independent of the definition adopted for $\sin^2 \theta_w$. Differences in physical results, obtained from different schemes, can occur only through terms of higher order, because the perturbative series is truncated at a given order. Actually $\sin^2 \theta_w$ is used only as a reference quantity. What really matters for precision tests is the prediction of the different measured quantities P in terms of the input parameters; or, in practice, for fixed m_Z, the value $P = P(m_t, m_H)$ of each given physical quantity in the allowed range of m_t and m_H.

At tree level the relation between $\sin^2\theta_w$ and m_Z is obtained from Eqs. (5.1) and (5.2) (with $\Delta r = 0$):

$$\sin^2\theta_w \cos^2\theta_w = (\pi\alpha/\sqrt{2}\,G_F)/(\varrho_0 m_Z^2)\,. \tag{8.1}$$

Beyond the tree level, the quantity Δr [52] is introduced by radiative corrections:

$$\sin^2\theta_w \cos^2\theta_w = (\pi\alpha/\sqrt{2}\,G_F)/[\varrho_0 m_Z^2(1-\Delta r)]\,, \tag{8.2}$$

where $\Delta r \equiv \Delta r(\alpha, \alpha_s, G_F, m_Z, m_f, m_H)$ is of course different for different definitions of $\sin^2\theta_w$. In the following, some particularly interesting definitions of $\sin^2\theta_w$ will be discussed. We now set $\varrho_0 = 1$. Let us first consider the usual definition [53]:

$$\sin^2\theta_w \equiv s_W^2 = 1 - (m_W^2/m_Z^2)\,. \tag{8.3}$$

From now on, the symbol s_W^2 will always refer to this specific definition of $\sin^2\theta_w$. Clearly, given m_Z from SLC or LEP, s_W^2 is directly equivalent to m_W. In this case Δr specifies the relation between m_Z and m_W:

$$(1 - m_W^2/m_Z^2)\, m_W^2 = (\pi\alpha/\sqrt{2}\,G_F)/(1-\Delta r)\,. \tag{8.4}$$

The value of Δr as a function of the input parameters has been studied in great detail (Fig. 6). In particular, the following results are obtained [48, 52, 54]:

$$1/(1-\Delta r) = \alpha(m_Z)/\alpha \cdot 1/[1 + (c_W^2/s_W^2)\delta\varrho] + \text{'small'} = 1/(1-\delta\alpha)\cdot 1/[1 + (c_W^2/s_W^2)\delta\varrho] + \text{'small'}\,, \tag{8.5}$$

where $c_W^2 = 1 - s_W^2$; $\delta\alpha$ was defined in Eqs. (7.1) and (7.2), and $\delta\varrho \to \delta\varrho_t$ for large m_t, with $\delta\varrho_t$ given by [50, 55]

$$\delta\varrho_t = (3G_F m_t^2/8\pi^2\sqrt{2}) + (3G_F m_t^2/8\pi^2\sqrt{2})^2\,(19 - 2\pi^2)/3 + O[(G_F m_t^2)^3]\,. \tag{8.6}$$

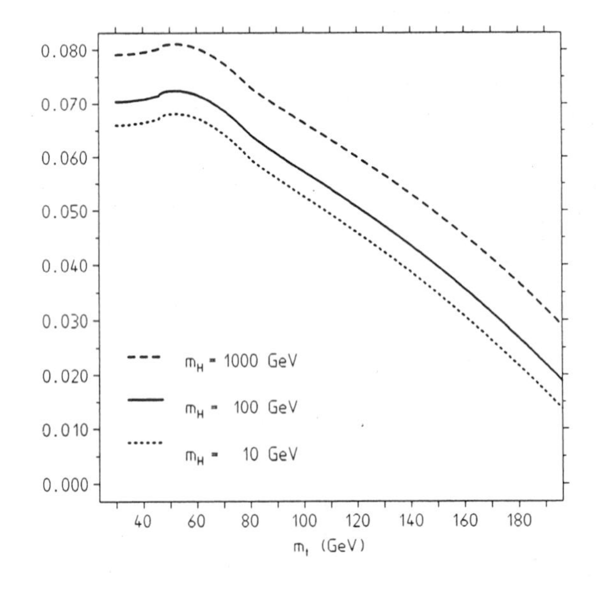

Fig. 6 Δr as a function of the t-quark mass for various m_H (m_Z = 91 GeV) [52].

By 'small' in Eq. (8.5), we mean terms (which at one-loop accuracy are known) without large logs and/or leading powers of $G_F m_t^2$; $\delta\varrho_t$ is the dominant term, for large m_t, of the famous ϱ-parameter first studied in Refs. [50]. The two-loop term was obtained in Ref. [55] and the geometric series resummation was recently advocated in Ref. [54].

Going back to the Z-exchange amplitude near the resonance and the parameters ϱ_{if}, k_i, and k_f defined in Eq. (5.5) with the definition of $\sin^2\theta_w \equiv s_W^2$, we obtain [48]

$$\varrho_{if} = 1 + \delta\varrho + \text{'small'} \tag{8.7}$$

and

$$k_f = 1 + (c_W^2/s_W^2)\,\delta\varrho + \text{'small'} \tag{8.8}$$

(here $f \neq b$; the b-quark will be reconsidered in the following), where the 'small' terms are non-universal (i.e. process-dependent). Note that k_f contains additional 'large' terms with respect to those included in Δr and ϱ_{if}. As these 'large' terms are also universal, this suggests that ks_W^2 could be a better effective $\sin^2\theta_w$ than s_W^2 for physics at the Z pole. Before going into this matter, we will add some comments on $\delta\varrho$. For any weak isospin fermion doublet (e.g. new heavy quarks or leptons), the quantity '$3m_t^2$' in $\delta\varrho_t$ (at one loop) becomes [50, 52]

$$\text{'}3m_t^2\text{'} = N_c\,[m_u^2 + m_d^2 - 2m_u^2 m_d^2/(m_u^2 - m_d^2)\ln m_u^2/m_d^2]\,, \tag{8.9}$$

where N_c is the number of colours. For negligible $m_{d,u}$, '$3m_t^2$' $\to N_c m_{u,d}^2$, whilst for $m_u = m_d + \epsilon$: '$3m_t^2$' $\to N_c\,{}^4\!/_3\epsilon^2$ to leading order in ϵ (the result vanishes for unsplit doublets). Similarly, many more kinds of broken weak isospin multiplets can contribute to '$3m_t^2$' [52, 56] (squarks and sleptons [57], charged Higgses [58], etc. The upper limit on 'm_t' $\lesssim$ 210 GeV (see Section 10) together with the direct lower limit on the t-quark mass, $m_t \gtrsim$ 77 GeV, leave little space for heavy multiplets, except for nearly degenerate ones.

The contribution of the neutral Higgs mass to $\delta\varrho$ is only logarithmic at one loop:

$$\delta\varrho_{\text{Higgs}} \simeq -\,(11G_F m_W^2/24\pi^2\sqrt{2})\,\mathrm{tg}^2\,\theta_w \ln(m_H^2/m_W^2) + \ldots\,. \tag{8.10}$$

There are no m_H^2 terms but only logs, because the 'custodial' SU(2) symmetry is not broken in the Higgs sector [26]. Power terms appear only at two loops [59], but their effect is sizeable only for $m_H \gtrsim$ 1 TeV.

We now consider a different class of definitions of $\sin^2\theta_w$. Assume that we fix $k_f = 1$ [defined by Eq. (5.5)] for one given Z vertex (e.g. $k_f = 1$ in $Z \to e^+e^-$ for on-shell Z). Then it follows from the previous discussion that δk is 'small' for all neutral-current processes (for $f \neq b$) near the Z pole. Let us introduce the notation

$$\sin^2\theta_w\,(\text{from } Z \to e^+e^-) \equiv \bar{s}_W^2\,. \tag{8.11}$$

In this case, Eq. (5.5) becomes

$$M_{fi} = [\sqrt{2}\,G_F m_Z^2/D(s)]\,\varrho_{if}\,(J_3^i - 2\bar{s}_W^2 J_{em}^i)\,(J_3^f - 2\bar{s}_W^2 J_{em}^f) + \text{'small'}\,, \tag{8.12}$$

where ϱ_{if} is given by Eq. (8.7) in terms of $\delta\varrho$, which is specified in Eqs. (8.6) and (8.10). In the present case,

$$\bar{s}_W^2\bar{c}_W^2 = (\pi\alpha/\sqrt{2}\,G_F)/[m_Z^2(1 - \Delta\bar{r})]\,, \tag{8.13}$$

and we find [48]

$$\bar{s}_W^2 \simeq s_W^2 + c_W^2 \delta\varrho + \text{'small'} . \tag{8.14}$$

In other words, apart from 'small' terms, we have

$$\bar{s}_W^2 \simeq 1 - (m_W^2/\varrho m_Z^2) \tag{8.15}$$

with $\varrho \simeq 1 + \delta\varrho$. It is interesting to note that $\bar{s}_W^2$ is less dependent than s_W^2 on m_t^2. In fact [48, 52],

$$\Delta r \simeq \delta\alpha - (c_W^2/s_W^2)\,\delta\varrho , \tag{8.16}$$

$$\Delta\bar{r} \simeq \delta\alpha - \delta\varrho , \tag{8.17}$$

so that the amplifying factor c_W^2/s_W^2 in front of $\delta\varrho$ is missing in $\Delta\bar{r}$. Note that given m_Z, $\bar{s}_W^2$ is known with an accuracy of about ± 0.002 when m_t varies between 60 and 210 GeV (see Fig. 7). There is a class of definitions of $\sin^2\theta_w$ in the literature that correspond to $\bar{s}_W^2$ if 'small' terms are neglected: $\bar{s}_W^2$ of Hollik [60]; $s^{*2}(m_Z^2)$ of Lynn and Kennedy [61]; $(\sin^2\theta_w)_{\overline{MS}}(m_Z^2)$ [62, 63], etc. Note that because of Eqs. (7.1) and (8.17) one can also write:

$$\bar{s}_W^2 \bar{c}_W^2 = [\pi\alpha(m_Z)/\sqrt{2}\,G_F]/\varrho m_Z^2 . \tag{8.18}$$

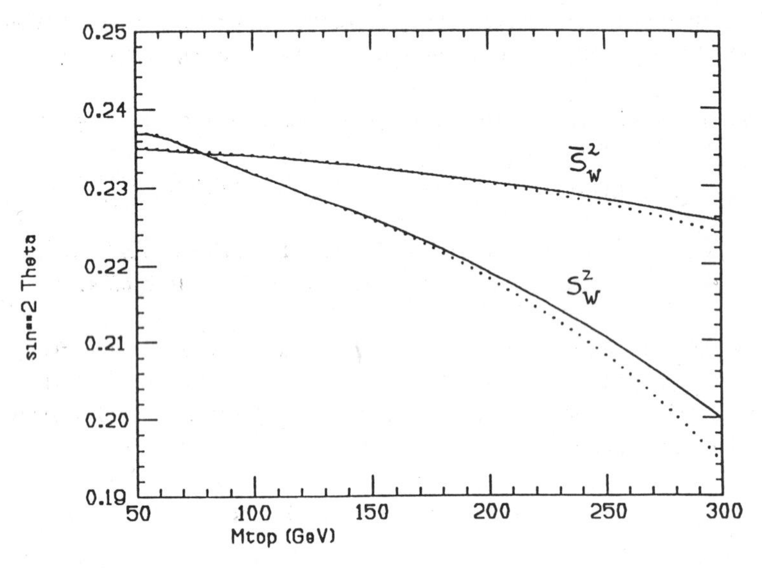

Fig. 7 The behaviour of s_W^2 and $\bar{s}_W^2$ [defined by Eqs. (8.3) and (8.11), respectively] as functions of m_t for $m_Z = 91$ GeV, $m_H = 100$ GeV. The solid [dashed] lines are computed to $O(\alpha^2)$ [$O(\alpha)$]. From Ref. [48].

9. IMPROVED BORN APPROXIMATION

For precision tests of the electroweak theory, the complete one-loop radiative corrections (plus higer-order/exponentiated purely photonic corrections) are mandatory and are indeed available for the processes of practical relevance. However, in many cases a less accurate estimate can be enough. For this purpose, it is useful to know [48] that formulae as simple as those of the Born approximation can be written in a way that takes all 'large' corrections into account. For $e^+e^- \to f\bar{f}$, with $f \neq e,b$, the amplitude for γ and Z exchange near the resonance can be written in the form

$$M_{f\bar{f}} = Q_e Q_f\,[4\pi\alpha(m_Z)/s]\,J^e_{em}J^f_{em} + \sqrt{2}\,G_F\varrho m_Z^2/(s - m_Z^2 + is\,\Gamma_Z/m_Z)\,\bar{J}^e\bar{J}^f\,, \tag{9.1}$$

where $\varrho = 1 + \delta\varrho$, $\delta\varrho$ being given by Eqs. (8.6) and (8.10), and

$$J^f_{em} = \gamma_\mu\,, \qquad \bar{J}^f = \gamma_\mu\,[I_3^f(1 - \gamma_5) - 2Q_f\bar{s}_W^2]\,, \tag{9.2}$$

with Q_f and I_3^f being respectively the electric charge and the third component of the weak isospin (e.g. for $f = \mu^-$, $Q = -1$, $I_3 = -1/2$). The important features are the replacement of α-fixed with α-running, the inclusion of the s-dependence for the total width Γ_Z in the resonant denominator, the presence of the factor of ϱ multiplying G_F, and the use in the Z couplings of the effective $\sin^2\theta_w = \bar{s}_W^2$, introduced in the previous section. Clearly for $f = e$, i.e. for Bhabha scattering, the t-channel exchange is also to be included. The improved Born approximations include the real parts of self-energies (sometimes [61] called 'oblique' corrections). All large logs and all $G_F m_t^2$ terms are included. What are left out are 'small' corrections from imaginary parts of self-energies, vertices, and boxes.

For $f = b$ there are additional large terms from the vertex corrections [60, 64] of the type shown in Fig. 8. The longitudinal W modes are, also in this case, responsible for the presence of quadratic mass terms. We can simply modify the improved Born approximation in order to include these terms as well. The recipe [48, 65] is as follows: in Eqs. (9.1) and (9.2), replace ϱ by

$$\sqrt{\varrho\varrho_b}\,, \qquad \varrho_b \equiv \varrho\,(1 - 4/3\,\delta\varrho)\,, \tag{9.3}$$

and $\bar{s}_W^2$ by

$$\bar{s}_W^2\,(1 + 2/3\,\delta\varrho) \equiv \bar{s}_W^2\,k_b\,. \tag{9.4}$$

Note that whilst many sorts of heavy particles can contribute do $\delta\varrho$, only the t-quark (or a new t′) contributes to $\delta\varrho_b$, because the W nearly always turns a b-quark into a t (or a t′) quark.

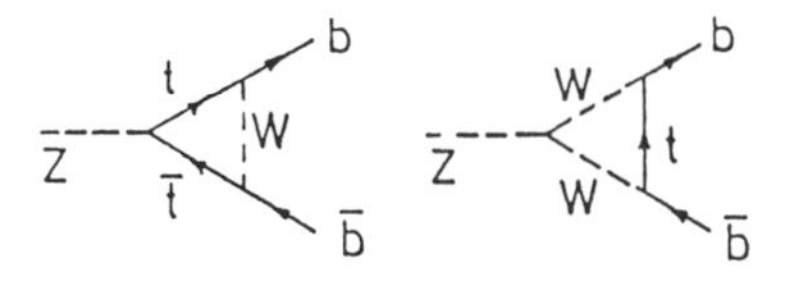

Fig. 8 Vertex corrections that contain large $G_F m_t^2$ terms.

As a final example, the improved Born approximation for the inclusive widths $\Gamma[Z \to f\bar{f}(\gamma,g)]$ (including photons and gluons in the final state), is given by

$$\Gamma[Z \to f\bar{f}(\gamma,g)] = N_c\,(G_F\varrho m_Z^3/24\pi\sqrt{2})\,[1 - (1 - 4|Q_F|\bar{s}_W^2)^2]\,, \tag{9.5}$$

where

$$N_c = \begin{cases} 1\,[1 + (3\alpha/4\pi)\,Q_f^2] & \text{(leptons)}\,, \\ 3\,[1 + (3\alpha/4\pi)\,Q_f^2]\,[1 + \alpha_s(m_Z)/\pi] & \text{(quarks)}\,. \end{cases} \tag{9.6}$$

In the same approximation, the total width is given simply by

$$\Gamma_Z = \Sigma_f \Gamma[Z \rightarrow f\bar{f}(\gamma,g)] + \text{rare decays}\,, \tag{9.7}$$

where the rare decays [66] (e.g. $Z \rightarrow H\mu^+\mu^-$ for a light Higgs) are numerically unimportant in the Standard Model (for $m_H > 10$ GeV) [67]. For $\Gamma(Z \rightarrow b\bar{b})$, replace [48, 65] ϱ by ϱ_b and $\bar{s}_W^2$ by $\bar{s}_W^2 k_b$, where ϱ_b and k_b are given in Eqs. (9.3) and (9.4). Note that for $f \neq b$, Eq. (9.5) can also be written in the form [by using Eq. (8.18):

$$\Gamma[Z \rightarrow f\bar{f}(\gamma,g)] = N_c\,[\alpha(m_Z)m_Z/(48\,\bar{s}_W^2\bar{c}_W^2)][1 - (1 - 4|Q_F|\bar{s}_W^2)^2]\,, \tag{9.8}$$

10. STATUS OF PRECISION TESTS OF THE ELECTROWEAK THEORY

After a period of stagnation, a big step forward in precision tests of the electroweak theory has been accomplished with the data presented this summer. Usually, one starts by presenting a table of results on s_W^2 extracted from different processes, with radiative corrections computed assuming a given value for m_t and m_H. Over the years, the assumed value of m_t started from 15 GeV, was then set at 30 GeV, then again increased to 45 GeV, and more recently to 60 GeV. This was a sensible procedure, because, for m_t not too large, the radiative corrections are not dominated by m_t. A moderate change of m_t in a range around and below 60 GeV does not affect the radiative corrections very much. This is, for example, evident from Fig. 6, where the value of Δr is plotted as a function of m_t. We now know from new data from CDF [68], UA1 [69], and UA2 [70] that m_t is probably > 70–80 GeV. For large m_t, the quadratic terms of order $G_F m_t^2$ become a driving term for the radiative corrections, which, as a result, depend strongly on m_t. It seems to me that a better strategy is to discuss the experimental results and the associated radiative corrections needed to extract $\sin^2\theta_w$ as functions of m_t, with m_t varying in the whole range of allowed values. The requirement that the electroweak theory be consistent with the data, restricts the allowed range of m_t significantly. We shall see that not only is an upper bound obtained on m_t (which is an old result), but also a lower bound can now be obtained. This limit [see Eq. (10.7)] is interesting in itself, although the limits from production experiments are more directly reliable. (Clearly, a lower limit on m_t from radiative corrections can be avoided if heavy new particles contribute to vacuum polarization diagrams, whilst the upper limit is *a fortiori* valid.)

The most important results are the new measurements of the weak gauge-boson masses. The old and new data on m_W and m_Z are summarized in Table 1. The precise determination of m_Z by MARK II [71] (and CDF [72]) are among the highlights of this summer's results. The CDF result is quite impressive if we consider the difficulty of such a measurement at hadron colliders. The MARK II determination of m_Z is a great achievement, not only in itself but also as a promise for the future of e^+e^- linear colliders.

By combining the CDF and MARK II results, we obtain

$$m_Z = 91.12 \pm 0.16\,\text{GeV}\,. \tag{10.1}$$

There is no theoretical error associated with the measurement of m_Z in e^+e^- annihilation, and also the derivation of s_W^2 (or $\bar{s}_W^2$) from m_Z is very clean, i.e. the theory of Δr (or $\Delta\bar{r}$) has no ambiguities in the minimal Standard Model, other than our ignorance of m_t and m_H. However, the value of s_W^2 obtained from m_Z depends strongly on m_t through Δr. This is seen in Fig. 9 [73], where the value of s_W^2 corresponding to a given value of m_Z is plotted versus m_t. As discussed in Section 8, the value of $\bar{s}_W^2$ extracted from m_Z is less sensitive to m_t. We shall come back later to the determination of $\bar{s}_W^2$.

The measurement of m_W/m_Z at $p\bar{p}$ colliders is also quite clean, although not as much as that of m_Z. The error on the energy scale drops when taking the ratio. The theoretical error is confined to

Table 1

The numbers in italics refer to errors on the energy scale.
Most of these errors cancel when taking the ratio m_W/m_Z.

	m_W (GeV)	m_Z (GeV)
Old data		
UA1	82.7 ± 1.0 ± *2.7*	93.1 ± 1.0 ± *3.1*
UA2	80.2 ± 0.6 ± *1.4*	91.5 ± 1.2 ± *1.7*
CDF	80.0 ± 3.3 ± *2.4*	
New data		
CDF [71]	80.0 ± 0.2 ± 0.5 ± *0.3*	90.9 ± 0.3 ± *0.2*
MARK II [72]		91.17 ± 0.18
UA2 [74]	80.0 ± 0.4 ± 0.4 ± *1.2*	90.2 ± 0.6 ± *1.4*

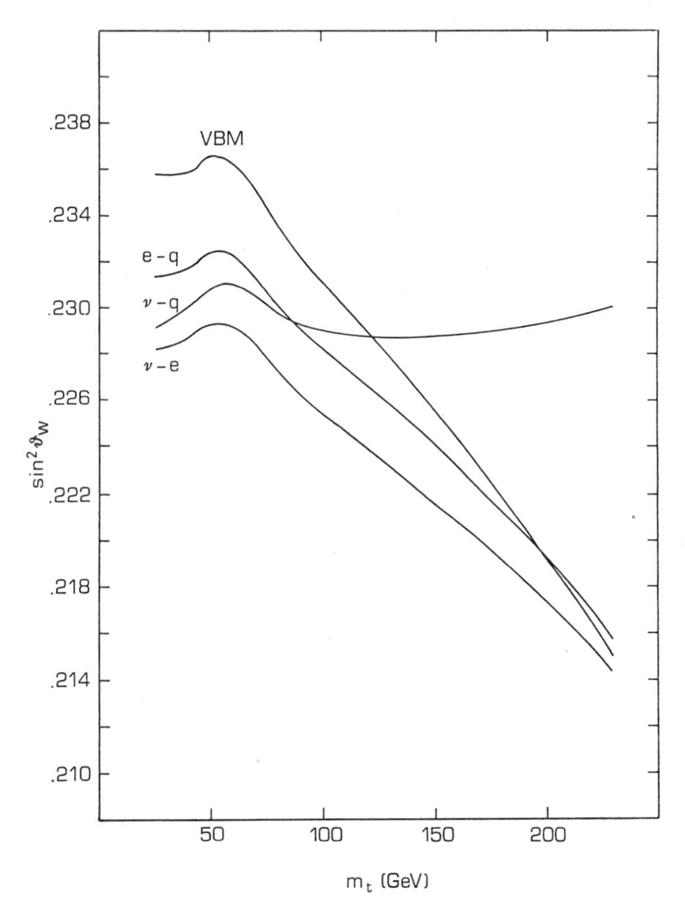

Fig. 9 Dependence on m_t of s_W^2 [defined in Eq. (8.3)] extracted from $m_{W,Z}$ (VBM) and from the neutral-current e–q, ν–q, or ν–e sectors (Ref. [74]).

some small uncertainties in the W transverse momentum distribution (e.g. structure functions, Λ_{QCD}, and related problems), which somewhat affect the determination of m_W from the Jacobian peak in the electron distribution. The values obtained by CDF [72] and UA2 [74] are in excellent mutual agreement:

$$\begin{aligned} m_W/m_Z &= 0.887 \pm 0.009 \quad \text{(UA2)}, \\ &= 0.880 \pm 0.007 \quad \text{(CDF)}. \end{aligned} \tag{10.2}$$

In this case, there are clearly no problems in obtaining s_W^2: the ratio m_W/m_Z determines s_W^2 directly, by definition. The above results correspond to

$$\begin{aligned} (s_W^2)_{m_W/m_Z} &= 0.213 \pm 0.015 \quad \text{(UA2)}, \\ &= 0.225 \pm 0.012 \quad \text{(CDF)}. \end{aligned} \tag{10.3}$$

The resulting combined value $s_W^2 = 0.220 \pm 0.009$ is reported in Fig. 10. Clearly, the relatively low value of s_W^2 obtained from m_W/m_Z when compared with m_Z, pushes the value of m_t towards large

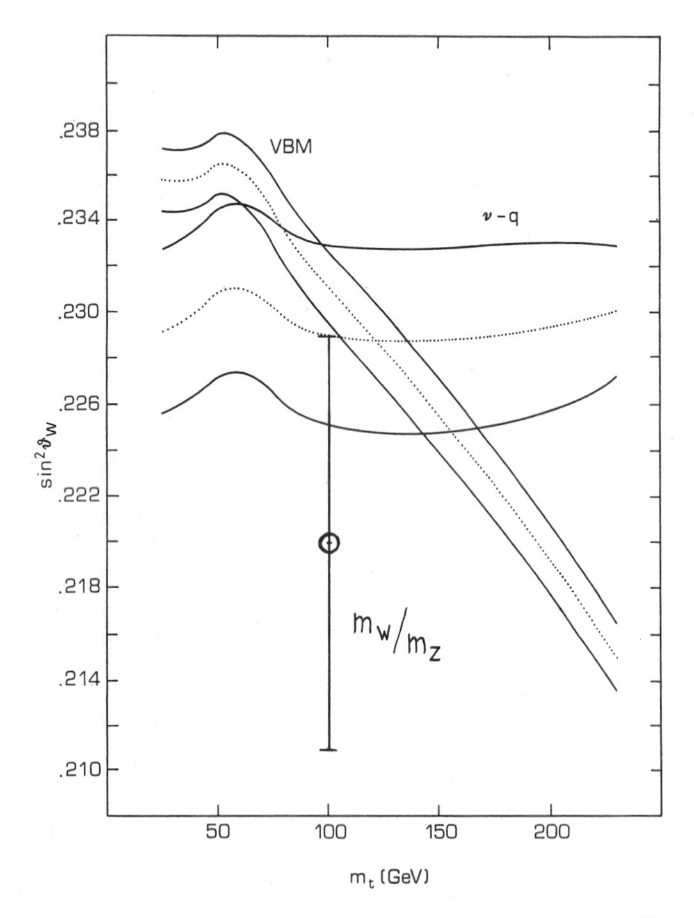

Fig. 10 s_W^2 [defined in Eq. (8.3)] obtained from m_Z, m_W/m_Z, and $R_\nu = \sigma_{\nu N}^{NC}/\sigma_{\nu N}^{CC}$ as a function of m_t. The error shown for R_ν is only a part of the total; it corresponds to a fixed charm mass m_c = 1.45 GeV and m_H = 100 GeV (courtesy of G.L. Fogli).

values (within the limited accuracy of the m_W/m_Z measurement). It is important to note that an upper limit on m_t around 260 GeV is already implied by the vector-boson masses alone.

We now consider neutral-current data. The most precise determination of s_W^2 is obtained from the ratio R_ν of neutral- to charged-current deep-inelastic neutrino scattering. It turns out that R_ν imposes an important constraint on the electroweak parameters when compared with m_W or m_Z because s_W^2, derived from R_ν, is nearly independent of m_t in the range of interest [75]. This near independence is apparent in Figs. 9 and 10. It arises because of a largely accidental cancellation of the m_t dependence introduced by defining s_W^2 from m_W/m_Z, and because of the m_t dependence induced by the ϱ parameter appearing in the neutral- to charged-current ratio. The most precise measurements of s_W^2 are due to the CHARM [76] and CDHS [77] Collaborations at CERN:

$$\begin{aligned} (s_W^2)_{\nu N} &= 0.236 \pm 0.007 \quad \text{(CHARM)}\,, \\ &= 0.228 \pm 0.007 \quad \text{(CDHS)}\,. \end{aligned} \tag{10.4}$$

Note that both $(s_W^2)_{\nu N}$ and $(s_W^2)_{m_W/m_Z}$ are independent [or nearly so for $(s_W^2)_{\nu N}$] of m_t. As a consequence, the direct comparison of Eq. (10.4) with Eq. (10.3) provides a quite remarkable and successful consistency check, independent of m_t, for the electroweak theory. Thus R_ν is a good measure of m_W/m_Z (as was recently stressed again in Ref. [78]). Once R_ν is fixed, if m_Z is decreased, then m_W has also to be decreased, which means that Δr is decreased, and finally that m_t is increased. The fact that the measured value of m_Z is below the previous world average of about 92 GeV, pushes m_t toward the large values.

The values of s_W^2 obtained from m_Z, m_W/m_Z, and R_ν as functions of m_t with the corresponding experimental errors are displayed in Fig. 10. Clearly, a lower and an upper limit on m_t are implied. For a quantitative determination of the allowed range of m_t and s_W^2, it must be stressed that the measurement of R_ν in neutrino deep-inelastic scattering and the extraction of s_W^2 from R_ν are obviously affected by much larger theoretical uncertainties than those connected with m_Z and m_W/m_Z. In νN scattering, we must cope with uncertainties connected with the validity of the parton model (i.e. the importance of higher twist and other pre-asymptotic corrections), with our relative ignorance of sea densities, with the effects of the onset of the charm threshold and the validity of the slow rescaling procedure (the related uncertainties are roughly parametrized in terms of an effective 'charmed mass' parameter m_c), with the modelling of the effect of experimental cuts, and so on. The errors stated by the experiments and quoted in Eq. (10.4) also include an estimate of theoretical errors. The error of 0.007 arises from adding, in quadrature, the experimental and theoretical errors according to $\delta s_W^2 = 0.012(m_c - 1.5) \pm 0.005$ (exp.) $\pm\ 0.003$ (theor.), with $m_c \approx 1.2$–1.8 GeV. It should, however, be kept in mind that the determination of s_W^2 from neutrino–nucleon scattering is quite involved, and that the evaluation of the related theoretical error is to some extent debatable. With the new data on m_W/m_Z and especially on Z physics from SLC/LEP 1, R_ν will increasingly become the weak link in the chain.

There are new important results on other neutral-current processes. First, I want to recall the recent measurement of parity violation in atomic caesium that was made at Boulder [79] (by a team of only three physicists!). They quote the remarkably accurate value

$$(s_W^2)_{Cs} = 0.219 \pm 0.019 \tag{10.5}$$

obtained for m_t light ($m_t \lesssim 60$ GeV). For larger m_t, the value of $(s_W^2)_{Cs}$ decreases rapidly, as is seen from the curve labelled 'e–q' in Fig. 9. The value of Eq. (10.5) is in agreement with previous, less precise, results on Cs obtained by the Paris group [80]. A decade of continuous progress on atomic 'table-top' experiments has allowed the validity of the electroweak theory at very low energies to be

Table 2
Results on ν–e scattering

	ν_μe	$\bar{\nu}_\mu$e	$\sin^2\theta_W$ (no rad. corr.)
CHARM 1	83	112	0.211 ± 0.035 ± 0.011
E734	160	97	0.195 ± 0.018 ± 0.013
CHARM 2	762	1017	0.233 ± 0.012 ± 0.008

established with a precision not far from that of experiments, at higher energies, performed with large detectors at big accelerators.

New data on $\overset{(-)}{\nu}_\mu$e scattering have been presented this summer, for the first time, by the CHARM 2 Collaboration [81]. As shown in Table 2 (where the numbers of observed events are also displayed), the CHARM 2 values are considerably more precise than in previous experiments. Also, the central value of s_W^2 is now in better agreement with the results from other processes. The CHARM 2 result, including radiative corrections with $m_t = 100$ GeV (the behaviour at larger values of m_t is shown in Fig. 9), is given by

$$(s_W^2)_{\overset{(-)}{\nu}_\mu e} = 0.232 \pm 0.014\,. \tag{10.6}$$

The experiment is still running, and the present stage of the analysis is far from achieving the final aim of reducing the error on s_W^2 to $\delta s_W^2 = \pm(6\text{–}7) \times 10^{-3}$.

Going back to Fig. 9 showing the dependence, on m_t, of s_W^2 obtained from different processes, it is seen that on the one hand ν-q is similar to m_W/m_Z, leading to a value of s_W^2 nearly independent of m_t), whilst on the other hand m_Z, e–q, and $\overset{(-)}{\nu}_\mu$e lead to s_W^2 with a common trend in m_t. Figure 11

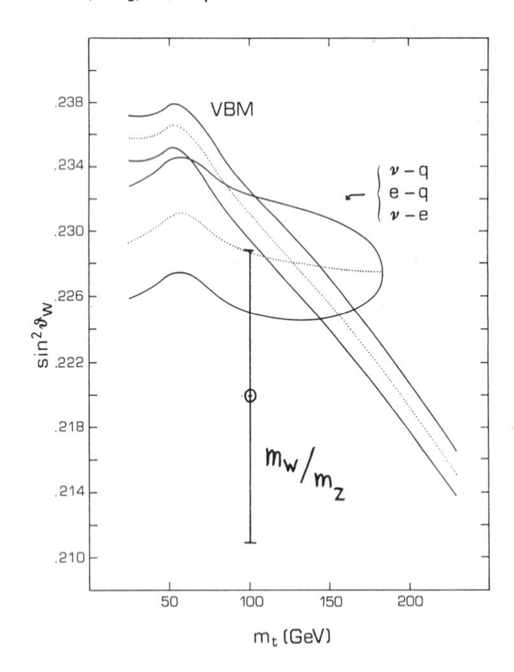

Fig. 11 Same as Fig. 10, but with the addition of the data on the e–q and ν–e sectors (courtesy of G.L. Fogli).

differs from Fig. 10 because the constraints from e–q and ν–e have been added to those from ν–q in order to complete the information from neutral-current processes.

Taking all the information together, we can conclude that the ranges of m_t and s_W^2 indicated by the present data on the electroweak theory are

$$m_t = 130 \pm 50 \text{ GeV}, \quad (10.7)$$

$$s_W^2 \equiv 1 - (m_W^2/m_Z^2) = 0.227 \pm 0.006 . \quad (10.8)$$

These values are in agreement with the results of a number of complete detailed analyses of the data [73, 82–84] (see, for example, Fig. 12 taken from Ref. [83]), although the errors quoted here are in some cases more conservative. In particular, the upper (lower) limit on the top mass from Eq. (10.7) is about $m_t \leqslant 210$ GeV ($m_t > 50$ GeV) at 90% CL. Note that the uncertainty in s_W^2 is actually much smaller now that it was before this summer. In fact, the present error is for all allowed values of m_t and m_H (for m_H we take the range 10 GeV to 1 TeV), whilst in the past the quoted errors referred to assumed fixed values of these masses.

A different question is, What is, at present, the best estimate of $\bar{s}_W^2$? i.e. the effective $\sin^2\theta_w$ for the on-shell Z couplings introduced in Section 8, Eqs. (8.11) to (8.15). We recall that the value of $\bar{s}_W^2$ obtained from m_Z is less dependent on m_t and m_H than is s_W^2. If we take for m_Z the value and the error given in Eq. (10.1), for m_t the range in Eq. (10.7), and for m_H the mass range 10 GeV < m_H < 1 TeV, we obtain [63, 78]

$$\begin{aligned} \bar{s}_W^2 &= 0.233 \pm 0.002 \pm 0.0074\,\delta m_Z\,(\text{GeV}) \\ &= 0.233 \pm 0.002 \pm 0.0012 . \end{aligned} \quad (10.9)$$

We see that indeed the Z couplings can be predicted with more accuracy from $\bar{s}_W^2$ than we would naïvely obtain by going through s_W^2. Note that $\bar{s}_W^2 \approx (\sin^2\theta_w)_{\overline{MS}}$ (i.e. apart from 'small' terms), is also the relevant quantity to compare with GUT predictions. The value in Eq. (10.9) is in disagreement with naïve SU(5) but in agreement with the minimal supersymmetric extension of SU(5) [85, 86].

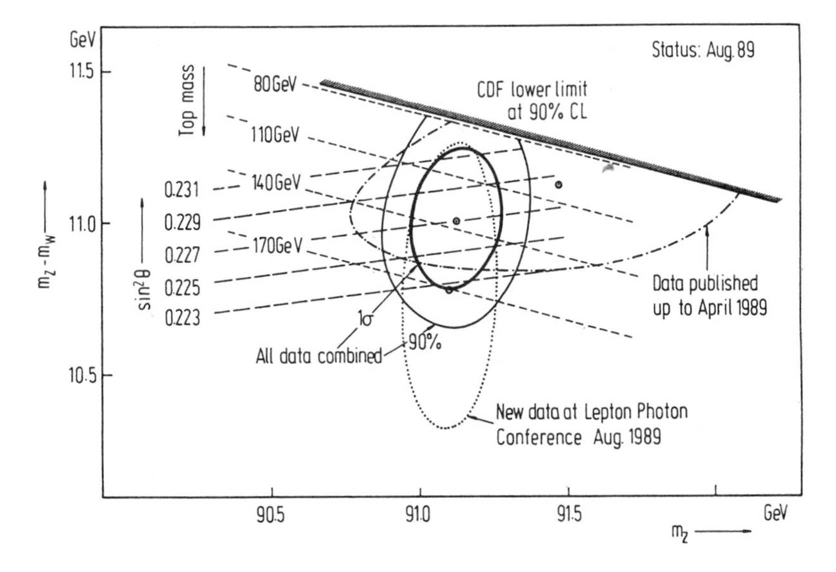

Fig. 12 Summary of the present status of electroweak tests. Given m_Z, $m_Z - m_W = m_Z(1 - m_W/m_Z)$ is equivalent to m_W/m_Z or s_W^2 (courtesy of D. Haidt).

11. THE Z LINE-SHAPE

The forthcoming experiments at the SLC, and especially at LEP, will certainly mark a big step forward in the domain of precision tests of the standard theory. The strategy at LEP 1/SLC is first to study the line shape and to measure m_Z, Γ, and the partial widths, to perform the counting of neutrinos, and to look for possible surprises in the final state. This is the part of the programme that has already started at the SLC and has led to the important results presented this summer. At LEP 1 the same measurements will presumably by completed very soon with considerably more precision*). The exploratory phase will be followed by a long period of running at the Z peak. The goals are to measure, as precisely as possible, a number of asymmetries that allow the most stringent tests of the theory, and to search for interesting rare Z decays, the most important being those involving the Higgs boson (i.e. $Z \to H\mu^+\mu^-$ and $Z \to H\gamma$). A thorough and updated review of Z physics in e^+e^- annihilation can be found in Ref. [47]. In the following, I will discuss very briefly some of the highlights of the physics programme outlined above.

A considerable amount of work has deservedly been devoted to the theoretical study of the Z line-shape [67]. The experimental accuracy on m_Z, which is planned at LEP, is $\delta m_Z = \pm 50$ MeV. The error on m_Z can eventually go down to $\delta m_Z = \pm 20$ MeV if the transverse polarization of the beams measured by adequate polarimeters will be used to calibrate the energy by spin-resonance methods [87]. Similarly, a measurement of the total width to an accuracy $\delta\Gamma = \pm 30$ MeV is now considered feasible. The prediction of the Z line-shape in the Standard Model to such an accuracy has posed a formidable challenge to theory, which has been successfully met. For the inclusive process $e^+e^- \to f\bar{f}X$, with $f \neq e$ (for simplicity, we leave Bhabha scattering aside) and X including γ's and gluons, the physical cross-section can be written in the form of a convolution [67]:

$$\sigma(s) = \int_{z_0}^{1} dz\, \hat{\sigma}(zs) G(z,s) , \tag{11.1}$$

where $\hat{\sigma}$ is the reduced cross-section, and G(z,s) is the radiator function that describes the effect of initial-state radiation; $\hat{\sigma}$ includes the purely weak corrections, the effect of final-state radiation (of both γ's and gluons), and also non-factorizable terms (initial- and final-state radiation interferences, boxes, etc.) which, being small, can be treated in lowest order and effectively absorbed in a modified $\hat{\sigma}$. The radiator G(z,s) has an expansion of the form [67]

$$G(z,s) = \delta(1 - z) + \alpha/\pi\,(a_{11}L + a_{10}) + (\alpha/\pi)^2\,(a_{22}L^2 + a_{11}L + a_{20}) + \ldots + (\alpha/\pi)^n \sum_{i=0}^{n} a_{ni}L^i , \tag{11.2}$$

where $L = \ln s/m_e^2 \simeq 24.2$ for $\sqrt{s} \simeq m_Z$. All first- and second-order terms are known exactly. The sequence of leading and next-to-leading logs can be exponentiated (closely following [88] the formalism of structure functions in QCD). For $m_Z \approx 91$ GeV, the convolution displaces the peak by +110 MeV, and reduces it by a factor of about 0.74. The exponentiation is important in that it amounts to a shift of about 14 MeV in the peak position.

A model-independent analysis [89] of the reduced cross-section $\hat{\sigma}$ leads to the following general expression (here $m \equiv m_Z$):

$$\hat{\sigma}(s) = [12\pi\Gamma_e\Gamma_f/|D(s)|^2]\,[s/m^2 + R_f\,(s - m^2)/m^2) + \Gamma/m\, I_f + \ldots] + [4\pi\alpha^2(m^2)Q_f^2N_c/3s] , \tag{11.3}$$

*) This has indeed been the case. At the moment of this write-up, after three weeks of running at LEP, the combined values from the four LEP experiments are $m_Z = 91.10 \pm 0.033 \pm 0.045$ GeV $= 91.10 \pm 0.06$ GeV, $\Gamma_Z = 2.584 \pm 0.075$ GeV, and $N_\nu = 3.17 \pm 0.21$.

with N_c given by Eq. (9.6) and

$$D(s) = s - m^2 + im\Gamma\,[s/m^2 + \epsilon\,(s - m^2)/m^2] + \dots . \tag{11.4}$$

This form of the resonant term was obtained by starting from a general renormalizable field theory (the Standard Model being a particular case). Near the resonance, $(s - m^2)/m^2 \approx \Gamma/m$ is of order α (or α_W). As only a perturbative calculation of $\hat{\sigma}$ is possible in α or α_W, at the same level of accuracy we can expand vertices, propagators, etc., in $(s - m^2)/m^2$ near the resonance. For example, for the inverse propagator, we can write the expansion

$$\begin{aligned} D(s) &= s - m^2 + \Pi(s) \\ &= s - m^2 + \mathrm{Re}\,\Pi(m^2) + (s - m^2)\,\mathrm{Re}\,\Pi'(m^2) + i\,\mathrm{Im}\,\Pi(m^2) + i(s - m^2)\,\mathrm{Im}\,\Pi'(m^2) + \dots \\ &= [1 + \mathrm{Re}\,\Pi'(m^2)]\{s - m^2 + im\Gamma\,[s/m^2 + \epsilon\,(s - m^2)/m^2 + \dots]\}\,, \end{aligned} \tag{11.5}$$

with

$$m\Gamma = \mathrm{Im}\,\Pi(m^2)/[1 + \mathrm{Re}\,\Pi'(m^2)]\,, \tag{11.6}$$

and $\mathrm{Re}\,\Pi(m^2) = 0$ because of the specific definition of $m \equiv m_Z$ that is adopted. The overall factor $1 + \mathrm{Re}\,\Pi'(m^2)$ is reabsorbed in the numerator. The parameter ϵ measures the deviation from the scaling behaviour of the s-dependent width; ϵ is noticeably different from zero only if in the final state of Z decays there are important channels with massive particles. For example, for $Z \to A\bar{A}$, $\epsilon_A \approx (4m_A^2/m^2)B(Z \to A\bar{A})$. The following identity is valid:

$$\begin{aligned} D(s) &= s - m^2 + im\Gamma\,[s/m^2 + \epsilon\,(s - m^2)/m^2 + \dots] \\ &= [1 + i\gamma(1 + \epsilon)](s - \bar{m}^2 + i\bar{m}\bar{\Gamma})\,, \end{aligned} \tag{11.7}$$

with $\gamma = \Gamma/m$ and

$$\bar{m} = m[1 - (\gamma^2/2)(1 + \epsilon) + \dots]\,, \tag{11.8}$$

$$\bar{\Gamma} = \Gamma[1 - (\gamma^2/2)(1 + 3\epsilon) + \dots]\,. \tag{11.9}$$

As the factor $1 + i\gamma(1 + \epsilon) \simeq \exp i\gamma(1 + \epsilon)$ cannot be observed from the absolute square of D(s), we see that, on the one hand, at $\epsilon = 0$ the replacement of the constant Γ by the s-dependent width $s\Gamma/m^2$ leads to a variation of the apparent mass by

$$\delta m = -\tfrac{1}{2}\,\gamma^2 m \simeq -34\ \mathrm{MeV}\,, \tag{11.10}$$

which is clearly an important effect [90, 91]. On the other hand, the additional effect from ϵ,

$$\delta m_\epsilon = -\tfrac{1}{2}\,\gamma^2 \epsilon m \simeq -34\epsilon\ \mathrm{MeV}\,, \tag{11.11}$$

is certainly small because ϵ cannot exceed 10% or so at most. Note that the effect of ϵ cannot be disentangled from m in the fit, so that it leads to an ambiguity in the determination of m. In conclusion, ϵ can be safely neglected at the price of allowing an error of a few MeV on m and Γ.

Of the two parameters R_f and I_f which appear in Eq. (11.3) for $\hat{\sigma}$ (in addition to the inclusive partial widths Γ_e and Γ_f), I_f, like ϵ, depends only on the spectrum of particles below the Z. In fact, I_f

is determined by absorptive parts of vertices and boxes. Thus large deviations from the Standard Model value cannot occur for I_f. In the Standard Model, I_f is very small [for $\mu^+\mu^-$, $I_\mu \simeq -(1\text{–}2) \times 10^{-2}$], and as it is multiplied by Γ/m, it can be neglected. The main contribution to R_f is already present at the Born level and arises from γ–Z interference. Higher-order corrections are relatively important, especially for muons. New physics, for example a heavy Z′, can modify R_f because the non-resonating background is changed. In the Standard Model, R_f is small [for muons, $R_\mu \simeq (4.5\text{–}6) \times 10^{-2}$, for hadrons $R_h \simeq (0.75\text{–}1) \times 10^{-1}$]. Determining R_f from a fit is difficult. In first approximation, it can be fixed at its Standard Model value. Realistic deviations from the Standard Model would not appreciably affect the determination of m and Γ.

In conclusion, a model-independent analysis of the reduced cross-section proves that a modified Breit–Wigner with s-dependent width plus photon exchange and interference is a perfectly adequate basis for the experimental study of the line shape. Whilst in the Minimal Standard Model all the parameters Γ, Γ_e, Γ_f, R_f, and I_f can be obtained from m_Z, given m_t and m_H, the general expression of $\hat{\sigma}$ will, in principle, allow a model-independent measurement of the Z parameters. Starting from Eq. (11.3) for $\hat{\sigma}$, approximate analytic solutions of the convolution integral can be found [89, 92–94]. The resulting compact analytic expressions are sufficiently precise for most applications.

An indicative set of theoretical predictions on the line shape for $m_Z \simeq 91$ GeV is reported in Tables 3 to 5 taken from Ref. [67]. The agreement on the line shape, obtained by different calculations using slightly different procedures or schemes, is better than 0.1% for $\mu^+\mu^-$, corresponding to an error of a few MeV on m_Z. For hadrons, the uncertainty on Γ introduced by $\alpha_s(m_Z) = 0.11 \pm 0.01$ [12] is $\delta\Gamma = \pm 6$ MeV. We see from Tables 4 and 5 that the visible cross-section at the peak for $N_\nu = 3$ is expected to be about 35 nb (for reasonable cuts on Bhabha events near the forward direction). This implies that the number of visible Z is expected to be

$$N_Z = \sigma^{\text{peak}}_{\text{visible}} \int L\,dt = 3.5 \times 10^6 \left(\int L\,dt/10^{38}\ \text{cm}^{-2}\right). \tag{11.12}$$

At the project luminosity of LEP, $L = 1.7 \times 10^{31}$ cm^{-2} s^{-1}, in a 'year' of 10^7 s, with an overall efficiency $\epsilon \sim 1/2$, we expect to have for each experiment

$$N_Z/\text{year} \simeq 3.5 \times 10^6 \times 1.7 \times 0.5 \simeq 3 \times 10^6\,. \tag{11.13}$$

Table 3

Partial and total Z widths for $m_Z = 91$ GeV and $\alpha_s = 0.12$ obtained from Ref. [67].

m_t (GeV)	m_H (GeV)	$\sin^2\theta_w$	Γ_Z (MeV)	$\Gamma_{Z\to\nu\bar{\nu}}$ (MeV)	$\Gamma_{Z\to e^+e^-}$ (MeV)	$\Gamma_{Z\to u\bar{u}}$ (MeV)	$\Gamma_{Z\to d\bar{d}}$ (MeV)	$\Gamma_{Z\to b\bar{b}}$ (MeV)
60	100	0.2366	2460	165.0	82.7	292.7	378.3	376.0
90	10	0.2312	2463	164.9	82.7	293.4	379.0	375.8
90	100	0.2328	2466	165.3	82.9	293.6	379.3	376.2
90	1000	0.2360	2458	165.0	82.7	292.3	377.9	374.8
150	100	0.2257	2479	166.0	83.3	295.9	382.0	375.5
200	100	0.2181	2493	166.9	83.7	298.6	385.2	374.5
230	10	0.2106	2501	167.2	83.8	300.4	387.2	373.3
230	100	0.2123	2504	167.6	84.0	300.6	387.6	373.7
230	1000	0.2158	2497	167.3	83.8	299.4	386.2	372.5

Table 4

The total cross-section including all corrections for $e^+e^- \rightarrow \mu^+\mu^-$, using a Z mass of 91 GeV and a minimum for $\sqrt{zs}$ of 0.2 GeV. The results are obtained from Ref. [67].

m_t (GeV)	m_H (GeV)	Γ_Z (MeV)	σ_{max} (nb)	$\sqrt{s}_{max}$ (GeV)	$\sqrt{s}_-$ (GeV)	$\sqrt{s}_+$ (GeV)
60	100	2460	1.492	91.090	89.802	92.656
90	10	2463	1.490	91.091	89.801	92.657
90	100	2466	1.492	91.091	89.800	92.659
90	1000	2458	1.493	91.090	89.804	92.654
150	100	2479	1.493	91.091	89.794	92.668
200	100	2493	1.497	91.091	89.786	92.678
230	10	2501	1.499	91.092	89.782	92.682
230	100	2504	1.500	91.092	89.781	92.685
230	1000	2497	1.500	91.091	89.785	92.680

Table 5

The total cross-section including all corrections for $e^+e^- \rightarrow$ hadrons, using a Z mass of 91 GeV and a minimum for $\sqrt{zs}$ of 10 GeV. The results are obtained from Ref. [67].

m_t (GeV)	m_H (GeV)	Γ_Z (MeV)	σ_{max} (nb)	$\sqrt{s}_{max}$ (GeV)	$\sqrt{s}_-$ (GeV)	$\sqrt{s}_+$ (GeV)
60	100	2460	30.652	91.091	89.820	92.637
90	10	2463	30.653	91.092	89.819	92.639
90	100	2466	30.660	91.092	89.817	92.641
90	1000	2458	30.658	91.091	89.821	92.635
150	100	2479	30.699	91.092	89.811	92.649
200	100	2493	30.761	91.093	89.805	92.659
230	10	2501	30.818	91.093	89.801	92.665
230	100	2504	30.817	91.093	89.800	92.667
230	1000	2497	30.796	91.093	89.803	92.662

At LEP—if everything works well—neutrino counting [95] can be completed very quickly. The fastest procedure is to extract N_ν from the peak value of the cross-section assuming the Standard Model. This method, which is already followed by MARK II [71], and similar ones based on the partial widths, would allow us to measure N_ν in the next few months, the expected accuracy being $\delta N_\nu = \pm 0.2$ with 3 pb^{-1}. A similar precision, $\delta N_\nu = 0.25$, can also be obtained by collecting a few pb^{-1} at a number of GeV above the resonance and looking for $e^+e^- \rightarrow \gamma$ + nothing.

12. FUTURE PERSPECTIVE ON PRECISION TESTS

The goals of LEP and the SLC over a longer term are to measure several asymmetries that provide the optimal precision tests at the Z peak. With no longitudinal polarization of the beam, the interesting quantities are the forward–backward asymmetries: A_{FB}^{f} for $e^+e^- \rightarrow f\bar{f}$, with $f = \mu$, b, c, ..., and the τ-polarization asymmetry A_{pol}^{τ}, measured in $e^+e^- \rightarrow \tau^+\tau^-$ by reconstructing the τ helicity from the $\tau \rightarrow \pi\nu$ decay. At LEP with $\gtrsim 100\ pb^{-1}$ of integrated luminosity, the accuracies reported in Table 6 can be expected [96–98]. Note that the quoted error on $\sin^2\theta_w$ actually refers to the quantity $\bar{s}_W^2$ defined in terms of effective couplings at the Z. A more significant representation of the planned accuracy is in terms of plots such as those in Figs. 13 and 14, where the predicted values

Table 6

Experimental errors on asymmetries and $\bar{s}_W^2$ expected at LEP with $> 100\ pb^{-1}$ of integrated luminosity

Measurement	Error	$\delta\bar{s}_W^2$
A_{FB}^{μ}	0.0035	0.0017
A_{FB}^{s}	0.007	0.0012
A_{FB}^{c}	0.007	0.0015
A_{FB}^{b}	0.005	0.0009
A_{FB}^{u}	0.010	0.0020
A_{pol}^{τ}	0.011	0.0014

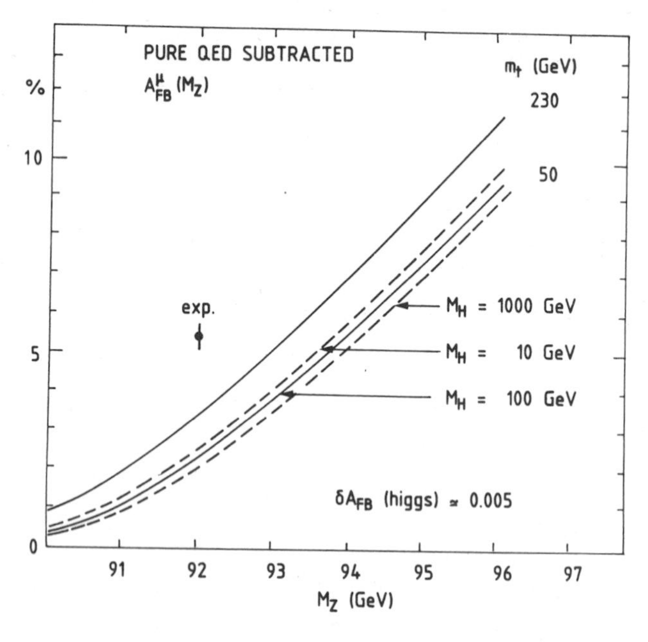

Fig. 13 A_{FB}^{μ} versus m_Z for given m_t and m_H (from Ref. [97]). The experimental error expected at LEP is also shown for comparison.

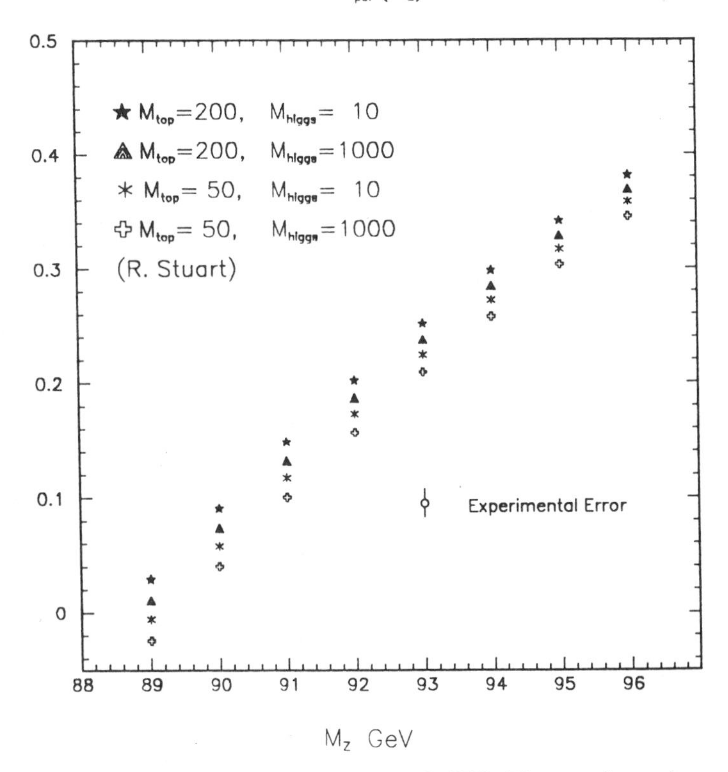

Fig. 14 A^{τ}_{pol} versus m_Z for given m_t and m_H (from Ref. [98]). The experimental error expected at LEP is also shown for comparison.

of A^{μ}_{FB} and A^{τ}_{pol} are shown as functions of m_Z for different m_t and m_H. The curves are obtained by using the most up-to-date and complete sets of electroweak radiative corrections. In fact, approximations like the improved Born formulae of Section 9 are not sufficient for precision tests at the level of these asymmetries. The pure QED corrections, being detector-dependent, have to be subtracted away [the asymmetries shown are computed from the analogues of the reduced cross-section $\hat{\sigma}$ in Eq. (11.1)]. The size of the experimental error is also shown. We see that the expected accuracy is of the order of the effect corresponding to a variation of m_H in the range 10 GeV $\leqslant m_H \leqslant$ 1 TeV. This means that the measurements are indeed probing the fine structure of radiative corrections.

Going back to Table 6, we note that the forward–backward asymmetry for the b-quark is really promising. The error shown in Table 6 for A^{b}_{FB} (which requires $\sim 10^7$ Z events) also includes the uncertainty associated with the correction [65] demanded by B^0–$\bar{B}^0$ mixing. In particular, the mixing parameter averaged over all produced b particles, has to be precisely measured at LEP (e.g. from the ratio of equal- to opposite-sign dileptons) in order to use this information for extracting A^{b}_{FB}. An updated analysis can be found in Ref. [65]. Note that A^{b}_{FB} is also important in another respect. The other asymmetries all measure essentially the same vacuum polarization effects. For example, if we plot A^{τ}_{pol} versus A^{μ}_{FB}, the resulting curve is nearly independent of m_t, as can be seen from Fig. 15. On the contrary, A^{b}_{FB} is also strongly dependent on large corrections from vertex diagrams, as was discussed in Section 9.

It is well known that having impletent longitudinally polarized beams is a great advantage for precision tests of the electroweak theory. This is because the left–right asymmetry A_{LR} has the same

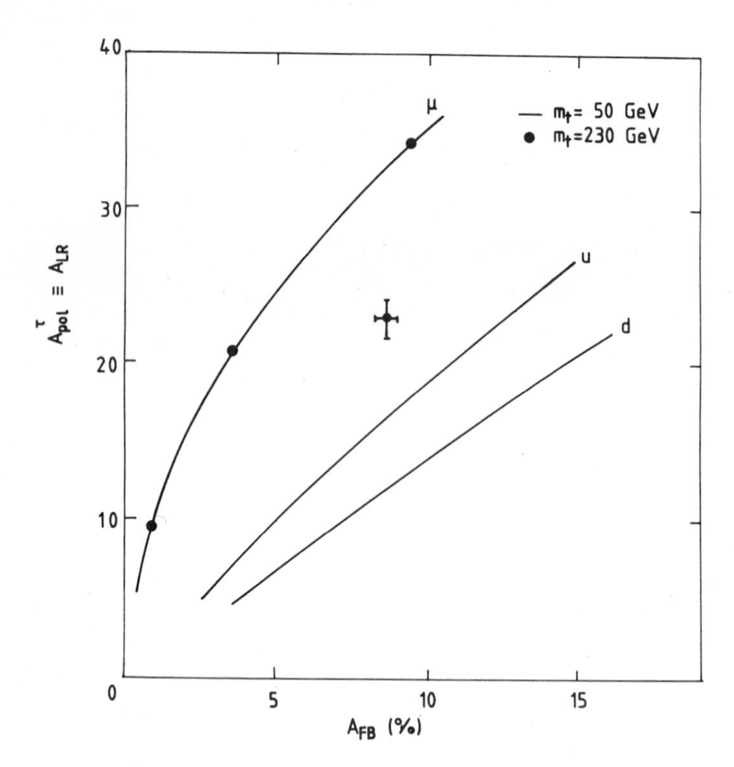

Fig. 15 A^{τ}_{pol} versus A_{FB} (for μ, u, and d final states). The curves for $m_t = 50$ GeV and 230 GeV are almost coincident (from Ref. [97]).

sensitivity as A^{τ}_{pol} to weak effects but is much easier to measure [96]: one simply measures the difference of the totally inclusive cross-sections from left- and right-handed electrons and then divides by the sum. The structure of radiative corrections is also much simpler. At LEP with 40 pb^{-1} of integrated luminosity, and polarization $P \simeq 0.5$ fixed with an accuracy of $\delta P \simeq 1\%$, it is possible to obtain [96] $\delta A_{LR} \simeq \pm 0.003$, corresponding to $\delta \bar{s}^2_W = \pm 0.0004$. This is a very exciting possibility, and longitudinal polarization is actually under serious consideration at CERN. But implementing longitudinally polarized beams at LEP is difficult (because of the length of time needed to build up the polarization, compared with the beam lifetime, and because of depolarizing effects due to the spin time and the energy spread, etc.). Hence, such implementation is costly. Also, the necessary polarization rotators have to be provided for all four LEP experiments. On the contrary, polarized beams are more easily obtained at the SLC. Thus the most obvious development of the experimental programme at the SLC with the SLD detector is to implement polarization and measure A_{LR}. Even with only 10^4 Z, one can expect [96] $\delta A_{LR} \simeq \pm 0.025$, corresponding to $\delta \bar{s}^2_W = \pm 0.003$, which is a quite respectable result.

At LEP 2, the measurement of m_W with a precision of $\delta m_W = \pm 100$ MeV will be possible [99]. Hadron colliders can hardly approach a comparable accuracy. In any case, the improvements needed, in terms of integrated luminosity and/or detectors, are such that the time spent on carrying them out would be more or less as long as the wait for LEP 2. Given m_Z, a precision of $\delta m_W \simeq \pm 100$ MeV is equivalent to $\delta s^2_W = \pm 0.002$. This accuracy is not only of the same order as that obtained from the measurement of asymmetries; it also has the additional advantage of carrying information of a very different qualitative nature.

We repeat here that the search for the Higgs boson [100] (see Section 4) is one of the most important goals of LEP. There is no better accelerator than LEP for discovering a Higgs with mass

$m_H \leqslant$ 40–50 GeV at LEP 1 and $m_H \leqslant$ 70–90 GeV at LEP 2. Actually, the Higgs search is a reference problem. More generally, one looks for a way to solve the problem of the origin of the Fermi scale of mass $\sim G_F^{-1/2}$. Indeed, LEP 2 is a formidable discovery machine, and I am confident that the low-lying fringes of the rich spectroscopy associated with all conceivable scenarios for new physics will already be observed at LEP 2.

In conclusion, a very exciting period for phenomenology lies ahead of us. We all hope that the wealth of data that are about to be produced at the CERN $p\bar{p}$ Collider and LEP, at the SLAC SLC and the FNAL Tevatron, at the DESY HERA, etc., will lead to new important breakthroughs in our understanding of particle physics.

REFERENCES

[1] S.L. Glashow, Nucl. Phys. **22** (1961) 579.
S. Weinberg, Phys. Rev. Lett. **19** (1967) 1264.
A. Salam, Proc. 8th Nobel Symposium, Aspenäsgården, 1968, ed. N. Svartholm (Almqvist and Wiksell, Stockholm, 1968), p. 367.

[2] See, for example, the following textbooks
I. Aitchison and A. Hey, Gauge theories in particle physics: a practical introduction (A. Hilger, Bristol, 1982).
D. Bailin, Introduction to gauge field theory (A. Hilger, Bristol, 1986).
E.D. Commins and P.H. Bucksbaum, Weak interactions of leptons and quarks (Cambridge Univ. Press, Cambridge, 1983).
L.B. Okun', Leptons and quarks (North-Holland Publ. Co., Amsterdam, 1982).

[3] Particle Data Group, Phys. Lett. **204B** (1988) 1.

[4] N. Cabibbo, Phys. Rev. Lett. **10** (1963) 531.

[5] S.L. Glashow, J. Iliopoulos, L. Maiani, Phys. Rev. **D2** (1970) 1285.

[6] J.H. Christenson, J.W. Cronin, V.L. Fitch and R. Turlay, Phys. Rev. Lett. **13** (1964) 138.

[7] C.N. Yang and R. Mills, Phys. Rev. **96** (1954) 191.

[8] E.S. Abers and B.W. Lee, Phys. Rep. **9** (1973) 1.

[9] J. Bjorken and S. Drell, Relativistic quantum fields (McGraw-Hill, New York, 1965).

[10] M. Consoli, S. Lo Presti and L. Maiani, Nucl. Phys. **B223** (1983) 474.
F. Jegerlehner, Z. Phys. **C32** (1986) 425.
Yu.D. Bardin, S. Riemann and T.Z. Riemann, Z. Phys. **C32** (1986) 121.

[11] G. Altarelli, Phys. Rep. **81** (1982) 1.

[12] G. Altarelli, Annu. Rev. Nucl. Part. Sci. **39** (1989), in press.

[13] G. Altarelli, Acta Phys. Austriaca, Suppl. **24** (1982) 229.

[14] J. Ellis et al., Annu. Rev. Nucl. Part. Sci. **32** (1982) 48.

[15] J.H. Kühn et al., Nucl. Phys. **B272** (1986) 560.

[16] P.W. Higgs, Phys. Lett. **12** (1964) 132; Phys. Rev. **13** (1964) 508; Phys. Rev. Lett. **145** (1966) 1156.
P.W. Anderson, Phys. Rev. **130** (1963) 439.
F. Englert and R. Brout, Phys. Rev. Lett. **13** (1964) 321.
G.S. Guralnik, C.R. Hagen and T.W.B. Kibble, Phys. Rev. Lett. **13** (1964) 585.

[17] S.L. Glashow and S. Weinberg, Phys. Rev. **D15** (1977) 1968.
E.A. Paschos, Phys. Rev. **D15** (1977) 1966.

[18] T.D. Lee, Phys. Rep. **9C** (1974) 143.
S. Weinberg, Phys. Rev. Lett. **37** (1976) 657.

[19] S.L. Adler, Phys. Rev. **177** (1969) 2426.
J.S. Bell and R. Jackiw, Nuovo Cimento **51** (1969) 47.
See also, C. Bouchiat, J. Iliopoulos and Ph. Meyer, Phys. Lett. **38B** (1972) 519.
D. Gross and R. Jackiw, Phys. Rev. **D6** (1972) 477.

[20] For reviews, see for example:
P. Fayet and S. Ferrara, Phys. Rep. **32** (1977) 251.
P. van Nieuwenhuizen, Phys. Rep. **68** (1981) 191.
J. Wess and J. Bagger, Supersymmetry and supergravity (Princeton Univ. Press, New York, 1983).
P. West, Introduction to supersymmetry and supergravity (World Scientific, Singapore, 1986).
N.P. Nilles, Phys. Rep. **110** (1984) 1.
H.E. Haber and G. Kane, Phys. Rep. **117** (1985) 75.

[21] See, for example, D.J. Gross, Proc. 24th Int. Conf. on High-Energy Physics, Munich, 1988, eds. R. Kotthaus and J. Kühn (Springer-Verlag, Berlin–Heidelberg, 1989), p. 310.

[22] E. Cremmer et al., Phys. Lett. **116B** (1982) 231.
R. Barbieri, S. Ferrara and A. Savoy, Phys. Lett. **119B** (1982) 343.

[23] E. Fahri and L. Susskind, Phys. Rep. **74** (1981) 277.

[24] H. Georgi et al., Phys. Lett. **143B** (1984) 152.
M.J. Dugan et al., Nucl. Phys. **B254** (1985) 299.

[25] L.F. Abbott and E. Fahri, Phys. Lett. **101B** (1981) 69; Nucl. Phys. **B189** (1981) 547.

[26] M. Veltman, Acta Phys. Polon. **B8** (1977) 475.
B.W. Lee, C. Quigg and H.B. Thacker, Phys. Rev. **D16** (1979) 1519.

[27] J. Ellis, M.K. Gaillard and D.V. Nanopoulos, Nucl. Phys. **B106** (1976) 292.
R. Barbieri and T.D. Ericson, Phys. Lett. **57B** (1975) 270.

[28] T.N. Pham and D.G. Sutherland, Phys. Lett. **151B** (1985) 444.
R.S. Willey and H.L. Yu, Phys. Rev. **D26** (1982) 3287.

[29] R.S. Willey, Phys. Lett. **173B** (1986) 480.

[30] M. Narain (CUSB Collab.), paper submitted to the 24th Int. Conf. on High-Energy Physics, Munich, 1988.

[31] F. Wilczek, Phys. Rev. Lett. **39** (1977) 1304.

[32] M.I. Vysotsky, Phys. Lett. **97B** (1980) 159.
See also:
J. Ellis et al., Phys. Lett. **158B** (1985) 417.
P. Nason, Phys. Lett. **175B** (1986) 223.

[33] For reviews, see for example:
A.D. Linde, Rep. Prog. Phys. **42** (1979) 389.
A.A. Ansel'm, N.G. Ural'tsev and V.A. Khoze, Sov. Phys.-Usp. **28** (1985) 113.

[34] A.D. Linde, Sov. Phys.–JETP Lett. **23** (1976) 64.
S. Weinberg, Phys. Rev. Lett. **36** (1976) 294.

[35] R.A. Flores and M. Sher, Phys. Rev. **D27** (1983) 1679.

[36] L. Maiani et al., Nucl. Phys. **B136** (1978) 115.
N. Cabibbo et al., Nucl. Phys. **B158** (1979) 295.

[37] M.A.B. Bég et al., Phys. Rev. **52** (1984) 883.
D.J. Callaway, Nucl. Phys. **B233** (1984) 189.
R. Dashen and H. Neuberger, Phys. Rev. Lett. **50** (1983) 1897.
K.S. Babu and E. Ma, Phys. Rev. **D31** (1984) 2861.
E. Ma, Phys. Rev. **D31** (1985) 322.

[38] M. Lindner, Z. Phys. **C31** (1986) 295.

[39] K. Decker, I. Montvay and P. Weisz, Nucl. Phys. **B268** (1986) 362.
J. Jersak, Proc. Conf. on Advances in Lattice Gauge Theory, Tallahassee, 1985, eds. D.W. Duke et al. (World Scientific, Singapore, 1985), p. 241.
D.J. Callaway and R. Petronzio, Nucl. Phys. **B267** (1986) 253.
A. Hasenfratz and P. Hasenfratz, Nucl. Phys. **B270** (1986) 687, and references therein.
[40] M.S. Chanowitz, Proc. 23rd Int. Conf. on High Energy-Physics, Berkeley, 1986 (World Scientific, Singapore, 1987), Vol. I, p. 445.
[41] See, for example, J. Ellis and R. Peccei (eds.), Physics at LEP (CERN 86-02, Geneva, 1986).
[42] See, for example, Proc. ECFA Workshop on LEP 200, Aachen, 1986, eds. A. Böhm and W. Hoogland (CERN 87-08, Geneva, 1987).
[43] J. Mulvey (ed.), Proc. Workshop on Physics at Future Accelerators, La Thuile and Geneva, 1987 (CERN 87-07, Geneva, 1987).
[44] R. Kleiss and W.J. Stirling, Phys. Lett. **200B** (1988) 193.
[45] J. Mulvey (ed.), The feasibility of experiments at high luminosity at the Large Hadron Collider (CERN 88-02, Geneva, 1988).
[46] R.N. Cahn et al., Proc. Workshop on Experiments, Detectors, and Experimental Areas for the Supercollider, Berkeley, 1987, eds. R. Donaldson and M.D. Gilchriese (World Scientific, Singapore, 1988), p. 20.
[47] G. Altarelli, R. Kleiss and C. Verzegnassi (eds.), Z Physics at LEP 1, CERN 89-08 (1989), Vols. 1-3.
[48] M. Consoli, W. Hollik and F. Jegerlehner, ibid., Vol. 1, p. 7.
[49] See, for example: F. Antonelli and L. Maiani, Nucl. Phys. **B186** (1981) 269.
S. Bellucci, M. Lusignoli and L. Maiani, Nucl. Phys. **B189** (1981) 329.
A recent discussion of 'large' terms can be found in G. Gounaris and D. Schildknecht, Z. Phys. **C42** (1989) 107.
[50] M. Veltman, Nucl. Phys. **B123** (1977) 89.
M.S. Chanowitz, M.A. Furman and I. Hinchliffe, Phys. Lett. **78B** (1978) 285.
[51] T. Appelquist and J. Carazzone, Phys. Rev. **D11** (1975) 2856.
[52] G. Burgers et al., *in* Ref. [47], Vol. 1, p. 55.
[53] A. Sirlin, Phys. Rev. **D22** (1980) 971.
W.J. Marciano and A. Sirlin, Phys. Rev. **D22** (1980) 2695 and **D29** (1984) 75, 945.
[54] M. Consoli, W. Hollik and F. Jegerlehner, preprint CERN-TH.5395/89 (1989).
[55] J.J. van der Bij and F. Hoogeveen, Nucl. Phys. **B283** (1987) 477.
[56] B.W. Lynn, M. Peskin and R.G. Stuart, *in* Ref. [41], Vol. 1, p. 90.
[57] R. Barbieri and L. Maiani, Nucl. Phys. **B224** (1983) 32.
L. Alvarez-Gaumé, J. Polchinski and M. Wise, Nucl. Phys. **B221** (1983) 495.
[58] W. Hollik, Z. Phys. **C37** (1988) 569.
[59] J. van der Bij and M. Veltman, Nucl. Phys. **B231** (1984) 205.
[60] W. Hollik, preprint DESY 88-188 (1988).
[61] D.C. Kennedy et al., Nucl. Phys. **B321** (1989) 83.
D.C. Kennedy and B.W. Lynn, Nucl. Phys. **B322** (1989) 1.
B.W. Lynn, Stanford report SU-ITP-867 (1989).
[62] W.J. Marciano and A. Sirlin, Phys. Rev. Lett. **46** (1981) 163.
J. Sarantakos, W.J. Marciano and A. Sirlin, Nucl. Phys. **B217** (1983) 84.
[63] A. Sirlin, preprint CERN-TH.5506/89 (1989).
[64] W. Beenakker and W. Hollik, Z. Phys. **C40** (1988) 141.
[65] J.H. Kühn et al., *in* Ref. [47], Vol. 1, p. 267.
[66] E.W.N. Glover et al., *in* Ref. [47], Vol. 2, p. 1.
[67] F.A. Berends et al., *in* Ref. [47], Vol. 1, p. 89.

[68] P. Sinervo, to appear in Proc. 14th Int. Symp. on Lepton and Photon Interactions at High Energies, Stanford, 1989.
[69] K. Eggert, ibid.
[70] L. Di Lella, ibid.
[71] G. Feldman, ibid. See also these proceedings.
[72] M.K. Campbell, ibid.
[73] J. Ellis and G.L. Fogli, preprints CERN-TH.5457/89 and CERN-TH.5511/89 (1989).
[74] A. Weidberg, as Ref. [68].
[75] R.G. Stuart, Z. Phys. **C34** (1987) 445.
[76] J.V. Allaby et al., Phys. Lett. **177B** (1986) 446; Z. Phys. **C36** (1987) 611.
[77] H. Abramowicz et al., Phys. Rev. Lett. **57** (1986) 298.
A. Blondel et al., preprint CERN-EP/89-101 (1989).
[78] A. Blondel, preprint CERN-EP/89-84 (1989).
[79] M.C. Noecker, B.P. Masterson and C.E. Wieman, Phys. Rev. Lett. **61** (1988) 310.
[80] M.A. Bouchiat et al., J. Phys. **47** (1986) 1709.
[81] J. Panman, to appear in Proc. 14th Int. Symp. on Lepton and Photon Interactions at High Energies, Stanford, 1989.
[82] P. Langacker, Univ. Pennsylvania preprint UPT-0400T (1989).
[83] D. Haidt, preprint DESY 89-073 (1989) (updated: private communication).
[84] Z. Hioki, Tokushima Univ. preprint 89-05 (1989).
[85] U. Amaldi et al., Phys. Rev. **D36** (1987) 1385.
G. Costa et al., Nucl. Phys. **B297** (1988) 244.
[86] W.J. Marciano, Brookhaven report BNL-41498 (1988).
[87] See, for example, G. Alexander et al. (eds.), Polarization at LEP, CERN 88-06 (1988).
[88] E.A. Kuraev and V.S. Fadin, Sov. J. Nucl. Phys. **41** (1985) 466.
G. Altarelli and G. Martinelli, *in* Ref. [41], Vol. 1, p. 47.
[89] A. Borrelli et al., preprint CERN-TH.5441/89 (1989).
[90] W. Wetzel, Nucl. Phys. **B227** (1983) 1, and *in* Ref. [41], Vol. 1, p. 40.
[91] D.Y. Bardin et al., Phys. Lett. **206B** (1988) 539.
[92] R.N. Cahn, Phys. Rev. **D36** (1987) 2666.
[93] F. Aversa and M. Greco, Frascati preprint LNF-89/025 (PT) (1989).
[94] S.N. Ganguli, A. Gurtu and K. Mazumdar, Tata preprints TIFR-EHEP/89/2 and 89/3 (1989).
[95] L. Trentadue et al., *in* Ref. [47], Vol. 1, p. 129.
[96] D. Treille, *in* Ref. [87], Vol. 1, p. 265.
[97] M. Böhm and W. Hollik, *in* Ref. [47], Vol. 1, p. 203.
[98] S. Jadach and Z. Wąs *in* Ref. [47], Vol. 1, p. 235.
[99] P. Roudeau et al., Proc. ECFA Workshop on LEP 200, Aachen, 1986, eds. A. Böhm and W. Hoogland (CERN 87-08, Geneva, 1987), Vol. 1, p. 49.
[100] P.J. Franzini and P. Taxil, *in* Ref. [47], Vol. 2, p. 58.

THE GEOMETRICAL PRINCIPLES OF STRING THEORY CONSTRUCTIONS

Jan Govaerts

Instituut voor Theoretische Fysica

Katholieke Universiteit Leuven, B-3030 Leuven, Belgium

1 INTRODUCTION

In their quest for a quantum consistent unified theory of all fundamental interactions among elementary particles, theoretical particle physicists in recent years have been entertaining the idea[1] that such a theory may be constructed within the framework of superstring theories[1]. As opposed to the previous attempts in the past fifteen years, all formulated within the framework of ordinary quantum field theory, superstring theories seem to possess the very same desirable properties that stimulated previous attempts as well as the resolution of problems which led to the dismissal of these attempts[2,3,4].

One of the most appealing features, not only from a particle physicist's point of view, is that string theories provide a unique unification between gravity and all other interactions. Indeed, *all* fundamental matter particles and gauge quanta mediating *all* interactions, including the graviton, are nothing but the same fundamental object, a relativistic string in different states of excitation. This idea resolves the centuries old dream of ultimate unification: the Universe and all that it contains being made of only one fundamental object, with dynamics so rich that it leads to this infinitely large variety of physical phenomena that we observe at all energy scales in our Universe.

Obviously, our present understanding of these theories is still far too poor to be able to make any definite conclusion concerning such fascinating speculations. About four years ago[5], when string theory became the prime candidate for fundamental unification, uniqueness was regarded as a very compelling argument. Indeed, only a handful of different theories were known at the time[1]. Since then however, countless new string theories have been constructed in space-time dimensions ranging from ten to four, with a few of the four dimensional theories leading to a massless particle content resembling that of the quarks and leptons in the Standard Model.

Furthermore, some four years ago, optimism was high that realistic particle phenomenology would easily be achieved withing string theory. This optimism was based on the study of the corresponding low-energy effective field theories[2,3,4]. Recent investigations however, at the level of string theory itself, show that the problems are much more subtle, with entirely new mechanisms coming into play[4,6].

The prevailing attitude nowadays[7], is to consider that all these theories are different vacua of a few underlying fundamental theories still to be formulated - the hope being that non-perturbative phenomena would determine which among all possible string constructions are actually dynamically generated, together with the ensuing particle content and phenomenology.

Clearly, such a state of affairs makes it even more difficult to make any experimental contact with string theories. On the other hand, this is not a good enough reason to dismiss this approach to unification altogether. After all, string theories are the best and only (so far) consistent formulation for such a quantum unification, including gravity. Moreover, we should not forget that it took many years before the basic ideas of the Standard Model were widely accepted by the scientific community, and that the detailed experimental tests of the Standard Model are only just beginning at the new big accelerators[8].

From a different point of view, the study of string theories is also responsible for a strong revival of interest in two-dimensional conformal field theories[9] and related mathematical structures.In recent years, this interest has led to fundamental developments[9] in the field of two-dimensional statistical systems at and off criticality. New results point to deep relationships between so far unrelated fields in pure mathematics[9], with completely new insights into different problems[10]. These are truely fascinating subjects in their own right.

All these reasons make the effort involved in getting acquainted with string theory worthwhile. One could even go as far as to consider that some familiarity with the field should now be part of common knowledge for any (theoretical) particle physicist, in the same manner that such familiarity is now assumed for grand unified theories, supersymmetry, and so on. Herein lies the main motivation for these lectures.

It is of course not possible to discuss in such lectures all the aspects of the subject which have emerged over the last years. What will not be discussed for example, are the following points:

- the calculation of string scattering amplitudes, at any order of perturbation theory, either through path-integral methods[11], or through operatorial methods such as the infinite Grassmannian approach[12],
- string field theory[13], i.e. the second quantization of string theory,
- particle phenomenology from string theory[4].

What will be discussed are the "principles of string theory construction". By first studying in detail the case of the bosonic string, we shall understand the origin and the meaning of the structure and of the consistency constraints which appear when quantizing the system. Having identified the relevant structures, we shall then show how they easily generalize to the case of spinning strings, and how these general constraints lead to the formulation of "principles for string theory construction".

Hopefully , by starting at an introductory level, newcomers to the subject will not be put off by the jargon which will be introduced in a pedestrian manner as we go along. Interested readers should find these notes a useful starting point for their own research.

Basic references may be found in Refs. 1,14,15,16, where some historical comments concerning the development of the subject from a theory for strong interactions to a theory of fundamental unification are also given. Here, we shall follow more closely Refs. 17,18,19. We shall also use units in which $\hbar = 1 = c$, and we shall take space-time to be D-dimensional, with the Minkowski metric $\eta^{\mu\nu}(\mu, \nu = 0, 1, ...D-1))$ of signature mostly "+" signs.

2 RELATIVISTIC BOSONIC STRINGS

2.1 Classical Lagrangian Description

2.1.1 The action principle

Although we shall discuss right away the relativistic bosonic string, the interested reader may find it useful to first consider the case of the relativistic scalar particle, through an analysis similar to the one followed in this chapter. For more details, see Refs. 17-20.

A bosonic string, being a 1-dimensional object, sweeps out a 2-dimensional surface, its world-sheet, as it propagates freely through space-time. To represent this space-time trajectory, one introduces coordinates $x^\mu(\sigma,\tau)$ giving the space-time position of the string. These quantities are D functions of two (dimensionless) world-sheet coordinates σ and τ, parametrizing the two-dimensional surface. Both for open and for closed strings, we shall take $0 \leq \sigma \leq \pi$; τ is considered as the time-evolution parameter of the system. Clearly, having both the observer's time coordinate x^o and the world-sheet parameter τ present in this formulation is a redundant feature. This is the price to pay for having an explicit space-time covariant description.

In order that the physical properties of the system do not depend neither on the observer's reference frame in space-time nor on the world-sheet parametrization (σ,τ), it is necessary that the action describing the dynamics of the bosonic string be a space-time scalar and a world-sheet scalar. The total world-sheet area between some initial and final string configurations $x_i^\mu(\sigma) = x^\mu(\sigma,\tau=\tau_i)(i=1,2)$, properly rescaled by a dimensionful quantity, is the obvious candidate for such an action.

There are two ways of expressing such an action. The first, by measuring the area using the metric induced on the world-sheet by the space-time Minkowski metric. This corresponds to the Nambu-Goto action[21], leading to non-linear equations of motion. The second way to express the action is by coupling the 2-dimensional fields $x^\mu(\sigma,\tau)$, which transform as space-time vectors and world-sheet scalars, to an intrinsic world-sheet metric. This corresponds to the so-called Polyakov action[22], originally discussed in Ref. 23, which leads to linear equations of motion.

In the following, we shall use the notation $\xi^\alpha(\alpha=0,1)$ for world-sheet coordinates, with $\xi^0=\tau$ and $\xi^1=\sigma$. As usual, partial derivatives with respect to τ (resp.σ) will be denoted by a dot (resp. a prime) above quantities.

With these notations, from the invariant space-time line element

$$ds^2 = \eta^{\mu\nu}dx_\mu dx_\nu = \eta^{\mu\nu}\partial_\alpha x_\mu \partial_\beta x_\nu d\xi^\alpha d\xi^\beta, \tag{2.1}$$

it is clear that the induced world-sheet metric $\gamma_{\alpha\beta}$ is given by the components

$$\gamma_{\alpha\beta} = \eta^{\mu\nu}\partial_\alpha x_\mu \partial_\beta x_\nu. \tag{2.2}$$

The Nambu-Goto action[21] then reads

$$S[x^\mu] = \frac{-1}{2\pi\alpha'}\int_{\tau_1}^{\tau_2} d\tau \int_0^\pi d\sigma(-det\gamma_{\alpha\beta})^{1/2} = \int_{\tau_1}^{\tau_2} d\tau \int_0^\pi d\sigma \mathcal{L}(\dot{x},x'), \tag{2.3}$$

with

$$\mathcal{L}(\dot{x},x') = \frac{-1}{2\pi\alpha'}\sqrt{(\dot{x}x')^2 - \dot{x}^2 x'^2}. \tag{2.4}$$

Here, α' is a dimensionful quantity, of dimension a lenght square, which thus determines the physical scales in the system, such as the scale of the mass spectrum of the bosonic string. The choice of sign in the factor $(-det\gamma_{\alpha\beta})^{1/2}$ is due to the fact that the induced metric $\gamma_{\alpha\beta}$ has a world-sheet signature $(-+)$. This also guarantees that no point of the string moves faster than the speed of light.[14,17]

Introducing an intrinsic world-sheet metric $g_{\alpha\beta}$, of signature $(-+)$, the Polyakov action[22,23] is given as:

$$S[x^\mu, g_{\alpha\beta}] = \frac{-1}{4\pi\alpha'} \int_{\tau_1}^{\tau_2} d\tau \int_0^\pi d\sigma \sqrt{-det g_{\alpha\beta}}\; g^{\alpha\beta}\partial_\alpha x^\mu \partial_\beta x^\nu \eta_{\mu\nu}. \tag{2.5}$$

A priori, one could also add to it the usual Einstein-Hilbert action term

$$\frac{1}{4\pi\kappa}\int d^2\xi \sqrt{-g} R^{(2)}, \tag{2.6}$$

or a world-sheet cosmological term

$$\frac{-1}{4\pi\alpha'}\mu^2 \int d^2\xi \sqrt{-g}. \tag{2.7}$$

However, the former term is a topological invariant in two dimensions (up to surface terms), which thus does not contribute to the equations of motion (in 2 dimensions, the Einstein tensor $G_{\alpha\beta} = R_{\alpha\beta} - \frac{1}{2}g_{\alpha\beta}R^{(2)}$ vanishes identically). The latter term is consistent with the classical equations of motion only if $\mu^2 = 0$.

At the quantum level, using a path-integral formulation[11], it may be seen that the Einstein-Hilbert action sets the strength of string interactions, whereas the cosmological term is induced by 1-loop short-distances singularities on the world-sheet[24]. It then has to be properly renormalized[24] so as to maintain the local Weyl symmetry of the classical Polyakov action (see below).

Let us remark here that the action (2.5) may also be used to represent the propagation of the bosonic string in a curved space-time background of metric $G_{\mu\nu}(x)$. This is easily done by replacing $\eta_{\mu\nu}$ in (2.5) by $G_{\mu\nu}(x)$, leading to the usual action of a non-linear sigma model in two dimensions. Such a formulation is the starting point of a much studied approach to the problem of space-time compactification in string theory[25].

By construction, both the Nambu-Goto and Polyakov actions are invariant under space-time Poincaré transformations and world-sheet reparametrizations. From the 2-dimensional field theoretical point of view, the former symmetry is a global internal symmetry, whereas the latter is a world-sheet gauge symmetry.

By Noether's theorem, associated to space-time translations and generalized rotations (i.e. rotations in space and Lorentz boosts), there exist locally conserved energy- and angular-momentum world-sheet currents, whose conserved charges, giving the total energy- and angular-momentum of the bosonic string, generate Poincaré transformations of the system. Clearly, the existence of such quantities follows from the fact that the bosonic string is propagating in a space-time with isometry precisely the Poincaré group. For a generic curved space-time, with arbitrary metric $G_{\mu\nu}$, such quantities would not exist, unless the space-time metric has some special symmetries.

In contradistinction, the world-sheet reparametrization gauge symmetry is always present for any string theory. Actually, it is precisely this gauge invariance, when imposed at the quantum level, which is responsible for all the profound and beautiful properties of string theory (this should become clear as we proceed with the discussion). Also, as we shall see in chapter 4, the formulation of "principles for string theory construction" immediately follows from the requirement of this gauge invariance in the quantized theory.

Among world-sheet gauge transformations, we have local world-sheet reparametrizations in the same connected component of the diffeomorphism group as the identity transformation. These reparametrizations are generated by repeated applications of infinitesimal reparametrizations, thus preserving the orientation of the world-sheet. Correspondingly, there should exist two quantities, associated to the two arbitrary functions $\tilde{\xi}^\alpha(\xi)$ specifying such transformations, which generate infinitesimal reparametrizations and which vanish at all times for physical solutions. In other words, associated to world-sheet reparametrization invariance, we have two constraints generating the local gauge symmetry of the system. As we shall see, imposing this local gauge invariance at the quantum level leads to a constraint on the number of world-sheet degrees of freedom used in the construction of string theories.

There also exist world-sheet gauge transformations which are not connected to the identity transformation, so called disconnected or global gauge transformations. Depending on the world-sheet topology, there may exist orientation preserving global reparametrizations. These symmetries lead to a new type of constraint at the quantum level, not existing in ordinary field theory, which goes by the name of modular invariance. One may also impose invariance under orientation reversing global reparametrizations, thus describing unoriented string theories (indeed, both the linear and non-linear actions above are invariant under such transformations). At the quantum level, this leads to further constraints on the physical spectrum of the corresponding theories. Both these constraints and modular invariance will be discussed in chapter 4.

When compared to the Nambu-Goto action, the Polyakov action (2.5) actually possesses one more local symmetry (when $\mu^2 = 0$) under local Weyl rescalings of the intrinsic metric $g_{\alpha\beta}$:

$$g_{\alpha\beta}(\xi) \longrightarrow e^{\phi(\xi)} g_{\alpha\beta}(\xi). \tag{2.8}$$

These transformations lead to local shifts in the conformal mode of the metric $g_{\alpha\beta}$, by the arbitrary world-sheet function $\phi(\xi)$. Hence, one expects one more constraint, involving the generator of local Weyl rescalings. Actually, this constraint combines with the previous two generators of local reparametrizations into the equation of motion for the metric $g_{\alpha\beta}$ which is an auxiliary field:

$$T_{\alpha\beta} = 0. \tag{2.9}$$

Here, $T_{\alpha\beta}$ is the world-sheet energy-momentum tensor of the scalar fields x^μ, with:

$$T_{\alpha\beta} = \partial_\alpha x^\mu \partial_\beta x_\mu - \frac{1}{2} g_{\alpha\beta} g^{\gamma\delta} \partial_\gamma x^\mu \partial_\delta x_\mu. \tag{2.10}$$

The constraint $T_{\alpha\beta} = 0$ is thus a consequence of local reparametrization and Weyl invariance. Its solution expresses the intrinsic metric $g_{\alpha\beta}$ as a Weyl rescaling of the induced metric $\gamma_{\alpha\beta}$. When substituted in the Polyakov action, one then recovers the Nambu-Goto action, thus showing that at the classical level, these two formulations of the relativistic bosonic string are indeed entirely equivalent.

This need not be true at the quantum level however, due to the conformal anomaly in two dimensions. In a quantization approach which explicitly preserves reparametrization gauge invariance but not Weyl invariance, such as Polyakov's path-integral approach[22], one finds that the conformal mode of the metric $g_{\alpha\beta}$ couples dynamically unless a specific condition on the number of degrees of freedom is met. In the case of the bosonic string discussed here, this condition fixes the number of scalar fields x^μ (or the dimension of space-time) as $D = 26$.

In a quantization approach which explicitly preserves Weyl invariance but not reparameterization invariance, such as the canonical approach which we shall use in these lectures[20], this latter symmetry is not realized at the quantum level unless the same condition ensuring decoupling of the conformal mode in Polyakov's path-integral is met[26]. Hence, when this condition is satisfied, both the linear and non-linear approaches lead to equivalent descriptions of the system, also at the quantum level. In these lectures, we shall only consider the construction of string theories in which the conformal mode decouples, i.e. reparametrization *and* Weyl invariant string theories. It is also possible to construct reparametrization but not Weyl invariant string theories, in which the conformal mode is dynamical[27]. Recently, more progress has been made in that direction[28].

For these reasons, we shall now concentrate on the non-linear Nambu-Goto action, and use canonical Hamiltonian methods to quantize the system. The same results actually also follow for the linear action when applying the same methods[20].

Considering the Nambu-Goto action (2.3), the conserved world-sheet currents associated to space-time translations and generalized rotations are easily found to be respectively

$$P^\alpha_\mu = \frac{\partial \mathcal{L}}{\partial(\partial_\alpha x^\mu)}, \quad M^\alpha_{\mu\nu} = P^\alpha_\mu x_\nu - P^\alpha_\nu x_\mu, \tag{2.11}$$

with

$$\partial_\alpha P^\alpha_\mu = 0, \quad \partial_\alpha M^\alpha_{\mu\nu} = 0. \tag{2.12}$$

We have:

$$P^0_\mu = \frac{-1}{2\pi\alpha'}\left((\dot{x}x')^2 - \dot{x}^2 x'^2\right)^{-1/2}[(\dot{x}x')x'_\mu - x'^2\dot{x}_\mu], \tag{2.13a}$$

$$P^1_\mu = \frac{-1}{2\pi\alpha'}\left((\dot{x}x')^2 - \dot{x}^2 x'^2\right)^{-1/2}[(\dot{x}x')\dot{x}_\mu - \dot{x}^2 x'_\mu]. \tag{2.13b}$$

The corresponding conserved charges are

$$P_\mu = \int_0^\pi d\sigma P^0_\mu, \quad M_{\mu\nu} = \int_0^\pi d\sigma M^0_{\mu\nu}, \tag{2.14}$$

giving the total energy -and angular-momentum of the string for any physical solution.

The variation of the action under infinitesimal variations $\delta x^\mu(\sigma,\tau)$ in the coordinates, leaving the initial and final configurations fixed (i.e. $\delta x^\mu(\sigma,\tau_i) = 0, i = 1,2$), leads to the equations of motion for the system and to boundary conditions in σ. The equations of motion precisely express the local conservation of the energy-momentum world-sheet current P^α_μ. The boundary conditions in σ appear since we are dealing with a two-dimensional field theory on a space with boundary.

For a simply-connected space-time, two choices of boundary conditions are possible, respectively:

$$\text{- open strings} : \quad P^1_\mu(\sigma = 0,\pi,\tau) = 0, \tag{2.15a}$$

$$\text{- closed strings} : \quad x^\mu(\sigma = 0,\tau) = x^\mu(\sigma = \pi,\tau). \tag{2.15b}$$

Note that the boundary conditions for open strings imply that there is no flow of energy -nor angular-momentum at the end points.

Finally, from the expressions for P^α_μ, it is straightforward to check that the following equations

$$[P^0_\mu \pm \frac{\partial_\sigma x_\mu}{2\pi\alpha'}]^2 = 0, \tag{2.16a}$$

$$[P^1_\mu \pm \frac{\partial_\tau x_\mu}{2\pi\alpha'}]^2 = 0, \tag{2.16b}$$

are identically satisfied. Actually, these are the constraints which we expect as consequences of local reparameterization invariance. Indeed, it may be seen, by performing an arbitrary infinitesimal reparametrization

$$\xi^\alpha \to \tilde{\xi}^\alpha = \xi^\alpha - \eta^\alpha(\xi), \tag{2.17}$$

that these constraints follow from reparametrization invariance of the Nambu-Goto action, and that the quantities in (2.16) are the corresponding generators.

However, from our previous discussion we would expect to have only two constraints whereas we seem to obtain four. Actually, the constraints (2.16a) and (2.16b) are not independent. They are related to one another by orientation preserving global reparameterizations exchanging the role of σ and τ. Later, we shall need to consider the constraints (2.16a) only.

Let us remark here that in the case of the open string, due to the boundary conditions (2.15a), the constraints (2.16b) reduce to

$$\dot{x}^2(\sigma = 0, \pi, \tau) = 0. \tag{2.18}$$

This implies that the end points move at the speed of light for any classical physical solution.

2.1.2 The conformal gauge

The general solution to the non-linear equations of motion of the Nambu-Goto action does not seem to be obtained easily. Moreover, it would involve two arbitrary functions of the world-sheet coordinates, due to reparameterization gauge invariance.

There exists however, a choice of world-sheet parametrization, i.e. of gauge-fixing, in which the equations of motion simply reduce to those of massless two-dimensional scalar fields. Indeed, locally on the world-sheet it is always possible to choose a parameterization such that

$$\gamma_{01} = 0 \quad , \quad \gamma_{00} + \gamma_{11} = 0; \tag{2.19a}$$

$$\text{or } (\dot{x} \pm x')^2 = 0. \tag{2.19b}$$

Geometrically, these conditions mean that the tangent vectors in the σ and τ directions are orthonormal (with respect to the space-time Minkowski metric) up to a local scale factor. (Strictly speaking, such a choice of parametrization always exist for a world sheet metric of Euclidean signature, but otherwise requires some additional technicalities in the case of a Minkowski signature[20]. This point is not essential in these lectures). These gauge-fixing conditions define the conformal or orthonormal gauge (the meaning of conformal will become clear shortly).

Using these conditions in the definition for P^α_μ, one finds:

$$P^0_\mu = \frac{\dot{x}_\mu}{2\pi\alpha'}, \; P^1_\mu = \frac{-x'_\mu}{2\pi\alpha'}, \tag{2.20}$$

so that the equations of motion reduce to

$$(\partial_\tau^2 - \partial_\sigma^2)x^\mu(\sigma,\tau) = 0. \tag{2.21}$$

These are indeed the Klein-Gordon equations for D massless scalar fields in two dimensions. The general solution is of the form

$$x^\mu(\sigma,\tau) = x^\mu_{(1)}(\tau-\sigma) + x^\mu_{(2)}(\tau+\sigma), \tag{2.22}$$

thus clearly showing the separation into right- and left-moving modes in the conformal gauge. Imposing the boundary conditions (2.15a) for open strings, one finds that the functions $x^\mu_{(1)}$ and $x^\mu_{(2)}$ are essentially equal and 2π-periodic (more precisely these are the properties of their derivatives). This is the expression of the fact that modes are reflected back at the end points of open strings. In the case of closed strings, one finds that the two functions $x^\mu_{(1)}$ and $x^\mu_{(2)}$ are completely independent. Thus, for closed string theories, right- and left-moving modes are entirely decoupled and "transparent" to each other. This is the basic property which allows for the vast richness in string theory constructions, since one may then use completely different structures in each moving sector.

Although in the conformal gauge the equations of motion have reduced to those of free massless scalar fields, the system has not really simplified to such a set of fields. Any solution to the equations of motion has to satisfy the gauge-fixing conditions (2.19), thus leading to constraints on the integration constants defining the solution.

In the case of the open string, the general solution to (2.21) and the corresponding boundary conditions in σ may be parametrized as:

$$x^\mu(\sigma,\tau) = \sqrt{2\alpha'}(q^\mu + \alpha_0^\mu\tau + i\sum_n{}' \frac{1}{n}\alpha_n^\mu e^{-in\tau} cos n\sigma), \tag{2.23a}$$

where

$$\alpha_n^\mu = \alpha_{-n}^\mu, \tag{2.23b}$$

$$\alpha_0^\mu = \sqrt{2\alpha'}P^\mu. \tag{2.23c}$$

($\sum_n'$ means summing over all positive and negative integers, but excluding $n = 0$). The gauge-fixing conditions are then equivalent to the constraints

$$L_n^{(\alpha)} = 0, \tag{2.24}$$

where the bosonic Virasoro generators are given by

$$L_n^{(\alpha)} = \frac{1}{2}\sum_m \alpha_{n-m}^\mu \alpha_m.$$

In particular, the zero-mode constraint $L_0^{(\alpha)} = 0$ leads to the classical mass formula

$$\alpha' M^2 = \sum_{n=1}^{\infty} \alpha_{-n}^\mu \alpha_{n\mu} = N. \tag{2.25}$$

In the case of closed strings, we have

$$x^\mu(\sigma,\tau) = \sqrt{2\alpha'}\{q^\mu + (\alpha_0^\mu + \overline{\alpha}_0^\mu)\tau + \frac{1}{2}i\sum_n{}' \frac{1}{n}(\alpha_n^\mu e^{-2in(\tau-\sigma)} + \overline{\alpha}_n^\mu e^{-2in(\tau+\sigma)})\}, \tag{2.26a}$$

where

$$\alpha_n^{\mu *} = \alpha_{-n}^\mu, \overline{\alpha}_n^{\mu *} = \overline{\alpha}_{-n}^\mu, \tag{2.26b}$$

$$\alpha_0^\mu = \frac{1}{2}\sqrt{2\pi\alpha'}P^\mu = \overline{\alpha}_0^\mu. \tag{2.26c}$$

The gauge-fixing conditions then reduce to

$$L_n^{(\alpha)} = 0 \quad , \quad \overline{L}_n^{(\alpha)} = 0, \tag{2.27a}$$

where

$$L_n^{(\alpha)} = \frac{1}{2}\sum_m \alpha^\mu_{n-m}\alpha_{m\mu} \quad , \quad \overline{L}_n^{(\alpha)} = \frac{1}{2}\sum_m \overline{\alpha}^\mu_{n-m}\overline{\alpha}_{m\mu}. \tag{2.27b}$$

The zero-mode constraints $L_0^{(\alpha)} = 0, \overline{L}_0^{(\alpha)} = 0$ lead to the classical mass formulae:

$$\frac{1}{2}\alpha' M^2 = N + \overline{N}, \tag{2.28a}$$

$$N = \overline{N}, \tag{2.28b}$$

with

$$N = \sum_{n=1}^{\infty} \alpha^\mu_{-n}\alpha_{n\mu} \quad , \quad \overline{N} = \sum_{n=1}^{\infty} \overline{\alpha}^\mu_{-n}\overline{\alpha}_{n\mu}. \tag{2.28c}$$

Note that except for the bosonic zero-modes q^μ, α_0^μ, and $\overline{\alpha}_0^\mu$ giving the space-time position of the center-of-mass of the string, we have twice as many non-zero modes α_n^μ and $\overline{\alpha}_n^\mu$ for the closed string as compared to the open string, corresponding to each moving sector. As said above, such a factorization in characteristic of all closed string theories and is the essential property which allows for a large variety of constructions in the conformal gauge.

As the reader has been suspecting since the beginning of this section, the conformal gauge-fixing conditions (2.19) actually do not completely fix the world-sheet parametrization. This may be seen in many ways.

We made it plausible that the constraints (2.16a) are the generators of the local reparametrization gauge invariance of the system. In the conformal gauge, these constraints precisely reduce to the gauge-fixing conditions, whose Fourier modes in σ are the Virasoro generators. The fact that these latter quantities must vanish for physical solutions thus points to the existence of a remaining gauge invariance in the conformal gauge, generated by the Virasoro generators.

As was noted above, the common scale of the tangent vectors $\dot{x}^\mu$ and x'^μ is not fixed by the conformal gauge conditions (2.19). Actually, there exist local reparametrizations $\tilde{\xi}^\alpha(\xi)$(i.e. non-singular and orientation preserving transformations) preserving the conformal gauge-fixing conditions. Such transformations must satisfy the equations

$$\frac{\partial\tilde{\tau}}{\partial\tau} = \frac{\partial\tilde{\sigma}}{\partial\sigma}, \frac{\partial\tilde{\tau}}{\partial\sigma} = \frac{\partial\tilde{\sigma}}{\partial\tau}. \tag{2.29}$$

For a world-sheet metric of Euclidean signature, the second of these relations would involve a minus sign. These relations would then correspond to the Cauchy conditions for analytic or conformal transformations on the complex plane. The condition (2.29) thus corresponds to (pseudo)-conformal transformations, the remaining local gauge invariance of the system in the conformal gauge (hence the name).

The general solution to (2.29) is of the form

$$\tilde{\tau} + \tilde{\sigma} = f_+(\tau + \sigma), \tilde{\tau} - \tilde{\sigma} = f_-(\tau - \sigma). \tag{2.30}$$

It may be checked that the world-sheet metric $\gamma_{\alpha\beta}$(or $g_{\alpha\beta}$) is then indeed rescaled by a local factor.

Depending on the world-sheet topology, the functions f_+ and f_- above have to satisfy some boundary conditions in σ. These functions are associated to the existence of conformal Killing vectors on the world-sheet. In the case of free closed and open strings, such conformal transformations may be shown to exist. Hence, the solutions given above in the conformal gauge are not the most general solutions to the two-dimensional massless Klein-Gordon equation. The general solution is obtained as in (2.23) and (2.26), with σ and τ given through two functions f_+ and f_- satisfying the appropriate boundary conditions, as in (2.30). Of course, all these solutions correspond to the same physical configuration of the system.

In conclusion, in the conformal gauge, the system has reduced to a set of conformal invariant "free" massless scalar fields on the world-sheet. The existence of this conformal symmetry follows from a remaining reparametrization gauge invariance of the system in the conformal gauge. This is also the reason why physical solutions must satisfy the Virasoro conditions $L_n^{(\alpha)} = 0, \overline{L}_n^{(\alpha)} = 0$. Indeed, as will be shown, these quantities are precisely the generators of conformal transformations in the conformal gauge.

2.1.3 The light-cone gauge

It is possible to completely fix the local reparametrization invariance of the system in the conformal gauge, by introducing an additional condition fixing the local scale of the world-sheet metric. However, as this amounts to solving explicitly the Virasoro constraints, such a complete gauge fixing cannot be achieved in an explicitly space-time covariant manner.

Since conformal reparametrizations (2.29) also satisfy the massless Klein-Gordon equation, it may be possible to fix this local scale by relating σ or τ to some linear combination of $x^\mu(\sigma,\tau)$ in the conformal gauge. Due to the boundary conditions in σ, both for the open and for the closed string, this may actually be done only for the time evolution parameter τ. Moreover, to maintain explicit space-time convariance as much as possible, let us introduce some constant space-time vector n^μ to define the additional gauge-fixing condition.

Thus finally, complete gauge-fixing of reparameterization invariance of the bosonic string is obtained through the three conditions,

$$(\dot{x} \pm x')^2 = 0, \quad n_\mu x^\mu(\sigma,\tau) = 2\alpha' n_\mu p^\mu \tau, \tag{2.31}$$

the first two being the conformal gauge conditions, the third fixing the scale of the world-sheet metric. The proportionality constant in this last condition is determined by relating it to the total momentum of the string. Let us also remark that use was made of the invariance under constant shifts in τ (these are indeed conformal transformations) to remove any constant which could appear in the last condition.

That the conditions (2.31) completely fix the world-sheet parametrization may best be seen from their geometrical meaning. A condition of the form $\eta_\mu x^\mu$ = constant defines a $(D-1)$-hyperplane. Hence, the third gauge fixing condition determines for each value of τ, a hyperplane intersecting the world-sheet. This then determines the parametrization in σ for constant τ, up to a constant shift in σ for the closed string. By the conformal gauge-fixing conditions, the parametrization in τ is then also determined.

The additional non-covariant gauge-fixing condition explicitly breaks the space-time Poincaré group down to the little group of the vector n_μ (for a résumé, see Ref. 17), which is still an explicit space-time symmetry of the gauge-fixed system. A priori, any vector n^μ may be used[29]. The most convenient choice turns out to be a light-like vector[30]

$n^\mu, n^2 = 0$, in which case the little group is isomorphic to the $(D-2)$-dimensional Euclidean group, having as subgroup the group $SO(D-2)_n$ of space rotations transverse to the vector n^μ. Such a choice corresponds to the light-cone gauge.

By convention, one takes for the light-like vector:

$$n^\mu = (\frac{1}{\sqrt{2}}, 0, \ldots, 0, \frac{-1}{\sqrt{2}}), \tag{2.32}$$

and one introduces the corresponding light-cone coordinates for any vectors u^μ, v^μ:

$$u^\pm = \frac{1}{\sqrt{2}}(u^0 \pm u^{D-1}), \quad u^i \qquad i = 1, \ldots D-2, \tag{2.33}$$

with

$$u.v = -u^+v^- - u^-v^+ + \sum_{i=1}^{D-2} u^i v^i. \tag{2.34}$$

The generators of the connected part of the little group of n^μ are then M^{+i}, M^{ij}, with M^{ij} being the generators of $SO(D-2)_n$. The remaining generators of the Lorentz group, M^{-i} and $M^{0(D-1)}$, are not explicit symmetries in the light-cone gauge. Their algebra needs to be checked explicitly, especially at the quantum level where anomalies may arise.

With the use of light-cone gauge-fixing conditions, it is possible to explicitly solve the constraints of reparametrization invariance and express the gauge degrees of freedom in terms of the physical ones[30]. One then finds that the latter degrees of freedom are the transverse coordinates $x^i(\sigma,\tau)$ (and their associated conjugated momenta $P^{oi}(\sigma,\tau)$), whereas the gauge degrees of freedom are the light-cone components given by:

$$x^+(\sigma,\tau) = 2\alpha' P^+ \tau, P^{o+}(\sigma,\tau) = \frac{1}{\pi} P^+, \tag{2.35}$$

and

$$P^{o-}(\sigma,\tau) = \frac{\pi}{2P+}[(P^{oi})^2 + (\frac{x'^i}{2\pi\alpha'})^2], \tag{2.36}$$

$$x'^-(\sigma,\tau) = \frac{\pi}{P+} P^{oi}(\sigma,\tau) x'^i(\sigma,\tau). \tag{2.37}$$

The solution to (2.37) is

$$x^-(\sigma,\tau) = \sqrt{2\alpha'} q^- + 2\alpha' P^- \tau + [1 - \frac{1}{\pi}\int_0^\pi d\sigma]. \int_0^\sigma d\sigma' \frac{\pi}{P+} P^{oi}(\sigma',\tau) \frac{\partial x^i}{\partial \sigma'}(\sigma',\tau), \tag{2.38}$$

with q^- an integration constant, and P^- given by

$$P^- = \int_0^\pi d\sigma P^{o-}(\sigma,\tau). \tag{2.39}$$

Hence, given the physical degrees of freedom x^i, P^{oi}, q^- and P^+, the gauge degrees of freedom $x^\pm, P^{o\pm}$ are uniquely determined and so is the trajectory of the bosonic string in space-time.

The relations above may also be expressed through mode expansions as in the conformal gauge. For open strings, this gives

$$x^i(\sigma,\tau) = \sqrt{2\alpha'}(q^i + \alpha_0^i \tau + i \sum_n{}' \frac{1}{n} \alpha_n^i e^{-in\tau} \cos n\sigma), \tag{2.40a}$$

with

$$\alpha_0^i = \sqrt{2\alpha'} P^i. \tag{2.40b}$$

For the light-cone components, one obtains:

$$q^+ = 0, \quad \alpha_n^+ = \sqrt{2\alpha'} P^+ \delta_{n,0}, \tag{2.41a}$$

$$q^- = q^-, \quad \alpha_n^- = \frac{1}{\sqrt{2\alpha'} P^+} L_n^\perp, \tag{2.41b}$$

with the transverse Virasoro generators

$$L_n^\perp = \frac{1}{2} \sum_m \alpha_{n-m}^i \alpha_m^i. \tag{2.42}$$

This leads to the mass formula:

$$\alpha' M^2 = N^\perp \equiv \sum_{n=1}^{\infty} \alpha_{-n}^i \alpha_n^i, \tag{2.43}$$

thus showing that the classical mass spectrum is indeed positive definite, a property not so obvious in the conformal gauge.

For the closed string, the corresponding results are:

$$x^i(\sigma, \tau) = \sqrt{2\alpha'}[q^i + (\alpha_0^i + \overline{\alpha}_0^i), \tau + \frac{1}{2} i \sum_n{}' \frac{1}{n} (\alpha_n^i e^{-2in(\tau-\sigma)} + \overline{\alpha}_n^i e^{-2in(\tau-\sigma)})], \tag{2.44a}$$

with

$$\alpha_0^i = \frac{1}{2}\sqrt{2\alpha'} P^i = \overline{\alpha}_0^i, \tag{2.44b}$$

and

$$q^+ = 0 \quad , \quad \alpha_n^+ = \frac{1}{2}\sqrt{2\alpha'} P^+ \delta_{n,o} = \overline{\alpha}_n^+, \tag{2.45a}$$

$$q^- = q^- \quad , \quad \alpha_n^- = \frac{2}{\sqrt{2\alpha'} p^+} L_n^\perp, \; \overline{\alpha}_n^- = \frac{2}{\sqrt{2\alpha'} P^+} \overline{L}_n^\perp, \tag{2.45b}$$

with

$$\alpha_0^- = \frac{1}{2}\sqrt{2\alpha'} P^- = \overline{\alpha}_0^-, \tag{2.45c}$$

$$L_n^\perp = \frac{1}{2} \sum_m \alpha_{n-m}^i \alpha_m^i, \; \overline{L}_n^\perp = \frac{1}{2} \sum_m \overline{\alpha}_{n-m}^i \overline{\alpha}_m^i. \tag{2.46}$$

The mass formulae are then:

$$\frac{1}{2}\alpha' M^2 = N^\perp + \overline{N}^\perp, \tag{2.47a}$$

$$N^\perp = \overline{N}^\perp, \tag{2.47b}$$

with

$$N^\perp = \sum_{n=1}^{\infty} \alpha_{-n}^i \alpha_n^i, \overline{N}^\perp = \sum_{n=1}^{\infty} \overline{\alpha}_{-n}^i \overline{\alpha}_n^i. \tag{2.47c}$$

As we shall see, the constraints $\alpha_0^- = \overline{\alpha}_0^-$, or equivalently $N^\perp = \overline{N}^\perp$, is the consequence of the fact that in the case of closed strings, the light-cone gauge-fixing conditions fix the parametrization in σ only up to a constant.

2.2 Hamiltonian Quantization

As is well-known, two different but essentially equivalent methods are available for quantizing a given system. One is the operatorial or canonical quantization approach. The other is the path-integral approach. In these lectures, we shall only consider the former approach for the quantization of string theories.

For regular system, i.e. systems without constraints, canonical quantization consists in the following procedure. Given a description of the system through an action principle, with a Lagrange function depending on coordinates and velocities, the Hamiltonian formulation is derived through the usual Legendre transform. The system is then described by a phase-space, i.e. the space of coordinates and their conjugate momenta, with time-evolution determined by a Hamiltonian function through its Poisson brackets. These Poisson brackets define a symplectic structure on phase-space. Canonical quantization of the system then proceeds by the correspondence principle. Phase-space degrees of freedom now correspond to quantum operators acting on a space of states. The algebra of these operators is given from the Poisson brackets : the value of the commutator (or anticommutator for Grassmannian quantities) of two operators is obtained as $(i\hbar)$ times the result of the corresponding Poisson bracket (for example $\{q, p\} = 1$ leads to $[q, p] = i\hbar$, for commuting degrees of freedom q and p). In addition, the space of states is equipped with an inner product such that the relevant adjointness properties of the quantum operators are satisfied. Finally, time-evolution on the space of states is determined by the Schrödinger equation, with the quantum Hamiltonian operator given by the classical Hamiltonian function, through a specifically chosen ordering of quantum operators.

It is then also possible to obtain a phase-space path-integral representation for any quantum matrix element, which is unambiguously defined in so far that the quantum system itself is uniquely determined.

If the integral over conjugate momenta may be completed, one then obtains the configuration space path-integral quantization approach for the same quantum system.

This is the approach which we shall adopt here. The presence of constraints however, renders the analysis more involved. The general Hamiltonian formulation of constrained systems has been given by Dirac[31] (for a résumé see Refs. 17,32) but we shall not need to consider this discussion in full generality here. Within that formalism, there are two possible ways of quantizing a constrained system.

The first, where the constraints are not solved for but are imposed on quantum physical states, typically leads to an explicitly covariant quantization of the system with states of negative norm. In the case of QED, this corresponds to the Gupta-Bleuler quantization procedure.

In the second procedure, one first solves for the constraints by introducing gauge-fixing conditions. The ensuing reduced phase-space[33] then leads to the gauge-fixed quantized system, in the manner explained above. Typically, the corresponding space of states has a positive-definite inner product, but explicit covariance is lost. For example, this is the case of QED quantized in the Coulomb gauge.

2.2.1 The Hamiltonian formalism[17–20].

In the case of the bosonic string in the Nambu-Goto description, associated to the coordinates $x^\mu(\sigma, \tau)$, we have the conjugate momenta defined by

$$\pi_\mu(\sigma, \tau) = \frac{\partial \mathcal{L}}{\partial \dot{x}^\mu(\sigma, \tau)} = P^0_\mu(\sigma, \tau). \tag{2.48}$$

The corresponding phase-space is then equipped with a symplectic structure defined through the Poisson brackets (we always only give the non-vanishing ones):

$$\{x^\mu(\sigma,\tau),\pi_\nu(\sigma',\tau)\} = \delta^\mu_\nu \delta(\sigma-\sigma'). \tag{2.49}$$

All these phase-space degrees of freedom are not independent however. They must satisfy the primary[31] constraints:

$$\phi_+^{(\alpha)} = 0 \quad , \quad \phi_-^{(\alpha)} = 0, \tag{2.50a}$$

where

$$\phi_\pm^{(\alpha)} = \frac{1}{2}\pi\alpha' \phi_\pm^\mu \phi_{\pm\mu} \quad , \quad \phi_\pm^\mu = \pi^\mu \pm \frac{\partial_\sigma x^\mu}{2\pi\alpha'}. \tag{2.50b}$$

They satisfy the following relations:

$$\{x^\mu(\sigma,\tau),\phi_\pm^{(\alpha)}(\sigma',\tau)\} = \pi\alpha'\phi_\pm^\mu(\sigma',\tau)\delta(\sigma-\sigma') \tag{2.51a}$$

$$\{x^\mu(\sigma,\tau),\phi_\pm^{(\alpha)}(\sigma',\tau)\} = \pm\frac{1}{2}\phi_\pm^\mu(\sigma',\tau)\partial_\sigma\delta(\sigma-\sigma') \tag{2.51b}$$

$$\{\phi_\eta^{(\alpha)}(\sigma,\tau),\phi_{\eta'}^{(\alpha)}(\sigma',\tau)\} = \eta\delta_{\eta\eta'}[\phi_\eta^{(\alpha)}(\sigma,\tau)+\phi_{\eta'}^{(\alpha)}(\sigma',\tau)]\partial_\sigma\delta(\sigma-\sigma'), \eta,\eta' = +,- \tag{2.52}$$

Time-evolution of the system in phase-space is generated by an Hamiltonian, through its Poisson brackets. The canonical Hamiltonian density

$$\mathcal{H}_0 = \dot{x}^\mu\pi_\mu - \mathcal{L} = 0 \tag{2.53}$$

vanishes identically, as a consequence of reparametrization invariance. However, one may also add to it an arbitrary linear combination of the primary constraints[31], and thus generate time-evolution (or translations in τ) with the Hamiltonian

$$H = \int_0^\mu d\sigma[\lambda^+\phi_+^{(\alpha)} + \lambda^-\phi_-^{(\alpha)}], \tag{2.54}$$

where λ^+, λ^- are functions of the world-sheet coordinates. These functions should be such that the constraints $\phi_\pm^{(\alpha)} = 0$ are consistently preserved through time-evolution[31], namely

$$\dot{\phi}_\pm^{(\alpha)} = \{\phi_\pm^{(\alpha)}, H\} = 0 \text{ for } \phi_\pm^{(\alpha)} = 0 \tag{2.55}$$

From (2.52), one sees that this is always the case. Hence, the functions λ^+ and λ^- may be chosen arbitrarily, and the full set of constraints consists of the primary and first-class[31] constraints $\phi_+^{(\alpha)}$ and $\phi_-^{(\alpha)}$, whose algebra is given in (2.52).

The corresponding equations of motion are:

$$\dot{x}^\mu = \{x^\mu, H\} = \pi\alpha'[\lambda^+\phi_+^\mu + \lambda^-\phi_-^\mu], \tag{2.56a}$$

$$\dot{\pi}^\mu = \{\pi^\mu, H\} = \frac{1}{2}\partial_\sigma[\lambda^+\phi_+^\mu - \lambda^-\phi_-^\mu]. \tag{2.56b}$$

Together with the constraint, they also follow from the first-order action on phase-space:

$$S[x^\mu,\pi_\mu;\lambda^+,\lambda^-] = \int_{\tau_1}^{\tau_2} d\tau \int_0^\pi d\sigma[\dot{x}^\mu\pi_\mu - \lambda^+\phi_+^{(\alpha)} - \lambda^-\phi_-^{(\alpha)}]. \tag{2.57}$$

Hence, λ^+, λ^- are the Lagrange multipliers for the constraints.

Solving for π^μ from (2.56a), the action (2.57) reduces to the linear action (2.5), with $g_{\alpha\beta}$ give by

$$g_{\alpha\beta} = e^{\phi}\frac{2}{\lambda^+ + \lambda^-}\begin{bmatrix} -\lambda^+\lambda^- & \frac{1}{2}(\lambda^+ - \lambda^-) \\ \frac{1}{2}(\lambda^+ - \lambda^-) & 1 \end{bmatrix}. \tag{2.58}$$

Here, ϕ is an arbitrary function of ξ^α, which does not appear in the linear action due to its Weyl invariance.

From the transformations generated by the first-class constraints to be discussed below, one may actually show[20] that this matrix indeed transforms as a world-sheet metric tensor under world-sheet reparametrizations. Hence, the Lagrange multipliers λ^+ and λ^- are also components of an intrinsic world-sheet metric, with ϕ being the conformal mode. We thus recover our previous result, that the equations of motion of the world-sheet metric field are the constraints $\phi_\pm^{(\alpha)} = 0$ of local reparametrization invariance. This should convince the reader that the same Hamiltonian analysis applied to the linear action reduces exactly[20] to the one considered here, which follows from the non-linear action.

The first-order action (2.57) is invariant under infinitesimal transformations generated by the first-class constraints. Let us define the quantity

$$\phi_\epsilon^{(\alpha)} = \int_0^\pi d\sigma[\epsilon^+\phi_+^{(\alpha)} + \epsilon^-\phi_-^{(\alpha)}], \tag{2.59}$$

where ϵ^+, ϵ^- are two arbitrary (infinitesimal) functions of (σ, τ), satisfying appropriate boundary conditions for open and for closed strings (for details, see Ref.20). Then, the following variations leave the action (2.57) invariant (up to surface terms which vanish for the appropriate boundary conditions):

$$\begin{aligned} \delta_\epsilon x^\mu &= \{x^\mu, \phi_\epsilon^{(\alpha)}\} = \pi\alpha'[\epsilon^+\phi_+^\mu + \epsilon^-\phi_-^\mu], && (2.60a) \\ \delta_\epsilon \pi^\mu &= \{\pi^\mu, \phi_\epsilon^{(\alpha)}\} = \frac{1}{2}\partial_\sigma[\epsilon^+\phi_+^\mu - \epsilon^-\phi_-^\mu], && (2.60b) \end{aligned}$$

and

$$\begin{aligned} \delta_\epsilon \lambda^+ &= \partial_\tau\epsilon^+ - \lambda^+\partial_\sigma\epsilon^+ + \partial_\sigma\lambda^+\epsilon^+, && (2.61a) \\ \delta_\epsilon \lambda^- &= \partial_\tau\epsilon^- + \lambda^-\partial_\sigma\epsilon^- - \partial_\sigma\lambda^-\epsilon^-. && (2.61b) \end{aligned}$$

Defining η^α by

$$\epsilon^\pm = \lambda^\pm\eta^0 \pm \eta^1, \tag{2.62}$$

and solving for π^μ, it may easily be seen that the variation $\delta_\epsilon x^\mu$ precisely reduces to the variation $\delta_\eta x^\mu = \eta^\alpha\partial_\alpha x^\mu$ induced in the Lagrangian formalism by the infinitesimal reparametrization (2.17). The matrix (2.58) may then also be seen to transform as a world-sheet metric tensor.

Thus, the constraints $\phi_+^{(\alpha)}$ and $\phi_-^{(\alpha)}$ are the generators of local connected gauge transformations, in the Hamiltonian formalism. Their Poisson brackets (2.52) define the algebra of two-dimensional diffeomorphisms.

The previous discussion shows that the arbitrariness in the choice of Lagrange multipliers λ^+, λ^- precisely corresponds to the arbitrariness in the choice of world-sheet parametrization in the Lagrangian formalism. Solutions to the Hamiltonian equations of motion depend on the two arbitrary functions λ^+ and λ^-. Hence, gauge-fixing of the system requires at least some choice for λ^+ and λ^-.

The conformal gauge corresponds to the choice

$$\lambda^+ = 1, \quad \lambda^- = 1. \tag{2.63}$$

Indeed, the equations of motion then reduce to:

$$\dot{x}^\mu = 2\pi\alpha'\pi^\mu, \quad \dot{\pi}^\mu = \frac{x''^\mu}{2\pi\alpha'}, \tag{2.64}$$

and the constraints $\phi_\pm^{(\alpha)} = 0$ then become equivalent to $(\dot{x}\pm x')^2 = 0$. The same solutions as in the Lagrangian approach are thus recovered. The Hamiltonian in the conformal gauge is

$$H = \int_0^\pi d\sigma[\phi_+^{(\alpha)} + \phi_-^{(\alpha)}].$$

The conformal gauge is not a complete gauge-fixing of the system however. There are transformations generated by $\phi_\pm^{(\alpha)}$ which leave the conformal gauge conditions (2.62) invariant but induce a transformation in phase-space. Such transformations correspond to zero-modes of the equation

$$(\partial_\tau \mp \partial_\sigma)\epsilon^\pm(\sigma,\tau) = 0. \tag{2.65}$$

With the use of (2.62) in the conformal gauge, we precisely recognize the infinitesimal (pseudo)conformal reparametrizations discussed previously, corresponding to the remaining gauge invariance of the system in the conformal gauge. The associated quantities ϕ_ϵ are thus the generators of conformal transformations in this gauge.

In the case of open strings, we have the boundary condition[17–20]:

$$(\epsilon^+ - \epsilon^-)(\sigma = 0, \pi, \tau) = 0, \tag{2.66}$$

so that the solutions to (2.65) are

$$\epsilon^+(\sigma,\tau) = \epsilon(\tau+\sigma), \epsilon^-(\sigma,\tau) = \epsilon(\tau-\sigma), \tag{2.67a}$$

with

$$\epsilon(\tau+2\pi) = \epsilon(\tau). \tag{2.67b}$$

Using the mode expansion

$$\epsilon(\tau) = \sum_n \epsilon_n e^{in\tau}, \tag{2.68}$$

we then have

$$\phi_\epsilon^{(\alpha)} = \sum_n \epsilon_n e^{in\tau}, \tag{2.69a}$$

where

$$L_n^{(\alpha)} = \int_0^\pi d\sigma[e^{in(\tau+\sigma)}\phi_+^{(\alpha)} + e^{in(\tau-\sigma)}\phi_-^{(\alpha)}]. \tag{2.69b}$$

These quantities are precisely the Virasoro generators introduce above. They satisfy the conformal algebra

$$\{L_n^{(\alpha)}, L_m^{(\alpha)}\} = -i(n-m)L_{n+m}^{(\alpha)}. \tag{2.70}$$

Note that the generator of translations in τ, namely the Hamiltonian, is the Virasoro zero-mode $L_0^{(\alpha)}$.

In the case of closed strings, we have the boundary conditions[17–20]

$$\epsilon^{\pm}(\sigma=\pi,\tau)=\epsilon^{\pm}(\sigma=0,\tau), \tag{2.71}$$

so that the solutions to (2.65) are

$$\begin{aligned} \epsilon^{\pm}(\sigma,\tau) &= \epsilon_{\pm}(\tau\pm\sigma), && (2.72a)\\ \text{with} \quad \epsilon_{\pm}(\tau+\pi) &= \epsilon_{\pm}(\tau). && (2.72b) \end{aligned}$$

Using the mode expansion

$$\epsilon_{\pm}(\tau)=\sum_n \epsilon_n^{\pm}e^{2in\tau}, \tag{2.73}$$

we then have

$$\begin{aligned} \phi_{\epsilon}^{(\alpha)} &= 2\sum_n[\epsilon_n^{+}\overline{L}_n^{(\alpha)}+\epsilon_n^{-}L_n^{(\alpha)}] && (2.74a)\\ \text{where} \quad \overline{L}_n^{(\alpha)} &= \frac{1}{2}\int_0^{\pi} d\sigma e^{2in(\tau+\sigma)}\phi_+^{(\alpha)},\ L_n^{(\alpha)}=\frac{1}{2}\int_0^{\pi} d\sigma e^{2in(\tau-\sigma)}\phi_-^{(\alpha)}. && (2.74b) \end{aligned}$$

These Virasoro generators, precisely the quantities introduced above, satisfy the two-dimensional conformal algebra

$$\begin{aligned} \{L_n^{(\alpha)},L_m^{(\alpha)}\} &= -i(n-m)L_{n+m}^{(\alpha)}, && (2.75a)\\ \{\overline{L}_n^{(\alpha)},\overline{L}_m^{(\alpha)}\} &= -i(n-m)\overline{L}_{n+m}^{(\alpha)}, && (2.75b)\\ \{L_n^{(\alpha)},\overline{L}_n^{(\alpha)}\} &= 0. && (2.75c) \end{aligned}$$

Note that the generator of translations in τ is now given by $2(L_0^{(\alpha)}+\overline{L}_0^{(\alpha)})$. The generator of translations in σ is easily seen to be $2(L_0^{(\alpha)}-\overline{L}_0^{(\alpha)})$. Hence, the constraint $L_0^{(\alpha)}=\overline{L}_0^{(\alpha)}$, or equivalently $N=\overline{N}$, is consequence of the fact that the origin in σ has no physical meaning for physical states of closed strings.

These results show, as was claimed previously, that the Virasoro generators $L_n^{(\alpha)},\overline{L}_n^{(\alpha)}$ are indeed the generators of conformal invariance, the remaining gauge invariance of the system in the conformal gauge. Thus, they must vanish for any physical solution, since otherwise the solution would depend on which conformal parametrization is used to describe it. Requiring that this local conformal invariance is realized at the quantum level leads to restrictions on the system, as we shall discuss.

To obtain the Hamiltonian formalism in reduced phase-space, two additional gauge-fixing conditions must be introduced, in order to completely fix the remaining gauge invariance of the conformal gauge. Corresponding to the light-cone gauge, these two conditions are[34] :

$$\begin{aligned} \Omega_1(\sigma,\tau) &\equiv x^{+}(\sigma,\tau)-2\alpha' P^{+}\tau=0, && (2.76a)\\ \Omega_2(\sigma,\tau) &\equiv \pi^{+}(\sigma,\tau)-\frac{1}{\pi}P^{+}=0, && (2.76b) \end{aligned}$$

with P^+ being some integration constant.

The full set of constraints $\phi_+^{(\alpha)},\phi_-^{(\alpha)},\Omega_1,\Omega_2$ then becomes second-class[31], so that by the use of Dirac brackets[31], one may explicitly solve for these constraints. The gauge degrees of freedom $x^{\pm}(\sigma,\tau), \pi^{\pm}(\sigma,\tau)$ are then determined in terms of the physical degrees of freedom $x^i(\sigma,\tau), q^-$ and their conjugate momenta $\pi^i(\sigma,\tau), P^+$,as was explained previously. The symplectic structure defined on this reduced phase-space , which is consistent with the solutions to the constraints, is given by the Dirac brackets:

$$\{\sqrt{2\alpha'}q^-, P^+\}_D = -1, \tag{2.77a}$$
$$\{x^i(\sigma,\tau), \pi^j(\sigma',\tau)\}_D = \delta^{ij}\delta(\sigma-\sigma'). \tag{2.77b}$$

Time-evolution is then generated through these brackets by the light-cone gauge Hamiltonian:

$$H_{l.c.} = \pi\alpha' \int_0^{\pi} d\sigma[(\pi^i)^2 + (\frac{x'^i}{2\pi\alpha'})^2] = 2\alpha' P^+ P^-. \tag{2.78}$$

2.2.2 Old covariant quantization

The so-called "old covariant quantization" corresponds to canonical quantization in the conformal gauge. By the correspondence principle, the phase-space degrees of freedom become quantum operators, whose commutation relations are given from their Poisson brackets by

$$[x^\mu(\sigma,\tau), \pi_\nu(\sigma',\tau)] = i\hbar\delta^\mu_\nu\delta(\sigma-\sigma'). \tag{2.79}$$

In terms of the mode expansions in the conformal gauge, we have equivalently ($\hbar = 1$):

- open string:

$$[\sqrt{2\alpha'}q^\mu, P^\nu] = i\eta^{\mu\nu}, \quad [\alpha^\mu_n, \alpha^\nu_m] = n\eta^{\mu\nu}\delta_{n+m,0}, \tag{2.80}$$

- closed string:

$$[\sqrt{2\alpha'}q^\mu, P^\nu] = i\eta^{\mu\nu}, \quad [\alpha^\mu_n, \alpha^\nu_m] = n\eta^{\mu\nu}\delta_{n+m,0} = [\overline{\alpha}^\mu_n, \overline{\alpha}^\nu_m], \tag{2.81}$$

with the adjointness properties:

$$(q^\mu)^+ = q^\mu, (P^\mu)^+ = P^\mu, (\alpha^\mu_n)^+ = \alpha^\mu_{-n}, (\overline{\alpha}^\mu_n)^+ = \overline{\alpha}^\mu_{-n}. \tag{2.82}$$

These algebras are very familiar. They are tensor product algebras of:

- position-momentum algebras for q^μ and P^μ, for each value of μ,
- harmonic oscillator algebras with creation operator α^μ_{-n} or $\overline{\alpha}^\mu_{-n}$ and annihilation operator $\alpha^\mu_n, \overline{\alpha}^\mu_n$ (and the wrong sign of the commutator for time components), for each value of μ and $n \geq 1$.

Note that for the closed string, the algebra of non-zero modes is the tensor product of that of the open string with itself. This tensor product structure extends to the space of states, and is characteristic of closed string theories.

The corresponding spaces of states are abstract linear representation spaces of these algebras, equipped with an inner product consistent with the properties (2.82). Hence, the space of states is obtained by acting with the creation operators α^μ_{-n} (for open strings) or α^μ_{-n} and $\overline{\alpha}^\mu_{-n}$ (for closed strings) on Fock vacua or ground states. These ground states are, say, eigenstates of the momentum operator P^μ, and are annihilated by the operators $\alpha^\mu_n, \overline{\alpha}^\mu_n (n \geq 1)$, namely:

$$P^\mu|\Omega; p> = p^\mu|\Omega; p>, \tag{2.83a}$$
$$\alpha^\mu_n|\Omega; p> = 0, \overline{\alpha}^\mu_n|\Omega; p> = 0, n \geq 1. \tag{2.83b}$$

They are normalized by

$$<\Omega; p|\Omega; p'> = \delta^{(D)}(p-p'). \tag{2.83c}$$

Due to the opposite sign for the time components of the oscillator algebra, many states actually are of negative norm. For example, the states $\alpha^{\mu}_{-n}|\Omega; p>$ $(n \geq 1)$ are all of negative norm for $\mu = 0$. The presence of such states is the price to pay for having explicit space-time covariance in the quantized theory. To avoid any problems with unitarity and causality, none of these negative-norm states should be a physical state.

To define physical states, we have to consider Virasoro generators, which are composite operators. Hence, we first need to specify a normal ordering for the fundamental operators. In the conformal gauge this is easy, since we are only dealing with free fields. As usual, the ordering chosen is defined by bringing all the position and creation operators to the left of all momentum and annihilation operators, which is denoted by double dots. From now on, all quantum operators are understood to be normal-ordered, so that for example:

$$L_n^{(\alpha)} = \frac{1}{2}\sum_n : \alpha^{\mu}_{n-m}\alpha_{m\mu} : . \tag{2.84}$$

Note that among Virasoro generators, normal ordering may only affect the zero-mode operators $L_0^{(\alpha)}, \overline{L}_0^{(\alpha)}$.

Since only matrix elements of the Virasoro generators need to vanish between physical states, it is enough to define quantum physical states by the following Virasoro constraints:

- open strings:

$$[L_0^{(\alpha)} - a]|\psi >= 0, \;\; L_n^{(\alpha)}|\psi >= 0, n \geq 1, \tag{2.85}$$

- closed string:

$$[L_0^{(\alpha)} + \overline{L}_0^{(\alpha)} - 2a]\psi = 0, \;\; [L_0^{(\alpha)} - \overline{L}_0^{(\alpha)}]|\psi >= 0, \tag{2.86a}$$

$$L_n^{(\alpha)}|\psi >= 0, \;\; \overline{L}_n^{(\alpha)}|\psi >= 0, n \geq 1 \tag{2.86b}$$

Here, "a" is a subtraction constant, so far not specified, following from normal ordering. From the constraint involving the zero-modes, we obtain the quantum mass formulae:

- open string:

$$\alpha' M^2 = N - a, \qquad N = \sum_{n=1}^{\infty} \alpha^{\mu}_{-n}\alpha_{n\mu}, \tag{2.87}$$

- closed string:

$$\frac{1}{2}\alpha' M^2 = N + \overline{N} - 2a, \;\; N = \overline{N}, \tag{2.88a}$$

$$N = \sum_{n=1}^{\infty} \alpha^{\mu}_{-n}\alpha_{n\mu}, \;\; \overline{N} = \sum_{n=1}^{\infty} \overline{\alpha}^{\mu}_{-n}\overline{\alpha}_{n\mu}. \tag{2.88b}$$

Note that Fock vacua are always physical states of positive norm, but that they are tachyonic, both for open and closed strings, when $a > 0$.

Generally, normal ordering would also affect the conformal algebra, as follows

$$[L_n^{(\alpha)}, L_m^{(\alpha)}] = (n-m)L_{n+m}^{(\alpha)} + \frac{1}{12}c(n)\delta_{n+m,0}, \tag{2.89a}$$

$$[\overline{L}_n^{(\alpha)}, \overline{L}_m^{(\alpha)}] = (n-m)\overline{L}_{n+m}^{(\alpha)} + \frac{1}{12}c(n)\delta_{n+m,0}, \tag{2.89b}$$

$$[L_n^{(\alpha)}, \overline{L}_m^{(\alpha)}] = 0. \tag{2.89c}$$

Here, $c(n)$ is a *central extension* due to normal ordering, or short distance singularities in the two dimensional field theory describing the string theory in the conformal gauge. This central extension corresponds to a conformal anomaly[9,22], or a Schwinger term, in the algebra of the world-sheet energy-momentum tensor of the system in the conformal gauge. By the Jacobi identity, the central extension is always of the form[35]

$$c(n) = cn^3 + \beta n \tag{2.90}$$

where "c" is the *central charge* of the Virasoro algebra[9] (2.89a) and β may be modified by constant shifts in $L_0^{(\alpha)}$.

Normal ordering however, does not affect explicit space-time covariance. Indeed, it may be checked that the Poincaré algebra is satisfied by the quantum operator P^μ and $M^{\mu\nu}$. This implies that the space of states transforms as an infinite set of irreducible representations of the Poincaré group. Moreover, since the Virasoro generators and the excitation level operators N and $\overline{N}$ commute with P^μ and $M^{\mu\nu}$, all physical states at each excitation level fall into such irreducible representations.

Hence, physical states may be classified according to their mass and to their "spin". By "spin" , we mean here the representation of the rotation subgroup of their little group[17]. For massless states, "spin" is thus given by a representation of $SO(D-2)$, and for massive states (with $m^2 > 0$) by a representation of $SO(D-1)$. For example in four dimensions, we then have representations of $SO(2)$ (helicity) and $SO(3)$(spin) respectively.

This discussion completes the characterization of the quantum system and its physical states. We have to make certain however, that none of the physical states is of negative norm, as otherwise unitarity would be spoiled in amplitudes where such states contribute as intermediate states. The analysis of this question requires first of all an analysis of the Virasoro algebra (2.89), which is most efficiently done by using techniques of conformal field theory in two dimensions (see for example Refs. 9,15,16,17,19).

The result of such an analysis is that the Virasoro algebra for the bosonic string is given by:

$$[L_n^{(\alpha)}, L_m^{(\alpha)}] = (n-m)L_{n+m}^{(\alpha)} + \frac{1}{12}Dn(n^2-1)\delta_{n+m,0}, \tag{2.91}$$

or

$$c(n) = Dn^3 - Dn. \tag{2.92}$$

Hence, each scalar field $x^\mu(\sigma,\tau)$ in two dimensions contributes a unit $(+1)$ to the central charge of the Virasoro algebra.

Considering then the problem of negative norm physical states, the no-ghost theorem[15,36] asserts that:
the necessary and sufficient conditions such that none of the physical states is of negative norm are:

- $a \leq 1$
- if $a = 1 : D \leq 26$,
- if $a < 1 : D < 26$ i.e. $D \leq 25$.

The interested reader may check (but not prove) these statement by solving the problem for the first few excitation levels (see Ref.17).

Although these conditions are necessary for quantum consistency, they may not be sufficient. Indeed, when studying 1-loop amplitudes, one finds[37] (see for example Refs.

15,38) that the correct pole and cut singularities due to intermediate states can only be obtained for $D = 26$. Thus, the necessary conditions for the absence of negative-norm physical states and 1-loop unitarity require that

$$D = 26, \quad a = 1 \tag{2.93}$$

(as we shall comment upon later on, these are still not sufficient conditions for quantum consistency).

Under the conditions (2.93), the statement of the no-ghost theorem is actually more detailed[15,36]. For the open string for example, we then have that any physical state $|\psi>$ is necessarily of the form

$$|\psi>=|\phi_0> + L_{-1}|\chi_1> + (L_{-2} + \frac{3}{2}L_{-1}^2)|\chi_2>, \tag{2.94}$$

where $|\psi_0>$ is a positive-norm transverse state (in 1-to-1 correspondence with a state in the light-cone gauge, to be discussed below), and $L_{-1}|\chi_1>, (L_{-2} + \frac{3}{2}L_{-1}^2)|\chi_2>$ are zero-norm physical states, such that

$$L_0|\chi_1>=0, (L_0+1)|\chi_2>=0, \ L_n|\chi_{1,2}>=0, n \geq 1. \tag{2.95}$$

Unphysical and zero-norm states are necessary for explicit space-time covariance. Although zero-norm states are physical by the definitions above, they actually (must) decouple in physical amplitudes (this point will be discussed further in chapter 4). The situation is analogous to the one of QED in the Lorentz gauge, where the longitudinal photon is precisely such a zero-norm physical state which decouples from amplitudes.

With the critical values (2.93) , let us describe the first few excitation levels of bosonic strings, beginning with the open string.

The physical ground states are states $|\Omega; p>$, with $N = 0$, such that $\alpha' m^2 = -1$. These are space-time scalar tachyonic states, thus spelling disaster for the quantum theory which is then not unitary.

At the next excitation level, we have states $\epsilon_\mu \alpha_{-1}^\mu |\Omega; p>$ with $N = 1$, hence $\alpha' m^2 = 0$. These massless states must satisfy the transversality condition $\epsilon_\mu p^\mu = 0$, which has two types of solutions. One, consisting of the 24 linearly independent vectors ϵ_μ such that $\epsilon^2 = 1, \epsilon p = 0$, corresponds to a positive-norm massless transverse state transforming as the vector representation $\Box_{SO(24)}$. The other, with $\epsilon^\mu = p^\mu$, corresponds to a zero-norm massless scalar state, transforming as a singlet under $SO(24)$, namely the longitudinal component of the massless vector. Note that this state is given as $L_{-1}|\Omega; p>$ (see (2.94)).

At the second excitation level $N = 2$, one finds three states; a positive-norm 2-index symmetric traceless transverse massive state, transforming as $\Box\Box_{SO(25)}$ and having 324 components, a zero-norm transverse massive vector state transforming as $\Box_{SO(25)}$, and a zero-norm massive scalar state transforming as a singlet under $SO(25)$. These states are respectively given by $\frac{1}{\sqrt{2}}\lambda_{\mu\nu}\alpha_{-1}^\mu \alpha_{-1}^\nu |\Omega; p>$, $L_{-1}\epsilon_\mu \alpha_{-1}^\mu |\Omega; p>$ and $(L_{-2} + \frac{3}{2}L_{-1}^2)|\Omega; p>$, where the polarization tensors satisfy $\lambda_{\mu\nu}\lambda^{\mu\nu} = 1, \lambda_{\mu\nu} - \lambda_{\nu\mu} = \lambda_\mu^\mu = p^\mu \lambda_{\mu\nu} = 0$ and $\epsilon^2 = 1, \epsilon p = 0$ (compare with (2.94)).

From the mass formula (2.87), it is clear that all states fall on "Regge trajectories" of slope α'. Indeed, if space-time were four dimensional, spin would be characterized by a single number J, and the mass formula would then lead to the linear dependence $J = \alpha' m^2 +$ constant.

Similar results easily follow for closed strings as well. From the mass formulae (2.88), the slope of "Regge trajectories " is now $\frac{1}{2}\alpha'$. The physical ground states $|\Omega; p>$, with

$\frac{1}{2}\alpha' m^2 = -2$, are space-time scalar tachyons in this sector of the theory as well. At the first excitation level $N = \overline{N} = 1$, we have massless states, which for those of positive norm combine into a 2-index symmetric transverse traceless tensor of $SO(24)$ with 299 components, a 2-index antisymmetric transverse tensor of $SO(24)$ with 276 components, and a scalar state of $SO(24)$ with 1 component. These states correspond respectively to a massless "spin 2" state or graviton, a massless antisymmetric tensor, and a massless scalar or dilaton.

One of the remarkable features of these spectra is the appearance of massless "spin 1" and "spin 2" states. Usually, the existence of such particle is associated with some local gauge invariance, a Yang-Mills gauge invariance for "spin 1" states and a space-time reparametrization invariance for "spin 2" states. Actually, the low-energy behavior of these string states is precisely that of such gauge bosons[39].

The profound origin of such space-time local symmetries in string theories is still not really understood. Nevertheless, such a relationship between world-sheet and space-time local symmetries seems to be a generic feature of string theory constructions. Here, world-sheet reparametrization invariance is related to space-time reparametrization invariance for closed strings and space-time Yang-Mills invariance for open strings. Later, we shall see how this extends to world-sheet supersymmetry and space-time supersymmetry, and to world-sheet current algebra and space-time Yang-Mills invariance.

2.2.3 Light-cone gauge quantization

Quantization in the light-cone gauge[30] proceeds through the correspondence principle from the Dirac brackets in the reduced phase-space Hamiltonian formulation in that gauge. Hence, we have the commutation relations

$$[\sqrt{2\alpha'}q^-, P^+] = -i\hbar, \;\; [x^i(\sigma,\tau), \pi^j(\sigma',\tau)] = i\hbar\delta^{ij}\delta(\sigma-\sigma'), \tag{2.96}$$

or equivalently in terms of the mode expansion ($\hbar = 1$):

- open string:

$$[\sqrt{2\alpha'}q^-, P^+] = -i, \;\; [\sqrt{2\alpha'}q^i, P^j] = i\delta^{ij}, \tag{2.97a}$$
$$[\alpha^i_n, \alpha^j_m] = n\delta^{ij}\delta_{n+m,0}, \tag{2.97b}$$

- closed string:

$$[\sqrt{2\alpha'}q^-, P^+] = -i, \;\; [\sqrt{2\alpha'}q^i, P^j] = \delta^{ij}, \tag{2.98a}$$
$$[\alpha^i_n, \alpha^j_m] = n\delta^{ij}\delta_{n+m,0} = [\overline{\alpha}^i_n, \overline{\alpha}^j_m], \tag{2.98b}$$

with the adjointness properties:

$$(q^-)^+ = q^-, (P^+)^+ = P^+, (q^i)^+ = q^i, (P^i)^+ = P^i, \tag{2.99a}$$
$$(\alpha^i_n)^+ = \alpha^i_{-n}, (\overline{\alpha}^i_n)^+ = \overline{\alpha}^i_{-n}. \tag{2.99b}$$

The structure of these algebras, and of the corresponding spaces of states should now be obvious. Fock vacua $|\Omega, p^i, p^+ >$ are eigenstates of the momentum operators P^i and P^+, , are annihilated by the operators $\alpha^i_n, \overline{\alpha}^i_n (n \geq 1)$, and are normalized by $< \Omega, p^i, p^+ | \Omega, p'^i, p'^+ >= \delta^{(D-2)}(p^i - p'^i)\delta(p^+ - p'^+)$. All states are obtained from these vacua by the action of the creation operators $\alpha^i_{-n}, \overline{\alpha}^i_{-n} (n \geq 1)$. Note again the structure for closed strings as compared to that of open strings.

In contradistinction to old covariant quantization, the present space of states does not have any state of negative norm. All states are physical ones, and correspond to transverse string excitations only.

The light-cone coordinate operators are determined from the constraints. Hence, we have the operational relations:

- open string:

$$q^+ = 0, \quad \alpha_n^+ = \sqrt{2\alpha'}P^+\delta_{n,0}, \tag{2.100a}$$

$$q^- = q^-, \quad \alpha_n^- = \frac{1}{\sqrt{2\alpha' P^+}}(L_n^\perp - a\delta_{n,0}), \tag{2.100b}$$

- closed string:

$$q^+ = 0, \ \alpha_n^+ = \frac{1}{2}\sqrt{2\alpha'}P^+\delta_{n,0} = \overline{\alpha}_n^+, \tag{2.101a}$$

$$q^- = q^-, \alpha_n^- = \frac{2}{\sqrt{2\alpha' P^+}}[L_n^\perp - a\delta_{n,0}], \overline{\alpha}_n^- = \frac{2}{\sqrt{2\alpha' P^+}}[\overline{L}_n^\perp - a\delta_{n,0}], \tag{2.101b}$$

$$\text{with} \quad \alpha_0^- = \overline{\alpha}_0^- = \frac{1}{2}\sqrt{2\alpha'}P^-. \tag{2.101c}$$

It is understood that the transverse Virasoro generators $L_n^\perp, \overline{L}_n^\perp$ have been normal ordered, with the same choice of normal ordering as in the conformal gauge. Here, "a" is the ensuing normal ordering subtraction constant for the Virasoro zero-modes.

Note that the transverse Virasoro generators satisfy the Virasoro algebra with a central extension $c(n) = (D-2)n(n^2-1)$ to which only transverse fields $x^i(\sigma, \tau)$ contribute.

From the previous relations, we have the mass formulae

- open string:

$$\alpha' M^2 = N^\perp - a, \ N^\perp = \sum_{n=1}^{\infty} \alpha^i_{-n}\alpha^i_n,$$

- closed string:

$$\frac{1}{2}\alpha' M^2 = N^\perp + \overline{N}^\perp - 2a, \ N^\perp = \overline{N}^\perp, \tag{2.102a}$$

$$N^\perp = \sum_{n=1}^{\infty} \alpha^i_{-n}\alpha^i_n, \ \overline{N}^\perp = \sum_{n=1}^{\infty} \overline{\alpha}^i_{-n}\overline{\alpha}^i_n. \tag{2.102b}$$

For the closed string the level matching condition for the right-and left-moving sectors follows from the fact that the light-cone gauge fixing conditions do not fix the origin in the parametrization in σ.

Since the light-cone gauge does not preserve space-time covariance explicitly, it may happen that quantization and normal ordering are not compatible with implicit Poincaré covariance. Since the little group of the light-like vector n^μ is an explicit symmetry in the light-cone gauge, only the algebra involving the generators M^{-i} and $M^{0(D-1)}$ needs to be checked. Actually, since only $\alpha_n^-, \overline{\alpha}_n^-$ are given by normal ordered composite operators, only the commutator $[M^{-i}, M^{-j}]$ could be anomalous, whereas it should vanish to maintain implicit space-time Poincaré covariance. An explicit and tedious calculation shows[30] that this commutator is indeed anomalous, unless we have:

$$D - 2 = 24, \qquad a = 1. \tag{2.103}$$

It is only under these conditions that physical states may be characterized by their mass and their "spin". Since in the light-cone gauge, all physical states transform explicitly in representations of the $SO(D-2)_n$ rotation subgroup of the little group of the vector n^μ, it is straightforward to identify the "spin" for massless states. For massive states however, with $m^2 > 0$, only under the conditions (2.103) do the $SO(D-2)_n$ representations combine into representations of $SO(D-1)$, which is the rotation subgroup of the corresponding little group.

This type of consideration alone actually leads to a restriction for the first excitation level only[40]. For example, in the case of the open string, we then have the states $\alpha_{-1}^i|\Omega>$ with $\alpha' m^2 = (1-a)$. These $(D-2)$ states may transform covariantly under the full Lorentz group, and not only under $SO(D-2)_n$, only if they transform as a massless vector state in D dimensions.Hence , Lorentz covariance requires[40] $a = 1$, which implies in turn that the physical ground state is a tachyon. For states at higher levels however, the counting of physical components is always such that the corresponding states *could* be combined into irreducible representations of $SO(D-1)$. Only the explicit calculation of the full Lorentz algebra can ensure that indeed the space of states is organized into irreducible but non-linear representations of the Poincaré group, and that this occurs only if $a = 1$ and $D = 26$.

There exists however an heuristic argument[40] for the value $D = 26$, which does not require the full calculation. The value for "a" follows from normal ordering in the transverse Virasoro zero-modes, and is thus given by the formal series:

$$a = -\frac{1}{2}(D-2)\sum_{n=1}^{\infty} n. \tag{2.104}$$

Using ζ-function regularization, we have

$$\sum_{n=1}^{\infty} n = \sum_{n=1}^{\infty} n^{-s}_{|s=-1} = \zeta(-1) = \frac{-1}{12}, \tag{2.105}$$

leading to

$$a = \frac{D-2}{24}. \tag{2.106}$$

Therefore, the necessary condition for Lorentz covariance $a = 1$ implies the critical value $D = 26$. Clearly, such an argument does depend on which regularization of the series is used. Only the full calculation of the commutator $[M^{-i}, M^{-j}]$ gives an unambiguous answer. Nevertheless, ζ-function regularization happens to give the correct result, and is used as an easy check on results. Later in chapter 4, we shall justify the use of this regularization from the point of view of modular invariance.

With the critical values (2.103), let us briefly describe the first few excitation levels, first for the open string. From the mass formula , we find again the physical ground state to be a scalar tachyon. At the next excitation level, we have the massless states $\alpha_{-1}^i|\Omega; p^i, p^+>$ transforming as $\Box_{SO(24)}$, i.e. as a massless space-time vector state, with

24 components. At the second excitation level $N^{\perp} = 2$, we have massive states with $\alpha' m^2 = 1$, corresponding to $\alpha^i_{-2}|\Omega>$ and $\alpha^i_{-1}\alpha^j_{-1}|\Omega>$ and transforming as $(\Box + \quad + \cdot)_{SO(24)}$. Clearly, they combine into a massive 2-index symmetric traceless tensor of $S0(25)$, with 324 components.

For the closed string, the physical ground-state is also a scalar tachyon with $\frac{1}{2}\alpha' m^2 = -2$. At the next level, we have the massless states $\alpha^i_{-1}\overline{\alpha}^j_{-1}|\Omega>$ transforming as $(\Box + \quad + \cdot)_{SO(24)}$, and thus corresponding to the graviton, the antisymmetric tensor state and the dilaton, with respectively 299, 274 and 1 components.

Clearly, positive norm physical states in the conformal and in the light-cone gauge quantizations are in 1-to-1 correspondence, as guaranteed by the no-ghost theorem (see (2.94)).

It is possible to introduce a function which counts the number of positive norm physical states at each excitation level, i.e. a generating function for level degeneracies. Let us first consider the open bosonic string.

In the light-cone gauge, we have the mass formula

$$\alpha' M^2 = N^{\perp} - 1 \tag{2.107}$$

Let us then introduce the generating function for the degeneracies d_n at the excitation level $N^{\perp} = n$:

$$f_O(q) = \sum_{n=0}^{\infty} d_n q^{2(n-1)} = \sum_{n=0}^{\infty} d_n q^{2\alpha' m_n^2}, \tag{2.108}$$

where q is a complex or real number such that $|q| < 1$. We have:

$$f_O(q) = Tr q^{2(N^{\perp}-1)} \tag{2.109}$$

where the trace is taken only on the non-zero modes $\alpha^i_n (n \neq 0)$. The calculation of the trace is straightforward[17] (the reader may like to first do the calculation for a single harmonic oscillator) , and one finds:

$$f_O(q) = \frac{1}{q^2}\left[\prod_{n=1}^{\infty}\frac{1}{(1-q^{2n})}\right]^{D-2} = q^{\frac{D-26}{12}}\eta^{2-D}(\tau), \tag{2.110}$$

with $\eta(\tau)$ being the Dedekind η-function of number theory and the theory of modular forms:

$$\eta(\tau) = q^{\frac{1}{12}}\prod_{n=1}^{\infty}(1-q^{2n}), \qquad q = e^{i\pi\tau}. \tag{2.111}$$

Expanding (2.110) in powers of q for $D = 26$, one obtains

$$f_O(q) = \eta^{-24}(\tau) = \frac{1}{q^2}[1 + 24q^2 + 324q^4 + 3200q^6 + 25650q^8 + ...], \tag{2.112}$$

thus giving the degeneracies at each excitation level. These numbers may of course be checked by explicit construction of the associated physical states. It may be shown[15] that the degeneracy numbers increase exponentially as the level increases.

For the closed bosonic string, the degeneracy number at excitation level $N^{\perp} = \overline{N}^{\perp} = n$ is simply given by d_n^2. The corresponding generating function is thus:

$$f_{\mathcal{X}}(|q|) = \sum_{n=0}^{\infty} d_n^2 |q|^{4(n-1)} = \sum_{n=0}^{\infty} d_n^2 |q|^{\alpha' m_n^2}, \tag{2.113}$$

where $|q|$ is a real number such that $|q| < 1$. For convenience, let us introduce the complex number

$$q = |q|e^{i\theta/2}, \qquad -\pi \leq \theta \leq \pi. \tag{2.114}$$

The generating function may then be expressed in terms of the corresponding function for the open string, as:

$$f_c(|q|) = \int_{-\pi}^{\pi} \frac{d\theta}{2\pi} |f_0(q)|^2 = \int_{-\pi}^{\pi} \frac{d\theta}{2\pi} |q^{\frac{D-26}{12}} \eta^{2-D}(\tau)|^2. \tag{2.115}$$

Hence, for $D = 26$, we have

$$f_c(|q|) = \int_{-\pi}^{\pi} \frac{d\theta}{2\pi} |\eta(\tau)|^{-48}. \tag{2.116}$$

We shall make use of these results when discussing modular invariance of 1-loop partition functions in chapter 4.

2.3 BRST Quantization

As was discussed extensively, conformal symmetry is the remaining reparametrization gauge invariance of the system in the conformal gauge. Any breakdown of conformal symmetry would thus imply breakdown of reparametrization invariance, and lead to a depence on the world-sheet parametrization used to compute physical quantities.

It seems that such a breakdown of conformal invariance is indeed obtained, since the central extension, or conformal anomaly, of the Virasoro algebra does not vanish, not even for the critical values $D = 26$ and $a = 1$. In other words, reparametrization invariance, which is the fundamental symmetry leading to the constraints $\phi_{\pm}^{(\alpha)} = 0$, which guarantee at the quantum level the decoupling of unphysical and zero-norm states from amplitudes, seems to be broken by quantum anomalies! Does the theory still make sense?

Since the days of Feynman, de Witt, Faddeev and Popov[41], we know that such a question can be meaningfully addressed only when taking into account the ghost sector[41] following from gauge fixing. This is usually done by considering the configuration space path-integral and its gauge-fixing, and using the Faddeev-Popov trick[41]. Such an approach however, does not always give a meaningful answer. In the general case of open algebras of constraints[42], i.e. when the structure coefficients of that algebra are functions on phase-space, one has to resort to more general Hamiltonian[43] or Lagrangian[44] methods developed by Batalin, Fradkin, Vilkovisky and others.

In the spirit of the Hamiltonian approach adopted in these lectures, we shall apply the general Hamiltonian methods leading to BRST quantization[45] of gauge invariant systems, namely the Batalin-Fradkin-Vilkovisky (BFV) formalism[32,42,43].

Actually, although we shall not need this method in its full glory, let us briefly discuss it independently of string theory, not only to give a flavor of its fundamental aspects but also because it provides today the most general approach to the quantization of gauge invariant systems. Moreover, such a brief discussion should help to demystify BRST quantization.

2.3.1 The BFV Hamiltonian formalism

For simplicity, we shall consider the case of a gauge-invariant system with bosonic phase-space degrees of freedom only, and a set of irreducible bosonic first-class constraints

ϕ_α (i.e. constraints which are linearly independent locally on phase-space, and with vanishing Poisson brackets when the constraints are imposed). The generalization to Grassmann (or anticommuting) degrees of freedom and/or to reducible constraints is rather straightforward, and follows the same general approach as the one discussed below[32,42,43].

Time-evolution on phase-space is obtained from a first-class Hamiltonian[31]

$$H_T = H + \lambda^\alpha \phi_\alpha, \tag{2.117}$$

where H is a specific first-class Hamiltonian, and λ^α are Lagrange multipliers for the constraints ϕ_α. We have the algebra of constraints

$$\begin{aligned} \{\phi_\alpha, \phi_\beta\} &= C_{\alpha\beta}^{\ \ \gamma}\phi_\gamma, & (2.118a) \\ \{H, \phi_\alpha\} &= V_\alpha^{\ \beta}\phi_\beta, & (2.118b) \end{aligned}$$

where generally $C_{\alpha\beta}^{\ \ \gamma}$ and $V_\alpha^{\ \beta}$ are structure functions on phase-space.

The corresponding Hamiltonian equations of motion also follow from a first-order action. This action is invariant under infinitesimal variations generated by linear combinations $\epsilon^\alpha\phi_\alpha$ of the constraints. For phase-space degrees of freedom, these variations are given by the corresponding Poisson brackets, and for the Lagrange multipliers, we have[32,42]:

$$\delta_\epsilon \lambda^\alpha = \dot{\epsilon}^\alpha + \lambda^\gamma \epsilon^\beta C_{\beta\gamma}^{\ \ \alpha} - \epsilon^\beta V_\beta^{\ \alpha}. \tag{2.119}$$

The algebra of constraints thus characterizes, in the Hamiltonian formalism, the algebra of the local gauge invariance of the system. (The reader may like to compare this general discussion with the previous analysis for the bosonic string).

In the BFV approach, the original phase-space is extended in two steps[32,42,43]. First, to render the Lagrange multipliers λ^α dynamical degrees of freedom, one introduces bosonic conjugate momenta π_α, with the Poisson brackets:

$$\{\lambda^\alpha, \pi_\beta\} = \delta_\beta^\alpha. \tag{2.120}$$

We then have the set of constraints $G_a = (G_{\alpha_1}, G_{\alpha_2})$, with

$$G_{\alpha_1} = \pi_{\alpha_1}, \qquad G_{\alpha_2} = \phi_{\alpha_2}, \tag{2.121}$$

and the algebra:

$$\begin{aligned} \{G_a, G_b\} &= C_{ab}^{\ \ c} G_c, & (2.122) \\ \{H, G_a\} &= V_a^{\ b} G_b. & (2.123) \end{aligned}$$

In order to compensate for the use of degrees of freedom which are not all independent, one then introduces a set of degrees of freedom of opposite statistics, i.e. Grassmann variables. (In the general case, for Grassmann odd constraints, the corresponding Lagrange multipliers are Grassmann odd, hence also their conjugate momenta, and the additional degrees of freedom which one introduces are then Grassmann even, i.e. bosonic). These BFV ghosts come in conjugate pairs $\eta^a, \mathcal{P}_a$, with the Poisson brackets

$$\{\eta^a, \mathcal{P}_b\} = -\delta_b^a, \tag{2.124}$$

and the properties

$$(\eta^a)^* = \eta^a, \qquad (\mathcal{P}_a)^* = -\mathcal{P}_a. \tag{2.125}$$

The BFV extended phase-space (EPS) thus consists of all these degrees of freedom, namely the original phase-space variables z_A together with $\lambda^\alpha, \pi_\alpha, \eta^a$ and $\mathcal{P}_a$. On this EPS, one also introduces an integer grading, by defining the ghost number as follows:

degree of freedom	$z_A, \lambda^\alpha, \pi_\alpha$	η^a	$\mathcal{P}_a$
ghost number	0	+1	-1

(2.126)

By counting Grassmann even and Grassmann odd degrees of freedom with plus and minus signs respectively, the reader may check that indeed the total number of locally independent degrees of freedom on EPS is the same as in a reduced phase-space approach.

One of the fundamental results of the Hamiltonian BFV formalism[42,43] is the following. Associated to the algebra of constraints ϕ_α, there exists a quantity Q_B, the BRST charge, which is uniquely determined[42], up to canonical transformations in EPS, by the properties:

- $Q_B^* = Q_B$,
- Q_B is Grassmann odd,
- Q_B has ghost number (+1),
- Q_B is nilpotent , i.e. $\{Q_B, Q_B\} = 0$.

The expression of the BRST charge is of the form

$$Q_B = \eta^a G_a - \frac{1}{2}\eta^b \eta^c C_{cb}^{\ \ a} \mathcal{P}_a + \text{“more”}, \tag{2.127}$$

where “more“ stands for additional terms necessary for nilpotency when the structure coefficients are functions rather than constants.

The nilpotency property of Q_B is a very strong condition indeed, since Q_B is a Grassmann variable. This property embodies in one single equation all the information concerning the algebra of constraints and the associated Jacobi identities[42]. Moreover, it only depends on the algebra of constraints, and not on any particular choice of gauge-fixing conditions. Hence, corresponding to gauge transformations induced by the constraints on the original phase-space, we now have BRST transformations on EPS generated by the BRST charge. It is the nilpotency property which now characterizes the gauge invariance of the system.

Note also that the ghost sector and the BRST charge only depend on the algebra of constraints, and in no manner whatsoever on how this algebra is explicitly realized for a given system, i.e. for a given set of phase-space degrees of freedom z_A. Any gauge-invariant system with the same Hamiltonian gauge algebra has the same ghost sector and BRST charge, leading thus to an identical discussion of BRST quantization. This remark is fundamental in the construction of all string theories.

Time evolution on EPS is obtained from a BRST invariant effective Hamiltonian, i.e. having a vanishing Poisson bracket with the BRST charge, of the form:

$$H_{eff} = H_B - \{\Psi, Q_B\}. \tag{2.128}$$

Here, H_B is a particular BRST invariant Hamiltonian, of the form

$$H_B = H + \eta^a V_a{}^b \mathcal{P}_a + \text{“more”}, \tag{2.129}$$

where “more” stands for additional terms necessary for BRST invariance when the structure coefficients are not constants.

The function Ψ is a pure imaginary Grassmann odd function on EPS of ghost number (-1). Its role is that of gauge-fixing the system, and has thus to be chosen appropriately[46] according to the structure of the space of gauge orbits of the system[32,46]. However , the Fradkin-Vilkovisky theorem[42,43] (which is another fundamental result of the BFV approach) establishes that two functions Ψ, corresponding to gauge equivalent sections of the space of gauge orbits[46], lead to an identical physical description of the gauge invariant system (this *does not mean*[46], as is usually stated[42,43], that the description of the system is independent of Ψ).

In order to recover the same physical description as in the old covariant approach, it is necessary that the physical solutions to the equations of motion in EPS have zero ghost and BRST charges[32,42]. This requires a choice of BRST invariant and zero-ghost charge boundary conditions on EPS[32,42]. Clearly, the system may then also be described starting from some BRST invariant first-order action on EPS[32,42].

Finally, BRST quantization simply corresponds to the canonical quantization of the BFV Hamiltonian formalism. Quantum physical states are then defined as BRST invariant states of zero-ghost charge (up to possible ordering constants).

2.3.2 The BFV formalism for bosonic strings[19,20].

Let us apply this general discussion to the bosonic string. We have the phase-space degrees of freedom $x^\mu(\sigma,\tau)$ and $\pi_\mu(\sigma,\tau)$,with the Poisson brackets (2.49). Associated to the Lagrange multipliers $\lambda^\pm(\sigma,\tau)$, we now have the conjugate momenta $\pi_\pm(\sigma,\tau)$, with

$$\{\lambda^\eta(\sigma,\tau), \pi_{\eta'}(\sigma',\tau)\} = \delta^\eta_{\eta'}\delta(\sigma-\sigma'), \eta, \eta' = +, -. \tag{2.130}$$

The full set of constraints $G_a (a = 1,2,3,4)$ is thus

$$G_1 = \pi_-, G_2 = \pi_+; G_3 = \phi^{(\alpha)}_-, G_4 = \phi^{(\alpha)}_+. \tag{2.131}$$

The BFV ghost sector is then given by Grassmann odd fields $\eta^a(\sigma,\tau), \mathcal{P}_b(\sigma',\tau)$ $(a = 1,2,3,4)$, with

$$\{\eta^a(\sigma,\tau), \mathcal{P}_b(\sigma',\tau)\} = -\delta^a_b\delta(\sigma-\sigma'), \tag{2.132}$$

$$(\eta^a)^* = \eta^a, (\mathcal{P}_a)^* = -\mathcal{P}_a. \tag{2.133}$$

From the two-dimensional reparametrization algebra (2.52) and the expression (2.127), it is straightforward to obtain the BRST charge:

$$Q_B = \int_0^\pi d\sigma[\eta^1\pi_- + \eta^2\pi_+ + \eta^3\phi^{(\alpha)}_- + \eta^4\phi^{(\alpha)}_+ - \eta^3\partial_\sigma\eta^3\mathcal{P}_3 + \eta^4\partial_\sigma\eta^4\mathcal{P}_4]. \tag{2.134}$$

In particular, we then have

$$\{\mathcal{P}_3, Q_B\} = -\phi^{(T)}_-, \quad \{\mathcal{P}_4, Q_B\} = -\phi^{(T)}_+, \tag{2.135}$$

with

$$\phi^{(T)}_\pm = \phi^{(\alpha)}_\pm + \phi^{(c)}_\pm, \tag{2.136a}$$

$$\phi^{(c)}_- = -2\partial_\sigma\eta^3\mathcal{P}_3 - \eta^3\partial_\sigma\mathcal{P}_3, \tag{2.136b}$$

$$\phi^{(c)}_+ = +2\partial_\sigma\eta^4\mathcal{P}_4 + \eta^4\partial_\sigma\mathcal{P}_4. \tag{2.136c}$$

The quantities $\phi_\pm^{(c)}$ may be shown to satisfy the same algebra as the bosonic constraints $\phi_\pm^{(\alpha)}$. Hence,the total quantities $\phi_\pm^{(T)}$ are the Hamiltonian generators of world-sheet reparametrizations for all EPS degrees of freedom, with $\phi_\pm^{(\alpha)}$ generating such transformations for the bosonic coordinates x^μ and π_μ, and $\phi_\pm^{(c)}$ generating those for the BFV ghosts. In the conformal gauge, the Fourier modes of $\phi_\pm^{(T)}$ will thus also be the generators of conformal transformations, as was shown for the constrains $\phi_\pm^{(\alpha)}$.

In the present case, since the first-class Hamiltonian H vanishes identically, the BRST invariant effective Hamiltonian is simply given by

$$H_{eff} = -\int_0^\pi d\sigma\{\Psi, Q_B\}, \tag{2.137}$$

where Ψ is some function on EPS.

This is not the place to discuss the problem of the appropriate choice for Ψ[20]. Let us only mention that there does not exist any choice of Ψ such that a global and complete section of the space of gauge orbits is obtained[20]. In other words, the bosonic string suffers[20,47] from a Gribov[48] problem.

Actually, we are only interested here in gauge-fixing the system to the conformal gauge, which is done as follows[20]. Let us consider for the function Ψ:

$$\Psi = \frac{1}{\beta}\mathcal{P}_1(\lambda^- - 1) + \frac{1}{\beta}\mathcal{P}_2(\lambda^+ - 1) + \mathcal{P}_3\lambda^- + \mathcal{P}_4\lambda^+. \tag{2.138}$$

The corresponding effective Hamiltonian and first-order action are then easily obtained. In these expressions, redefine then $\pi_\pm, \mathcal{P}_1$ and $\mathcal{P}_2$ by $\beta\pi_\pm, \beta\mathcal{P}_1$ and $\beta\mathcal{P}_2$ respectively. Only then take the limit $\beta \to 0$.

Through this series of manipulations, the degrees of freedom $\lambda^\pm, \pi_\pm, \eta^{1,2}$ and $\mathcal{P}_{3,4}$ become auxiliary fields, which are determined by the dynamical degrees of freedom $x^\mu, \pi_\mu, \eta^3, \eta^4$ and $\mathcal{P}_1, \mathcal{P}_2$[19,20]. In particular, we have

$$\lambda^- = 1, \qquad \lambda^+ = 1, \tag{2.139}$$

showing that the system has indeed been gauge-fixed to the conformal gauge.

Usually, the choice $\Psi = \mathcal{P}_3 + \mathcal{P}_4$ is believed[49] to correspond to the conformal gauge.This however is not correct[32]. Indeed, it does not lead to the constraints $L_n^{(\alpha)} = 0$ for classical solutions[32]. Only the procedure described above[20,32] leads to conformal gauge-fixing within the BFV formalism.

On the other hand, the procedure just presented is actually singular, since it "squeezes out" part of EPS by taking the limit $\beta \to 0$. It may be shown however[20,32,46], also at the quantum level, that all the associated singularities in some sense "factor out" from the system, and may be ignored. Of course, one looses by this procedure any local information concerning the integration measure over the space of gauge orbits[20,32,46].

In order to make contact with the usual notation[15,16] , let us redefine the dynamical fields as follows:

$$c^- = \eta^3, c^+ = \eta^4, b_{--} = -2i\pi\mathcal{P}_1, b_{++} = -2i\pi\mathcal{P}_2. \tag{2.140}$$

With the procedure outlined above, we then obtain for the first-order action:

$$S_{eff} = \int_{\tau_1}^{\tau_2} d\tau \int_0^\pi d\sigma[\dot{x}^\mu\pi_\mu - (\phi_+^{(\alpha)} + \phi_-^{(\alpha)}) + \frac{i}{2\pi}b_{--}(\partial_\tau + \partial_\sigma)c^- + \frac{i}{2\pi}b_{++}(\partial_\tau - \partial_\sigma)c^+], \tag{2.141}$$

for the BRST charge:

$$Q_B = \int_0^{\pi} d\sigma[c^-\phi_-^{(\alpha)} + c^+\phi_+^{(\alpha)} + \frac{i}{2\pi}c^-\partial_\sigma c^- b_{--} - \frac{i}{2\pi}c^+\partial_\sigma c^+ b_{++}], \tag{2.142}$$

and for the ghost charge:

$$Q_c = \frac{1}{2\pi}\int_0^{\pi} d\sigma[c^- b_{--} + c^+ b_{++}]. \tag{2.143}$$

For the Poisson brackets, we now have:

$$\{x^\mu(\sigma,\tau), \pi_\nu(\sigma',\tau)\} = \delta^\mu_\nu \delta(\sigma - \sigma'), \tag{2.144}$$

$$\{c^-(\sigma,\tau), b_{--}(\sigma',\tau)\} = -2i\pi\delta(\sigma-\sigma') = \{c^+(\sigma,\tau), b_{++}(\sigma',\tau)\}, \tag{2.145}$$

with

$$\{b_{\pm\pm}, Q_B\} = -2i\pi\phi_\pm^{(T)}, \tag{2.146}$$

and

$$\phi_\pm^{(c)} = \frac{\mp i}{2\pi}[2\partial_\sigma c^\pm b_{\pm\pm} + c^\pm \partial_\sigma b_{\pm\pm}]. \tag{2.147}$$

The corresponding BRST invariant boundary conditions in σ are[20]:

- open strings:

$$\partial_\sigma x^\mu(\sigma = 0,\pi;\tau) = 0, \tag{2.148a}$$

$$(c^- - c^+)(\sigma = 0,\pi;\tau) = 0 = (b_{--} - b_{++})(\sigma = 0,\pi;\tau), \tag{2.148b}$$

- closed strings:

$$\text{periodicity in} \quad \sigma \to \sigma + \pi. \tag{2.149}$$

Note that the expression for the BRST charge now explicitly depends on the choice of the conformal gauge. Actually, when solving for the conjugate momenta π_μ, the action (2.141) and the BRST charge (2.124) precisely correspond to the Lagrangian BRST formulation of the theory, first discussed in Ref. 26, and also derived by the usual Faddeev-Popov path-integral approach[1,15]. The (b,c) ghost system is the corresponding Faddeev-Popov ghost sector due to reparametrization invariance. As emphasized previously, this ghost system always appears for any two dimensional reparametrization invariant system which is gauge-fixed in the conformal gauge.

To conclude this classical Hamiltonian BRST approach, let us give the solutions to the equations of motion in the conformal gauge. Obviously, the solutions in the bosonic sector are those given previously. Only the ghost sector needs to be considered.

For open strings, we have

$$c^\pm(\sigma,\tau) = \sum_n c_n e^{-in(\tau\pm\sigma)}, \tag{2.150a}$$

$$b_{\pm\pm}(\sigma,\tau) = \sum_n b_n e^{-in(\tau\pm\sigma)}. \tag{2.150b}$$

Then

$$\phi_\pm^{(k)} = \frac{1}{2\pi}\sum_n L_n^{(k)} e^{-in(\tau\pm\sigma)}, k = T, \alpha, c, \tag{2.151}$$

with $L_n^{(\alpha)}$ given previously, and

$$L_n^{(c)} = \sum_m (n-m) b_{n+m} c_{-m}, L_n^{(T)} = L_n^{(\alpha)} + L_n^{(c)}. \quad (2.152)$$

One also obtains:

$$Q_B = \sum_n [L_n^{(\alpha)} + \frac{1}{2} L_n^{(c)}] c_{-n}, \quad (2.153)$$

$$Q_c = \sum_n c_{-n} b_n, \quad (2.154)$$

and the corresponding Poisson brackets lead to

$$\{c_n, b_m\} = -i\delta_{n+m,0}, \quad (2.155)$$

$$\{b_n, Q_B\} = -iL_n^{(T)}. \quad (2.156)$$

Similarly, for closed strings, we have

$$c^-(\sigma,\tau) = \frac{1}{2}\sum_n c_n e^{-2in(\tau-\sigma)}, c^+(\sigma,\tau) = \frac{1}{2}\sum_n \bar{c}_n e^{-2in(\tau+\sigma)}, \quad (2.157a)$$

$$b_{--}(\sigma,\tau) = 4\sum_n b_n e^{-2in(\tau-\sigma)}, b_{++}(\sigma,\tau) = 4\sum_n \bar{b}_n e^{-2in(\tau+\sigma)}. \quad (2.157b)$$

One also obtains

$$\phi_-^{(k)} = \frac{2}{\pi}\sum_n L_n^{(k)} e^{-2in(\tau-\sigma)}, \qquad k = T, \alpha, c, \quad (2.158a)$$

$$\phi_+^{(k)} = \frac{2}{\pi}\sum_n \bar{L}_n^{(k)} e^{-2in(\tau+\sigma)}, \qquad k = T, \alpha, c, \quad (2.158b)$$

with the same expressions for $L_n^{(k)}, \bar{L}_n^{(k)}$ as for the open string, in terms of right–and left-modes. Finally, we have:

$$Q_B = \sum \{[L_n^{(\alpha)} + \frac{1}{2} L_n^{(c)}] c_{-n} + [\bar{L}_n^{(\alpha)} + \frac{1}{2}\bar{L}_n^{(c)}]\bar{c}_{-n}\}, \quad (2.159)$$

$$Q_c = \sum_n [c_{-n} b_n + \bar{c}_{-n}\bar{b}_n], \quad (2.160)$$

and

$$\{c_n, b_m\} = -i\delta_{n+m,0} = \{\bar{c}_n, \bar{b}_m\}, \quad (2.161)$$

$$\{b_n, Q_B\} = -iL_n^{(T)}, \{\bar{b}_n, Q_B\} = -i\bar{L}_n^{(T)}. \quad (2.162)$$

Hence, the separation into left-and right-modes obtained for the non-zero mode bosonic sector, with each moving sector identical to the structure which appears for the open string, extends to the ghost sector as well.

Finally, it should be clear that the Virasoro generator $L_n^{(k)}, \bar{L}_n^{(k)}$ satisfy the usual algebras (2.70) and (2.75) of conformal transformations.

2.3.3 BRST quantization

BRST quantization in the conformal gauge simply follows by canonical quantization from the previous classical BRST formulation. In the bosonic sector, this leads of course to the same quantum algebra of commutation relations for $q^\mu, P^\mu, \alpha_n^\mu, \overline{\alpha}_n^\mu$ as in the old covariant quantization. In this ghost sector, we now have the anticommutation relations:

$$\{c_n, b_m\} = \delta_{n+m,0}, \quad \{\overline{c}_n, \overline{b}_m\} = \delta_{n+m,0}, \tag{2.163}$$

with

$$c_n^+ = c_{-n} \quad , \quad \overline{c}_n^+ = \overline{c}_{-n}, \tag{2.164a}$$

$$b_n^+ = b_{-n} \quad , \quad \overline{b}_n^+ = \overline{b}_{-n}. \tag{2.164b}$$

Hence, except for the bosonic zero-mode sector, the quantum operator algebra for the closed string theory and thus also the space of states, is essentially the tensor product of those for the open string, with however the important feature that in the ghost sector the right-and left-moving ghost modes anticommute rather than commute with one another, since

$$\{c_n, \overline{c}_m\} = 0 \quad , \quad \{c_n, \overline{b}_m\} = 0, \tag{2.165a}$$

$$\{b_n, \overline{c}_m\} = 0 \quad , \quad \{b_n, \overline{b}_m\} = 0. \tag{2.165b}$$

For the sake of definiteness, let us restrict the discussion to the open string (or one of the two sectors of the closed string).

In the bosonic sector, the ground-states of the space of states are the states $|\Omega, p>$ introduced in the old covariant quantization. The space of states is now considerably enlarged through the ghost sector, which is represented as follows.

The algebra of ghost zero-modes

$$\{c_0, b_0\} = 1, \quad c_0^2 = 0, \quad b_0^2 = 0, \quad c_0^+ = c_0, \quad b_0^+ = b_0, \tag{2.166}$$

is represented on a two dimensional space spanned by two basis vectors $|->$ and $|+>$ such that:

$$c_0|->=|+> \quad , \quad c_0|+>=0, \tag{2.167a}$$

$$b_0|->=0 \quad , \quad b_0|+>=|->, \tag{2.167b}$$

$$<-|+>=<+|->=1 \quad , \quad <-|->=<+|+>=0. \tag{2.167c}$$

For the non-zero mode ghost algebra, let us introduce a ground state also denoted $|\Omega>$, such that:

$$b_n|\Omega>=0, \quad c_n|\Omega>=0, \quad n \geq 1, \tag{2.168a}$$

$$<\Omega|\Omega>=1. \tag{2.168b}$$

Thus, the space of states is spanned by the ground states $|\Omega, p, \pm>$ and their bosonic and ghost excitations, obtained through the action of the bosonic and ghost creation operators α_{-n}^μ, c_{-n} and $b_{-n} (n \geq 1)$.

The complete definition of the quantum system also requires a choice of normal ordering. In the bosonic sector, this choice is as before. In the ghost sector, one again chooses to bring all the creation operator $c_{-n}, b_{-n} (n \geq 1)$ to the left of all annihilation operators $c_n, b_n (n \geq 1)$. For the zero-modes, we must have

$$: c_0 b_0 := \frac{1}{2}[c_0 b_0 - b_0 c_0]. \tag{2.169}$$

This leads to

$$Q_B = \sum_n : (L_{-n}^{(\alpha)} + \frac{1}{2} L_{-n}^{(c)}) c_n : - a c_0, \tag{2.170}$$

$$Q_c = \frac{1}{2}(c_0 b_0 - b_0 c_0) + \sum_{n=1}^{\infty} [c_{-n} b_n - b_{-n} c_n], \tag{2.171}$$

$$L_n^{(T)} = \{b_n, Q_B\} = L_n^{(\alpha)} + L_n^{(c)} - a \delta_{n,0}, \tag{2.172}$$

where all composite operators $L_n^{(k)}$ are understood to be normal ordered quantities. Here again, "a" is a normal ordering constant. Obviously, similar expressions apply for the closed string, with left-moving modes then also contributing.

The question of quantum conformal invariance in the conformal gauge way now consistently be addressed, either by computing the central extension in the Virasoro algebra of the total Virasoro generator $L_n^{(T)}$, or equivalently by checking gauge invariance of the quantum system, namely by checking nilpotency of the quantum BRST charge.

Using techniques of conformal field theory (see for example Ref.19) , one obtains:

$$[L_n^{(\alpha)}, L_m^{(\alpha)}] = (n-m) L_{n+m}^{(\alpha)} + \frac{1}{12} D n (n^2 - 1) \delta_{n+m,0}, \tag{2.173a}$$

$$[L_n^{(c)}, L_m^{(c)}] = (n-m) L_{n+m}^{(c)} + \frac{1}{12}[-26 n^3 + 2n] \delta_{n+m,0}, \tag{2.173b}$$

$$[L_n^{(T)}, L_m^{(T)}] = (n-m) L_{n+m}^{(T)} + \frac{1}{12}[(D-26) n^3 + (2 + 24a - D) n] \delta_{n+m,0}, \tag{2.173c}$$

and

$$\begin{aligned} \{Q_B, Q_B\} &= 2 Q_B^2 = \sum_n \frac{1}{12}[(D-26) n^3 + (2 + 24a - D) n + 2(a-1)] : c_{-n} c_n : \\ &= \sum_n \{[L_n^{(T)}, L_m^{(T)}] - (n-m) L_{n+m}^{(T)}\} : c_{-n} c_n : . \end{aligned} \tag{2.174}$$

Hence, quantum conformal invariance and nilpotency of the quantum BRST charge are obtained under the same conditions, namely[26]

$$D = 26, \qquad a = 1. \tag{2.175}$$

That these two properties are equivalent may be seen from the relation (2.172), which leads to the following when $Q_B^2 = 0$:

$$[L_n^{(T)}, Q_B] = 0, \qquad [L_n^{(T)}, L_m^{(T)}] = (n-m) L_{n+m}^{(T)}. \tag{2.176}$$

Thus, nilpotency of Q_B or cancellation of the central extension in the total Virasoro algebra are one and the same expression of conformal invariance in the conformal gauge.

The critical values (2.175), which we found in the old covariant approach, are precisely those which ensure reparametrization invariance of the quantum system.

We still have to define physical states in the present formalism. First of all, since Q_B implements gauge transformations, two states which differ by a state of the form $Q_B|\chi>$ must be considered as representing the same quantum state. Thus, gauge equivalent states fall into cohomology classes of the BRST charge. Moreover, physical states must be BRST or gauge invariant states, namely

$$Q_B|\Psi>=0. \tag{2.177}$$

Hence, physical states correspond to BRST invariant cohomology classes of Q_B, or in other words the coset $KerQ_B/ImQ_B$. Since the null class is always BRST invariant for a nilpotent BRST charge, we only need to consider the non-trivial BRST invariant classes. The answer to this problem is the following[50].

All states in non-trivial BRST invariant classes are of the form

$$|\Psi>=|\Psi^-_{(\alpha)};->+|\Psi^+_{(\alpha)};+>, \tag{2.178}$$

(up to a state $Q_B|\chi>$ of course), where $|\Psi^{\pm}_{(\alpha)}>$ are states with bosonic excitations only and of strictly positive norm, i.e. transverse physical states in 1-to-1 correspondence with states in the light-cone gauge.

In order to restrict the choice further, it is useful to require that physical states have ghost number (-1/2):

$$Q_c|\Psi>=(-1/2)|\Psi>, \tag{2.179}$$

so that we finally have

$$|\Psi>=|\Psi^-_{(\alpha)};->. \tag{2.180}$$

The BRST invariant condition (2.177) then becomes

$$\{c_0[L_0^{(\alpha)}-1]+\sum_{n=1}^{\infty}c_{-n}L_n^{(\alpha)}\}|\Psi^-_{(\alpha)};->=0, \tag{2.181}$$

which clearly shows that the Virasoro conditions defining physical states in the old covariant approach are indeed recovered.

In conclusion, within BRST quantization, physical states correspond to BRST invariant cohomology classes of the BRST charge of ghost number (-1/2). Such non-trivial classes are in 1-to-1 correspondence with physical states of strictly positive norm both in the old covariant approach and in the light-cone gauge approach.

3 FERMIONIC STRINGS

Our discussion so far has been very detailed, and emphasized the fundamental features of the quantization of bosonic strings in the conformal and in the light-cone gauge. The reason is that these basic properties are generic of any string theory, so that it should now be easy to understand how any of their extensions and generalizations are to be used in the construction of any string theory. In this chapter, one such extension will be considered in some detail, not only because it provides an example of a generalization of the structures encountered so far, but also because it plays an important role in the construction of "realistic" string theories.

Bosonic strings obviously lack an important feature : their space-time spectrum does not contain any fermions. Such states can be obtained within string theory through two main approaches, where one introduces additional world-sheet degrees of freedom. In the first approach, the so-called spinning string[23] or Neveu-Schwarz-Ramond (NSR)[1,52] formulation, one introduces world-sheet Majorana spinors transforming as space-time vectors. In the second approach, the so-called superstring or Green-Schwarz formulation[53], one introduces world-sheet scalars transforming as space-time Majorana-Weyl spinors. In the latter approach, the construction of the theory explicitly realizes a space-time global supersymmetry, while in the former approach such a symmetry is only achieved by some specific projection of the spectrum[54], necessary for quantum consistency[55]. It is widely believed however, that the two approaches are actually equivalent[56], but this has not yet been proved in full generality.

By lack of space, we shall not consider here the superstring approach. Let us only mention that recent progress[57] has been made in the covariant quantization of these theories (for their quantization in the light-cone gauge, see for example Refs. 15,17), and that they have been used in the construction of four dimensional string theories[58].

As was shown in the previous chapter, reparametrization invariance and the corresponding constraints, which are the equations of motion for an intrinsic metric, are essential in obtaining a consistent quantum theory. In the conformal gauge, they are necessary in establishing the decoupling of unphysical and zero-norm states from physical amplitudes. In the light-zone gauge, they are essential for obtaining space-time covariance. Thus, by enlarging the system through the introduction of world-sheet spinors, we need at the same time to have an additional world-sheet local symmetry, leading to new constraints necessary to ensure consistency of the quantized system. The candidate for this additional symmetry is obviously a supersymmetry[23]. Hence, by extending the previous bosonic construction in the linear formulation to a two dimensional supergravity theory with local supersymmetry[23], we should expect that the necessary constraints would follow as the equations of motion of the supergravity sector.

Indeed, the action for the spinning string[23] is that of a two dimensional $N = 1$ minimal supergravity theory, with D matter multiplets $(x^\mu(\xi), \lambda^\mu(\xi))$ coupled to a supergravity multiplet $(e^a_\alpha(\xi), \psi_\alpha(\xi))$, where $e^a_\alpha(\xi)$ is a world-sheet zweibein or graviton, $\psi_\alpha(\xi)$ is a world-sheet Majorana gravitino and the $N = 1$ superpartners $\lambda^\mu(\xi)$ of the bosonic coordinates are world-sheet Majorana spinors. Here, $\alpha = 0, 1$ are the usual world-sheet coordinate indices, and $a = 0, 1$ are the corresponding tangent space indices. In the following, we shall use[17] the Majorana representation of the Dirac algebra in two dimensions, namely

$$\rho^0 = \begin{pmatrix} 0 & -1 \\ 1 & 0 \end{pmatrix}, \ \rho^1 = \begin{pmatrix} 0 & 1 \\ 1 & 0 \end{pmatrix}, \ \rho^5 = -\rho^0\rho^1 = \begin{pmatrix} 1 & 0 \\ 0 & -1 \end{pmatrix}, \tag{3.1}$$

with

$$\{\rho^a, \rho^b\} = 2\eta^{ab}, \tag{3.2}$$

$$\eta^{ab} = \begin{pmatrix} -1 & 0 \\ 0 & 1 \end{pmatrix}. \tag{3.3}$$

In this representation, a Majorana spinor is a real spinor.

As for the bosonic string, the spinning string action is not only invariant under local world-sheet diffeomorphisms and supersymmetry transformations, i.e. local super-reparametrizations, but also under local Weyl rescalings of the graviton supermultiplet and the associated fermionic transformation, i.e. super-Weyl transformations (for more details, see for example Refs. 15,17). As a consequence, the corresponding Weyl modes of the graviton and gravitino do not couple in the classical action, and will not appear in the canonical quantization of the theory. They do couple in a path-integral quantization[22] unless conditions analogous to those discussed for the bosonic string are satisfied. Moreover, the graviton multiplet is only an auxiliary multiplet in two dimensions, whose equations of motion are precisely the constraints[15,17] associated to super-reparametrization and super-Weyl invariance and generating these transformations.

These local gauge invariances allow for gauge-fixing the system to the superconformal gauge, defined by

$$e^a_\alpha(\xi) = e^{\phi(\xi)}\delta^a_\alpha \,, \ \psi_\alpha(\xi) = e^a_\alpha(\xi)\rho_a\chi(\xi) \,, \tag{3.4}$$

where $\chi(\xi)$ is a Majorana spinor. In the superconformal gauge, the system still possesses a local gauge invariance under superconformal transformations. The corresponding generators are then simply the Fourier modes of the constraints, as we showed in the case of the bosonic string. The important point which makes this possible is that the superconformal gauge-fixing preserves an explicit *global* $N = 1$ world-sheet supersymmetry.

3.1 Old Covariant Quantization

3.1.1 The superconformal gauge[15,17]

In the superconformal gauge, the action of the spinning string reduces to

$$S = \frac{-1}{4\pi\alpha'}\int d^2\xi \left[\eta^{\alpha\beta}\partial_\alpha x^\mu \partial_\beta x_\mu + i\overline{\lambda}^\mu \delta^\alpha_a \rho^a \partial_\alpha \lambda_\mu\right] \,, \tag{3.5}$$

where

$$\eta^{\alpha\beta} = \begin{pmatrix} -1 & 0 \\ 0 & 1 \end{pmatrix} \,, \tag{3.6}$$

$$\lambda^\mu = \begin{pmatrix} \lambda^\mu_- \\ \lambda^\mu_+ \end{pmatrix} \,, \tag{3.7}$$

and $\overline{\chi} = \chi^+\rho^0$ for a spinor χ.

This action is invariant under the $N = 1$ global supersymmetry defined by

$$\delta x^\mu = i\overline{\eta}\lambda^\mu \,, \ \delta\lambda^\mu = \delta^\alpha_a \rho^a \partial_\alpha x^\mu \eta \,, \tag{3.8}$$

where η is a constant infinitesimal Majorana spinor. For the Majorana representation given in (3.1), the action (3.5) becomes

$$S = \frac{-1}{4\pi\alpha'}\int d^2\xi \left[x^\mu(\partial^2_\tau - \partial^2_\sigma)x_\mu - i\lambda^\mu_+(\partial_\tau - \partial_\sigma)\lambda_{+\mu} - i\lambda^\mu_-(\partial_\tau + \partial_\sigma)\lambda_{-\mu}\right] \,, \tag{3.9}$$

thus leading to the equations of motion

$$(\partial^2_\tau - \partial^2_\sigma)x^\mu = 0 \tag{3.10}$$

$$(\partial_\tau - \partial_\sigma)\lambda^\mu_+ = 0 \,, \ (\partial_\tau + \partial_\sigma)\lambda^\mu_- = 0 \,. \tag{3.11}$$

Hence, as for the bosonic string, the system has reduced in the superconformal gauge to a set of "free" massless fields, which have to satisfy a set of constraints to be discussed below. Note that the separation into right- and left-movers extends to the fermionic sector, where λ_+^μ and λ_-^μ correspond to left- and right-moving Majorana-Weyl spinors respectively.

The boundary conditions in σ in the bosonic sector are as specified previously. For the fermionic matter fields, we have for open strings :

$$\lambda_-^\mu(\sigma=0,\tau) = \lambda_+^\mu(\sigma=0,\tau)\,, \tag{3.12a}$$
$$\lambda_-^\mu(\sigma=\pi,\tau) = \epsilon\lambda_+^\mu(\sigma=\pi,\tau)\,, \tag{3.12b}$$

where $\epsilon = +1$ corresponds to the Ramond (R) sector[51], and $\epsilon = -1$ to the Neveu-Schwarz (NS) sector[52] of the spinning string. For closed strings, we have

$$\lambda^\mu(\sigma=\pi,\tau) = \Lambda\lambda^\mu(\sigma=0,\tau) \tag{3.13}$$

with $\Lambda = -\mathbb{1}$, $\Lambda = \rho^5$, $\Lambda = -\rho^5$, $\Lambda = \mathbb{1}$ respectively for (N,NS), (R,NS), (N,R), (R,R) sectors, where the first entry corresponds to the right-moving modes and the second to the left-moving modes.

Actually, R boundary conditions preserve the global supersymmetry (3.8), whereas NS ones break it. As we shall see, such a breakdown of world-sheet supersymmetry is related to the presence of space-time tachyons in the physical spectrum.

Corresponding to space-time Poincaré invariance of the system, we have locally conserved world-sheet currents whose charges are the total energy- and angular-momentum of the string. In the superconformal gauge, on finds[17] for the energy-momentum current

$$P_\mu^0 = \frac{\dot{x}^\mu}{2\pi\alpha'}\,,\; P_\mu^1 = \frac{-x'^\mu}{2\pi\alpha'}\,,\; \partial_\alpha P_\mu^\alpha = 0\,, \tag{3.14}$$

thus showing that the total energy-momentum of the spinning string

$$P_\mu = \int_0^\pi d\sigma\, P_\mu^0\,, \tag{3.15}$$

is entirely carried by the bosonic coordinates. On the other hand, fermionic degrees of freedom do contribute to the angular-momentum of the system. Indeed, one finds[17]

$$M_{\mu\nu}^0 = \frac{1}{2\pi\alpha'}\left\{[\dot{x}_\mu x_\nu - \dot{x}_\nu x_\mu] + \frac{1}{2}i[\lambda_\mu^T\lambda_\nu - \lambda_\nu^T\lambda_\mu]\right\}\,, \tag{3.16a}$$
$$M_{\mu\nu}^1 = \frac{1}{2\pi\alpha'}\left\{[-x'_\mu x_\nu + x'_\nu x_\mu] + \frac{1}{2}i[\lambda_\mu^T\rho^5\lambda_\nu - \lambda_\nu^T\rho^5\lambda_\mu]\right\}\,, \tag{3.16b}$$

with

$$\partial_\alpha M_{\mu\nu} = 0\,, \tag{3.17}$$

and

$$M_{\mu\nu} = \int_0^\pi d\sigma\, M_{\mu\nu}^0\,. \tag{3.18}$$

Note that the boundary conditions in σ for open spinning strings are such that there is no flow of energy- nor angular-momentum at the end points.

Local world-sheet reparametrizations and supersymmetry transformations are generated by the two dimensional matter energy-momentum tensor $T_{\alpha\beta}$ and supercurrent J_α respectively, which are given in the superconformal gauge by [15,17].

$$\begin{aligned} T_{\alpha\beta} &= \partial_\alpha x^\mu\partial_\beta x_\mu + \frac{1}{4}i\bar{\lambda}^\mu[\rho_\alpha\partial_\beta + \rho_\beta\partial_\alpha]\lambda_\mu \\ &\quad -\frac{1}{2}\eta_{\alpha\beta}\left[\eta^{\gamma\delta}\partial_\gamma x^\mu\partial_\delta x_\mu + \frac{1}{2}i\bar{\lambda}^\mu\rho^\gamma\partial_\gamma\lambda_\mu\right]\,, \end{aligned} \tag{3.19}$$
$$J_\alpha = \partial_\beta x_\mu\rho^\beta\rho_\alpha\lambda^\mu\,, \tag{3.20}$$

where

$$\rho^\beta = \delta^\beta_b \rho^b \,, \; \rho_\alpha = \eta_{\alpha\beta}\delta^\beta_a \rho^a \,. \tag{3.21}$$

They obey the local conservation equation :

$$\partial_\alpha T^{\alpha\beta} = 0 \,, \; \partial_\alpha J^\alpha = 0 \,. \tag{3.22}$$

In fact, as a consequence of super-reparametrization and super-Weyl invariance, the equations of motion for the graviton multiplet simply reduce to the constraint equations[15,17] :

$$T_{\alpha\beta} = 0 \,, \; J_\alpha = 0 \,, \tag{3.23}$$

which must be satisfied by any solution to the free field equations (3.10) and (3.11).

In the case of open spinning strings, we have the same mode expansion for bosonic coordinates as in (2.23), and for fermionic coordinates :

$$\lambda^\mu_\pm(\sigma,\tau) \;=\; \sqrt{\alpha'} \sum_q c^\mu_q e^{-iq(\tau\pm\sigma)} \,, \tag{3.24a}$$

$$\text{with} \qquad (c^\mu_q)^* \;=\; c^\mu_{-q} \tag{3.24b}$$

Here, the fermionic modes c^μ_q are integer-moded and denoted $d^\mu_n = c^\mu_q$ ($q = n \in \mathbb{Z}$) for a R sector, and half-integer moded and denoted $b^\mu_r = c^\mu_q$ ($q = r \in \mathbb{Z} + 1/2$) for a NS sector.

The Fourier modes of the constraints then lead to the super Virasoro constraints

$$L^{(X)}_n = 0 \,, \; J^{(X)}_q = 0 \,, \tag{3.25}$$

where the super Virasoro generators are given by

$$L^{(X)}_n \;=\; L^{(\alpha)}_n + L^{(\lambda)}_n \,, \; L^{(\lambda)}_n = \frac{1}{2}\sum_q \left(q - \frac{1}{2}n\right) c^\mu_{n-q} c_{q\mu} \,, \tag{3.26}$$

$$J^{(X)}_q \;=\; \sum_m c^\mu_{q-m} \alpha_{m\mu} \,. \tag{3.27}$$

Here, the superscripts (X), (α) and (λ) stand for the contributions of the matter supermultiplet, of the bosonic modes and of the fermionic modes respectively. The supercurrent modes $J^{(X)}_q$ are integer moded and denoted $J^{(X)}_q = F^{(X)}_n$ for a R sector, and half-integer moded and denoted $J^{(X)}_q = G^{(X)}_r$ for a NS sector.

These quantities $L^{(X)}_n$ and $J^{(X)}_q$ are the generators of $N = 1$ superconformal transformations, which is the remaining super-reparametrization invariance of the system in the superconformal gauge.

For closed spinning strings, the mode expansion for bosonic coordinates is as in (2.26), and for fermionic coordinates we have :

$$\lambda^\mu_-(\sigma,\tau) = \sqrt{2\alpha'} \sum_q c^\mu_q e^{-2iq(\tau-\sigma)} \,, \; \lambda^\mu_+(\sigma,\tau) = \sqrt{2\alpha'} \sum_q \bar{c}^\mu_q e^{-2iq(\tau+\sigma)} \,, \tag{3.28}$$

with the following notation :

right-left sector	c^μ_q	$\bar{c}^\mu_q$	
(NS, NS)	b^μ_r	$\bar{b}^\mu_r$	$r \in \mathbb{Z} + 1/2$
(NS, R)	b^μ_r	$\bar{d}^\mu_n$	$n \in \mathbb{Z}$
(R, NS)	d^μ_n	$\bar{b}^\mu_r$	
(R, R)	d^μ_n	$\bar{d}^\mu_n$	

(3.29)

The Fourier modes of the constraints (3.23) lead to the super Virasoro constraints :

$$L_n^{(X)} = 0 \quad , \quad \overline{L}_n^{(X)} = 0 \, , \tag{3.30a}$$
$$J_q^{(X)} = 0 \quad , \quad \overline{J}_q^{(X)} = 0 \, , \tag{3.30b}$$

with the same notations and definitions as in (3.25), (3.26) and (3.27), now in terms of right- and left-moving modes. Note the by now familiar fact of doubling of fermionic and non-zero bosonic modes, when comparing closed strings to open strings.

The old covariant quantization in the superconformal gauge would require the corresponding Hamiltonian formulation of the system. Compared to the bosonic string, there are essentially two new features which arise here. The world-sheet Majorana spinors lead to second-class constraints, which may be solved by using Dirac brackets for these degrees of freedom. The set of first-class constraints generating the gauge symmetries of the theory now consist of bosonic and Grassmannian constraints. The associated Lagrange multiplets are respectively bosonic and Grassmannian variables, related to each other by local supersymmetry transformation and transforming as components of a supergravity multiplet. Superconformal gauge fixing amounts to setting the bosonic Lagrange multipliers λ^+ and λ^- equal to 1, and the Grassmann Lagrange multiplies equal to zero. It should thus be clear to the reader how such an Hamiltonian formulation of the spinning string could be obtained, and we shall not give any other details here.

In the superconformal gauge, we therefore have the same commutation relations as before for the bosonic modes (see (2.80), (2.81), (2.82)), both for open and for closed strings. For the fermionic matter fields $\lambda_\pm^\mu(\sigma,\tau)$, we obtain the anticommutation relations :

$$\{c_{q_1}^\mu, c_{q_2}^\nu\} = \eta^{\mu\nu}\delta_{q_1+q_2,0} \quad , \quad \{\overline{c}_{q_1}^\mu, \overline{c}_{q_2}^\nu\} = \eta^{\mu\nu}\delta_{q_1+q_2,0} \, , \tag{3.31a}$$
$$\text{with} \qquad (c_q^\mu)^+ = c_{-q}^\mu \quad , \quad (\overline{c}_q^\mu)^+ = \overline{c}_{-q}^\mu . \tag{3.31b}$$

These results allow for the description of the corresponding spaces of states. Let us first consider the case of the open spinning string.

In the Ramond sector, the Fock vacuum is a space-time spinor $|u(p)\rangle$, with $u(p)$ being a D-dimensional spinor, such that

$$\alpha_n^\mu|u(p)\rangle = 0 \, , \; d_n^\mu|u(p)\rangle = 0 \, , \; n \geq 1 \tag{3.32a}$$
$$\alpha_0^\mu|u(p)\rangle = \sqrt{2\alpha'}p^\mu|u(p)\rangle \, , \tag{3.32b}$$
$$d_0^\mu|u(p)\rangle = |\frac{1}{\sqrt{2}}i\gamma^\mu u(p)\rangle \, , \tag{3.32c}$$

where γ^μ are the D-dimensional Dirac matrices which satisfy the Clifford algebra

$$\{\gamma^\mu, \gamma^\nu\} = -2\eta^{\mu\nu} \quad , \quad \gamma^{\mu +} = \gamma^0\gamma^\mu\gamma^0 . \tag{3.33}$$

Indeed, the fermionic zero-modes d_0^μ satisfy the algebra

$$\{d_0^\mu, d_0^\nu\} = \eta^{\mu\nu} \quad , \quad \{d_0^\mu, d_n^\nu\} = 0 \quad , \quad n \neq 0 \, , \tag{3.34}$$

so that they are represented by

$$d_0^\mu = \frac{i}{\sqrt{2}}(-1)^{\sum_{n=1}^\infty d_{-n}^\mu d_{n\mu}}\gamma^\mu . \tag{3.35}$$

All other states are obtained by the action of the creation operators α_{-n}^μ, d_{-n}^μ ($n \geq 1$) on these Fock vacua. Clearly, all states in the Ramond sector are space-time fermions. In particular, the ground-state spinor transforms under Lorentz transformations as :

$$M^{\mu\nu}|u(p)\rangle = \sqrt{2\alpha'}[p^\mu q^\nu - p^\nu q^\mu]|u(p)\rangle + |\Sigma^{\mu\nu}u(p)\rangle \,, \tag{3.36a}$$

where

$$\Sigma^{\mu\nu} = -\tfrac{1}{4}i[\gamma^\mu, \gamma^\nu] \tag{3.36b}$$

is the generator of Lorentz transformations in the spinor representation. The first term in (3.36a) corresponds to the orbital contribution.

In the Neveu-Schwarz sector, since there are no fermionic zero-modes, the Fock vacua are momentum eigenstates $|\Omega;p\rangle$ annihilated by α_n^μ, b_r^μ ($n \geq 1, r \geq 1/2$), and transforming as space-time scalars. Hence, all states in the NS sector are space-time bosons.

In the case of closed spinning strings, except for the bosonic zero-mode sector, the space of states is essentially obtained as the tensor product of spaces of states of the open string, with the subtlety that the right- and left-moving fermionic modes anticommute. We thus have for the Fock vacua and the space-time spectra in each sector :

sector	Fock vacua	space-time spectrum
(NS, NS)	$\|\Omega, p\rangle$	space-time bosons
(R, NS)	$\|u(p)\rangle$	space-time fermions
(NS, R)	$\|\overline{u}(p)\rangle$	space-time fermions
(R, R)	$\|u(p), \overline{u}(p)\rangle$	space-time bosons

(3.37)

where $u(p)$ and $\overline{u}(p)$ are independent D-dimensional spinors associated to the right- and left-moving sectors respectively.

Let us remark that all these spaces of states, both for open and for closed strings, have states of negative norm, due to the fact that $\eta^{00} = -1$.

The definition of the quantum system is complete only when a normal ordering of quantum operators has been specified. As usual, all position and creation operators are brought to the left of all momenta and annihilation operators. For the fermionic zero-modes in the R sector, we must choose :

$$: d_0^\mu d_0^\nu := \frac{1}{2}(d_0^\mu d_0^\nu - d_0^\nu d_0^\mu) \,. \tag{3.38}$$

Defining then the super Virasoro generators by the same expressions as given before, but now understood to be normally ordered, it is possible to compute the corresponding $N = 1$ super-Virasoro algebras, both in R and NS sectors, using techniques of conformal field theory (see for example Refs. 17,19). We only give here the results for the open string or for the right-moving sector of the closed string. Similar results obviously hold for the left-moving modes.

Ramond sector

$$\{F_n^{(X)}, F_m^{(X)}\} \;=\; 2L_{n+m}^{(X)} + \frac{1}{2}Dn^2\delta_{n+m,0} \,, \tag{3.39a}$$

$$[L_n^{(X)}, F_m^{(X)}] = \frac{1}{2}n - m \quad F_{n+m}^{(X)}, \tag{3.39b}$$

$$[L_n^{(X)}, L_m^{(X)}] = (n-m)L_{n+m}^{(X)} + \frac{1}{8}Dn^3\delta_{n+m,0}, \tag{3.39c}$$

where

$$[L_n^{(\lambda)}, L_m^{(\lambda)}] = (n-m)L_{n+m}^{(\lambda)} + \frac{1}{12}D\left[\frac{1}{2}n^3 + n\right]\delta_{n+m,0}. \tag{3.40}$$

Neveu-Schwarz sector

$$\{G_r^{(X)}, G_s^{(X)}\} = 2L_{r+s}^{(X)} + \frac{1}{2}D(r^2 - \frac{1}{4})\delta_{r+s,0}, \tag{3.41a}$$

$$[L_n^{(X)}, G_r^{(X)}] = \left(\frac{1}{2}n - r\right)G_{n+r}^{(X)}, \tag{3.41b}$$

$$[L_n^{(X)}, L_m^{(X)}] = (n-m)L_{n+m}^{(X)} + \frac{1}{8}Dn(n^2-1)\delta_{n+m,0}, \tag{3.41c}$$

where

$$[L_n^{(\lambda)}, L_m^{(\lambda)}] = (n-m)L_{n+m}^{(\lambda)} + \frac{1}{24}Dn(n^2-1)\delta_{n+m,0}. \tag{3.42}$$

As for the bosonic string, we obtain central extensions to the $N = 1$ superconformal algebra, thus showing that short distance quantum effects in the matter sector of the theory (seem to) break the super-reparametrization invariance of the spinning string. Note that each Majorana-Weyl spinor contributes a value $(+1/2)$ to be central charge of the total Virasoro algebra.

It should be obvious that the quantization in the superconformal gauge explicitly preserves space-time Poincaré covariance. Indeed, one may show that the Poincaré algebra is obtained at the quantum level, and that it commutes with the super Virasoro algebra. Hence, all physical states may be characterized by their mass and by their "spin", as was explained for the bosonic string.

In a R sector, physical states are defined by the super-Virasoro constraints :

$$F_n^{(X)}|\psi\rangle = 0, \quad L_n^{(X)}|\psi\rangle = 0, \; n \geq 1, \tag{3.43a}$$

$$(F_0^{(X)} - \sqrt{a_+})|\psi\rangle = 0, \quad (L_0^{(X)} - a_+)|\psi\rangle = 0. \tag{3.43b}$$

In a NS sector, we have

$$G_r^{(X)}|\psi\rangle = 0, \quad L_n^{(X)}|\psi\rangle = 0, \quad r \geq 1/2, n \geq 1, \tag{3.44a}$$

$$(L_0^{(X)} - a_-)|\psi\rangle = 0 \tag{3.44b}$$

Here, a_+ and a_- are subtraction constants following from normal ordering in the super-Virasoro zero-modes. Note that we introduced a normal ordering constant for $F_0^{(X)}$. On the one hand, this does not seem necessary since $F_0^{(X)}$ does not suffer from an ordering ambiguity. On the other hand, it seems necessary since we have $(F_0^{(X)})^2 = L_0^{(X)}$. As it turns out, a_+ indeed vanishes (see the no-ghost theorem below). Note also that the constraint which involves $F_0^{(X)}$ generalizes Dirac's equation[51]. As usual, the constraints involving the Virasoro zero-modes lead the mass formulae. For open strings, we have:

$$\text{R sector} \quad : \quad \alpha' M^2 = N_R - a_+, \tag{3.45a}$$

where

$$N_R = \sum_{n=1}^{\infty} \alpha_{-n}^{\mu}\alpha_{n\mu} + \sum_{n=1}^{\infty} n d_{-n}^{\mu} d_{n\mu}. \tag{3.45b}$$

$$\text{NS sector} \quad : \quad \alpha' M^2 = N_{NS} - a_- , \tag{3.46a}$$

where

$$N_{NS} = \sum_{n=1}^{\infty} \alpha^{\mu}_{-n} \alpha_{n\mu} + \sum_{r=1/2}^{\infty} r b^{\mu}_{-r} b_{r\mu} . \tag{3.46b}$$

For closed strings, we have :

$$\frac{1}{2}\alpha' M^2 = (N - a) + (\overline{N} - \overline{a}) , \tag{3.47a}$$

$$N - a = \overline{N} - \overline{a} , \tag{3.47b}$$

where $N, \overline{N}$ are N_R, N_{NS}, $\overline{N}_R$, $\overline{N}_{NS}$, and a, $\overline{a}$ are $a_+, a_-, \overline{a}_+, \overline{a}_-$ depending on whether we have a R or a NS sector in each moving sector. The level matching constraint (3.47b) has the usual geometrical meaning in terms of constant shifts in σ.

For the spinning string, the results of the no-ghost theorem[59] lead to the critical values :

$$D = 10 , \quad a_+ = 0 , \quad a_- = \frac{1}{2} . \tag{3.48}$$

These are the necessary conditions for the absence of negative norm physical states and for the correct cut and pole singularities in one-loop amplitudes. The values (3.18) may be checked by explicitly constructing the first few physical excitation levels. Let us give those of strictly positive norm.

In the Ramond sector of the open spinning string, the physical ground-state is at excitation level $N_R = 0$. It corresponds to a massless space-time spinor $|u(p)\rangle$ with $\not{p}u|p\rangle = 0$. At the next excitation level $N_R = 1$, we have the states $\alpha^{\mu}_{-1}|u(p)\rangle$ and $d^{\mu}_{-1}|u(p)\rangle$, which lead to a massive Rarita-Schwinger ("spin $\frac{3}{2}$") state. All states in this sector are space-time fermions, with a double degeneracy of the Regge "trajectories" due to the space-time chirality of the spinor $u(p)$.

In the NS sector of the open string, the physical ground state is a space-time scalar tachyon $|\Omega; p\rangle$, with $\alpha' m^2 = 1/2$. At the next excitation level $N_{NS}\frac{1}{2}$, we have the massless states $b^{\mu}_{-1/2}|\Omega; p>$, giving a massless vector state. At the second excitation level $N_{NS} = 1$, we have the massive states $\alpha^{\mu}_{-1}|\Omega; p>, b^{\mu}_{-1/2}b^{\nu}_{-1/2}|\Omega; p>$, with $\alpha' m^2 = \frac{1}{2}$, leading to a 2-index antisymmetric tensor state. All states in this sector are space-time bosons, with two types of Regge "trajectories". Those corresponding to an even number of fermionic excitations, to which the tachyon belongs, and those with an odd number of fermionic excitations, to which the massless vector belongs.

For the closed spinning string, the discussion is very similar. In the (N,NS) sector, the physical ground-state $|\Omega, p\rangle$ is a scalar tachyon with $\frac{1}{2}\alpha' m^2 = -1$. At the next excitation level, we have a massless graviton, a massless 2-index antisymmetric transverse tensor and a massless dilaton, obtained from $b^{\mu}_{-1/2}\overline{b}^{\nu}_{-1/2}|\Omega; p\rangle$.

In the (R,NS) sector, the would-be spinor tachyon $|u(p)\rangle$ is not a physical state, due to the level matching condition. The physical ground-state has $N_R = 0$, $\overline{N}_{NS} = 1/2$, and thus corresponds to the massless states $\epsilon_{\mu}\overline{b}^{\mu}_{-1/2}|u(p)\rangle$, with $\not{p}u(p) = 0$ and $\epsilon \cdot p = 0$. Since the condition $\not{\epsilon}u(p) = 0$ is not obtained, we have a massless spinor ("spin 1/2") and a massless Rarita-Schwinger ("spin 3/2") state.

The discussion in the (NS, R) sector is similar, by exchanging the right- and left-moving sectors. Finally, in the (R, R) sector, the physical ground-state $|u(p), \overline{u}(p)\rangle$ is massless, with $\not{p}u(p) = 0.$, $\not{p}\overline{u}(p) = 0$, and corresponds to all completely antisymmetric space-time tensors, appearing in the tensor product of the two space-time spinors $u(p)$ and $\overline{u}(p)$.

Clearly, the Regge "trajectories" of the closed string also have a double "degeneracy", associated either to the chirality of the spinor in a R sector or to the parity of the fermion excitation number in a NS sector, in each of the two moving sectors.

3.1.2 Light-cone gauge quantization

Let us briefly describe how the same results are obtained in the light-cone gauge. For the spinning string, the light-cone gauge fixing conditions are[17] (in addition to the superconformal gauge conditions) :

$$x^+(\sigma,\tau) = 2\alpha' P^+ \tau \,, \quad \lambda^+(\sigma,\tau) = 0 \,. \tag{3.49}$$

The super-Virasoro constraints $L_n^{(X)} = 0$, $J_q^{(X)} = 0$ may then be solved for explicitly, by expressing the light-cone components $x^\pm(\sigma,\tau)$ and $\lambda^\pm(\sigma,\tau)$ (or equivalently $\alpha_n^\pm$, $\overline{\alpha}_n^\pm$ and $c_q^\pm$, $\overline{c}_q^\pm$) in terms of the transverse components $x^i(\sigma,\tau)$, $\lambda^i(\sigma,\tau)$, (or equivalently q^i, α_n^i, $\overline{\alpha}_n^i$, c_q^i, $\overline{c}_q^i$) and the two integration constants q^- and P^+.

From the corresponding Dirac brackets, we then have the usual commutation and anticommutation relations for the transverse modes q^i, P^i, α_n^i, $\overline{\alpha}_n^i$, c_q^i, $\overline{c}_q^i$ and q^-, P^+. Thus, the Fock vacua for a NS sector are simply states $|\Omega, p^i, p^+\rangle$ annihilated by α_n^i and b_r^i $(n \geq 1, r \geq 1/2)$. For a R sector, they are $SO(D-2)_n$ spinors $|u(p^i,p^+)\rangle$ annihilated by α_n^i and d_n^i $(n \geq 1)$, since the fermionic zero-modes d_0^i satisfy the Clifford algebra $\{d_0^i, d_0^j\} = \delta^{ij}$. By construction, all states have strictly positive norm, and correspond to transverse excitations only.

At the quantum level, when solving for the light-cone coordinates in terms of the transverse ones, normal ordering constants appear in the expressions for the zero-modes α_0^- and $\overline{\alpha}_0^-$. This then leads to the mass formulae. For open strings, we have :

$$\alpha' M^2 = N^\perp - a \,, \tag{3.50}$$

with

$$R \text{ sector}: \qquad N^\perp = \sum_{n=1}^{\infty} \alpha_{-n}^i \alpha_n^i + \sum_{n=1}^{\infty} n d_{-n}^i d_n^i \,, \quad a = a_+ \,, \tag{3.51a}$$

$$NS \text{ sector}: \qquad N^\perp = \sum_{n=1}^{\infty} \alpha_{-n}^i \alpha_n^i + \sum_{r=1/2}^{\infty} r b_{-r}^i b_r^i \,, \quad a = a_- \,. \tag{3.51b}$$

For closed strings, we obtain :

$$\frac{1}{2}\alpha' M^2 = (N^\perp - a) + (\overline{N}^\perp - \overline{a}) \,, \tag{3.52a}$$

$$N^\perp - a = \overline{N}^\perp - \overline{a} \,, \tag{3.52b}$$

with an obvious meaning for the notation (as in (3.47)). Here again, the level matching condition (3.52b) follows from the fact that the light-cone gauge fixing conditions do not fix the origin in σ parametrization for closed strings.

Due to the necessary normal ordering of composite operators, the Lorentz algebra may not be realized at the quantum level due to quantum anomalies. Here again, only the commutator $[M^{-i}, M^{-j}]$ needs to be checked, and can be shown to vanish only if we have[60] :

$$D = 10 \,, \quad a_+ = 0 \,, \quad a_- = 1/2. \tag{3.53}$$

The values for the subtraction constants are easily understood[40]. For definiteness, let us consider the open string. In the Ramond sector, the ground state $|u(p)\rangle$ has $2^{[D-2]/2}$ components and a mass $\alpha' m^2 = -a_+$. This state *may* transform covariantly under the full Lorentz group only if it is massless, i.e. only if $a_+ = 0$. In the NS sector, the same argument applies to the first excitation level $b^i_{-1/2}|\Omega; p\rangle$, which has $(D-2)$ components and a mass $\alpha' m^2 = 1/2 - a_-$. We must have $a_- = 1/2$, thus also implying that the physical ground state is a space-time scalar tachyon.

If we now use our favorite[40] ζ-function regularization for the evaluation of a_+ and a_-, we obtain :

$$-a_+ = (D-2)\left[\sum_{n=1}^{\infty}\frac{1}{2}n - \sum_{n=1}^{\infty}\frac{1}{2}n\right] = \frac{(D-2)}{2}[\zeta(-1) - \zeta(-1)] = 0, \tag{3.54}$$

$$-a_- = (D-2)\left[\sum_{n=1}^{\infty}\frac{1}{2}n - \sum_{n=1/2}^{\infty}\frac{1}{2}r\right] = \frac{(D-2)}{2}[\zeta(-1) - \frac{-1}{2}\zeta(-1)] = -\frac{D-2}{16}. \tag{3.55}$$

In each of these expressions, the first term is the contribution of the bosonic zero-point fluctuations, and the second term that of the fermionic ones. They cancel in the R sector but not in the NS sector, due to the global $N = 1$ supersymmetry which is preserved by R boundary conditions but broken by NS boundary conditions. This breakdown of supersymmetry is responsible for the appearance of a space-time tachyon in the NS sector. If we now set $a_- = 1/2$, we find indeed $D = 10$ from (3.55).

The reader is invited to check that all states of strictly positive norm obtained in the superconformal gauge, both for open and for closed strings, are indeed recovered in the light-cone gauge, and that all massive states indeed transform as irreducible representations of the covering group Spin(9) of $SO(D-1) = SO(9)$. Note also that the same types of "degeneracies" of Regge "trajectories" as in the covariant approach appear in the light-cone gauge.

3.1.3 GSO projection and space-time supersymmetry

The physical space-time spectrum of spinning strings contains bosonic and fermionic states. Actually, we have to think of the complete spectrum of these theories as the direct sum of the spectra in the different sectors.

We saw however that in the NS and (N,NS) sectors, one obtains space-time scalar tachyons, causing problems for unitarity. Moreover, the massless "spin 3/2" states also lead to problems, unless they couple to a conserved current of "spin 1/2", i.e. to a space-time supersymmetry current, in the same way that gravitons couple to a conserved energy-momentum tensor. Gliozzi, Olive and Scherk (GSO)[54] were the first to realize that by taking advantage of the double "degeneracy" of Regge "trajectories", there exists a consistent truncation of the spectrum of the theory, in which these problems are avoided, namely such that :

- tachyons are no longer physical states,
- the numbers of space-time bosons and fermions are equal at each mass level[54],
- the spectrum is space-time supersymmetric[61].

The corresponding truncation is called the GSO projection. In the next chapter, we shall find a more geometrical understanding of it, from the point of view of two dimensional reparametrization invariance. Actually, it is the only projection making the theory quantum consistent to all orders of perturbation theory[55].

To discuss the GSO projection, let us first consider the open spinning string. The quantity which distinguishes between the "degenerate" Regge "trajectories" is essentially the fermion number F, given in the R sector by

$$F = \sum_{n=1}^{\infty} d^{\mu}_{-n} d_{n\mu} \,, \tag{3.56}$$

and in the NS sector by

$$F = \sum_{r=1/2}^{\infty} b^{\mu}_{-r} b_{r\mu} \,. \tag{3.57}$$

Of course, the same definitions apply in the light-cone gauge, where one then sums only over transverse components.

Finally, let us introduce in the R sector the string chirality operator

$$\Gamma = \gamma_{11}(-1)^F \,, \tag{3.58}$$

where γ_{11} is the space-time chirality operator

$$\gamma_{11} = \gamma^{11} = \gamma^0\gamma^1 \ldots \gamma^9 \,. \tag{3.59}$$

(In the light-cone gauge, we would have $\Gamma = \gamma_9(-1)^F$ with $\gamma_9 = \gamma^1 \ldots \gamma^8$).

GSO projection of the open spinning string is then (not yet completely) defined by the conditions[54] :

$$\bullet\ R \text{ sector :} \qquad \Gamma|\psi\rangle = +|\psi\rangle \,, \ \ (\text{or } \Gamma = -1) \,, \tag{3.60a}$$

$$\bullet\ NS \text{ sector :} \qquad (-1)^F|\psi\rangle = -|\psi\rangle \,. \tag{3.60b}$$

The latter condition projects out in the NS sector all states having an half-integer value for $\alpha' m^2$, thus including the tachyon. Only states with integer values for $\alpha' m^2$ are retained. The physical ground state is then the massless vector state with 8 components. At the next excitation level with $\alpha' m^2 = 1$, we have the positive norm states among $b^{\mu}_{-3/2}|\Omega;p\rangle$, $b^{\mu}_{-1/2}b^{\nu}_{-1/2}b^{\rho}_{-1/2}|\Omega;p\rangle$ and $\alpha^{\mu}_{-1}b^{\nu}_{-1/2}|\Omega;p\rangle$ leading to the $SO(9)$ representations $\left(\begin{smallmatrix}\Box\\\Box\\\Box\end{smallmatrix} + \Box\Box\right)_{SO(9)}$, and having $(84 + 44 = 128)$ components respectively.

In the R sector, the ground state spinor $|u(p)\rangle$ decomposes into right- and left-handed components $u_{\pm}(p) = \frac{1}{2}(1 \pm \gamma_{11})u(p)$. The GSO projections retain states at all integer values of $\alpha' m^2$, but only half the original number. At the massless level, the condition $\Gamma = +1$ (resp. $\Gamma = -1$) implies that the physical ground-state of the truncated theory is a right-handed (resp. left-handed) massless Weyl spinor, with 16 real components.

A Majorana condition on $u(p)$ would reduce this number further to 8, i.e. the same number of components as in the bosonic sector, which is a necessary condition for having space-time supersymmetry. The remarkable fact is precisely that Majorana-Weyl spinors exist in Minkowski space-times only of dimension $D = 2(\text{mod } 8)$[54], which in particular includes $D = 10$. Hence, the complete GSO$projectionconditionsaredefinedby(3.60)tc$

In the R sector, the physical ground-state is then a right-handed massless Majorana-Weyl spinor, with 8 physical components. At the excitation level $\alpha' m^2 = 1$, we have the states $\alpha^\mu_{-1}|u_+(p)\rangle$ and $d^\mu_{-1}|u_-(p)\rangle$, combining into a massive Majorana Rarita-Schwinger "spin 3/2" state, with 128 components.

That the number of bosonic and fermionic physical states (of positive norm) is the same at each mass level may best be seen by computing the corresponding degeneracy generating functions[54]. In the R sector, we have :

$$f_R(q) = \sum_{n=0}^{\infty} d_R(n) q^{2\alpha' m_n^2} = \sum_{n=0}^{\infty} d_R(n) q^{2n} = d\, Tr\, q^{2N_R^\perp} \tag{3.61a}$$

$$= d \prod_{n=1}^{\infty} \left(\frac{1+q^{2n}}{1-q^{2n}} \right)^8 \tag{3.61b}$$

$$= d[1 + 16q^2 + 144q^4 + 960q^6 + \ldots], \tag{3.61c}$$

where q is a complex number such that $|q| < 1$, d_n is the degeneracy at excitation level $N_R^\perp = n$, d is the number of physical components of the ground-state spinor $u(p)$ ($d = 8$ for a Majorana-Weyl spinor on-shell), and the trace in (3.61a) is taken over all non-zero bosonic and fermionic modes.

In the NS sector, let us distinguish the states with an even and an odd number of fermionic excitations. We then have :

$$f^\pm_{NS} = Tr \frac{1}{2}(1 \pm (-1)^F) q^{2[N^\perp_{NS} - 1/2]} \tag{3.62a}$$

$$= \frac{1}{2q} \prod_{n=1}^{\infty} \frac{1}{(1-q^{2n})^8} \left\{ \prod_{n=1}^{\infty} (1+q^{2n-1})^8 \pm \prod_{n=1}^{\infty} (1-q^{2n-1})^8 \right\}, \tag{3.62b}$$

leading to the series expansions :

$$f^+_{NS}(q) = \frac{1}{q}[1 + 36q^2 + 402q^4 + 1964q^6 + \ldots], \tag{3.63a}$$

$$f^-_{NS}(q) = 8[1 + 16q^2 + 144q^4 + 960q^6 + \ldots]. \tag{3.63b}$$

Comparing (3.61) and (3.63), it becomes clear that only the GSO projection as defined above may lead to a space-time supersymmetric spectrum. To show the equality of $f_R(q)$ and $f^-_{NS}(q)$, it is useful to express these functions in terms of Jacobi theta functions defined as follows :

$$\theta_1(z|\tau) = i \sum_n (-1)^n q^{(n-1/2)^2} e^{2i\pi(n-1/2)z}$$

$$= 2q^{1/4} \sin \pi z \prod_{n=1}^{\infty} (1-q^{2n}) \prod_{n=1}^{\infty} (1 - q^{2n} e^{2i\pi z})(1 - q^{2n} e^{-2i\pi z}), \tag{3.64a}$$

$$\theta_2(z|\tau) = \sum_n q^{(n-1/2)^2} e^{2i\pi(n-1/2)z}$$

$$= 2q^{1/4} \cos \pi z \prod_{n=1}^{\infty} (1-q^{2n}) \prod_{n=1}^{\infty} (1 + q^{2n} e^{2i\pi z})(1 + q^{2n} e^{-2i\pi z}), \tag{3.64b}$$

$$\theta_3(z|\tau) = \sum_n q^{n^2} e^{2i\pi n z} = \prod_{n=1}^{\infty} (1-q^{2n}) \prod_{n=1}^{\infty} (1 + q^{2n-1} e^{2i\pi z})(1 + q^{2n-1} e^{-2i\pi z}), \tag{3.64c}$$

$$\theta_4(z|\tau) = \sum_n (-1)^n q^{n^2} e^{2i\pi n z} = \prod_{n=1}^{\infty} (1-q^{2n}) \prod_{n=1}^{\infty} (1 - q^{2n-1} e^{2i\pi z})(1 - q^{2n-1} e^{-2i\pi z}), \tag{3.64d}$$

where as before we set $q = e^{i\pi\tau}$. In the following, we shall only need the values $\theta_i(z = 0|\tau)$ $(i = 1,2,3,4)$, which will be denoted as θ_i $(i = 1,2,3,4)$. Then, one finds:

$$f_R(q) = \frac{1}{16} d \frac{\theta_2^4}{\eta^{12}}, \tag{3.65a}$$

$$f_{NS}^{\pm}(q) = \frac{1}{2} \frac{1}{\eta^{12}} [\theta_3^4 \pm \theta_4^4], \tag{3.65b}$$

where $\eta(\tau)$ is the Dedekind η-function defined in (2.111).

That the GSO projection leads indeed to a space-time spectrum with the same number of bosons and fermions at each mass level now follows[54] from the Riemann identity $(\theta_1 = 0)$:

$$\theta_1^4 + \theta_3^4 = \theta_2^4 + \theta_4^4 . \tag{3.66}$$

Moreover, it has been shown[61] that the spectrum then transforms under a $D = 10$ $N = 1$ supersymmetry algebra, with in particular the massless states falling into a $D = 10$ $N = 1$ Yang-Mills supermultiplet, i.e. a massless vector and a massless Majorana-Weyl spinor. From that point of view, the Riemann identity (3.66) is an expression of space-time supersymmetry.

For closed spinning strings, *GSO* projections are defined in a similar way[17]. Here however, we have two choices in the relative chirality of the two Majorana spinors in the two moving sectors, thus leading to two different theories. The first is the Type IIa fermionic string, defined by

$$\begin{array}{ll} \bullet & u(p), \overline{u}(p) \ : \text{Majorana spinors} \\ \bullet & \Gamma = +1 \ \ \overline{\Gamma} = -1 \ \text{(or vice-versa)} \\ \bullet & (-1)^F = -1\,, \ \ (-1)^{\overline{F}} = -1 \end{array} \tag{3.67}$$

and the second is the Type IIb fermionic string, defined by

$$\begin{array}{ll} \bullet & u(p), \overline{u}(p) \ : \text{Majorana spinors} \\ \bullet & \Gamma = +1\,, \ \ \overline{\Gamma} = +1 \ (\text{or } \Gamma = \overline{\Gamma} = -1) \\ \bullet & (-1)^F = -1\,, \ \ (-1)^{\overline{F}} = -1 \end{array} \tag{3.68}$$

Here, as usual, quantities with and without a bar correspond respectively to the left- and right-moving sectors.

It should be clear that in both cases, the numbers of space-time fermions and bosons are equal at each mass level, and are given by the square of the corresponding degeneracy numbers of the open spinning string. Moreover, in each theory, the space-time spectrum transforms under a space-time supersymmetry. For the Type IIa theory, this is the non-chiral $D = 10$ $N = 2a$ supersymmetry, and for the Type IIb theory, it is the chiral $D = 10$, $N = 2b$ supersymmetry, with the massless states falling into the corresponding graviton supermultiplets. In both cases, the (NS,NS) sector leads to the massless graviton, 2-index antisymmetric tensor and dilaton, transforming as $(\square\!\square + \begin{smallmatrix}\square\\\square\end{smallmatrix} + \bullet)_{SO(8)}$ and having respectively $(35 + 28 + 1 = 64)$ components. In the (R,R) sector, since the two Majorana spinors $u(p)$ and $\overline{u}(p)$ have different relative chiralities in each theory, we have different bosonic states. For the Type IIa theory, we have the states $(\square + \begin{smallmatrix}\square\\\square\\\square\end{smallmatrix})_{SO(8)}$ with $(8 + 56 = 64)$ components respectively. For the Type IIb theory, we have the states $(\bullet + \begin{smallmatrix}\square\\\square\end{smallmatrix} + \begin{smallmatrix}\square\\\square\\\square\\\square\end{smallmatrix}_{\pm})_{SO(8)}$ with $(1 + 28 + 35 = 64)$ components respectively, where $\begin{smallmatrix}\square\\\square\\\square\\\square\end{smallmatrix}_{\pm}$ is a self-dual or anti-selfdual tensor depending on the chirality of the spinors $u(p)$ and $\overline{u}(p)$. From the (R,NS) and (NS,R) sectors we obtain the massless fermions. In the type IIa theory, we have a Majorana spinor and a Majorana Rarita-Schwinger "spin 3/2" gravitino, none being Weyl fermions. In the Type IIb theory, we have a left-handed Weyl spinor and a right-handed Rarita-Schwinger "spin 3/2" gravitino, none being a Majorana fermion. In both cases, the number of components are $2(8+56) = 128$, respectively. Clearly, the spectrum of the Type IIa theory is non-chiral, whereas that of the Type IIb theory is indeed chiral.

3.2 BRST quantization

So far in our discussion, we have not taken into account the necessary ghost sector associated to superconformal gauge fixing. In principle, to be faithfull to the Hamiltonian approach adopted in these lectures, we should again use the BFV Hamiltonian formalism for the BRST quantization of the spinning string.

However, as long as one is not primarily interested in that aspect of the theory, such an analysis can be avoided by taking advantage[19] of the explicit world-sheet global $N = 1$ supersymmetry remaining after superconformal gauge fixing. We simply need to find out how to supersymmetrize the BRST quantization of the bosonic string, which is most efficiently done by means of world-sheet superfields[16,19].

Indeed, matter fields (x^μ, λ^μ) correspond to such superfields, and the ghost sector now follows from a ghost superfield[16,19]. The fermionic components of the ghost superfield are the (b, c) ghost system associated to the bosonic constraints of world-sheet reparametrization invariance. Its bosonic components are the ghost system associated to the fermionic constraints of world-sheet local supersymmetry, the so-called (β, γ) system of superghosts.

Since we have in the superconformal gauge, the action (3.9) for the matter fields, and the part of the BRST invariant action associated to the bosonic constraints and its ghost system (see (2.141)), it is rather easy to guess[19] what the BRST invariant action and the BRST charge for the spinning string should be in the superconformal gauge. By lack of space, and since we do not really need the details of BRST quantization in the remainder of these notes, we shall not develop these considerations any further, despite the fact that the superghost system plays an important role in string theory[16], and is essential in one general approach to string theory constructions[62,63]. The interested reader is invited to apply the approach sketched above and work out for himself the BRST quantization of the spinning string in the superconformal gauge. Details can be found in Refs. 15, 16, 19.

For later purposes, it suffices here to state some of the results following from such an analysis. As for the bosonic string, superconformal transformations of the spinning string in this BRST formulation are generated by total super-Virasoro generators $L_n^{(T)}$ and $J_q^{(T)}$. Now, they not only include the contributions $L_n^{(X)}$ and $J_q^{(X)}$ of the matter fields considered previously, but they also include the total contributions $L_n^{(B)}$ and $J_q^{(B)}$ of the ghost sector. In particular, the total ghost Virasoro generators $L_n^{(B)}$ are given by the sum of the (b, c) contribution $L_n^{(c)}$ obtained for the bosonic string, and a contribution $L_n^{(\gamma)}$ from the (β, γ) superghost system.

The moding of the ghost supercurrent modes $J_q^{(B)}$ is the same as that of the fermions, namely integer for a Ramond sector and half-integer for a Neveu-Schwarz sector. Indeed, the matter supercurrent J_α in (3.20) satisfies the same boundary conditions as the matter fermions, hence so do the superghosts (β, γ) associated to the gauge invariance generated by J_α, and the ghost supercurrent coupling to them. Hence, the contribution of the superghost system to the central extension of the total Virasoro algebra depends on the fermionic boundary condition. For the Virasoro algebra generated by $L_n^{(\gamma)}$, one finds[15,19] :

$$\bullet\, R \text{ sector}: \qquad c(n) = 11n^3 - 2n\,, \tag{3.69}$$

$$\bullet\, NS \text{ sector}: \qquad c(n) = 11n^3 + n\,. \tag{3.70}$$

If we now add up all contributions to the total central extension of the total Virasoro algebra, which, as for the bosonic string, includes the subtraction constants a_+ and a_- following from normal ordering, one obtains[15,19] :

$$\bullet\, R \text{ sector}: \qquad c(n) = \frac{3}{2}(D-10)n^3 + 24a_+ n\,, \tag{3.71}$$

$$\bullet\, NS \text{ sector}: \qquad c(n) = \frac{3}{2}(D-10)n^3 + \left[\frac{3}{2}(2-D) + 24a_-\right] n\,. \tag{3.72}$$

Hence, conformal invariance, and thus reparametrization invariance, is maintained at the quantum level only if we have the same critical values as obtained previously from different considerations

$$D = 10\,,\ a_+ = 0\,,\ a_- = \frac{1}{2}\,. \tag{3.73}$$

Recall that within the present quantization approach, super-Weyl invariant is explicit, since the superconformal modes of the world-sheet graviton and gravitino decouple. In Polyakov's path-integral approach[22], super-reparametrization invariance is explicitly preserved, but super-Weyl invariance is only realized if we have the conditions (3.73)[22]. Otherwise, the superconformal modes are dynamical fields.

Actually, with the critical values (3.73), not only does the central extension of the total Virasoro algebra vanish, but so do all the central extensions of the total $N = 1$ super-Virasoro algebra generated by $L_n^{(T)}$ and $J_q^{(T)}$[15,19]. Moreover, the total quantum BRST charge is then also nilpotent[64], thereby ensuring gauge invariance, namely super-reparametrization invariance, of the quantum spinning string.

That all these considerations are related and all lead to the critical values (3.73) is not surprising, considering the fact that the super-Virasoro generators $L_n^{(T)}$ and $J_q^{(T)}$ are given by anticommutators and commutators of the b and β ghosts with the BRST charge, as was the case also for the bosonic string. This is the reason why it is enough to consider for example, the total central extension of the Virasoro algebra to determine the critical values for gauge invariance.

As a last comment, let us say that physical states are again defined as BRST invariant states of definite ghost charge. Indeed, as for the bosonic string, the non-trivial BRST invariant cohomology classes of the BRST charge, both in Ramond and Neveu-Schwarz sectors, are in 1-to-1 correspondence with positive norm physical states in the old covariant or light-cone gauge quantization, with however a degeneracy due to ghost and superghost zero-modes[64,65]. In the NS sectors, this degeneracy is two-fold since only the (b, c) system has zero-modes. But in the R sector, this two-fold degeneracy is itself infinitely degenerate[16] due to the (β, γ) bosonic zero-modes. Hence the necessary restriction on the ghost number, to obtain a 1-to-1 correspondence with the physical content of the theory.

4 THE PRINCIPLES OF STRING THEORY CONSTRUCTION

Now that we have described and understood the structure, the origin and the meaning of the restrictions which appear for bosonic and spinning strings, we may actually infer a general description of the requirements that any string theory should meet. So far, we have concentrated on the restrictions following from local gauge invariance, i.e. invariance under infinitesimal gauge transformations. As we shall discuss, additional requirement also follow from global gauge invariance, i.e. invariance under gauge transformations not connected to the identity transformation.

The formulation of the basic rules for string theory construction that we are about to present, actually grew out from the construction of ten dimensional heterotic strings with space-time supersymmetry[66], and their toroïdal compactifications[67].

4.1 Quantum Consistency Constraints

4.1.1 Local world-sheet invariances

As is now clear, string theories are two dimensional field theories of quantum gravity coupled to matter fields, which are reparametrization invariant in the world-sheet. The constraints which follow from this gauge invariance are the vanishing of the generators of local diffeomorphisms, i.e. local world-sheet reparametrizations connected to the identity.

This world-sheet gauge invariance may be extended to some larger symmetry, including reparametrization invariance, as is the case for example for the spining string where we have world-sheet super-reparametrization invariance under a local $N = 1$ supersymmetry. Other extensions are also possible.

Such local world-sheet symmetries are essential in obtaining a consistent quantum string theory. In the conformal gauge, the existence of such symmetries at the quantum level guarantees that negative and zero-norm states decouple from physical states and amplitudes, as is necessary for unitarity (actually, we only established this property at the level of the physical spectrum, through the no-ghost theorem, but it may be shown to be also true for tree level amplitudes[15]). In the light-cone gauge, the existence of local world-sheet symmetries is essential for the space-time Poincaré covariance of the space of states.

Moreover, *world-sheet* local gauge invariances seem to be related in a very profound and natural way to *space-time* local symmetries. Indeed, string spectra always contain the associated massless gauge bosons, such as the graviton, Yang-Mills bosons and gravitinos, and the corresponding quantum field theories are known to be consistent only if such massless particles couple to the associated conserved space-time currents.

Hence, it is vital for quantum consistency of a string theory that its local world-sheet symmetries be preserved at the quantum level. As we know, this leads to restrictions on the number of world-sheet degrees of freedom, since the associated quantum anomalies must vanish.

Gauge invariances of string theories require gauge-fixing, with the ensuing ghost sectors. In the conformal gauge, as was discussed in detail, the ghost sector associated to world-sheet reparametrization invariance, which is a symmetry for any string theory, is the (b, c) ghost system. Its contribution to the central extension of the total Virasoro algebra is

$$c(n) = -26n^3 + 2n. \tag{4.1}$$

If in addition we have local $N = 1$ supersymmetry in the world-sheet, the corresponding (β, γ) superghost system contributes the following values to the total central extension.

$$\text{R sector: } c(n) = +11n^2 - 2n, \tag{4.2}$$
$$\text{NS sector: } c(n) = +11n^3 + n. \tag{4.3}$$

Similarly, for any other local world-sheet gauge invariance of the theory, one must take into account the corresponding contributions of the associated ghost sector to the central extension in the conformal gauge.

In this gauge, the remaining world-sheet degrees of freedom, namely the matter fields which couple to the two dimensional quantum gravity sector, define a conformal field theory, with the corresponding Virasoro algebra and central extension. Local reparametrization invariance at the quantum level, requires that the total central extension of the total Virasoro algebra, which includes the subtraction constant due to normal ordering, and the contributions from all the ghost sectors, vanishes identically. Clearly, this determines the value of the total central charge for the matter fields, and the value of the normal ordering constant. The same values should ensure that any other local world-sheet symmetries of the system are then also realized at the quantum level. This may be checked by computing the central extensions of the full symmetry algebra in the conformal gauge, or equivalently, by checking that the corresponding BRST charge is indeed nilpotent.

In order to describe a string theory in a D dimensional flat Minkowski space-time, we must have as world-sheet matter fields D scalar fields $x^\mu(\sigma, \tau)$, giving the space-time position of the string. In the conformal gauge, each of these scalar fields contributes to the central extension of the Virasoro algebra by

$$c(n) = n^3 - n. \tag{4.4}$$

For string theories with space-time fermions in the NSR formulation, we need a world-sheet local $N = 1$ supersymmetry such that the D scalar fields have as superpartners D world-sheet Majorana-Weyl spinors $\lambda^\mu_\pm(\sigma, \tau)$. In the conformal gauge, each of these spinors contributes as follows to the central extension of the Virasoro algebra:

$$\text{R sector:} \quad c(n) = \frac{1}{2}n^3 + n, \tag{4.5}$$
$$\text{NS sector:} \quad c(n) = \frac{1}{2}n^3 - \frac{1}{2}n. \tag{4.6}$$

The remaining world-sheet degrees of freedom, if any, are then considered as internal degrees of freedom, which in the conformal gauge define a conformal field theory contributing to the Virasoro algebra with the corresponding central extension:

$$c(n) = c_{int}n^3 + \beta_{int}n. \tag{4.7}$$

Hence, for a theory having only world-sheet reparametrization invariance, so called $N = 0$ supersymmetry, quantum gauge invariance requires that the central charge of the internal conformal field theory and the subtraction constant in the total Virasoro algebra be given by:

$$c_{int} = 26 - D, \tag{4.8}$$
$$a = \frac{1}{24}[(D-2) - \beta_{int}]. \tag{4.9}$$

For a theory having $N = 1$ world-sheet supersymmetry, the corresponding values are:

$$c_{int} = \frac{3}{2}(10 - D), \tag{4.10}$$

$$\text{R sector}: \quad a_+ = \frac{1}{24}[-\beta^+_{int}], \tag{4.11}$$

$$\text{NS sector}: \quad a_- = \frac{1}{24}[\frac{3}{2}(D-2) - \beta^-_{int}]. \tag{4.12}$$

Clearly, the value for the subtraction constant determines the lowest mass values of the physical spectrum. We have the following contributions to this normal ordering constant: $+\frac{1}{24}$ from a scalar degree of freedom, $-\frac{1}{24}$ from a Majorana-Weyl spinor in the Ramond sector and $+\frac{1}{48}$ from a Majorana-Weyl spinor in the Neveu-Schwarz sector. These are also the values given by ζ-function regularization of the corresponding infinite series. Note again the breakdown of $N = 1$ supersymmetry for NS boundary conditions.

The same restrictions on c_{int} and the subtraction constant may be obtained in the light-cone gauge[68]. Indeed, the transverse string coordinates together with the internal degrees of freedom define a conformal field theory, with a central extension such that the full Lorentz algebra is realized at the quantum level, i.e. such that the commutator $[M^{-i}, M^{-j}]$ vanishes identically. For example, the total central charge of this conformal field theory must be $c = 24$ for a string theory with $N = 0$ supersymmetry[30], and $c = 12$ for a theory with $N = 1$ supersymmetry[60]. These restrictions then lead again to (4.8) and (4.10). The same applies to the normal ordering constants, which however may also be determined by considering the would-be massless physical states, as was explained for bosonic and spinning strings[40].

With the string theories discussed in the previous two chapters, we have a situation where there are no internal degrees of freedom. From (4.8) to (4.12), we then recover of course the same critical values. Historically, the values $D = 26$ and $D = 10$ were believed to correspond to critical values of the dimension of space-time for which these theories could be consistently quantized. Now, they are seen to correspond to upper bounds on the dimension of space-time. Indeed, in order that the internal conformal field theory be unitary, we must have[9,69] $c_{int} \geq 0$.

However, in the present formulation of string theory, there is no dynamical argument which could single out any particular value of D. String theories may be constructed for space-time dimensions ranging from 26 to 2. In particular, the case $D = 4$ requires $c_{int} = 22$ for $N = 0$ supersymmetry, and $c_{int} = 9$ for $N = 1$ supersymmetry. If the classification of the corresponding conformal field theories were known, one could then give the complete list of all possible four-dimensional string theories.

From the present point of view, the problem of decoupling of conformal modes of the two dimensional gravity sector becomes somewhat academic. Indeed, in the conformal gauge, when these modes are dynamical fields, they simply correspond to some conformal field theory, and as such they may be regarded as some internal degrees of freedom. The study of the corresponding conformal field theory may be nonetheless important.[27,28]

The reader may have been wondering why we restricted all our attention to $N = 0$ and $N = 1$ world-sheet supersymmetry, and did not consider local $N \geq 2$ supersymmetries. The reason is that the corresponding possible[70] dimensions of space-time are $D \leq 2$, which does not leave much room for describing strings vibrating in transverse space dimensions! Nevertheless, the corresponding conformal structures find important applications elsewhere[9], and also in string theory as we shall comment upon later.

In conclusion, the requirements discussed in this section determine the necessary and sufficient conditions for maintaining , at the quantum level, the local world-sheet gauge invariances of string theory, namely local reparametrization invariance and its extensions. They are also necessary, but not necessarily sufficient as we shall discuss in the next section, for full quantum consistency of the theory.

The corresponding world-sheet structures can be used in the construction of open and of closed string theories. For closed strings however, we have the additional feature that different such structures may be used in each moving sector. Except for the space-time bosonic zero-modes q^μ and P^μ giving the space-time position of the center-of-mass and the total space-time momentum of the string, all other modes indeed separate into right-and left-moving modes, including the internal degrees of freedom. Hence, we have the following broad classification of closed string theories[7], depending on the number of world-sheet supersymmetries in each moving sector:

	sector		
string theory	right	left	dimension
bosonic	N=0	N=0	$D \leq 26$
heterotic	N=0	N=1	$D \leq 10$
Type II	N=1	N=1	$D \leq 10$

4.1.2 Global world-sheet invariances

Global world-sheet reparametrizations, which are not continuously connected to the identity transformation, but should also be symmetries of the quantum string theory, fall into two classes: orientation preserving and orientation reversing diffeomorphisms. If only the former symmetries are imposed on the system, one is describing an oriented string theory. Otherwise, we have an unoriented string theory. Let us first discuss this latter situation.

In the case of a closed string, an orientation reversing reparametrization, such as $\sigma \to \pi - \sigma$, clearly exchanges the right-and left-moving sectors. Hence, all states which are symmetric under this transformation remain physical for unoriented closed strings. For example, the massless 2-index antisymmetric tensor is always projected out from the spectrum.

For open strings, the orientation reversing diffeomorphism ($\sigma \to \pi - \sigma$) clearly changes the phases of bosonic and fermionic modes. These phase factors may be compensated for by coupling open strings to gauge degrees of freedom, by introducing Chan-Paton factor[71]. These Chan-Paton factors may intuitively be thought to correspond to charges attached at the end points of the open string. This was originally suggested by Chan and Paton for $U(n)$ groups[71], by viewing these charges as a quark and an anti-quark in the fundamental representation of $U(n)$.

Actually, this idea is easily generalized[72] by associating to each state of the open string some representation of some algebra. However, this may not be done arbitrarily, since the correct factorization of tree-level scattering amplitudes must be obtained[72]. Indeed, the gauge quantum numbers of intermediate states must be such that they may couple to those of the external states. This tree-level consistency requirement may be solved[72,73], with the following result:

- only the Lie algebras associated to $U(n), SO(n)$ and $USp(n)$ are allowed,

- all mass levels even in $\alpha' m^2$ must be in the adjoint representation,

- all mass levels odd in $\alpha' m^2$ must be in the following representations:
  - $U(n)$: adjoint representation
  - $SO(n)$: $\square\!\square + \bullet$
  - $USp(n)$: $\begin{smallmatrix}\square\\ \square\end{smallmatrix} + \bullet$

Moreover, $U(n)$ gauge degrees of freedom may only be introduced for oriented open strings, and $SO(n)$ or $USp(n)$ ones for unoriented open strings (this result agrees with one's intuition).

When considering string interactions, where open strings interact at their end points, it is clear that both open and closed string sectors of a same theory must be included in the description of the spectrum. Indeed, a single open string may interact at its own end points to form a closed string, or a closed string may break to form an open string. Hence, the possible situations are:

- open and closed oriented strings,
- open and closed unoriented strings,
- closed oriented strings,
- closed unoriented strings.

Therefore, there exist the following $D = 26$ bosonic theories:

- open and closed oriented strings, with gauge group $U(n)$,
- open and closed unoriented strings, with gauge group $SO(n)$ or $USp(n)$,
- closed oriented strings,
- closed unoriented strings.

The first two classes correspond to the Veneziano Model[74], the third to the Extended Shapiro-Virasoro Model[75] and the fourth to the Restricted Shapiro-Virasoro Model[75]. All these theories have space-time scalar tachyons, a feature incompatible with unitarity.

Since the open spinning string has only one space-time supersymmetry, we have the following $D = 10$ supersymmetric string theories[72]:

- Type I open and closed unoriented superstring, with gauge group $SO(n)$ or $USp(n)$,
- Type I closed unoriented superstring,
- Type IIa closed oriented superstring,
- Type IIb closed oriented superstring.

The closed string sector of the first two classes is obtained by truncation of the the type IIb superstring under the orientation reversing diffeomorphism symmetry. Indeed, the space-time supercharges associated to the two moving sectors must have the same chirality. Hence, the first two cases have $N = 1, D = 10$ space-time supersymmetry, whereas the last two cases have $N = 2a$ or $N = 2b, D = 10$ supersymmetry.

These are thus the restrictions following from the requirement of quantum invariance under orientation reversing world-sheet diffeomorphisms, by considering quantum consistency at tree-level. By considering higher loop amplitudes however, further restrictions are imposed.

In the case of open strings, this is well known[5]. Indeed, when considering 1-loop amplitudes with external open string states of zero-norm, such amplitudes must vanish identically, i.e. all null states must decouple. In the case of Type I superstrings, this question has been investigated for external massless gauge bosons[5] and gravitons[76], with the conclusion that such a decoupling occurs only for the gauge group $SO(32) = SO(2^{10/2})$. Although such a decoupling has not been established yet for all other null states, including massless and massive fermions, it will most probably also occur for the same gauge group only.

In the case of open bosonic strings, the same question has been investigated for arbitrary external open string states, with the conclusion[38,77] that such a decoupling can never occur because of the open string tachyon, but that otherwise, except for this non-vanishing tachyon contribution, decoupling would require the gauge group $SO(2^{26/2})$[78], independently of any regularization[38,77].

Hence, higher loop unitarity restricts open string theories to those of unoriented open and closed strings with gauge group $SO(2^{D/2})$, up to possible tachyon contributions[38,77], in which case the theory is ill-defined anyway.

In the case of unoriented closed strings, not much is known along these lines. In the case of oriented closed strings however, it is widely believed that modular invariance (which is discussed below) and the absence of tachyons are sufficient conditions to guarantee decoupling of all null states. This has been established explicitly for massless bosonic states in the case of superstrings, and from the point of view of the corresponding low-energy effective field theory[63]. General arguments support this conjecture.[63]

It so happens that the $SO(2^{D/2})$ unoriented open string theories can be obtained from oriented closed string theories by a $\mathbb{Z}_2$-orbifold construction in the *world-sheet*[79]. Thus, this leads to the idea[38,77] that all consistent open string theories are unoriented string theories which are obtained through such a construction from an oriented closed string theory. Moreover, quantum consistency of these open string theories, with in particular decoupling of all null states to all orders of perturbation theory, would then simply follow from modular invariance to all orders of the corresponding oriented closed string theory[38,77,79]. Since a $\mathbb{Z}_2$-orbifold structure may be defined in two ways, one may believe that one of these structures would lead to a theory of unoriented open and closed strings, and the other structure to a theory of unoriented closed strings only. In addition, more complicated theories may possibly be obtained through other abelian (and non-abelian?) orbifold structures on the world-sheet, thus corresponding to theories of strings with more than two end points[38,77].

Hence, this opens the way to the construction of new open (and more complicated?) string theories in less than 26 or 10 dimensions, including 4 dimensions[80], by considering any consistent closed oriented string theory. This is certainly an interesting possibility, in particular with regards to the possible gauge symmetries. Much however remains to be done in this direction.

In the remainder of these lectures, we shall concentrate on the construction of oriented closed strings, i.e. theories for which the world-sheet has a well-defined orientation. In string perturbation theory, scattering amplitudes are obtained by considering a world-sheet corresponding to a surface of given topology, to which the external states are attached[11]. The topology of the surface is characterized by its genus g, which counts the number of holes in the surface, or the number of handles attached on the sphere (of genus g=0) to obtain the surface. In the case of oriented closed strings, the space of global diffeomorphisms is non-trivial when the world-sheet has a topology more complex than that of the sphere, namely when its genus is greater than zero, $g > 0$.

Let us introduce some terminology[11]. $Diff_0$ denotes the space of all world-sheet diffeomorphisms continuously connected to the identity transformations; they thus preserve the orientation. $Diff$ denotes the space of all orientation preserving diffeomorphisms. The quotient $Diff/Diff_0 = \Omega_g$ then forms a group, the group of all disconnected components of $Diff$, known as the modular (or mapping class) group. This group is defined for each genus g separately.

By imposing the constraints discussed in the previous section, we are assured that $Diff_0$ is a symmetry of the quantized theory. However, requiring that string theory is fully reparametrization invariant still imposes the additional constraint of modular invariance under Ω_g to all orders in the loop expansion.

Actually, it is enough to impose this invariance for vacuum amplitudes only. Modular invariance for scattering amplitudes should then follow by factorization. Moreover, it seems to be a fact that modular invariance of the 1-loop vacuum amplitude and the correct spin-statistics relation (namely, that space-time bosons and fermions contribute with opposite signs in loops) are sufficient conditions for modular invariance to all orders[55,81].

For this reason, we shall restrict the discussion to 1-loop vacuum amplitudes, or partition functions, for closed oriented string theories. The corresponding world-sheet has genus $g = 1$; this is the topology of a torus. By using local diffeomorphisms and Weyl rescalings, it is always possible to map this torus into a parallelogram, with a flat metric, in the upper-half complex plane $\mathbb{C}^+$, with opposite sides identified, and defined by two vectors attached at the origin and ending at 1 and $\tau = \tau_1 + i\tau_2 (Im\tau_2 > 0)$ respectively (the parameter τ should not be confused with the world-sheet coordinate τ). There is thus a torus associated to every point in $\mathbb{C}^+$. The opposite however, is not true. A same torus may correspond to different points in $\mathbb{C}^+$. Indeed, modular transformations leave the torus invariant but act non-trivially on $\mathbb{C}^+$.

The modular group of a given compact Riemann surface is generated by so-called Dehn twists[82,83]. For the genus $g = 1$ case, there are two such Dehn twists, obtained as follows. For each of the two homology cycles of the torus, cut the surface along this cycle, twist it by 2π and glue it back together. Clearly, this is a global diffeomorphism, leaving the torus invariant, but not the associated parameter τ. Indeed, cutting along the cycle corresponding to σ-parametrization, we have the Dehn twist acting on τ as:

$$T : \tau \to \tau + 1. \tag{4.13}$$

For the Dehn twist along the other cycle, we have

$$S : \tau \to -1/\tau. \tag{4.14}$$

Hence, the modular group $\Omega_g = 1$ of the torus is generated by these two transformations S and T. It is isomorphic to the group $PSL(2\mathbb{Z}) = SL(2, \mathbb{Z})/\mathbb{Z}_2$, since it acts on τ as:

$$\tau \to \frac{a\tau + b}{c\tau + d}, \quad a, b, c, d \in \mathbb{Z}; \quad ad - bc = 1. \tag{4.15}$$

Let us introduce here some more terminology[11,82]. For a Riemann surface of given topology g, let us consider the space of all metrics (of Euclidean or Minkowski signature) defined on this surface. The quotient of this space of metrics by the semi-direct product $Diff_0\times$ Weyl, with "Weyl" representing Weyl transformations, is then called Teichmüller space. Generally, the modular group Ω_g acts non-trivially on Teichmüller space, whose quotient by Ω_g is called modular space. Modular space parametrizes all possible conformal, complex or geometrical structures of the corresponding Riemann surface. In particular, we just saw that for $g = 1$, Teichmüller space is the upper-half complex plane $\mathbb{C}^+$ and that modular space is simply $\mathbb{C}^+/PSL(2\mathbb{Z})$. Let us choose the following fundamental domain of the modular group in Teichmüller space, to represented modular space:

$$F = \{\tau \in \mathbb{C} : |\tau| > 1, -\frac{1}{2} \leq Re\tau < \frac{1}{2}, Im\tau > 0\}. \tag{4.16}$$

Modular invariance thus requires that the transformations (4.13) and (4.14) leave 1-loop vacuum amplitudes invariant. To express these amplitudes, all the machinery of path-integrals[11,24,82] is not really necessary, if one recalls that the 1-loop vacuum amplitude of a single relativistic bosonic or fermionic degree of freedom in ordinary quantum field theory is given by:

$$\mp\frac{1}{2}ln\ det[-\Box + m^2] = \mp\frac{1}{2}trln[-\Box + m^2], \tag{4.17}$$

where the upper (resp.lower) sign applies to a boson (resp. fermion). Through a proper-time representation, this quantity is given (up to a factor) by:

$$\Lambda(m^2) = \pm\frac{1}{2}\int_0^\infty \frac{dt}{t}(4\pi t)^{-D/2}e^{-t\frac{m^2}{\mu^2}} \tag{4.18}$$

where t measures the total proper-time in the loop, and μ^2 is some mass scale. Note that in this expression, the trace over the zero-modes, namely the loop momentum, has already been taken.

For a mass distribution $\rho(m^2)$ of states of the same statistics, given as

$$\rho(m^2) = \sum_{n=0}^{\infty} d_n\delta(\frac{m^2}{\mu^2} - (\alpha n + \beta)), \tag{4.19}$$

where α, β are rational numbers, and the degeneracies d_n are defined by the function

$$f(q) = \sum_{n=0}^{\infty} d_n q^{2(\alpha n+\beta)}, q \in \mathbb{C}, |q| < 1, \tag{4.20}$$

the total contribution to the 1-loop vacuum amplitude is then:

$$\Lambda = \pm\frac{1}{2}\int_0^\infty \frac{d\tau_2}{\tau_2}(8\pi^2\tau_2)^{-D/2}f(e^{-\pi\tau_2}). \tag{4.21}$$

Here, we set $t = 2\pi\tau_2$.

Let us now consider the case of a closed string. Typically, for bosonic and fermionic states, we have the mass formulae:

$$\frac{1}{2}\alpha' m^2 = (\alpha n + \beta_R) + (\alpha m + \beta_L), \tag{4.22a}$$
$$\alpha n + \beta_R = \alpha m + \beta_L, \tag{4.22b}$$

where α, β_R, β_L are rational numbers, such that $\alpha^{-1}(\beta_R - \beta_L)$ is an integer, and n, m are positive integers (it is always possible to choose the factor α to be the same in each moving sector). The values αn and αm are the eigenvalues of the excitation level operators N and $\overline{N}$ (or $N^\perp$ and $\overline{N^\perp}$) in each moving sector, namely the Virasoro zero-modes with the contribution of the bosonic zero-modes subtracted away. To each moving sector, we associate a partition function:

$$f_R(q) = Trq^{2(N^\perp+\beta_R)} = \sum_{n=0}^{\infty} d_R(n) q^{2(\alpha n+\beta_R)}, \tag{4.23a}$$

$$f_L(q) = Trq^{2(\overline{N}^\perp+\beta_R)} = \sum_{n=0}^{\infty} d_L(n) q^{2(\alpha n+\beta_L)}, \tag{4.23b}$$

with the traces taken over all non-zero modes. As was explained for the closed bosonic string after (2.113), the level matching condition (4.22b) is easily implemented in the partition function for the theory, which then has the expression:

$$f(q) = \int_{-\pi}^{\pi} \frac{d\theta}{2\pi} f_R(|q|e^{i\theta/2\alpha}) \overline{f_L(|q|e^{i\theta/2\alpha})}. \tag{4.24}$$

When substituted in (4.21) , we thus obtain for the corresponding 1-loop vacuum amplitude:

$$\Lambda = \pm\frac{1}{2}\alpha(8\pi^2)^{-D/2} \int_0^\infty \frac{d\tau_2}{\tau_2^2} \int_{-1/2\alpha}^{1/2\alpha} d\tau_1 \frac{1}{\tau_2^{D/2-1}} f_R(e^{i\pi\tau}) \overline{f_L(e^{i\pi\tau})}, \tag{4.25}$$

where

$$\tau = \tau_1 + i\tau_2, \quad \theta = 2\pi\alpha\tau_1. \tag{4.26}$$

On the other hand, we have:

$$f_R(e^{i\pi\tau}) \overline{f_L(e^{i\pi\tau})} = Tre^{i\pi\tau_1 P} e^{-\pi\tau_2 H'}, \tag{4.27}$$

where

$$P = 2[(N^\perp + \beta_R) - (\overline{N}^\perp + \beta_L)] = 2[(L_0^\perp + \beta_R) - (\overline{L}_0^\perp + \beta_L)], \tag{4.28a}$$

$$H' = 2[(N^\perp + \beta_R) + (\overline{N}^\perp + \beta_L)] = 2[(L_0^\perp + \beta_R) + (\overline{L}_0^\perp + \beta_L)] - \alpha'(P^i)^2. \tag{4.28b}$$

Hence, P is the generator of translations in σ, and H' the generator of translations in τ with the bosonic zero-modes subtracted out. The physical meaning of the expression (4.25) is thus clear. As the closed string propagates freely before annihilating itself, it propagates for a total proper-time value (proportional to) $(\pi\tau_2)$ and twists on itself by an angle (proportional to) $(\pi\tau_1)$. The factor (4.27) precisely computes the corresponding partition function. We recognize in τ_1 and τ_2 respectively the real and imaginary parts of the modular parameter τ of the associated torus topology. When multiplied by the factor $\tau_2^{1-D/2}$, which corresponds to the trace over the bosonic zero-modes (P^i) , the partition function (4.27) is then integrated in (4.25) over all possible values of τ_1 and τ_2, i.e. over all possible geometries of the torus associated to this 1-loop vacuum amplitude.

We thus conclude that the 1-loop vacuum amplitude of a closed string theory is given by the expression (up to a factor):

$$\int \frac{d\tau_1 d\tau_2}{\tau_2^2} \tau_2^{1-D/2} Trq^{2(N^\perp+\beta_R)} \overline{q}^{2(\overline{N}^\perp+\beta_L)}, \tag{4.29}$$

where

$$q = e^{i\pi\tau}, \quad \tau = \tau_1 + i\tau_2, \tag{4.30}$$

and the integration is performed over some domain of Teichmüller space. Modular invariance requires on the one hand, that the integrand in (4.29) be modular invariant, namely invariant under the S and T transformations of the modular parameter, and on the other hand, that the integral in (4.29) be performed over the fundamental domain F defined in (4.16), as otherwise the same geometrical configuration would be included more than once in the amplitude.

It is easy to see that invariance under T in (4.13) only requires that

$$N + \beta_R = \overline{N} + \beta_L \quad (mod\ 1) \tag{4.31}$$

This is the constraint expressing invariance under constants shifts in $(\sigma \to \sigma + \sigma_0)$. For the transformation S in (4.14), it is easy to show that we have

$$\frac{d\tau_1 d\tau_2}{\tau_2^2} \quad \to \quad \frac{d\tau_1 d\tau_2}{\tau_2^2}, \tag{4.32}$$

$$\tau_2 \quad \to \quad \frac{\tau_2}{|\tau|^2}. \tag{4.33}$$

Hence, invariance under $\tau \to -1/\tau$ requires that the factor

$$\tau_2^{1-D/2} Tr q^{2(N^\perp + \beta_R)} \overline{q}^{2(\overline{N}^\perp + \beta_L)} \tag{4.34}$$

be modular invariant.

By restricting the integral (4.29) to the fundamental domain F, the usual short-distance divergence at $\tau_2 \to 0$ is excluded. Thus, string theories have a much better short-distance behaviour than ordinary quantum field theories of particles. This restriction due to modular invariance is one of the reasons for the possible finiteness of string theories. However, the point $\tau \to 0$, a short-distance limit, is related to the point $\tau \to i\infty$, a large-distance limit, through the modular transformation $\tau \to -1/\tau$. This is the simplest example of a duality property between large distance and small distance physics in string theories.

The fact that the large distance point $\tau_2 \to i\infty$ is included in the integral (4.29) implies that its expression is ill-defined for tachyonic theories. Indeed, we have

$$q^{m^2} = e^{i\pi\tau_1 m^2} e^{-\pi\tau_2 m^2}. \tag{4.35}$$

Actually, this divergence has no physical meaning[38,77]. It is only a consequence of the fact that the proper-time representation used in (4.18) is not adequate[38,77,78] when $m^2 < 0$. It may be shown[38,77] that the correct result is obtained by inserting in (4.29) a factor $e^{-\pi\tau_2\gamma}$, with $\gamma \geq -m^2_{\text{tachyon}}$, and then taking $\gamma \to 0$ once the integral is completed. However, modular invariance is then explicitly broken. But after all, this is not so surprising. Modular invariance is believed to be essential in the quantum consistency of string theories. For example, it should guarantee that all null states decouple to all orders of perturbation theory, which is necessary for unitarity. For a tachyonic theory, unitarity is certainly not obtained, and as we discussed, null states do not decouple either. Thus, if modular invariance indeed implies unitarity, a tachyonic theory can never be modular invariant. For these reasons of unitarity and modular invariance, a tachyonic theory can never be considered to be a consistent quantum theory.

Finally, let us remark here that the factorization of the partition function (4.29) into right-and left-moving contributions generally extends to all orders of perturbation theory. This property is known as holomorphic factorization.[11]

4.2 Examples of Modular Invariance

4.2.1 The D=26 closed bosonic string

As a first example of the principles of string theory construction at work, let us reconsider the simplest of all string theories, namely the closed oriented bosonic theory studied in chapter 2. In this case, all world-sheet matter fields are bosonic coordinates $x^\mu(\sigma,\tau)$, taking their values in some Minkowski space-time of D dimensions. From our discussion above, we know that quantum local reparametrization invariance in the world-sheet requires the values (see (4.8) and (4.9)):

$$D = 26, \qquad a = \frac{D-2}{24} = 1. \tag{4.36}$$

These values immediately lead to the description of the space-time spectrum which we gave in chapter 2.

Let us now consider the global constraint of modular invariance, and, as an exercise, assume that the values (4.36) are not specified. The contribution of one scalar degree of freedom $x^\mu(\sigma,\tau)$ to the partition function (4.29) in one moving sector is simply:

$$\frac{1}{q^{2\beta^0}} \prod_{n=1}^{\infty} \frac{1}{(1-q^{2n})} = \left[q^{2(\beta^0-\frac{1}{24})}\eta(\tau)\right]^{-1}, \tag{4.37}$$

where β^0 is the contribution to the ordering constant "a" from one scalar field in one moving sector, and $\eta(\tau)$ is the Dedekind η-function defined in (2.111). Hence, the 1-loop vacuum amplitude is:

$$\int \frac{d\tau_1 d\tau_2}{\tau_2^2} \tau_2^{1-D/2} \left[q^{2(\beta_-^0-\frac{1}{24})}\bar{q}^{2(\beta_+^0-\frac{1}{24})}\right]^{(2-D)} |\eta(\tau)|^{2(2-D)}. \tag{4.38}$$

Under modular tranformations, the $\eta(\tau)$ function transforms as follows:

$$T: \tau \to \tau+1 \qquad \eta(\tau) \to e^{i\pi/12}\eta(\tau), \tag{4.39}$$

$$S: \tau \to -1/\tau \qquad \eta(\tau) \to (-i\tau)^{1/2}\eta(\tau). \tag{4.40}$$

Hence, modular invariance under $\tau \to \tau+1$ requires

$$\beta_-^0 = \beta_+^0 \ (mod\ 1), \tag{4.41}$$

whereas invariance under $\tau \to -1/\tau$ requires

$$\beta_-^0 = \frac{1}{24} = \beta_+^0. \tag{4.42}$$

Therefore, modular invariance requires that each scalar field contributes a value $1/24$ to the subtraction constant. This is the value also following from local reparametrization invariance, and from ζ-function regularization. From this point of view, one may say that this type of regularization is compatible with modular invariance, so that it may consistently be used to obtain the correct values for the subtraction constants. Intuitively, the reason for this compatibility is that ζ-function regularization does not require any mass scale, and thus should preserve conformal , Weyl and modular invariance[24]. Note however that modular invariance does not fix the value for D; the quantity (4.38) is modular invariant for any value D, provided we have the values (4.42) for the ordering constants.It is only local reparametrization invariance, through the requirement of a vanishing total central charge of the Virasoro algebra, which fixes the value $D = 26$.

In conclusion, the 1-loop partition function for the oriented closed string in 26 dimensions is:

$$\int_F \frac{d\tau_1 d\tau_2}{\tau_2^{14}} |\eta(\tau)|^{-48}. \tag{4.43}$$

Thus, modular invariance does not impose any additional constraints than those following from local reparametrization or conformal invariance. From that point of view, the closed bosonic string is a consistent theory. However, we know that the values (4.42) lead to the existence of a tachyon, which manifests itself in (4.43) by the fact that the factor $|\eta(\tau)|^{-48}$ diverges as $(e^{4\pi\tau_2})$ when $\tau \to i\infty$. This divergencess indeed corresponds to a state with $\frac{1}{2}\alpha' m^2 = -2$. As we discussed in the last section, such a situation leads to the breakdown of modular invariance, a consequence of the fact that the theory is not unitary.

4.2.2 The D=10 closed spinning string

Let us now consider the case of the spinning string discussed in chapter 3. The matter fields are then D $N = 1$ $d = 2$ chiral supermultiplets (x^μ, λ^μ). Local super-reparametrization invariance requires

$$D = 10, \;\; a_R = 0, \;\; a_{NS} = \frac{1}{2}. \tag{4.44}$$

In our analysis of modular invariance, we shall use these values, although the reader may like to repeat the argument with D, a_R and a_{NS} unspecified.

The contribution of a single scalar degree of freedom to the partition function in one moving sector is:

$$Trq^{2[N^\perp - \frac{1}{24}]} = \frac{1}{q^{1/12}} \prod_{n=1}^{\infty} \frac{1}{(1-q^{2n})} = \frac{1}{\eta(\tau)}. \tag{4.45}$$

Here, we used the fact that the corresponding subtraction constant is $\frac{1}{24}$. For the Majorana-Weyl degrees of freedom, we have to distinguish between Ramond and Neveu-Schwarz boundary conditions, with the corresponding subtraction constants $(-\frac{1}{24})$ and $(+\frac{1}{48})$ respectively. In the Ramond sector, we then have the following contribution to the partition function for one Majorana-Weyl degree of freedom

$$Trq^{2[N^\perp + \frac{1}{24}]} = q^{\frac{1}{12}} \prod_{n=1}^{\infty} (1+q^{2n}) = \frac{1}{\sqrt{2}} \left[\frac{\theta_2}{\eta}\right]^{1/2}. \tag{4.46}$$

In the Neveu-Schwarz sector, we obtain

$$Trq^{2[N^\perp - \frac{1}{48}]} = q^{-\frac{1}{24}} \prod_{n=1}^{\infty} (1+q^{2n-1}) = \left[\frac{\theta_3}{\eta}\right]^{1/2}, \tag{4.47}$$

and if the contributions are weighted by the fermion number $(-1)^F$, we have

$$Tr(-1)^F q^{2[N^\perp - \frac{1}{48}]} = q^{-\frac{1}{24}} \prod_{n=1}^{\infty} (1-q^{2n-1}) = \left[\frac{\theta_4}{\eta}\right]^{1/2}. \tag{4.48}$$

In these expressions, η is the Dedekind η-function, θ_i are the theta functions $\theta_i(0|\tau)$ defined in (3.64), and the quantities $N^\perp$ and F represent the contribution of the corresponding degrees of freedom to the respective quantum operators, with the trace taken accordingly over the associated subspace of the space of states.

It should then be clear that the corresponding partition function has the form:

$$\int \frac{d\tau_1 d\tau_2}{\tau_2^2} \frac{1}{\tau_2^4} |\eta(\tau)|^{-24} \left[\frac{1}{16} d_R \theta_2^4 - \alpha_R \theta_3^4 - \beta_R \theta_4^4\right] \overline{\left[\frac{1}{16} d_L \theta_2^4 - \alpha_L \theta_2^4 - \beta_L \theta_4^4\right]} \tag{4.49}$$

Here, d_R and d_L count the degeneracies of the ground state spinors in the respective Ramond sectors. This degeneracy is 8 for an on-shell massless Majorana-Weyl spinor, 16 for a massless Majorana or Weyl spinor and 32 for a massless Dirac spinor. The quantities $\alpha_R, \alpha_L, \beta_R$ and β_L correspond to the "projectors" $\frac{1}{2}\left[\alpha_R + \beta_L(-1)^F\right]$ and $\frac{1}{2}\left[\alpha_R + \beta_L(-1)^{\overline{F}}\right]$ in the respective Neveu-Schwarz sectors, with $F, \overline{F}$ being the fermion numbers. For $\beta_R = 0$ or $\beta_L = 0$, we have no projection, hence a possible tachyon state. The case $(\alpha + \beta)_{R,L} = 0$ corresponds to GSO projection; the tachyon is then avoided. In order to count correctly the number of states at each mass level, we must have $|\alpha_{R,L}| + |\beta_{R,L}| = 1$. Finally, the relative minus sign between R and NS contributions is due to the fact that space-time bosons and fermions should contribute with opposite signs to the 1-loop vacuum amplitude. Thus, the correct spin-statistics relation is obtained when α_R and α_L are positive (with d_R,d_L positive).

Under modular transformations, we have:

$$T : \tau \to \tau + 1 \qquad \eta(\tau) \to e^{i\pi/12}\eta(\tau), \tag{4.50}$$

$$\theta_2 \to e^{i\pi/4}\theta_2, \tag{4.51a}$$

$$\theta_3 \to \theta_4, \tag{4.51b}$$

$$\theta_4 \to \theta_3. \tag{4.51c}$$

$$S : \tau \to -1/\tau \qquad \eta(\tau) \to (-i\tau)^{1/2}\eta(\tau), \tag{4.52}$$

$$\theta_2 \to (-i\tau)^{1/2}\theta_4, \tag{4.53a}$$

$$\theta_3 \to (-i\tau)^{1/2}\theta_3, \tag{4.53b}$$

$$\theta_4 \to (-i\tau)^{1/2}\theta_2. \tag{4.53c}$$

Hence, invariance of (4.49) under T requires that:

$$\alpha_R + \beta_R = 0, \qquad \alpha_L + \beta_L = 0. \tag{4.54}$$

This precisely corresponds to GSO projection in the respective NS sectors. With the requirement that $|\alpha_{R,L}| + |\beta_{R,L}| = 1$, (4.54) implies that all $\alpha_{R,L}, \beta_{R,L}$ take the values $\pm 1/2$. The sign of these quantities is now fixed by modular invariance under S, since it requires:

$$-\beta_R = \frac{d_R}{16}, \qquad -\beta_L = \frac{d_L}{16}. \tag{4.55}$$

Therefore, modular invariance of the 1-loop partition function implies:

$$d_R = d_L = 8, \qquad \alpha_R = \alpha_L = \frac{1}{2} = -\beta_R = -\beta_L. \tag{4.56}$$

These are precisely[55] the conditions imposed in the GSO projection[54]. They show that this projection of the oriented closed spinning string indeed leads to a consistent string theory. Note that we also obtain the correct spin-statistics relation [55]. In conclusion, the 1-loop modular invariant partition function of the type II spinning string is:

$$\int \frac{d\tau_1 d\tau_2}{\tau_2^6} \left| \frac{1}{2\eta^{12}} \left[\theta_2^4 - \theta_3^4 + \theta_4^4\right] \right|^2. \tag{4.57}$$

Note first of all that this quantity is finite, since the limit $\tau \to i\infty$ is not singular. Indeed, the would-be tachyon in the (N_S, N_S) sector is projected out in the difference $(\theta_3^4 - \theta_4^4)$. Moreover, this quantity actually vanishes exactly, due to the Riemann identity (3.66). This is a consequence of the space-time supersymmetry in each moving sector.

Note also that the discussion of modular invariance considered each moving sector independently from the other. In particular, this implies that the chiralities of the two Majorana-Weyl spinors in the R sectors are not correlated. These chiralities may be the same or different, corresponding respectively to the Type IIb and Type IIa superstrings. In lower dimensions, there exist examples of string theories where the right-and left-moving sectors are related in a non-trivial manner through modular invariance.

4.3 Spin Structures

It is convenient to express these last results concerning the spinning string in an other language, by introducing a new geometrical concept: spin structures[55]. In the quantization of world-sheet spinors, we considered two sectors associated to Ramond and Neveu-Schwarz boundary conditions, corresponding respectively to having periodic (P) or anti-periodic (A) fermions when taken around the closed string in the σ-direction on the world-sheet.

Clearly, when the world-sheet has the topology of a torus, since fermions may then be taken around the two homology cycles of the torus, we have $2^2 = 4$ possible choices of boundary conditions. Thus, NS (resp.R) spinors then correspond to anti-periodic (resp. periodic) fermions around the homology cycle associated to σ-parametrization, with periodic or anti-periodic boundary conditions on the other homology cycle associated to τ-parametrization. We shall denote these choices by ${}_{A,P}\underset{A,P}{\square}$, where the horizontal line corresponds to the boundary condition in the σ-cycle, and the vertical line to the boundary condition in the τ-cycle. Hence, NS fermions correspond to ${}_{A}\underset{A}{\square}$ and ${}_{P}\underset{A}{\square}$, and R fermions to ${}_{A}\underset{P}{\square}$ and ${}_{P}\underset{P}{\square}$. These four possible choices correspond to so-called spin structures of the torus[55,82,83]. In the same way that a torus is endowed with a geometry (or equivalently a complex or a conformal structure) by specifying the modular parameter τ, the torus is endowed with a spin structure, which is necessary to describe fermions on it, by specifying these boundary conditions on a basis of the homology group of the torus. Let us only remark here that this notion of spin structure extends to compact Riemann surfaces of higher genus, once a basis of the associated homology group is given[82,83].

Under modular transformations of the torus, it should be clear that spin structures are mapped into one another as follows:

$T : \tau \to \tau + 1$

$$
\begin{aligned}
&P\underset{P}{\Box} \to P\underset{P}{\Box}\,,\\
&A\underset{P}{\Box} \to A\underset{P}{\Box}\,,\\
&A\underset{A}{\Box} \to P\underset{A}{\Box}\,,\\
&P\underset{A}{\Box} \to A\underset{A}{\Box}\,,
\end{aligned}
\tag{4.58}
$$

$S : \tau \to -1/\tau$

$$
\begin{aligned}
&P\underset{P}{\Box} \to P\underset{P}{\Box}\,,\\
&A\underset{P}{\Box} \to P\underset{A}{\Box}\,,\\
&A\underset{A}{\Box} \to A\underset{A}{\Box}\,,\\
&P\underset{A}{\Box} \to A\underset{P}{\Box}\,,
\end{aligned}
\tag{4.59}
$$

Therefore, when considering modular invariant partition functions of string theories with world-sheet fermions, one must not only sum over the different possible geometries of the torus, by integrating over τ in the fundamental domain F, but one must also sum over all possible spin structures of the torus[55]. From this point of view, the GSO projection of the previous section corresponds to summing over all geometrical data of the torus topology of the world-sheet for the 1-loop vacuum amplitude. At genus $g = 1$, the $P\underset{P}{\Box}$ sector does not mix with the other sectors. At a higher genus, this is not the case anymore, so that it indeed participates non-trivially to the sum over spin structures necessary for modular invariance[55].

The transformations of spin structures under the modular group are very reminiscent of the associated transformations of theta functions (see (4.51) and (4.53)). The relationship can be made exact. The ordinary partition function of a Majorana-Weyl fermion in two dimensional quantum field theory corresponds to having anti-periodic boundary conditions in the time direction[82,83]. Thus, to the spin structure $A\underset{A}{\Box}$, we must associate the contribution (4.47) in the NS sector, namely:

$$
A\underset{A}{\Box} : \left[\frac{\theta_3}{\eta}\right]^{1/2} , \tag{4.60}
$$

whereas the spin structure $P\underset{A}{\Box}$ is associated to (4.48), since the insertion of the fermion number $(-1)^F$ precisely corresponds to changing the boundary condition in time:

$$
P\underset{A}{\Box} : \left[\frac{\theta_4}{\eta}\right]^{1/2} . \tag{4.61}
$$

This association is indeed compatible with the modular transformation under T of these quantities. From the transformation under S, we then have for the spin structure $A\underset{P}{\Box}$:

$$
A\underset{P}{\Box} : \left[\frac{\theta_2}{\eta}\right]^{1/2} . \tag{4.62}
$$

Finally, we may thus expect to have for the remaining spin structure

$$\mathrm{P}\underset{\mathrm{P}}{\square} : \left[\frac{\theta_1}{\eta}\right]^{1/2} . \tag{4.63}$$

This is indeed correct. The first three spin structures are known as even spin structures, whereas the last one is an odd spin structure, since the corresponding partition function vanishes identically (see(3.64)). This is due to the existence of fermion zero modes for the associated boundary conditions[83].

In terms of this notion of spin structures, the partition function (4.57) of the Type II superstring may be rewritten symbolically as:

$$\left[\frac{1}{2}\left(\mathrm{A}\underset{\mathrm{P}}{\square}+\mathrm{P}\underset{\mathrm{P}}{\square}\right)-\frac{1}{2}\left(\mathrm{A}\underset{\mathrm{A}}{\square}-\mathrm{P}\underset{\mathrm{A}}{\square}\right)\right]\overline{\left[\frac{1}{2}\left(\mathrm{A}\underset{\mathrm{P}}{\square}+\eta\,\mathrm{P}\underset{\mathrm{P}}{\square}\right)-\frac{1}{2}\left(\mathrm{A}\underset{\mathrm{A}}{\square}-\mathrm{P}\underset{\mathrm{A}}{\square}\right)\right]} \tag{4.64}$$

where the relative sign $\eta = \pm 1$ in the Ramond sectors corresponds to the Type IIa ($\eta = -1$) or to the Type IIb ($\eta = +1$) superstring partition function. In (4.64), the notation for the boundary conditions means that all 8 Majorana-Weyl spinors in each moving sector (in the light-cone gauge) have the same boundary conditions.

Remark

To conclude, let us remark that the notion of spin structure can be extended to arbitrary twisted fermions, where the phase factors that fermions pick up when taken around homology cycles are now arbitrary rather than simply (± 1) factors. The corresponding partition functions may then be expressed in terms of general theta functions with characteristics. For more details, see Ref.83.

5 COMMENTS AND CONCLUSIONS

Unfortunately due to lack of space, we have to leave our discussion of string theory constructions here. It would be interesting to apply the general principles of the last chapter to the torus compactification of bosonic strings[67], or to the construction of all ten dimensional heterotic string theories[55,66,85], either in a bosonic or in a fermionic formulation, i.e. where the internal conformal field theory is realized in terms of free scalars or free spinors on the world-sheet. This is left as an exercise for the interested reader. He may find in Ref. 63 a useful discussion of torus compactification of bosonic strings, and in Refs. 15, 17, 82 a description of $D = 10$ heterotic strings, both in a bosonic or a fermionic formulation. For example, he may like to construct the $D = 10$ $N = 1$ supersymmetric heterotic strings[66] with gauge group $E_8 \times E_8$ or Spin $(32)/\mathbb{Z}_2$, and the $D = 10$ non-supersymmetric tachyon-free heterotic string[55,85] with gauge group $SO(16) \times SO(16)$, by considering all possible choices of spin structures for the internal world-sheet fermions and imposing modular invariance.

A new fundamental structure that these theories realize is that of a world-sheet current algebra, namely a Kac-Moody algebra[35], which leads to the existence of massless vector states, corresponding to the gauge bosons of a space-time gauge symmetry of the theory. In a bosonic formulation, this structure is obtained through the so-called Frenkel-Kac construction of level 1 representations of Kac-Moody algebras, using string vertex operators (for more details, see Refs. 15, 17, 35, 66). In a fermionic formulation, the current algebra structure is obtained through a quark model type of construction (for more details, see Refs. 15, 35, 86). Hence, we have here another example of the general and profound relationship which exists between space-time and world-sheet local symmetries.

Despite the necessity of ending these lecture notes here, hopefully the reader will have found in the above discussion a useful introduction to and understanding of the principles of string theory construction. This should allow him to understand the literature on the subject and develop his own research. Clearly, the "name of the game" has become the following : pick your favorite dimension D of space-time (usually $D = 4$!), and use for the internal degrees of freedom some conformal field theory with central charge $c = 26 - D$ or $c = 3(10 - D)/2$, corresponding respectively to $N = 0$ or $N = 1$ local supersymmetry in the associated moving sector. This internal conformal field theory should be such that the complete 1-loop partition function of the theory is modular invariant, and consistent with the usual spin-statistics connection (and also free of tachyon-related divergences). Then, work out the physical spectrum and its space-time symmetries, and decide if it has a chance of resembling, in the massless sector, the spectrum of quarks, leptons and gauge bosons of the $SU(3) \times SU(2) \times U(1)$ Standard Model, or some Grand Unification of it. Clearly, when such a string model is found, there is still a very large amount of work to be done[3,4,6] before calling it a "realistic string theory".

Originally, since string theory became the most prominent candidate for a unified quantum theory right after the wave of popularity of Kaluza-Klein theories[87], people tended to think of the internal degrees of freedom as compactified degrees of freedom. For example, starting from a ten dimensional string theory, by compactifying six bosons and fermions on some internal compact space, such as a torus, a four dimensional theory may be obtained[1,72]. When this was considered at the level of the low-energy effective field theory in ten dimensions, it led[2] to compactifications on Calabi-Yau manifolds[2,4,15] when requiring $N = 1$ $D = 4$ global supersymmetry. But clearly, this is only one possible approach. More generally, an arbitrary construction would not have such a simple geometrical interpretation (for example, when the right- and left-moving sectors are completely different), although conformal field theory in two dimensions is so particular that such an interpretation may sometimes be possible nevertheless.

From this point of view, very large numbers of string theories in 4 dimensions have been constructed over the last years, using many different formulations for the internal degrees of freedom which do not always lead to different theories. Let us only mention here toroïdal compactifications[67], covariant lattice constructions[62,63], orbifolds[6,88], group manifolds[89], fermionic constructions[7,90], tensor product constructions[68,91,92], etc....

Clearly, the approach to string theory construction can be considered from a much more general point of view. As was one of the main leit-motives of these lectures, string theories *in the conformal gauge*, and thus *also in the light-cone gauge*, are nothing but conformal field theories, where the internal degrees of freedom define a $N = 0$ or $N = 1$ conformal field theory (CFT) with a given central charge[7,68]. Thus, if a complete list of these CFT and of their partition functions was known, irrespective of explicit realizations of those theories in terms of two dimensional fields, a complete classification of consistent string theories could be given. This is one among many other reasons for the very rapidly expanding present research activity on the classification of conformal field theories in two dimensions[9]. Let us remark here that from such a point of view, the calculation of string scattering amplitudes to any order of perturbation theory becomes an exercise in the calculation of correlation functions in conformal field theory[9].

As should be emphasized, the present approach to string theory construction does not give any understanding as to why space-time should be four dimensional and flat. Only requirements of quantum consistency are imposed, which only lead to the upper bound $D \leq 10$. The hope is that, by classifying all possible string theories, we could not only find out if there are potential "realistic" theories, but moreover, such an approach could bring some new insight into the little understanding that we have concerning the dynamics of these theories, especially in non-perturbative regimes[7,68]. In this context, all present constructions should be viewed as the constructions of different classical string vacua of what could possibly be only a few fundamental string theories, still to be formulated. For example, all toroïdal compactifications of bosonic strings are related to one another, via expectation values of massless physical states[93]. Dynamics in the spaces of moduli (or parameters) describing these different classical vacua, induced for example by possible non-perturbative effects, could lead to some unique "ground state" string theory. This is the general present "philosophy" behind the approach discussed in these lectures.

Clearly, since the number of possibilities for constructing string theories is so large, one may be less ambitious, and only consider the classification of possible "realistic" string theories, by imposing some additional general "phenomenological" requirements. The obvious one is to choose $D = 4$. One may also require that the gauge symmetry group of the theory contains the group $SU(3) \times SU(2) \times U(1)$, with massless chiral fermions transforming as a triplet and a doublet under $SU(3) \times SU(2)$. Under very general conditions, it has been shown[94] that such a requirement excludes all Type II string constructions, leaving thus only heterotic strings as possible candidates for "realistic" theories.

One further constraint which one may like[2] to enforce is that the four dimensional heterotic string theory has one global space-time supersymmetry. It has been shown[95] that in *the conformal gauge*, such a requirement imposes the existence of a *global* $N = 2$ superconformal invariance in at least one moving sector of the theory.

Hence, from such a point of view,"realistic" string theories should be obtained through heterotic string constructions using $N = 2$ superconformal field theories with central charge $c = 9$. The simplest such theories are obtained by tensor products of the corresponding members of the unitary discrete series[9]. Indeed, they lead to consistent string theories, which share many of the properties of Calabi-Yau compactifications of ten dimensional heterotic strings, or the associated orbifold constructions[92,96]. Other constructions of $N = 2$ superconformal theories with $c = 9$ are also possible[97], leading to new types of string theories[98]. Thus, the classification of all $N = 2$ superconformal field theories (with $c = 9$) is very important in that respect, and it has indeed grown into a research field of its own over the past months[99].

Clearly, although the main research activity in string theory emphasizes nowadays more the conformal field theoretical aspects of these theories, it is a field of research which is going to be with us for many more years, and as such any young theoretician should be familiar with its basic structure and properties. Not only is two dimensional conformal field theory a research field in its own right, it also has important connections, many probably still to be discovered, with statistical mechanics in two dimensions and pure mathematics. Moreover, after more than five years since string theory is back at the fore-front of theoretical particle physics, with so many people having worked on it from very different perspectives, not one single argument has been put forth which could make these theories less an appealing or even inconsistent framework for the unification of all interactions.

As there is no other alternative known today, string theory still remains a fascinating and promising speculative avenue for the ultimate quantum unification of all our understanding of particle physics. The beauty of the mathematics involved makes it even more rewarding.

Acknowledgements

It is a pleasure to thank the organizers, and especially R. Gastmans, for their invitation to lecture on this subject at such an interesting and pleasant Summer School. Many thanks are also due to all participants for their interest and their stimulating questions.

References

These references do not claim to provide a complete list of all the work done on a given subject. They are only to be considered as indicative, and their aim is to help the interested reader in finding his own way in the vast literature of string theory.

My apologies are offered here to any author who feels unjustifiably unmentioned.

1. For many references to original work, see
   "Superstrings, The First 15 Years of Superstring Theory", ed. J.H. Schwarz (World Scientific, Singapore, 1985), 2 volumes.

2. P. Candelas, G.T. Horowitz, A. Strominger and E. Witten, Nucl. Phys. **B258** (1985) 46.

3. E. Witten, Phys.Lett. **B155** (1985) 151; Nucl. Phys. **B258** (1985) 75.
   M. Dine, R. Rohm, N. Seiberg and E. Witten, Phys. Lett. **B156** (1985) 55.
   J.- P. Derendinger, L.E. Ibáñez and H.P. Nilles, Nucl. Phys. **B267** (1986) 365.
   N. Seiberg, in Proc. 3rd Jerusalem Winter School, 1985, eds. T. Piran and S. Weinberg (World Scientific, Singapore, 1987).
   B.R. Greene, K.H. Kirklin, P.J. Miron and G.G. Ross, Nucl.Phys. **B278** (1986) 667; Nucl. Phys. **B292** (1987) 606; Phys. Lett. **B180** (1986) 69.

4. For a recent discussion and references, see:
   G.G. Ross, Lectures at the Banff Summer Institute (CAP) on Particles and Fields, 1988, preprint CERN-TH.5109/88 (December 1988).

5. M.B. Green and J.H. Schwarz, Phys. Lett. **B149** (1984) 117; Nucl. Phys. **B255** (1985) 93; Phys. Lett. **B151** (1985) 21.

6. A. Font, L.E. Ibáñez, F. Quevedo and A. Sierra, preprint CERN-TH.5326/89, LAPP-TH.241/89 (March 1989), and references therein.

7. J.H. Schwarz, Int. J. Mod.Phys. **A2** (1987) 593.

8. See the Lectures by G. Altarelli, J.J Aubert , J.D. Bjorken, G. Feldman, F. Pauss, J. Steinberg and J. Stirling in these Proceedings.

9. See the Lectures by C. Itzykson in these Proceedings.

10. E. Witten, Comm. Math. Phys. **121** (1989) 351; Nucl. Phys. **B322** (1989) 629; and references therein.

11. For a review and references , see:
E. D'Hoker and D.H. Phong, Rev. Mod. Phys. **60** (1988) 917.

12. L. Alvarez-Gaumé, P. Nelson, C. Gomez, G. Sierra and C. Vafa, Nucl.Phys. **B311** (1988) 333; and references therein.

13. E. Witten, Nucl. Phys. **B268** (1986) 253; Nucl. Phys. **B276** (1986) 291;
H. Hata, K. Itoh, T.Kugo, H. Kunitomo and K. Ogawa, Phys. Lett. **B172** (1986) 186, 195; Phys. Rev. **D34** (1986) 2360; Nucl. Phys. **B283** (1987) 433;
A. Neveu and P. West, Nucl. Phys. **B278** (1986) 601;
M. Kaku, Int. J. Mod. Phys. **A2** (1987) 1;
and references therein.

14. J. Scherk, Rev. Mod. Phys. **47** (1975) 123.
J. H. Schwarz, Physics Reports **89** (1982) 223.

15. M.B. Green, J.H. Schwarz and E. Witten, "Superstring Theory" (Cambridge, University, Cambridge, 1987) , 2 volumes.

16. S. Shenker, in "Unified String Theories", eds. M. Green and D. Gross (World Scientific , Singapore, 1986), p.141-161;
D. Friedan, in "Unified String Theories", eds. M. Green and D. Gross (World Scientific, Singapore, 1986), p. 162-213;
D. Friedan, E. Martinec and S. Shenker, Nucl.Phys. **B271** (1986)93;
M. Peskin, Lectures at the 3rd Advanced Study Institute in Elementary Particle Physics, Santa-Cruz, June 23- July 19, 1986, preprint SLAC-PUB-4251 (March 1987)

17. J. Govaerts, in Proc. 2nd Mexican School of Particles and Fields, Cuernavaca-Morelos, 1986, eds. J.-L. Lucio and A. Zepeda (World Scientific, Singapore, 1987), p. 247-442.

18. J. Govaerts, Lectures at the 1st International Workshop on Mathematical Physics, Bujumbura, September 28-October 10, 1987, preprint CERN-TH.4953/88 (January 1988).

19. J. Govaerts, in Proc. X Autumn School "Physics Beyond the Standard Model", Lisbon, October 10-14, 1988, eds. G.C. Branco and J.C. Romão (Nuclear Physics B- Proceedings Supplements Section), to appear; preprint CERN-TH.5306/89 (February 1989).

20. J. Govaerts, Int. J. Mod. Phys. **A4** (1989) 173.

21. Y. Nambu, Lecture at the Copenhagen Symposium, 1970, unpublished;
T. Goto, Prog. Theor. Phys. **46** (1971) 1560.

22. A.M. Polyakov, Phys. Lett. **B103** (1981) 207, 211.

23. S. Deser and B. Zumino, Phys. Lett. **B65** (1976) 369.
L. Brink, P. Di Vecchia and P. Howe, Phys. Lett. **B65** (1976) 471.

24. J. Polchinski, Comm. Math. Phys. **104** (1986) 37.

25. For recent reviews and references , see:
A.A. Tseythin, Int. J. Mod. Phys. **A4** (1989) 1257;
Int. J. Mod. Phys. **A4** (1989) 4279;
Trieste preprint ICTP-IC/89/90.

26. M. Kato and K. Ogawa, Nucl. Phys. **B212** (1983) 443.

27. A. Bilal and J.L. Gervais, Nucl. Phys. **B284** (1987) 397; Phys. Lett. **B187** (1987) 39; Nucl. Phys. **B293** (1987) 1.

28. A.M. Polyakov, Mod. Phys. Lett. **A2** (1987) 893;
V.G. Knizhnik, A.M.Polyakov and A.B. Zamolodchikov, Mod. Phys. Lett. **A3** (1988) 819;
A. M. Polyakov and A.B. Zamolodchikov, Mod. Phys. Lett. **A3** (1988) 1213;
A.M. Chamseddine and M. Reuter, Nucl. Phys. **B317** (1989) 757;
K. Fujikawa, N. Nakazawa and H. Suzuki, Phys. Lett. **B221** (1989) 289;
Y. Matsuo, Phys. Lett. **B227** (1989) 209;
P. Forgács, A. Wipf, J. Balog, L. Fehér and L. O'Raifeartaigh, Phys. Lett. **B227** (1989) 214;
J. Distler, Z. Hlousek and H. Kawai, Cornell preprints CLNS 88/854, CLNS 88/871;
and references therein.

29. P. Goddard, A.J. Hanson and G. Ponzano, Nucl. Phys. **B89** (1975) 76.

30. P. Goddard, J. Goldstone, C. Rebbi and C.B. Thorn, Nucl. Phys. **B56** (1973) 109.

31. P.A.M. Dirac, "Lectures on Quantum Mechanics" (Yeshiva University, New York, 1964).

32. J. Govaerts, "Hamiltonian Quantization and Constrained Dynamics" (Leuven University Press), in preparation.

33. P.A.M. Dirac, Phys. Rev. **114** (1959) 924.
L.D. Faddeev, Theor. Math. Phys. **1** (1969) 3.

34. K. Sundermeyer, "Constrained Dynamics", Lecture Notes in Physics 169 (Springer Verlag, Berlin, 1982).

35. See for example
P. Goddard and D. Olive, Int. J. Mod. Phys. **A1** (1986) 303.

36. R.C. Brower, Phys. Rev. **D6** (1972) 1655;
P. Goddard and C.B. Thorn, Phys. Lett. **B40** (1972) 235.

37. C. Lovelace, Phys. Lett. **B34** (1971) 500.

38. J. Govaerts, Int. J. Mod. Phys. **A4** (1989) 4353.

39. T. Yoneya, Progr. Theor. Phys. **51** (1974) 1907;
J. Scherk and J.H. Schwarz, Nucl. Phys. **B81** (1974) 118.

40. L. Brink and H.B. Nielsen, Phys. Lett. **B45** (1973) 332.

41. L.D. Faddeev and V.N. Popov, Phys. Lett. **B25** (1967) 29; and references therein.

42. M. Henneaux, Physics Reports **126** (1985) 1.

43. I.A. Batalin and E.S. Fradkin, Riv. Nuovo Cim. **9** (1986) 1; and references therein.

44. I.A. Batalin and G.A. Vilkovisky, Phys. Rev. **D28** (1983) 2567; and references therein.

45. C. Becchi, A. Rouet and R.Stora, Phys. Lett. **B52** (1974) 344;
Ann. Phys. **98** (1976) 287;
I.V. Tyutin, Lebedev preprint FIAN No.39 (1975), unpublished.

46. J. Govaerts, Int. J. Mod. Phys. **A4** (1989) 4487.

47. T.P. Killingback, Comm. Math. Phys. **100** (1985) 267;
M.A. Solov'ev, JETP Lett. **44** (1986) 469;
Z. Jaskolski, J. Math. Phys. **29** (1988) 1035.

48. V.N. Gribov, Nucl. Phys. **B139** (1978) 1.
I.M. Singer, Comm. Math. Phys. **60** (1978) 7.

49. S. Hwang, Phys. Rev. **D28** (1983) 2614.

50. M.D. Freeman and D.I. Olive, Phys. Lett. **B175** (1986) 151;
M. Henneaux, Phys. Lett. **B177** (1986) 35.

51. P. Ramond, Phys. Rev. **D3** (1971) 2415.

52. A. Neveu and J.H. Schwarz, Nucl. Phys. **B31** (1971) 86.

53. M.B. Green and J.H. Schwarz, Phys. Lett. **B109** (1982) 444;
Phys. Lett. **B136** (1984) 367;
Nucl. Phys. **B243** (1984) 285.

54. F. Gliozzi, J. Scherk and D.Olive, Nucl. Phys. **B122** (1977) 253.

55. N. Seiberg and E. Witten, Nucl. Phys. **B276** (1986) 272.

56. R. Nepomechie, Phys. Lett. **B178** (1986) 207;
K. Arakawa and M. Maeno, Progr. Theor. Phys. **77** (1987) 145.

57. R. Kallosh, Phys. Lett. **B224** (1989) 273; Phys. Lett. **B225** (1989) 49;
S.J. Gates, M.T. Grisaru, M. Roček, W. Siegel, P. van Nieuwenhuizen and A.E. van de Ven, Phys. Lett. **B225** (1989) 44;
M.B. Green and C.H. Hull, Phys. Lett. **B225** (1989) 57.

58. A.H. Chamseddine and J.-P. Derendinger, Nucl. Phys. **B301** (1988) 381;
A.H. Chamseddine, J.-P. Derendinger and M. Quirós, Nucl. Phys. **B311** (1988) 140;
Phys. Lett. **B227** (1989) 41.

59. P. Goddard, C. Rebbi and C.B. Thorn, Nuovo Cim. **12** (1972) 425;
R.C. Brower and K.A. Friedman, Phys. Rev. **D7** (1973) 535;
J.H. Schwarz, Nucl. Phys. **B46** (1972) 61.

60. Y. Iwasaki and K. Kikkawa, Phys. Rev. **D8** (1973) 440.

61. M.B. Green and J.H. Schwarz, Nucl. Phys. **B181** (1981) 502.

62. W. Lerche, D. Lüst and A.N. Schellekens, Nucl. Phys. **B287** (1987) 477.

63. For a review and references, see
W. Lerche, A.N. Schellekens and N.P. Warner, Physics Reports **177** (1989) 1.

64. N. Ohta , Phys. Lett. **B179** (1986) 347; Phys. Rev. **D33** (1986) 1681.

65. M. Henneaux, Phys. Lett. **B183** (1987) 59.

66. D. Gross, J. Harvey, E. Martinec and R. Rohm, Nucl. Phys. **B256** (1985) 253;
Nucl. Phys. **B267** (1986) 75.

67. K. Narain, Phys. Lett. **B169** (1986) 41;
K. Narain, M. Sarmadi and E. Witten, Nucl.Phys. **B279** (1986) 369.

68. J.H. Schwarz , Int. J. Mod. Phys. **A4** (1989) 2653.

69. D. Friedan, Z. Qiu and S. Schenker, Phys. Rev. Lett. **52** (1984) 1575;
Comm. Math. Phys. **107** (1986) 535.

70. M. Ademollo et al., Nucl.Phys. **B111** (1976) 77;
M. Ademollo et al., Nucl. Phys. **B114** (1976) 297;
M. Ademollo et al., Phys. Lett. **B62** (1976) 105;
L. Brink and J.H. Schwarz, Nucl. Phys. **B121** (1977) 285.

71. J.E. Paton and H.M. Chan, Nucl. Phys. **B10** (1969) 516.

72. J.H. Schwarz, Physics Reports **89** (1982) 223.

73. N. Marcus and A. Sagnotti, Phys. Lett. **B119** (1982) 97.

74. G. Veneziano, Nuovo Cim. **A57** (1968) 190.

75. M.A. Virasoro, Phys. Rev. **177** (1969) 2309;
J.A. Shapiro, Phys. Lett. **B33** (1970) 361.

76. M. Hayashi, N. Kawamoto, T. Kuramoto and K. Shigemoto, Nucl. Phys. **B294** (1988) 459;
Nucl. Phys. **B296** (1988) 373.

77. J. Govaerts, Phys. Lett. **B220** (1989) 77;
J. Govaerts, preprint CERN-TH.5149/88 (August 1988), to appear in the Proceedings of the XVIIth Intern.Colloquium on Group Theoretical Methods in Physics, June 27 - July 2, 1988, St. Adèle (Canada).

78. A. Neveu and P. West, Phys. Lett. **B194** (1987) 200.

79. A. Sagnotti, in "Non-Perturbative Methods in Field Theory, Cargèse 1987" eds. G. Mack et al. (Plenum, New York, 1988);
M. Bianchi and A. Sagnotti, Phys. Lett. **B211** (1988) 407;
G. Pradisi and A. Sagnotti, Phys. Lett. **B216** (1988) 59.

80. L. Clavelli, P.H. Cox and B.Harms, Phys. Rev. Lett. **61** (1988) 787;
L. Clavelli and S.T. Jones, Phys. Rev. **D39** (1989) 3795;
Z. Bern and D. Dunbar, Phys. Lett. **B203** (1988) 109;
D.C. Dunbar, Nucl. Phys. **B319** (1989) 72:
Z. Bern and D.C. Dunbar, Nucl. Phys. **B319** (1989) 104;
N. Ishibashi and T. Onogi, Nucl. Phys. **B318** (1989) 239;
P. Hořava, Prague preprint PRA-HEP-89/1 (February 1989).

81. A. Parkes, Phys. Lett. **B184** (1987) 19;
A.N. Schellekens, Phys. Lett. **B199** (1987) 427;
R. Bluhm, L. Dolan and P. Goddard, Nucl. Phys. **B309** (1988) 330;
and references therein.

82. L. Alvarez-Gaumé and P. Nelson, in "Supersymmetry, Supergravity and Superstrings '86", eds. B. de Wit, P. Fayet and M. Grisaru (World Scientific, Singapore, 1987), p. 419-510;
and references therein.

83. L. Alvarez-Gaumé, G. Moore and C.Vafa, Comm. Math. Phys. **106** (1986) 40.

84. H. Kikkawa and M. Yamasaki, Phys. Lett. **B149** (1984) 357;
N. Sakai and I. Senda, Progr. Theor. Phys. **75** (1986) 692;
V.P. Nair, A. Shapere, A. Strominger and F. Wilczek, Nucl. Phys. **B287** (1987) 402;
R. Dijkgraaf, E. Verlinde and H. Verlinde, Comm. Math. Phys. **115** (1988) 649;
A. Giveon, E. Rabinovici and G. Veneziano, Nucl. Phys. **B322**(1989) 167;
M. Dine, P. Huet and N. Seiberg, Nucl. Phys. **B322** (1989) 301;
J. Lauer , J. Mas and H. P. Nilles, Phys. Lett. **B226** (1989) 251;
J.M. Molera and B.A. Ovrut, Phys. Rev. **D40** (1989) 1146.

85. L. Alvarez-Gaumé, P. Ginsparg, G. Moore and C. Vafa, Phys. Lett. **B171** (1986) 155;
L.J. Dixon and J. A. Harvey, Nucl. Phys. **B274** (1986) 93.

86. P. Goddard, D.I. Olive and A. Schwimmer, Phys. Lett. **B157** (1985) 393.

87. For a review , see:
M.J. Duff, B.E.W. Nilsson and C.N. Pope, Physics Reports **130** (1986) 1.

88. L. Dixon, J.A. Harvey, C. Vafa and E. Witten, Nucl. Phys. **B261** (1985) 678; Nucl. Phys. **B274** (1986) 285;
M. Mueller and E. Witten, Phys. Lett. **B182** (1986) 28;
K. Narain , M.H. Sarmadi and C. Vafa, Nucl. Phys. **B288** (1987) 551;
L.J. Dixon, in "Superstrings, Unified Theories and Cosmology, 1987", eds. G. Furlan et al. (World Scientific , Singapore, 1988), p. 67-126.

89. D. Nemenschansky and S.Yankielowicz, Phys. Rev. Lett. **54** (1985) 620;
D. Gepner and E. Witten, Nucl. Phys. **B278** (1986) 493;
R. Bluhm, L. Dolan and P. Goddard, Nucl. Phys. **B289** (1987) 364.

90. H. Kawai, D.C. Lewellen and S.-H. H. Tye, Nucl. Phys. **B288** (1987) 1;
I. Antoniadis, C. P. Bachas and C. Kounnas, Nucl. Phys. **B289** (1987) 87;
I. Antoniadis and C. Bachas, Nucl. Phys. **B298** (1988) 586.

91. K. Bardacki, E. Rabinovici and B. Säring, Nucl. Phys. **B299** (1988) 151.

92. D. Gepner, Phys. Lett. **B199** (1987) 380; Nucl. Phys. **B296** (1988) 757.

93. P.Ginsparg, Phys. Rev. **D35** (1987) 648;
P. Ginsparg and C. Vafa, Nucl. Phys. **B289** (1987) 414.

94. L.J. Dixon, V.S. Kaplunovsky and C. Vafa, Nucl. Phys. **B294** (1987) 43.

95. T. Banks, L.J. Dixon, D. Friedan and E. Martinec, Nucl. Phys. **B299** (1988) 613.

96. D. Gepner, Lectures at the Spring School on Superstrings, Trieste, April 3-11, 1989, Princeton preprint PUPT-1121 (April 1989).

97. Y. Kazama and H. Suzuki, Phys. Lett. **B216** (1989) 112;
Nucl. Phys. **B321** (1989) 232;
Int. J. Mod. Phys. **A4** (1989) 235.

98. D. Gepner, Nucl. Phys. **B311** (1988) 191; Nucl. Phys. **B322**(1989) 65;
P. Zoglin, Phys. Lett. **B218** (1989) 444;
A. Font, L.E. Ibáñez and F. Quevedo, Phys. Lett. **B224** (1989)79;
A. Font, L.E. Ibáñez , M. Mondragón, F. Quevedo and G.G. Ross, Phys. Lett. **B227** (1989) 34;
K. Ito, Phys. Lett. **B226** (1989) 264;
C.A. Lütken and G.G. Ross, Phys. Lett. **B213** (1988) 152;
M. Lynker and R. Schimmrigk, Phys. Lett. **B215** (1988) 681;
T. Eguchi, H. Ooguri, A. Taormina and S.K. Yang, Nucl. Phys. **B315** (1989) 193.

99. A. Taormina, Lectures given at the Third Regional Conference in Mathematical Physics, Islamabad, February 18-25, 1989, preprint CERN-TH.5409/89 (May 1989);
G. Mussardo, G. Sotkov and M. Stanishkov, Phys. Lett. **B218** (1989) 191;
A.N. Schellekens and S. Yankielowicz, Phys. Lett. **B227** (1989) 387;

preprint CERN-TH.5440/89 (June 1989);
C. Vafa and N.P. Warner, Phys. Lett. **B218** (1989) 51;
C. Vafa, Mod. Phys. Lett. **A4** (1989) 1169;
W. Lerche, C. Vafa and N.P. Warner, Harvard preprint HUTP-88/A065;
L.J. Dixon, M.E. Peskin and J. Lykken, preprint SLAC-PUB-4884 (March 1989).

TWO TOPICS IN QUANTUM CHROMODYNAMICS*

J. D. Bjorken
Stanford Linear Accelerator Center
Stanford University, Stanford, CA 94309

INTRODUCTION

Quantum chromodynamics (QCD) has reached a level of credibility and maturity which deserves textbook status. Indeed, textbooks exist[1] and others are on the way.[2] Nevertheless, to my mind a textbook treatment of QCD is made much more difficult than that of quantum electrodynamics (QED) because of the confinement problem. Even perturbative QCD—which is all that will really be discussed here—suffers this problem. There is no S-matrix theory of quarks and gluons as there is for QED, as given in the LSZ formalism.[3] The concept of "on-mass-shell" or "asymptotic" quark and/or gluon is highly suspect. And the typical "Feynman diagram" used in perturbative QCD contains internal quark and gluon lines and external hadron lines. What does that really mean? How does one derive and justify Feynman-rules for such amplitudes in the absence of good control over the confinement question?

These issues are more matter-of-principle ones than operational ones. In general, I find no fault with what is being calculated, only that there is need for a more solid logical basis—as opposed to the intuitive, common-sense one—for what is done. The question is perhaps similar to, albeit much less profound than, the early days of quantum theory, where the calculations came fast and the real understanding of what they meant came more slowly.

In the last year, I have been lecturing on QCD at the University of Chicago, with these issues in mind. While I cannot claim much progress, the material which follows is influenced by the above concerns. It has also been most positively influenced by the students who patiently endured my gropings through this difficult subject and provided much help. Some are here at this school; to all, I give thanks.

⋆ Work supported by Department of Energy contract DE–AC03–76SF00515.

Particle Physics: Cargèse 1989
Edited by M. Lévy *et al.*
Plenum Press, New York, 1990

GUIDELINES FOR SETTING UP PERTURBATIVE QCD

Perturbative QCD is, at best, applicable only at short space-time intervals because of the "asymptotic freedom" property of the running coupling constant. What does this mean? Perturbative field theory is, essentially by definition, based on Feynman diagrams. Can one construct Feynman-diagram amplitudes whose ingredients depend only upon short distances? The answer appears affirmative, provided these diagrams are for Green's functions whose sources are restricted to small, contiguous space time regions. What, in turn, does this mean? I prefer to think of this restriction in terms of actual physical processes in principle localizable to small space-time regions. Tiny sources, of scale small compared to the confinement scale Λ_{QCD}^{-1}, create "beams" of quarks and gluons which interact, making reaction products which may be observed with tiny detectors, again of scale small compared to the confinement scale.[4] All of this should fit into space-time regions within which perturbation theory is really justifiable. Then (and only then?) can such processes be calculable by perturbative techniques alone. A *strictly-perturbative space-time region* can be defined as one which has the property that any straight line segment lying entirely within the region has an invariant length small compared to the confinement scale Λ_{QCD}^{-1} (whether or not the segment is spacelike or timelike). A little reflection should convince one that such strictly-perturbative domains are just the space-time regions adjacent to light cones. (For a light cone the only line-segments satisfying the criterion are null; so that the regions between hyperboloids $x^2 = a^2$ and $x^2 = -a^2$ evidently satisfy, for small enough a, the criterion for a strictly perturbative domain). While I haven't tried to prove it, it seems to me eminently reasonable that the relevant amplitudes and Green's functions within strictly perturbative domains really can be computed reliably using perturbation theory. Conversely, when the space-time region extends beyond such domains, it seems unavoidable that nonperturbative effects enter. It would be nice to sharpen these opinions further, but in these lectures it will only be done by example and not in generality.

What is the nature of convenient sources of quark and gluon "beams?" At large invariant distances the color-fields should be screened. A most economical way to guarantee this is that the *external sources of initial-state quark and gluon beams be local and color singlet.* For example, to obtain a beam of bottom quarks, first build a beam of Ws (of very high energy) and let them decay into $b\bar{c}$. Virtual photons are evidently an alternative. These are what we shall use in our examples, i.e., the "one-photon" and "two-photon" processes which form the lifeblood of e^+e^- collider physics experimentation. But, in general, we may assert the following:

1. Amplitudes for strictly perturbative processes shall be constructed from Green's functions

$$G(x_1,\ldots x_n) = \langle O|T(O_1(x_1)\ldots O_n(x_n))|O\rangle \quad , \tag{1}$$

   for which all operators $O_i(x_i)$ are local and color singlet.

2. After Fourier transformation, the momentum-space Green's functions $\tilde{G}(p_1\ldots p_n)$ will depend only on short distances, when the p_i^2 are all large and spacelike, and when the $p_i \cdot p_j$ $(i \neq j)$ are suitably restricted. (A sufficient restriction is that all p_i be "Euclidean" momenta, but this may not be necessary).

3. Given the confinement hypothesis, *all* information should be obtainable by analytic continuation of the Green's functions we have introduced. However, this does *not* imply that analytic-continuation of the approximate Green's functions constructed in perturbation theory provides this information.

The Green's functions we shall use in our examples involve only the electromagnetic current operator:

$$j^{\mu} = \sum_{i} e_i \, \overline{q}_i \gamma^{\mu} q_i \quad . \tag{2}$$

We first consider the two-point function for the vacuum polarization operator, and then the four-point function for the forward scattering amplitude for two virtual photons. These are sufficient for considering the e^+e^- total annihilation cross section into hadrons, and for the structure-functions for deep inelastic scattering of an electron from a virtual photon. Very interesting, but beyond the scope of these lectures is the question of how to describe the "final-state" properties of such processes, which, according to the lore of perturbative QCD, consist of sets of quark and gluon jets. Formally, these may be seen in the absorptive parts of the (appropriately analytically continued) Green's functions we have defined, as calculated in perturbation theory. Less formally they should be described in terms of the "physical" processes we have alluded to: tiny calorimeters placed "near the light-cone" pick up the quarks and gluons before they hadronize and measure the energy-momentum deposited into finite elements of solid angle $\Delta y \Delta \varphi$. How to link the formal description using the absorptive parts of the Green's functions to this "physical" picture is an interesting problem, well beyond the scope of these lectures. Some day I want to understand it better.

But this is more than enough of such general platitudes. The remainder of these lectures will be devoted to the e^+e^- total cross section and the virtual-photon's structure functions for deep-inelastic scattering. These examples will hopefully elucidate somewhat what I am driving at. Throughout this discussion, I assume the reader has some familiarity with a "standard" presentation of perturbative QCD as found in many places; the most immediate place is the fine set of lectures given by James Stirling in these proceedings.

ELECTRON-POSITRON ANNIHILATION INTO HADRONS

The total cross section for $e^+e^- \to$ *hadrons* normalized to the lowest-order cross section for $e^+e^- \to \mu^+\mu^-$, is given by the famous[5] function $R(s)$, which in the naïve lowest-order calculations is a constant equal to the sum of squares of charges of the participating quarks

$$R_{pert} = \sum_{i} e_i^2 \, [1 + \ldots] \quad , \tag{3}$$

and where the three dots denote perturbative-QCD radiative corrections, to be discussed later. Formally, R is related to the Fourier transform of a Green's function built from two electromagnetic currents

$$(q_\mu q_\nu - g_{\mu\nu} q^2) R(s) \; \alpha \int d^4x e^{iq\cdot x} < O | j_\mu(x) j_\nu(0) | O > \quad , \tag{4}$$

with $s = q^2$ timelike.

In general, is $R(s)$ a perturbatively calculable quantity for large s, according to our criteria? If so it should only depend upon the current correlation function at short space-time intervals. In the center-of-mass frame we deal with time intervals only. To test whether R is only sensitive to short time intervals, we may cut off the current correlation function at large times:

$$< O|j_\mu(x)j_\nu(0)|O > \rightarrow < O|j_\mu(x)j_\nu(0)|O > \exp\{-t^2/2\tau^2\} \quad , \tag{5}$$

and see the effect on $R(s)$. An easy calculation shows (for $\mu = \nu \neq 0$) that

$$sR(s) \rightarrow \int dE \; \exp\{-\tau^2 E^2/2\} \; s' R(s') \; \equiv \; s\overline{R}(s) \quad , \tag{6}$$

with $\sqrt{s'} = \sqrt{s} - E$.

In other words, $R(s)$ must be averaged[5] over an energy interval $\Delta\sqrt{s} \gg \tau^{-1}$ in order to be a strictly perturbative quantity. In particular, quarkonium resonances and sharp features of heavy-flavor thresholds must be smeared out over an energy scale large compared to $\Lambda_{QCD} \sim 200$ MeV. This is evidently just the uncertainty principle at work. It is amusing that, given a top quark mass in excess of the W mass, *physics* does the local averaging. The width of the top-quark decay $t \rightarrow Wb$ is large, in excess of 1 GeV. Thus there is no time available for toponium formation or even the formation of $T \equiv t\overline{q}$ mesons. Such processes of hadronization take place on a time scale long compared to the confinement time Λ_{QCD}^{-1}. Thus the threshold structure is already made smooth by the short t-quark lifetime. However, as pointed out by Fadin and Khose,[6] there are significant QCD radiative corrections near threshold which are numerically large, and which can be reliably estimated using a perturbative-QCD calculation, because everything happens within a strictly perturbative space-time domain.

A quantity related to $R(s)$ is the hadronic vacuum-polarization, evaluated at spacelike momenta. This is a sum of Feynman diagrams; one has:

$$(q_\mu q_\nu - g_{\mu\nu} q^2) \, \Pi(Q^2) \; \propto \; \int d^4x e^{iq\cdot x} < O|T(j_\mu(x)j_\nu(0))|O > \quad . \tag{7}$$

For large spacelike $q^2 = -Q^2$, the function $\Pi(Q^2)$ is necessarily smooth so that the averaging procedure is not needed (We will see this explicitly in what follows.) To lowest order in a perturbative calculation, $\Pi(Q^2)$ vanishes at $Q^2 = 0$ (after charge renormalization has been carried out) and grows logarithmically at large Q^2:

$$\Pi(Q^2) \propto \sum_i e_i^2 \; \ell n \frac{Q^2}{m_i^2} \quad . \tag{8}$$

A somewhat more convenient quantity for what will follow is the logarithmic derivative of $\Pi(Q^2)$, which we denote by $D(Q^2)$:

$$D(Q^2) \; = \; Q^2 \frac{d\Pi}{dQ^2} \; = \; \sum_i e_i^2 \; [1 + \ldots] \quad . \tag{9}$$

At the near-trivial parton level of calculation, D and R are, in fact, identical.

It should be clear that knowledge of R, i.e., of $< O|j_\mu(x)j_\nu(0)|O >$ implies knowledge of $\Pi(Q^2)$, hence of D. This is formally expressed in momentum space in the fact that $\Pi(Q^2)$ (or D) is an analytic function of Q^2 in the cut complex Q^2 plane, and that R is obtained from $\Pi(Q^2)$ by analytic continuation. In particular, R is the discontinuity of $\Pi(Q^2)$ across the branch cut. The formula is:

$$\Pi(Q^2) = Q^2 \int_0^\infty \frac{ds\ R(s)}{s(s+Q^2)} \quad . \tag{10}$$

Note that the threshold of the s-integral is at $4m_\pi^2$. Were R to be estimated perturbatively, the threshold would be somewhere else ($\sim 4m_q^2$) and $\sqrt{s}$ smearing is definitely called for.

We are now prepared to discuss the nature of the perturbative-QCD correction to $\overline{R}(s)$ [or $D(Q^2)$]. We again assume some familiarity with the workings of the theory and summarize what happens (in the MS or $\overline{MS}$ renormalization scheme)

1. In order to manage the divergences which appear, one cuts off the Feynman integrals by evaluating them in a space-time dimension D slightly less than four:

$$D = 4 - \epsilon \quad . \tag{11}$$

2. By dimensional analysis, the (bare) coupling constant α_0 multiplying the Feynman integrals acquires a nonvanishing dimensionality proportional to ϵ.
3. A new dimensionless coupling $\alpha_0(\mu^2)$ is defined by introducing an arbitrary scale-factor μ^2:

$$\alpha_0 \equiv \alpha_0(\mu^2)\mu^{4-D} \tag{12}$$

4. The theory is renormalized, removing the singular dependence on space-time dimension, i.e., on ϵ, of the Feynman-integrals.
5. The renormalized quantity $\overline{R}$ then depends upon three variables:

$\sqrt{s}$: the energy variable;

$\alpha_s(\mu^2)$: the renormalized (dimensionless) coupling constant; and

μ^2 : the arbitrary mass scale.

6. However, since $\overline{R}$ represents physics and the value of μ is an arbitrary choice, $\overline{R}$ cannot depend upon μ. This means:

$$O = \mu^2 \frac{d}{d\mu^2} \overline{R}(s,\ \alpha_s(\mu^2),\ \mu^2) = \mu^2 \frac{\partial \alpha_s}{\partial \mu^2}\frac{\partial \overline{R}}{\partial \alpha_s} + \mu^2 \frac{\partial \overline{R}}{\partial \mu^2} \quad . \tag{13}$$

Defining, somewhat unconventionally,

$$\beta(\alpha_s) = -\frac{\mu^2}{\alpha_s}\frac{\partial \alpha_s}{\partial \mu^2} \quad , \tag{14}$$

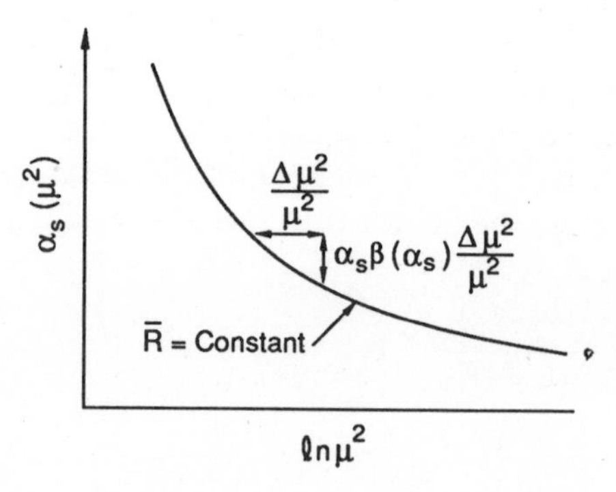

Figure 1. Behavior of the running coupling constant with scale factor μ.

we see that in $\alpha_s - \ell n\,\mu^2$ space (Fig. 1), there are lines along which $\overline{R}$ does not change. The local slope of such lines can be read off Eq. (13):

$$\frac{\mu^2}{\alpha_s}\frac{d\alpha_s}{d\mu^2} = -\beta(\alpha_s) \quad . \tag{15}$$

These curves $\alpha_s = \alpha_s(\mu^2)$, along which $\overline{R}$ remains constant, define the running coupling constant.

7. Only one of the curves will be consistent with experiment: for a given choice of μ^2, $\alpha_s(\mu^2)$ has to be chosen to agree with the data.
8. Dimensional analysis demands that $\overline{R}$ be a function only of α_s and of s/μ^2. Putting this constraint together with the argument that R be independent of μ^2 leads to the conclusion that R is a function of only a *single* variable, which must be $\alpha_s(s)$.
9. Because perturbative QCD allows a formal power series expansion in $\alpha_s(\mu^2)$, this implies the existence of a formal power series expansion in $\alpha_s(s)$.

We have left out details; the student is urged to consult Stirling's lectures and standard sources to fill them out. But the main point is that this line of argument is based on the renormalizability of the theory and is quite general. It therefore applies equally well to the vacuum polarization $\Pi(Q^2)$ or, better, $D(Q^2)$. We therefore have for the quantity experimentalists measure

$$\overline{R}(s) = \sum_i e_i^2 \left[1 + \sum_{m=1}^{\infty} r_m \left(\frac{\alpha_s(s)}{\pi}\right)^m\right] \quad , \tag{16}$$

and for the quantity theorists calculate

$$D(Q^2) = \sum_i e_i^2 \left[1 + \sum_{m=1}^{\infty} d_m \left(\frac{\alpha_s(Q^2)}{\pi}\right)^m\right] \quad . \tag{17}$$

These two quantities are related by a dispersion relation following directly from Eq. (10).

$$D(Q^2) = Q^2 \int_0^\infty \frac{ds R(s)}{(s+Q^2)^2} = \int_0^\infty \frac{dx R(xQ^2)}{(1+x)^2} \cong \int_0^\infty \frac{dx \overline{R}(xQ^2)}{(1+x)^2} \quad . \tag{18}$$

It turns out that interesting information emerges just from the fact that both D and $\overline{R}$ admit power-series expansions in α_s. We see from Eq. (18) that D is just a local average of $\overline{R}$, so that if $\overline{R}$ is slowly varying, D and $\overline{R}$ are essentially the same. But if we expand $\overline{R}$ as a power series in $\alpha_s(xQ^2)$, then it is possible to use the equation for the running coupling constant to express $\alpha_s(xQ^2)$ (perturbatively) in terms of $\alpha_s(Q^2)$ and thereby construct the power series for $D(Q^2)$. Evidently, D and $\overline{R}$ are not *identically* the same, so that the series are not the same. Indeed, in what follows we shall find evidence that the series expansion of D–$\overline{R}$ is almost certainly asymptotic, not absolutely convergent. The same statement probably applies for D and $\overline{R}$ separately, as well.

In order to make the connection one must know how the coupling constant runs. With our definition, the β-function admits a power-series expansion in α_s which begins in first order. Keeping only that contribution leads to the well-known expression:

$$\frac{1}{\alpha_s(Q^2)} = \frac{1}{\alpha_s(\mu^2)} + \frac{b}{\pi}\,\ell\text{n}\,\frac{Q^2}{\mu^2} \equiv \frac{b}{\pi}\,\ell n\,\frac{Q^2}{\Lambda^2_{QCD}} \quad , \tag{19}$$

with $b = (33-2n_f)/12 = 2.08\pm0.17$ for the effective number of flavors 4 ± 1. Expressing $\alpha_s(xQ^2)$ in terms of $\alpha_s(Q^2)$ will therefore lead to a power-series expansion in $\alpha_s(xQ^2)$ with the nth coefficient being (at most) an nth-order polynomial in $\ell\text{n}\, x$. (This is still true when the higher-order corrections to the β-function are included.) Thus the generic term in the convolution integral is of the form (note the symmetry under $x \leftrightarrow x^{-1}$):

$$\int_0^\infty \frac{dx(\ell\text{n}\, x)^n}{(1+x)^2} = \begin{cases} 0 & n \text{ odd} \quad , \\ 2\zeta(n)[1-2^{1-n}]n! & n \text{ even} \quad . \end{cases} \tag{20}$$

(It is gratifying to see a ζ-function appearing here, signalling perhaps some connection to conformal symmetry. It would be a terrible thing not to be at least slightly *au courant*. Long ago, Feynman observed[7] that there were then only two options open to theorists: to either form a group or to disperse. There seems to be even less choice nowadays: one must conform.)

With knowledge of how to do the integrals, it is only algebra to figure out the series for D–$\overline{R}$. Evidently, the order α_s/π term vanishes. Because the integral over one power of $\ell\text{n}\, x$ vanishes, the $(\alpha_s/\pi)^2$ term also vanishes. Only in third order do the two series begin to differ, and one easily finds

$$D - \overline{R} = \left(\frac{\pi^2 r_1 b^2}{3}\right)\left(\frac{\alpha_s}{\pi}\right)^3 + \ldots \approx 14\left(\frac{\alpha_s}{\pi}\right)^3 \quad , \tag{21}$$

where we use the known result that the first radiative correction to $\overline{R}$ is

$$r_1 = 1 \quad . \tag{22}$$

The perturbative series for $\overline{R}$ has been calculated through order $(\alpha_s/\pi)^3$, and the results for the higher order coefficients are

$$\begin{aligned} r_2 &= 1.53 \pm 0.12 \quad , \\ r_3 &= 66.10 \pm 1.24 \quad , \end{aligned} \tag{23}$$

where, again, we let $n_f = 4 \pm 1$ for this purpose. The large value of the $(\alpha_s/\pi)^3$ term, calculated recently[8] by Gorishny, Kataev, and Larin (hereafter GKL), has surprised many people. However, from the point of view of Eq. (21) for the difference D–$\overline{R}$ (which, by the way, is explicitly presented in the GKL paper; they, after all, calculate D and quote $\overline{R}$), this should not be surprising. There seems no particular reason why D should converge better than $\overline{R}$ (or vice versa), so that the estimate for the difference should reflect the behavior of the individual quantities.

But this is not the end of the story. The coefficient of the $(\alpha_s/\pi)^3$ term for the difference of D–$\overline{R}$ was very simple to calculate and used only lowest-order QCD calculations. Not only have higher order corrections to $\overline{R}$ been computed, but also higher-order corrections to the β-function; in our notation, we write

$$-\frac{\mu^2}{\alpha_s}\frac{d\alpha_s}{d\mu^2} = \beta(\alpha_s) = b\left(\frac{\alpha_s}{\pi}\right)\left[1+\sum_{n=1}^{\infty} b_n\left(\frac{\alpha_s}{\pi}\right)^n\right] , \tag{24}$$

with

$$\begin{aligned} b &= \frac{33-2n_f}{12} = 2.08 \pm 0.17 \\ b_1 &= 1.51 \pm 0.27 \\ b_2 &= 2.97 \pm 1.50 \quad . \end{aligned} \tag{25}$$

This allows computation of the next two orders of D–$\overline{R}$, out through the $(\alpha_s/\pi)^5$ contribution! One finds, after slightly more arduous algebra,

$$\begin{aligned} d_4 - r_4 &= \pi^2 b^2\left(r_2 + \frac{5}{6}r_1 b_1\right) = 43(1.5+1.3) = 119 \pm 17 \\ d_5 - r_5 &= \pi^2 b^2\left(2r_3 + \frac{7}{3}r_2 b_1 + r_1 b_2 + \frac{1}{2}r_1 b_1^2\right) + \frac{7\pi^4}{15}\, r_1 b^4 \\ &\cong 43(132+5+3+1)+855 \\ &= 6920 \pm 620 \quad . \end{aligned} \tag{26}$$

The largeness of these terms mostly reflects the largeness of b (note that for QED, the corresponding quantity is about seven times smaller). However, one also sees the beginnings of an asymptotic series emerging in the factors of $n!$ in numerators from the integrals of Eq. (20). To go to still higher orders is, in general, tedious, and requires unknown input. However, one observes that the contributions from higher order corrections to the β-function proportional to b_1 and b_2 were not especially significant numerically. This invites considering the approximation of neglecting all but the leading term for the β-function. It turns out that in this limit one can easily estimate the nth-order coefficient in the expansion of D–$\overline{R}$. To do this most efficiently, it is convenient to introduce the Borel transform of the perturbation series. One assumes that the functions D, $\overline{R}$, etc., which we generically call $F(\alpha_s)$, are obtainable as Laplace transforms in α_s^{-1} of another function $\tilde{F}(z)$ (the Borel transform):

$$F(\alpha_s) = \int_0^\infty dz e^{-z/\alpha_s}\,\tilde{F}(z) \quad . \tag{27}$$

Why this representation? If $\tilde{F}(z)$ admits a power series expansion in z (more or less),

$$\tilde{F}(z) = F_0\delta(z) + \sum_{n=1}^{\infty} \tilde{f}_n \, z^{n-1} \quad , \tag{28}$$

then one immediately finds:

$$F(\alpha_s) \equiv \sum_{n=1}^{\infty} f_n \alpha_s^n = F_0 + \sum_{n=1}^{\infty} (n-1)! \, \tilde{f}_n \, \alpha_s^n \quad . \tag{29}$$

In other words, the power series for $\tilde{F}$ is related to that of F and converges much better:

$$\tilde{f}_n = \frac{f_n}{(n-1)!} \quad . \tag{30}$$

For this reason, this Borel transform has been used by theorists[9] to investigate the convergence of the perturbation series. It is believed that $\tilde{F}(z)$ has a finite radius of convergence, with branch-point singularities on the real z-axis known in the trade as "renormalons." If this is truly the case, then the radius of convergence of the usual perturbation series for $F(\alpha_s)$ is zero with the nth term in the series eventually growing roughly as $n!$

The utility of the Borel transform, in our case, comes from the fact that the convolution relating D and $\overline{R}$, Eq. (18), factorizes. Introducing the Borel transform for $\overline{R}$ and using Eq. (19) for the running coupling constant yields:

$$\begin{aligned} D(\alpha_s) &= \int_0^{\infty} \frac{dx}{(1+x)^2} \int_0^{\infty} dz \tilde{R}(z) \, \exp\{-z/[\alpha_s(xQ^2)]\} \quad , \\ &= \int_0^{\infty} dz \, \exp\{-z/[\alpha_s(Q^2)]\} \, \tilde{R}(z) \int_0^{\infty} \frac{dx \, \exp\{-bz\ell n \, x/\pi\}}{(1+x)^2} \quad . \end{aligned} \tag{31}$$

The x-integral is a beta-function, and the remainder is in the Borel-transform format. Therefore, the result is simply:

$$\tilde{D}(z) = \tilde{R}(z) \cdot \left(\frac{bz}{\sin bz}\right) \quad . \tag{32}$$

One sees singularities appearing on the real z axis; these occur at the positions of the renormalons to which we alluded.

From this recursion relation, one may easily construct the power series expansion for $\tilde{D}$, hence of D, from that of $\overline{R}$. The easiest way to write this is:

$$\tilde{d}_n = \tilde{r}_n + \sum_{k=2,4\ldots} 2\zeta(k) \, (1 - 2^{1-k}) \, \tilde{r}_{n-k} \left(\frac{b}{\pi}\right)^k \quad , \qquad (n \geq 3) \quad . \tag{33}$$

Note there are no large coefficients in this expansion-nor small ones. However, upon returning from the Borel expansion coefficients to the original ones, we pick up a factor $(n-1)!$, which *is* large.

I have made some estimates of the nth coefficient, assuming:

$$r_n \gtrsim |d_n - r_n| \quad . \tag{34}$$

This leads to the values of $\delta_n = d_n - r_n$ quoted in Table I. One sees a remarkable growth in the coefficients. In Table II, the actual values for the nth term of the series expansion for δ_n and/or r_n are tabulated for a variety of choices of α_s/π. The entries above the line are secure, while the entries below depend on the guesswork we have introduced, namely, the approximate validity of the leading order expression, Eq. (19), for the running coupling constant, as well as the estimate for r_n in Eq. (34).

TABLE I Estimated coefficients for $\delta_n = d_n - r_n$.

$$\begin{aligned}
\delta_3 &= 15 \\
\delta_4 &= 120 \\
\delta_5 &= 6900 \\
\delta_6 &= 23{,}000 \\
\delta_7 &= 2{,}400{,}000 \\
\delta_8 &= 12{,}000{,}000 \\
\delta_9 &= 1{,}600{,}000{,}000
\end{aligned}$$

TABLE II N^{th} order contributions to $D - \overline{R}$ and $\overline{R}$.

		$\frac{\alpha_s}{\pi} = 0.3$		$\frac{\alpha_s}{\pi} = 0.1$		$\frac{\alpha_s}{\pi} = 0.05$ (PEP/PETRA)		$\frac{\alpha_s}{\pi} = 0.03$ (SLC/LEP)		$\frac{\alpha_s}{\pi} = 0.01$ (GUT)	
1		1.00		1.000		1.000		1.000		1.00	
$r_1(\frac{\alpha}{\pi})$		0.30		0.100		0.050		0.030		10^{-2}	
$r_2(\frac{\alpha}{\pi})^2$		0.13		0.015		0.004		0.001		1.5×10^{-4}	
$r_3(\frac{\alpha}{\pi})^3$	$\delta_3(\frac{\alpha}{\pi})^3$	1.8	0.4	0.069	0.015	0.008	0.002	0.002	4×10^{-4}	6.9×10^{-5}	1.5×10^{-5}
	$\delta_4(\frac{\alpha}{\pi})^4$		1.0		0.012		0.0008		1×10^{-4}		1.2×10^{-6}
	$\delta_5(\frac{\alpha}{\pi})^5$		17		0.069		0.002		1.7×10^{-4}		6.9×10^{-7}
?	$\delta_6(\frac{\alpha}{\pi})^6$		17		0.023		4×10^{-4}		1.7×10^{-5}		2×10^{-8}
?	$\delta_7(\frac{\alpha}{\pi})^7$		500		0.24		2×10^{-3}		5×10^{-5}		2×10^{-8}
?	$\delta_8(\frac{\alpha}{\pi})^8$		800		0.12		5×10^{-4}		8×10^{-6}		1.2×10^{-9}
?	$\delta_9(\frac{\alpha}{\pi})^9$		30,000		1.6		3×10^{-3}		3×10^{-5}		1.6×10^{-9}

I think the main lesson to be learned from this is something recognized for a long time by the experts: the perturbation series *is* asymptotic. The exercise we have done helps to give some feeling as to when the trouble appears. According to Table II, one should truncate the series at order $n \sim 1.5\alpha_s^{-1}$. For most "practical" energies (PEP/PETRA and above), this is still well beyond the calculations. However, there is perhaps a message for those who work on perturbative QCD at the interface with nonperturbative effects, i.e., at small mass scales, e.g., where $\alpha_s(\mu^2) \sim 1$. It is simply that one should probably *stop* at leading order and settle for roughly 30 to 50 percent agreement with experiment: attempts to "improve" the situation by calculating higher orders most likely only create confusion and worsen the situation. I think that perturbative-QCD theorists should find this welcome. It is some justification for laziness.

The reader may find it surprising that the apparently innocent relationship of D to an average of $\overline{R}$, as expressed in Eq. (18), should lead to such evidence for the asymptotic nature of the perturbation series. What happened? It is true that, for any α_s reasonably small, D and $\overline{R}$ are nearly equal. But because of the ubiquity of QCD logarithms, to obtain the difference to high accuracy, one needs to sample R over a dynamic range in $\log x$ which grows linearly with the perturbation order n. Thus when $n \gtrsim \ell n\, s/\Lambda^2_{QCD}$, one must sample the infrared region, where $\overline{R}$ is poorly defined by the perturbation series. But this condition is the same as the estimate $n \sim 1.5\alpha_s^{-1}$ quoted earlier.

As for the interpretation of renormalon poles, one may see from the defining equation for the Borel transform that a pole at $z = n$ is related to a power-law contribution to R (to leading order in the series expansion of the β-function):

$$\begin{aligned} F(\alpha_s) &= \sum_n \int_0^\infty \frac{dz \exp\{-z/\alpha_s\}}{bz - n\pi} \rightarrow \sum_n (Residues) \cdot \exp\{-n\pi/[b\alpha_s(Q^2)]\} \\ &\sim \sum_n (Residues) \cdot (Q^2/\Lambda^2_{QCD})^{-n} \quad . \end{aligned} \tag{35}$$

Thus one finds linkage between power-law contributions to $\overline{R}$, as computed using operator-product expansion techniques, and the location and structure of the renormalon singularities.

THE STRUCTURE FUNCTION OF A VIRTUAL PHOTON

Our second example of a strictly perturbative process is deep inelastic scattering of an electron from a virtual (spacelike) photon. The classic deep-inelastic process of electron-proton scattering does not qualify because the proton is evidently too big to fit into a strictly perturbative space-time domain near the light-cone. The virtual spacelike photon with squared mass $p^2 = -P^2$ *is* small, with transverse extension of order P^{-1}. It is produced from an electron or positron in the familiar two-photon process studied at e^+e^- colliders.

While the electromagnetic structure of the spacelike photon is amenable to a perturbative analysis, it would, of course, be nice to consider the real photon, not to mention the timelike photon as well; especially the extrapolation to the vector-boson ρ, ω, ϕ states. Indeed, the case of the real photon is a very interesting subject with a rich history. There was at one time considerable optimism that this process was an excellent test-bed for perturbative QCD, and might provide an accurate measurement of α_s. But there arose complications, to be described in more detail in what follows. By now, the optimism has waned considerably.[10] Nevertheless, the process is most interesting theoretically.

For a spacelike virtual photon with four-momentum p, there are several structure functions to consider. We shall restrict our attention to transverse photons only, and

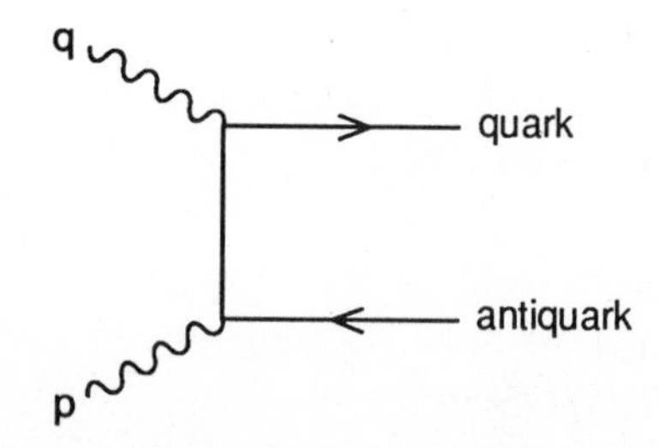

Figure 2. Lowest-order graph for "photon fusion" into quark–antiquark pairs.

only the helicity-independent contribution as well. The kinematic formulae can be found in many places[11] and will not be reproduced here.

Letting $Q^2 = -q^2$ be the (spacelike) squared mass of the probing photon, it should be clear that when both Q^2 and $P^2 = -p^2$ are large, the starting-point is the simple "photon-photon fusion" of quark pairs (Fig. 2), which gives what naïvely would be the leading contribution to the cross section. This was calculated long ago,[12] even before the advent of QCD. For $Q^2 \gg P^2 \gg \Lambda^2_{QCD}$, the relevant structure function (analogous to F_2 for nucleons) is easily computed to be:

$$F_2(x) \approx \left(\sum_i e_i^4 \right) x \left[x^2 + (1-x)^2 \right] \ell n\, Q^2/P^2 \quad . \tag{36}$$

For $P^2 \gg Q^2 \gg \Lambda^2_{QCD}$, the structure function vanishes rapidly, because the probing photon sees a small dipole of size P^{-1}, but only has resolving power Q^{-1}. Hence, there is a power-law scaling violation proportional to Q^2/P^2 (up to logarithmic factors).

The factors in Eq. (36) are reasonably easy to understand:

1. The amplitude is proportional to the square of the quark charge; hence the cross section to the fourth power.
2. The small x behavior follows the Regge-pole rule: exchange of spin J in the amplitude leads to an s^{2J-2} or x^{2-2J} behavior in the cross section. In this case, $J = 1/2$, the exchanged quark being treated as "elementary."
3. The factor $[x^2 + (1-x)^2]$ is the probability of finding a quark of momentum xp_μ in the photon of momentum p_μ (as $p_0 \to \infty$); it is just an Altarelli–Parisi "splitting function." (We assume the reader to be familiar with the basics of the Altarelli–Parisi formalism).[13]
4. The logarithm appears from a "collinear" singularity in the angular distribution of the quark pair relative to the photon direction, and is of the same nature that generates the leading logarithms in QCD.

It is of special interest that scaling in x is violated logarithmically (for $Q^2 \to \infty$; P^2 fixed), and that at large x the structure function is big: it approaches a constant as $x \to 1$. This is in sharp contrast to what is expected from vector-meson dominance, where the hadronic part of the photon is assumed to be a mixture of ρ, ω, $\phi \ldots$. The structure functions of such mesons are believed to vanish as $x \to 1$ in a manner similar to that of the pion or kaon, where the behavior is $(1-x)^n$, with n between 1 and 2. Therefore, observation of the photon structure function at large x ("large" meaning $x > 0.4$) is a good method of revealing its "pointlike" components.

The theoretical situation became considerably more interesting when Witten[14] considered the QCD radiative corrections to this parton-model calculation. He showed that at sufficiently large Q^2 even a real photon should exhibit the pointlike behavior, and that its structure function should violate scaling by one power of log Q^2 just as in Eq. (36). However, the *shape* of the function of the scaling variable x multiplying the log Q^2 term is changed in a calculable way by the higher-order QCD corrections. Thus it appeared that a measurement of this predicted shape would be a good test of QCD.

The complications began when next-to-leading order corrections to Witten's leading-order calculation were considered.[15] At small x, these had a very large effect—so large that the computed structure function went *negative,* a clearly inadmissible result. Subsequent work which traced down the origin of this phenomenon showed that one needed to include terms nonleading in $(\ell n\, Q^2)$ and link them carefully to the leading term in order to resolve this problem. The importance of the nonleading contributions, which have a Q^2 dependence typical of hadron structure functions instead of the $(\ell n\, Q^2)^1$ scaling-violation present in leading order, makes the phenomenology more complicated, and thus far has dampened the original optimism that this process was a good quantitative testing ground for perturtubative QCD.

It is our intention to review this somewhat confusing situation. We shall start with consideration of the structure function when $p^2 = -P^2$ is large and spacelike[16] so that by our criterion the process is strictly perturbative. (Historically, Witten considered real photons only, and while there is nothing technically wrong with his analysis, this choice has been a source of some of the confusion.) Once this case is worked out, we consider what happens when P^2 becomes null or timelike. The main tool to be used here is analytic continuation in P^2. The variable P^2 will here play a role quite similar to the momentum variable Q^2 used in the discussion of R and vacuum polarization. We will be able to see in a controlled way how the hadronic, nonperturbative aspects of the problem enter when P^2 is allowed to become small.

Within the QCD ideology, there are two major lines of attack on the structure-function problem. One uses the operator-product expansion plus renormalization group considerations to calculate the scaling violations of moments of the structure functions.[17] The other uses the Altarelli–Parisi evolution-equations for the structure-functions themselves. The former method is more rigorous, but also more abstract. The latter method allows some physical insight at the parton-model level into what is going on, but is harder to justify theoretically, especially when nonleading contributions are to be included. Indeed, the best justification for the Altarelli–Parisi approach is that it gives the same answers as the operator-product-expansion methodology.

In this discussion we shall use both methods, but begin with the Altarelli–Parisi approach. Their equation is schematically written as:

$$\begin{aligned} Q^2\,\frac{dF_2\,(x,Q^2)}{dQ^2} &= P_{qq}\otimes F_2 + P_{qG}\otimes G \quad , \\ Q^2\,\frac{dG}{dQ^2} &= P_{Gq}\otimes F_2 + P_{GG}\otimes G \quad , \end{aligned} \tag{37}$$

with the convolution being given by a ratio-kernel:

$$P \otimes F \equiv x \int_x^1 \frac{dy}{y} P(x/y, Q^2)\ F(y, Q^2) \quad , \tag{38}$$

and with the "splitting-functions" $P(z, Q^2)$ simple, known quantities.

The physics is that: (i) the importance of the QCD radiative corrections increases with Q^2 because of the increase in available phase-space; and (ii) the important contributions to leading logarithmic accuracy come from approximately collinear configurations of the initial-state and final-state quarks and/or gluons. Item (i) implies the integro-differential nature of the equation: the *change* with $\ell n\, Q^2$ in the parton distribution is given by the convolution. Item (ii) assures the survival of the parton-model interpretation despite the increase in transverse momentum of relevant constituents. The essential dynamics remains collinear, as required by parton-model ideology.

The standard Altarelli–Parisi equations, as written, are homogeneous. But for the photon structure-function there is an inhomogeneous driving team because the "bare" process in Fig. 2 has the linear dependence on log Q^2 as given by Eq. (36). For simplicity in what follows, we write down the modified Altarelli–Parisi equation omitting the gluon contributions; i.e., we consider a "nonsinglet" structure function. (Nothing essential is lost in this simplification, and in what follows we shall indicate the necessary modifications at the appropriate places). The equation then becomes:

$$Q^2 \frac{dF_2}{dQ^2} = P \otimes F_2 + f(x) \quad , \tag{39}$$

with

$$f(x) = \left(\sum_i e_i^4 \right) x\ [x^2 + (1-x)^2] \quad . \tag{40}$$

The solution of integral equations with ratio kernels is found with the aid of the Mellin transform. One defines moments of the structure-function and splitting function:

$$\begin{aligned} \tilde{F}(n, Q^2) &= \int_0^1 dx x^{n-2}\ F_2(x, Q^2) \quad , \\ \tilde{P}(n, Q^2) &= \int_0^1 dz z^{n-1}\ P(z, Q^2) \quad . \end{aligned} \tag{41}$$

Applying this to the integral equation unravels the convolution:

$$Q^2 \frac{d\tilde{F}(n, Q^2)}{dQ^2} = \tilde{P}(n, Q^2)\tilde{F}(n, Q^2) + \tilde{f}(n) \quad . \tag{42}$$

The general solution of this simple differential equation is a sum of a homogeneous solution and a particular solution. The homogeneous solution is the (hopefully) familiar one used in hadron structure-function analysis:

$$\tilde{F}(n,Q^2) = \tilde{F}(n,Q_0^2)\,\exp\left\{\int_{Q_0^2}^{Q^2}(d\sigma^2/\sigma^2)\tilde{P}(n,\sigma^2)\right\} . \tag{43}$$

The "anomalous dimension" $\tilde{P}(n,\sigma^2)$ is, to leading order, proportional to the running coupling constant $\alpha_s(\sigma^2)$; hence, proportional to $(\ell n\,\sigma^2/\Lambda_{QCD}^2)^{-1}$. When this dependence is introduced into Eq. (42), a short calculation leads to the behavior:

$$\frac{\tilde{F}(n,Q^2)}{\tilde{F}(n,Q_0^2)} = \left[\frac{\alpha_s(Q^2)}{\alpha_s(Q_0^2)}\right]^{d(n)} . \tag{44}$$

The exponent $d(n)$ is obtained from the appropriate moment of the splitting function and Eq. (19) for α_s. It is positive for large n (large x) and negative for small n. This behavior leads to the familiar pattern of scaling violations for the structure function F_2.

(If one were to consider the coupled equations, Eq. (39) would include the gluon structure functions, and one would find after taking moments a 2×2 matrix problem to solve. The answer involves a sum of two pieces behaving as $[\alpha_s(Q^2)]^{d^{\pm}(n)}$ with $d^{\pm}(n)$ the eigenvalues of a 2×2 "anomalous dimension" matrix. This modification will not alter in a significant way the discussion to follow.)

Inspection of Eq. (42) shows that a particular solution is, as already advertised, proportional to log Q^2, i.e., to α_s^{-1}.

$$\tilde{F}(n,Q^2) = [constant]\cdot[\alpha_s(Q^2)]^{d(n)} + \frac{\tilde{f}(n)\ell n\,Q^2/\Lambda_{QCD}^2}{1+d(n)} . \tag{45}$$

This follows because of the behavior of the α_s multiplying the splitting function $\tilde{P}$:

$$\tilde{P}(n,Q^2) \equiv \frac{-d(n)}{\ell n\,Q^2/\Lambda_{QCD}^2} . \tag{46}$$

The presence of the inhomogeneous term requires a special procedure for obtaining the normalization constant for the homogeneous term. It is found by observing that for virtual P^2 near Q^2 there is no QCD evolution at all, and that the lowest order "bare quark" term should represent, to this accuracy, the whole contribution. Also, we have been suppressing the P^2 dependence of $\tilde{F}$. Putting that in and applying the boundary condition produces the desired result:

$$\begin{aligned}\tilde{F}(n,Q^2,P^2) &= \frac{-\tilde{f}(n)(\ell n\,P^2/\Lambda_{QCD}^2)}{1+d(n)}\left[\frac{\alpha_s(Q^2)}{\alpha_s(P^2)}\right]^{d(n)} + \frac{\tilde{f}(n)(\ell n\,Q^2/\Lambda_{QCD}^2)}{1+d(n)} \\ &= \frac{\tilde{f}(n)(\ell n\,Q^2/\Lambda_{QCD}^2)}{1+d(n)}\left\{1-\left[\frac{\alpha_s(Q^2)}{\alpha_s(P^2)}\right]^{d(n)}\right\} .\end{aligned} \tag{47}$$

Thus the boundary condition for simple behavior when $Q^2 \sim P^2$ has provided linkage between the homogeneous ("hadronic") term, with its typical scaling-violation

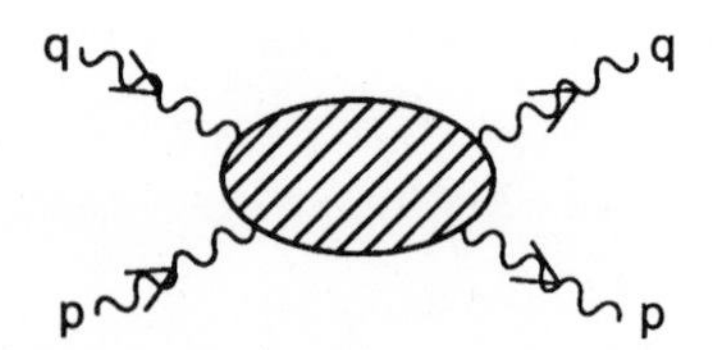

Figure 3. Photon-photon forward scattering amplitude.

behavior, and the inhomogeneous ("pointlike") term with single-log "Witten" behavior $\alpha_s(Q^2)^{-1}$.

Evidently, as $d(n)$ goes negative for small n, it is essential to have *both* terms present; keeping only the "leading" term would lead to an unphysical singularity. This is essentially what happened in the early days of confusion. An additional problem is evident if one wishes to let P^2 go to small or timelike values. If $[1 + d(n)] > 0$, it appears that the limit is stable, and only the leading "Witten" term survives. If $[1 + d(n)] < 0$, the conventional"hadronic" term looms up in importance. And in both cases, one must ask whether *nonperturbative* effects can infiltrate the results. To cope with these issues it is convenient to use analytic continuation in P^2, the probed photon squared mass, for fixed Q^2.

To understand that this is possible, it is convenient to review briefly the formal operator-product approach to deep-inelastic processes. We shall provide only a sketch, omitting almost all technical detail. One starts with the forward scattering amplitude $T(q,p)$ of current q from current p; diagrammatically shown in Fig. 3. It is a sum of Feynman diagrams, the Fourier transform of a four-point function:

$$T(q,p) = \int d^4x d^4y d^4z \, e^{iq\cdot(x-y)} \, e^{-ip\cdot z} \, < O|T(j(x)j(y)j(z)j(0))|O > \quad , \qquad (48)$$

where all spin indices, etc., have been omitted. For fixed $q^2 = -Q^2$ and $p^2 = -P^2$ spacelike, T is an analytic function of the energy variable $\nu = q \cdot p$, except for branch cuts on the real axis beginning at $(p \pm q)^2 = 4m_q^2$ or $4m_\pi^2$, depending upon whether one considers the perturbative approximant $\overline{T}$ or the exact amplitude T (for which there are also poles at $(p \pm q)^2 = m_\pi^2$). In the ν plane, the cuts therefore occur at:

$$\nu = \nu_0 \approx \pm\frac{P^2 + Q^2}{2} \quad . \qquad (49)$$

In terms of the scaling variable,

$$\omega = x^{-1} = \frac{2q \cdot p}{Q^2} \quad . \qquad (50)$$

T (or $\overline{T}$) satisfies a dispersion relation in the cut ω-plane with threshold

$$\omega_0 = 1 + \frac{P^2}{Q^2} \gtrsim 1 \quad . \qquad (51)$$

One has (for a "nonsinglet" structure function):

$$T(\omega, Q^2, P^2) = \int_{\omega_0}^{\infty} d\omega' \left[\frac{W(\omega', Q^2, P^2)}{\omega' - \omega} + \frac{\overline{W}(\omega', Q^2, P^2)}{\omega' + \omega} \right] \quad . \tag{52}$$

The discontinuities of T across the branch cuts (i.e., the absorptive part), which we call W or $\overline{W}$, are, up to suppressed kinematic factors, the cross sections for the photon-photon collision. These are essentially just the structure functions F_2.

The operator-product expansion expresses the product of currents $j(x)j(y)$ in Eq. (48) by a sum of local operators $O_n(x)$ multiplied by c-number coefficients $C_n(x-y)$, depending only on the coordinate differences, which are to be regarded as "small". In this application, it turns out to be enough to have the invariant distance $(x-y)^2$ small, i.e., near the light-cone.

Upon doing the y-integration in Eq. (48), one sees that all the dependence on q in the nth term of the expansion goes into the coefficient function. The remaining contribution depends only on p. This, however does not imply a complete factorization. Spin indices, brutally suppressed here, must be restored. Typically, the operator "O_n" has n tensor indices (give or take one or two) which are contracted into similar tensor indices possessed by the coefficient function C_n. This gives a structure for the nth term in the operator product expansion as follows:

$$n\text{th term} = A_n(Q^2)\ B_n(P^2)\ P_n(q \cdot p) \quad , \tag{53}$$

where the P_n is a *polynomial* in $(q \cdot p)$ of order n (give or take one or two). What this implies is that a series expansion of T in $q \cdot p$ (for fixed Q^2 and P^2) essentially allows identification of its nth term with the nth contribution to the operator-product expansion. Because of nonvanishing photon spin, gauge-invariance, etc., things are not quite so simple as sketched above. But the complications are essentially of a technical nature.

Because at large spacelike Q^2 and P^2 the amplitude T is analytic in a large neighborhood around $q \cdot p = 0$, the series-expansion is convergent. The coefficients are given in terms of the moments of the structure functions W and $\overline{W}$ already encountered in the Altarelli–Parisi approach. This can be seen from the dispersion-relation, Eq. (52). In the scaling limit, one finds (for $P^2 \ll Q^2$):

$$T = \sum_n T_n\ (2q \cdot p/Q^2)^n \quad , \tag{54}$$

with

$$T_n = \int_{\omega_0}^{\infty} \frac{d\omega'}{(\omega')^{n+1}} \left[W - (-)^n\ \overline{W}\right] = \int_0^1 dx x^{n-1} \left[W - (-)^n\ \overline{W}\right] \quad . \tag{55}$$

Here, the usual deep-inelastic scaling variable is given by:

$$x \equiv (\omega')^{-1} \quad . \tag{56}$$

Our main reason for going through all this is to emphasize that the moments of the structure functions W and $\overline{W}$ are related to a Green's function (sum of Feynman diagrams), with that Green's function being a finite number of derivatives with respect to $q \cdot p$ of T about $q \cdot p = 0$. Furthermore, $T(Q^2, P^2, 0)$ (and the derivatives thereof) has good analytic properties in P^2 at fixed spacelike Q^2 (or *vice versa*). This is most easily seen by examining, for any Feynman diagram, the expression obtained after "combining denominators" using Feynman's famous formula and doing the momentum-space integrations.[18] The resultant expression has the generic form.

$$I = \int \frac{d\alpha_1 \cdots d\alpha_n N(\alpha_1 \cdots \alpha_n)}{[\zeta_1(\alpha)Q^2 + \zeta_2(\alpha)P^2 + M^2(\alpha)]^m} \quad , \tag{57}$$

where ζ_1, ζ_2, and M^2 can be shown to be positive semidefinite irrespective of choice of $\alpha's$ or of Feynman-diagram. Thus one concludes that the moments of the structure functions are, for fixed spacelike q^2, analytic in the cut p^2 plane, with a branch cut with normal thresholds at positive p^2. Thus one has, at long last, the desired result.[19]

$$\tilde{F}(n, Q^2, P^2) \equiv \int_0^1 dx x^{n-2} \; F_2(x, Q^2, P^2) = \int_0^\infty \frac{d\sigma^2 \rho(n, Q^2, \sigma^2)}{(P^2 + \sigma^2)} \quad . \tag{58}$$

The dispersion integral should be convergent, since for large $P^2 (P^2 \gg Q^2)$, $\tilde{F}$ tends to zero, according to the discussion following Eq. (36).

What can one say about the weight function $\rho(\sigma^2)$? [We suppress for a while the Q^2 and n dependences]. At the level of the uncorrected $\gamma\gamma \to q\bar{q}$ "fusion" diagram, one has a log Q^2/P^2 behavior which is reproduced by:

$$\rho \approx \begin{cases} \textit{constant} & \sigma^2 < Q^2 \quad , \\ 0 & \sigma^2 > Q^2 \quad . \end{cases} \tag{59}$$

The QCD modifications to ρ need to be inferred from Eq. (47), which expresses the desired answer. What to do is perfectly clear[20]; one simply writes:

$$\rho \approx \begin{cases} \rho_0 \left[\dfrac{\alpha_s(Q^2)}{\alpha_s(\sigma^2)} \right]^{d(n)} & \sigma^2 < Q^2 \quad , \\ 0 & \sigma^2 > Q^2 \quad . \end{cases} \tag{60}$$

The *absorptive* part in P^2 of the moment of the structure function posses the scaling-violation pattern typical of ordinary "hadronic" structure functions. In this sense, it seems inappropriate to describe the dispersion integral over it as the difference of a "pointlike" part (contribution from the upper limit) and "hadronic" part (contribution from the lower limit). Just as for the case of vacuum-polarization and/or R ($e^+e^- \to$ *hadrons*), the separation into short-distance and long-distance pieces (aside from contributions of heavy-flavor thresholds) has to do with the large σ^2 and small σ^2 contributions of the absorptive part.

The importance of the low-σ^2 part of $\rho(n, \sigma^2, Q^2)$ evidently depends on n (or, if one likes, the "conjugate" variable x). For large n (large x), the perturbative-QCD correction *suppresses* the small-σ^2 contribution. However, for small n (small x), the opposite occurs and the low-σ^2, infrared region is enhanced. In this region in particular,

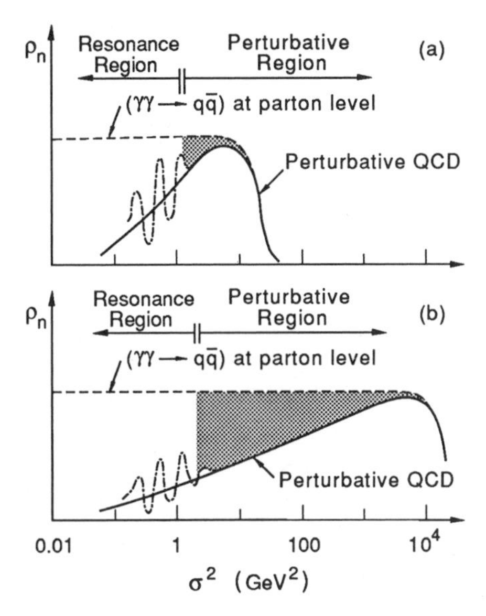

Figure 4. Sketch of the weight-function $\rho_n(\tau^2, Q^2)$ for $n \sim 4$ and (a) $Q^2 \sim 20$ GeV, and (b) $Q^2 \sim 20,000$ GeV.

there will be nonperturbative contributions from Regge-poles. Since Regge-poles are associated with the presence of *bound* $q\bar{q}$ states, they are indeed nonperturbative. And the perturbative calculations simply should not be trusted quantitatively at all when extended into this region. What they say is just that for small n and small x, there is a big infrared contribution, while at large n and large x, the infrared contributions are probably suppressed.

Where does this leave the matter experimentally? For $\Lambda^2_{QCD} \ll P^2 \ll Q^2$, the calculations are quite solid, but the sensitivity to the QCD corrections is poor until very large Q^2 (much larger than is now available) is attained. The real-photon structure-function at large x, according to our argument, can, in principle, be well estimated by the perturbative-QCD expression when Q^2 is sufficiently large. Note that the moments of the structure function are just (for $P^2 = 0$):

$$M(n, Q^2) = \int \frac{d\sigma^2}{\sigma^2} \rho(n, \sigma^2, Q^2) \quad ; \tag{61}$$

i.e., just the area under the curve when ρ is plotted versus log σ^2. This is done in Fig. 4 for $Q^2 = 20$ GeV^2 and $Q^2 = 20,000$ GeV^2. The region below, say, $\sigma^2 = M^2 \sim 2$ GeV^2 is no doubt *nonperturbative*. There are double-poles at ρ, ω, ϕ masses in the moments, leading to (approximate) $\delta'(\sigma^2 - m^2)$ singularities in ρ. So the perturbative estimate of the area below $\sigma^2 = M^2$ should not be trusted to a factor of two, if even that much. This leaves the "reliable" perturbative effect the shaded area as shown. For $Q^2 = 20$ GeV^2, the effect is very small; for $Q^2 = 20,000$ GeV^2, there seems to be some hope. It appears that if the *real-photon* structure function could be measured from, say, $Q^2 \sim 100$ GeV^2 to $Q^2 \gtrsim 10^4$ GeV^2, there might be some chance of getting a quantitative measure of the QCD correction. But, at present, what has been accomplished experimentally is observation of the presence of the large-x "pointlike" contribution from the $\gamma\gamma \to q\bar{q}$

process. This comes both from the shape of F_2 at large x and from its logarithmic Q^2 dependence.

CONCLUDING REMARKS

These discussions of R and of photon structure functions have been based on a conservative view of perturbative QCD. As expressed in the introduction, I feel that there is a need for developing this point of view further. I am encouraged by the fact that the first two examples attacked this way yielded something which at least I have found not uninteresting. In both cases, the use of Green's functions built from local color singlet operators, and especially their analyticity properties, was a central feature. While there are more things to do in generalizations along this line, I think there are also interesting questions of how to deal with the interpretation of the perturbative intermediate states which build the absorptive parts of the Green's functions, and how to relate them to the extant QCD jet calculus used in phenomenology. I hope to work on this in the future.

ACKNOWLEDGMENTS

Again, I thank my students at the University of Chicago for helping to clarify these topics. In addition, I thank Al Mueller for introducing me to the wonderful world of renormalons, and Jeff Owens for important help on photon structure functions. I also thank my colleagues at Fermilab, especially Bill Bardeen and Keith Ellis, for helpful discussions, and Michael Peskin for useful criticism.

REFERENCES

1. F. Yndurain, "Quantum Chromodynamics," Springer-Verlag, New York (1983).
2. R. Field, "Applications of Perturbative QCD," Addison-Wesley (1989);
   M. Peskin and D. Schroeder, in preparation.
3. As might be anticipated, I have in mind the line of argument presented in J. Bjorken and S. Drell, "Relativistic Quantum Fields," McGraw-Hill, New York (1965), chs. 16–17.
4. More discussion of essentially this point of view can be found in some old lectures of mind; *cf.* J. Bjorken, *Proc. of the SLAC Summer Institute on Particle Physics,* Stanford, CA (1979), A. Mosher, ed.; SLAC–REP–224; also preprint SLAC–PUB–2372.
5. E. Poggio, H. Quinn, and S. Weinberg, *Phys. Rev.* **D13**:1958 (1976).
6. V. Fadin, V. Khose, and T. Sjostrand, CERN Preprint CERN–TH–5394/89.
7. R. P. Feynman, *Proc. of the Int. Conf. on High Energy Physics,* Aix-en-Provence, vol. 2, E. Cremien-Alcan *et al.,* ed., Saclay, France (1961).
8. S. Gorishny, A. Kataev, and S. Larin, *Phys. Lett,* **B212**:238 (1988).
9. See, for example, G. 'tHooft, "The Whys of Subnuclear Physics, Erice 1977," A. Zichichi, ed., Plenum, New York (1977);
   G. Parisi, *Nucl. Phys.* **B150**:163 (1979);

A. Mueller, *Nucl. Phys.* **B250**:327 (1985);
G. West, Los Alamos preprint LA–UR–89–3785.

10. For a recent review, see S. Maxfield, *Proc. of the XXIV Int. Conf. on High Energy Physics, Munich 1988,* Springer-Verlag, Berlin, Heidelberg (1989), p. 661.
11. An especially complete treatment is given by G. Rossi, U.C. San Diego preprint UCSD–10P10–227.
12. T. Walsh and P. Zerwas, *Nucl. Phys.* **B41**:551 (1972); *Phys. Lett* **44B**:195 (1973); R. Kingsley, *Nucl. Phys.* **B60**:45 (1973).
13. G. Altarelli and G. Parisi, *Nucl. Phys.* **B126**:298 (1977); see also J. Stirling's lectures, these proceedings.
14. E. Witten, *Nucl. Phys.* **B120**:189 (1977).
15. W. Bardeen and A. Buras, *Phys. Rev.* **D20**:166 (1979); D. Duke and J. Owens, *Phys. Rev.* **D22**:2280 (1980).
16. T. Uematsu and T. Walsh, *Phys. Lett* **101B**:263 (1981); *Nucl. Phys.* **B199**:93 (1982).
17. Muta's book, reference 1, contains a thorough exposition.
18. More details can be found in Bjorken and Drell, Ref. 4, Chap. 18.
19. Analyticity in P^2 has recently been utilized by A. Gorski and B. Ioffe, University of Bern preprint BUTP–89/12.
20. At least to me. This follows either from inspection or from the realization that ρ satisfies the usual renormalization group equation *without* an inhomogeneous term.

HADRONIC MATRIX ELEMENTS AND WEAK DECAYS IN LATTICE QCD

Guido MARTINELLI

Dipartimento di Fisica, Università di Roma "La Sapienza"
P.le A.Moro 2, 00185 Roma, Italy
INFN - Sezione di Roma, Italy

INTRODUCTION

Many important progresses have been done in lattice QCD since the first pioneering numerical calculations were started about 10 years ago. Since then substantial advances have been possible thanks to a great theoretical effort in understanding the lattice quantum field theory and to the extraordinary increase in the available conputer power. Almost all aspects of particle physics are currently studied on the lattice, already with phenomenologically relevant results. Although we are not yet at the point to be able to predict with sufficient accuracy the mass of the proton, many steps forward have been done in the calculation of the hadronic matrix elements. Lattice QCD offers in fact the unique possibility, of computing with the same method many different matrix elements which are of interest in the phenomenology of the Standard Model. Among the other quantities I will report the most recent results obtained for the pion and proton structure functions and electromagnetic form factors, the nucleon σ-term, the meson decay constants and the weak hamiltonian matrix elements, including the semileptonic decays of D mesons. These lectures have been conceived to serve as an elementary introduction to this fascinating field and to offer to non-experts a survey of the most relevant results.

Particular attention has been devoted to discuss the systematic errors actually present in the lattice approach, expecially those coming from the "quenched" approximation and from effects due to the finitess of the lattice spacing a.

In Section 1 a short introduction to lattice QCD is given. In Section 2 the problems connected with the lattice regularization of the fermionic action are described. Sect.3 contains some basic notions on the calculations of operator matrix elements on the lattice. In Section 4 and 5 I discuss in length the calculation of the pion and nucleon structure functions. In Section 6 the problem of the σ-term and its connection to the systematic errors entailed the quenched approximation is presented. Section 7 is devoted to the weak decays of kaons. In Section 8 the calculation of the semileptonic decay of D mesons is discussed and finally in Section 9 a summary of the present situation and the perspectives in this field is given.

Particle Physics: Cargèse 1989
Edited by M. Lévy *et al.*
Plenum Press, New York, 1990

1. GENERALITIES ON LATTICE GAUGE THEORIES

1.1 *The Gauge Action on the Lattice*

The physics of strong interactions can be entirely derived starting from the action:

$$S = \int d^4x \, [- \frac{1}{2} \bar{q}(x) \gamma^{\mu} \overleftrightarrow{D}_{\mu} q(x) - m_q \bar{q}(x)q(x) - \frac{1}{4} F_{\mu\nu}(x)^2] \tag{1.1}$$

In eq.(1.1) $q(\bar{q})$ are the quark fields; $D_\mu = \partial_\mu + ig_0 A^a_\mu \lambda^a$ is the covariant derivative and $F^a_{\mu\nu} = \partial_\mu A^a_\nu - \partial_\nu A^a_\mu - g_0 f^{abc} A^b_\mu A^c_\nu$, where A^a_μ (a=1,...,8) are the gluon fields; g_0 is the bare coupling constant. The action in eq.(1.1) is a generalization to a non abelian group (SU(3) of colour for QCD) of the usual, abelian local gauge invariant action of electrodynamics[1].

Unfortunately the simple action of eq.(1.1) gives rise to a very complicated physics because the fields in the Lagrangian do not correspond to observable particles but are permanently confined to form hadrons (pions, protons, ...). Thus, unlike for weak and electromagnetic interactions, it is possible to use perturbation theory only for deep inelastic (very high energy) phenomena where asymptotic freedom is at work[2].

Actually the only formulation which is (will become) able to give quantitative predictions for hadron physics at low energy (masses, widths, decay amplitudes etc.) is lattice QCD. The lattice formulation of QCD on is a very natural way of introducing an ultraviolet cutoff (necessary to any renormalizable theory) in a gauge invariant way[3]. The continuum limit is obtained when the lattice spacing $a \rightarrow 0$ (i.e. the ultraviolet cutoff goes to infinity).

We proceed in two steps:

a) The quantum theory in a euclidean space time.

In the usual Minkowsky space-time any physical amplitude can be expressed in terms of the vacuum generating functional and its derivatives (vacuum expectation values of operators). They have the form:

$$Z(J) = \sum_{\{\phi\}} e^{i\int d^4x \, \mathcal{L}(\phi,\partial_\mu\phi,J)}$$

$$< O(\phi)>_{J=0} = \frac{1}{Z(0)} \sum_{\{\phi\}} O(\phi) e^{i\, S(\phi,\partial_\mu\phi,0)} \tag{1.2}$$

ϕ is any quantum field. $\sum_{\{\phi\}}$ indicates the integral over all the possible values of the fields, at any space-time point (this is formal definition to which it is possible to give a precise mathematical meaning at least in perturbation theory).

We make an analytic continuation in the time coordinate which defines the model in four dimensional Euclidean space[4]:

$$t \rightarrow i\, x_4 \tag{1.3}$$

Eqs.(1.2) now have the form:

$$Z(J) = \sum_{\{\phi\}} e^{-\int d^4x\, \mathcal{L}(\phi,\partial_\mu\phi,J)}$$

$$< O(\phi)>_{J=0} = \frac{1}{Z(0)} \sum_{\{\phi\}} O(\phi) e^{-S(\phi,\partial_\mu\phi,0)} \tag{1.4}$$

Z can be interpreted as the partition function of a four dimensional classical system whose Hamiltonian is given by the relation:

$$\beta H = S \tag{1.5}$$

We have thus transformed the quantum field theory in the study of the physics of a statistical system of classical fields.

b) The Theory on a lattice

To regularize the infinities of the theory we replace the continuum space-time by a mesh of discrete lattice points. The most simple case is a hypercubic lattice with point separation a. The lattice coordinates of a point are denoted by a four integer vector $n = (n_1, n_2, n_3, n_4)$ and the fields are defined only on the points of the lattice.

A very simple example is given by a d-dimensional free field scalar field:

$$S_{\text{continuum}} = \int d^d x \, \frac{1}{2} (\partial_\mu \phi)^2$$

$$\phi(x) \rightarrow \phi(n) \, ; \quad \partial_\mu\phi(x) \rightarrow \frac{\phi(n+\hat{\mu}) - \phi(n)}{a} \tag{1.6}$$

$$S_{\text{lattice}} = \frac{1}{2} \sum_{n,\hat{\mu}} [\phi(n+\hat{\mu}) - \phi(n)]^2$$

where $\phi(n) = a^{(d/2)-1}\, \phi(x)$ at $x = na$ and $\hat{\mu}$ is the unit vector in the μ-direction. It is trivial to see that:

$$\lim_{a\rightarrow 0} S_{\text{lattice}} = S_{\text{continuum}}$$

In the case of a gauge theory, we require that the lattice action satisfies two requisities[3]

i) Local (lattice) gauge invariance

ii) Formal (tree level) $a \rightarrow 0$ limit.

An example of a pure gauge (no quarks) action which satisfies i) and ii) was originally proposed by Wilson

$$S_G(U) = \frac{1}{g_0^2} \sum_P \text{tr}\, [U(P) + U^+(P)] \tag{1.7}$$

where U(P) is the product of link matrices belonging to an elementary plaquette as explained in Figs. 1a and 1b:

$$U(P) = U_\mu (n) U_\nu (n+\hat{\mu})\, U^+_\mu (n+\hat{\nu})\, U^+_\nu \,(n) \tag{1.8}$$

The link variable $U_\mu(n)$ is an element of the symmetry group and it is related to the gluon field A^a_μ by the equation:

$$U_\mu(n) = e^{i\, a\, g_0 A^a_\mu(n)\, \lambda^a} \tag{1.9}$$

where λ^a are the generators of the group. $S_G(U)$ is invariant under the local gauge transformation:

$$U_\mu(n) \rightarrow g(n)\, U_\mu(n)\, g^+(n+\hat{\mu}) \tag{1.10}$$

where g(n) is an element of the group. It is easy to verify that for $a \rightarrow 0$:

$$S_G(U) \rightarrow -\frac{1}{4g_0^2} \int F_{\mu\nu}(x)^2\, d^4x + 0(a^2) \tag{1.11}$$

which is the usual expression for the continuum action (up to a normalization of the gauge fields).

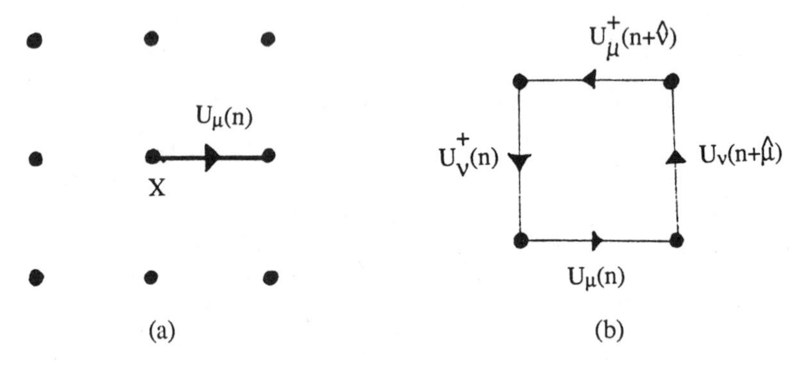

FIGURE 1
a) We associate to any point of the lattice and direction an oriented link variable $U_\mu(n)$ which connects the point n with the point $n+\hat{\mu}$. The relation between $U_\mu(n)$ and the corresponding gluon field is given in eq.(1.9); b) the ordered product of four links belonging to an elementary plaquette defines U(P).

Two comments are necessary at this point. The first comment is that there is an arbitrariness in defining the lattice action due to the extra $0(a^2)$, $0(a^4)$ ... terms present on the lattice. We could for example define an action made by the product of 6 links which still satisfies i) and ii) as illustrated in Fig. 2. The second comment is that going to the lattice formulation we have lost Lorentz invariance: only a residual symmetry under rotations of ninty degrees is left. In general, by putting the theory on the lattice (see also Section 2), we obtain symmetry properties which are different from the corresponding properties in the continuum. A proper $a \rightarrow 0$ limit of the lattice action should recover the symmetry properties of the original theory.

The vacuum functional for the lattice theory in absence of external currents can be written as

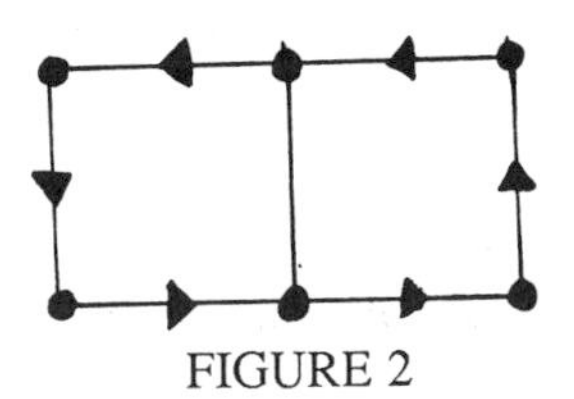

FIGURE 2

The action defined by the product of 6 links will tend to the continuum action for $a \to 0$. At finite a it differs from the definition shown in Fig. 1b.

$$Z = \int \prod_{n,\mu} d\,[U]_{n,\mu}\, e^{S_G(U)} \tag{1.12}$$

where d[U] is the invariant Haar measure with the properties:

$$\begin{aligned} &d[U] = 1 \\ &\int d[U]\, f(U) = \int d[U]\, f(gU) \end{aligned} \tag{1.13}$$

where g is an arbitrary element of the group.

1.2 *Removing the ultraviolet cut-off*

The full quantum continuum limit of the lattice action is a much more subtle problem than the formal limit given in eq.(1.11) and it involves the full restoration of the symmetries and eventually of the topological properties of the theory.

The only parameters which enter in the action of eq.(1.7) are the coupling constant g_o and (implicitely) the lattice spacing a. All masses and lengths may be expressed in terms of these parameters:

$$m_i = \frac{1}{a} f_i\,(g_o) \tag{1.14}$$

In the limit in which the lattice spacing becomes smaller and smaller we expect that it is possible to vary g_o and a in such a way that the physical properties of the system remain unaltered. Cut-off independence for masses implies:

$$a\frac{dm_i}{da} = 0 \tag{1.15}$$

(This is the condition that in the limit $a \to 0$ physics becomes independent of the ultraviole cut-off). At small g_o, because of asymptotic freedom, we expect:

$$- a\frac{dg_o}{da} = - \beta_o g_o^3 - \beta_1 g_o^5 + \ldots = \beta(g_o) \tag{1.16}$$

where $\beta_{o,1} > 0$ are universal coefficients that can be computed in perturbation theory.

Eqs.(1.15) and (1.16) imply that, for a $a \to 0$ all masses become proportional to a unique fundamental scale:

$$m_i = c_i \Lambda_{latt}$$
$$m_i/m_j = \text{constant} \qquad (1.17)$$

where c_i are dimensionless constants and also that:

$$a = \frac{1}{\Lambda_{latt}} \exp\left[\frac{-1}{2\beta_o g_o^2}\right] (\beta_o g_o^2)^{(-\beta_1/2\beta_o^2)} \left[1+0(g_o^2)\right] \qquad (1.18)$$

At Λ_{latt} fixed (i.e. masses fixed in physical units) eq.(1.18) shows that the lattice spacing a goes to zero exponentially in $1/g_o^2$ as $g_o \to 0$. Correspondingly all dimensioless correlation lenghts $\xi_i\, a^{-1} = 1/(m_i a)$ go to infinity: in the language of statistical mechanics the system undergoes a second order phase transition. The size of the physical particles becomes bigger and bigger in units of the lattice spacing as g_o is decreased so that we have a better and better description of the continuum physics as much in the same way one obtains a more and more accurate result in a numerical evaluation of an integral by increasing the number of subdivisions of the interval of integration.

In the continuum limit different lattice actions should produce the same continuum physics (universality principle). On one hand this means that ratios of masses (e.g. the mass of the proton over the mass of the ρ) must be independent of the action we used. On the other hand two different forms of lattice actions, which have different scale parameters Λ_{latt}, should correspond the same physical quantities:

$$m_i = c_i \Lambda_{latt} = c_i' \Lambda_{latt}'$$
$$R = \frac{\Lambda'_{latt}}{\Lambda_{latt}} = \frac{c_i}{c_i'} \qquad (1.19)$$

R can be predicted in perturbation theory and compared with the results for c_i and c_i' (obtained for example by Montecarlo techniques) as a test of universality.

We have seen that all the relevant physical information are given by expectation values of (gauge invariant) operators:

$$< O(U) > = \frac{\int d[U]\, e^{-S_G(U)}\, O(U)}{\int d[U]\, e^{-S_G(U)}} \qquad (1.20)$$

The computation of the functional integral in eq.(1.20) on an euclidean lattice is not easier than the original computation in the Minkowky continuum space-time. For most of the theories (including QCD) an exact evaluation is an impossible task. Several approximation techniques are available: in the following I briefly discuss only the Montecarlo techniques and their application to lattice QCD. Other approximations like strong coupling expansion, mean field theory, 1/N expansion, 1/d expansion etc. will not be covered in these lectures.

In a Montecarlo simulation one attempts a direct integration of the gauge fields on a *finite* lattice. The accuracy of the results is dictated by limitations imposed on the size of the lattice by computer memory and speed.

The situation is illustrated in Fig.3: at large values of g_o the size of the hadron is of the order the lattice spacing and strong effects due to the too small ultraviolet cutoff are present; at small g_o, with a limited number of lattice points, the hadron becomes as large as the lattice and it feels strong finite size (infrared) effects. The ideal situation is shown in Fig.3c: the hadron is much larger than the lattice spacing but much smaller than the lattice size. In actual simulations this situation is really never fully achieved; we will return on this point when discussing Montecarlo results. A usefull example of what occurs in practical cases is also given in Fig.4: the data with errors are measurements of some hadron mass by Montecarlo simulation; the full curve comes from strong coupling expansion; the dashed line corresponds, up to a multiplicative constant, to the expected asymptotic renormalization group behaviour of eq.(1.18); the dotted-dashed line is the spin-wave ($g_o \to 0$ on a finite lattice) prediction. If the lattice is not large enough, finite size effects could prevent us to see the region where masses scale according to eq.(1.18) (indicated as scaling region in Fig.4). Even in the most favourable case only the results on a small range of g_o can be used for the extrapolation to the continuum since correlation lenghts increase exponentially in $1/g_o^2$.

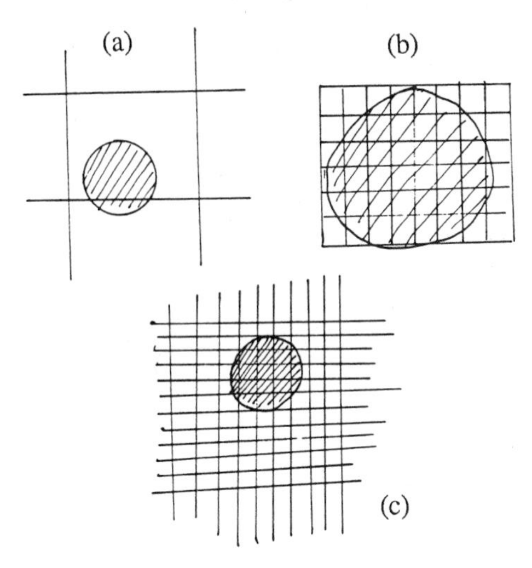

FIGURE 3
a) Hadron on the lattice at g_o large: the hadron is smaller than the lattice spacing; b) when $g_o \to 0$ the hadron becomes as large as the whole lattice; c) the hadron size is much smaller of the lattice and much larger of the lattice spacing.

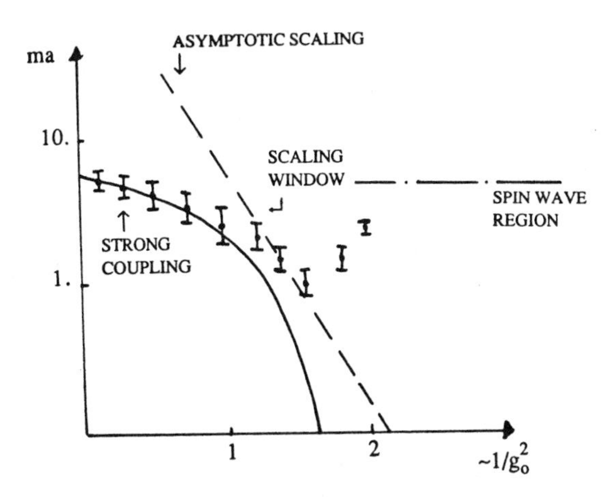

FIGURE 4
Schematic representation of the behaviour of some hadron mass as a function of $1/g_o^2$. The scales of the plot are arbitrary.

To avoid finite size effects, the lattice size must be increased linearly in the correlation length, that is the number of lattice points increases as the fourth power of the correlation length. Notice that on a lattice of 10^4 points link variables possess 3×10^5 degrees of freedom on which we have to integrate. It is clear that to make a multidimensional integral on 10^5 variables one needs to use a Montecarlo technique and important sampling. I will not discuss any of these technical aspects of the lattice approach furthermore.

1.3 *Fermions on the Lattice - The Fermion Doubling Problem*

A special problem arises when one tries to write on the lattice the fermionic part of the action given in eq.(1.1). The naïve transcription of the fermionic action on the lattice is given by:

$$S_\psi = \sum_{f,n} \{ \frac{1}{2} \sum_\mu [\bar{\psi}^f(n)\, \gamma_\mu\, U_\mu(n)\, \psi^f(n+\hat{\mu}) - \bar{\psi}^f(n+\hat{\mu})\, \gamma_\mu\, U^+_\mu(n)\, \psi^f(n)]$$

$$+ m_f\, \bar{\psi}^f(n)\, \psi^f(n) \} \qquad (1.21)$$

ψ^f is the quark field with flavour f and mass m_f and γ_μ are the Dirac matrices.

The above lattice action is obtained by the replacement

$$\partial_\mu f(x) = \frac{f(x+\hat{\mu}) - f(x)}{a} \qquad (1.22)$$

in the continuum theory: S_ψ has apparently the same chiral properties of the continuum action. In fact, in the limit in which the explicit mass term $m_f=0$, the action is invariant under the following (global) transformations:

$$\psi(x) \rightarrow e^{i\alpha}\, \psi(x) \qquad\qquad \psi(x) \rightarrow e^{i\alpha\gamma_5}\, \psi(x) \qquad (1.23)$$

More generally the lattice action is invariant under the global continuum symmetry groups:

$$SU(n_f)_V \times SU(n_f)_A \times U(1)_V \times U(1)_A \qquad (1.24)$$

exactly as it was for the continuum QCD action. In QCD the $U(1)_V$ symmetry corresponds to the conservation of the baryon number; the $SU(n_f)_A$ symmetry is expected to be dynamically broken giving rise to massless Goldstone bosons (pions, kaons, ...) while the $U(1)_A$ is broken because of the triangle anomaly (see Fig.5) and this should explain why the η' is not a light particle.

However, on the lattice something strange is happening: in fact the triangle anomaly is identically zero.

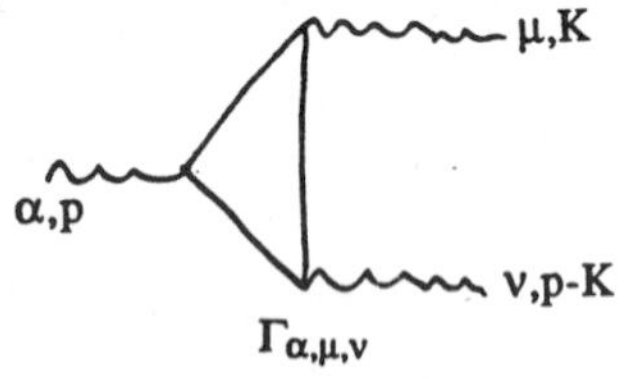

$$T_{\mu\nu} = p^\alpha\, \Gamma_{\alpha,\mu,\nu} = (\sum_i Q_i)\, I_{\mu\nu}$$

FIGURE 5
The Feynman diagram for the triangle anomaly

The reason is that the fermion spectrum on the lattice (if the corresponding continuum theory has only one flavour) is constituted by 16 different replicas of the original Dirac field, each of the replicas giving an opposite sign contribution to the triangular anomaly. The simplest explanation of the doubling of fermions can be found by looking at the free fermion propagator on the lattice:

$$\frac{1}{\frac{1}{a}\sum_{\mu} \gamma_{\mu} \sin p_{\mu} a} \tag{1.25}$$

The propagator in eq.(1.25) has 16 poles corresponding to p_{μ} a =0,π. The triangle anomaly is proportional to the sum of the axial charges of the looping fermions (see Fig.5) which have charges[5]:

$$Q(0,0,0,0) = +1, \qquad Q(\pi,0,0,0) = 1$$

and so on.

The unexpected properties of the lattice action considered above are general in character and have been states precisely by a "No go" theorem by Nielsen and Ninomiya[6] which proves that it is not possible to write a fermion action on the lattice which is local, has the same chiral properties of the continuum and avoids the fermion doubling problem.

Two approaches have been proposed to overcome the problem of the fermion doubling (never completely at finite a, because of the N.N. theorem). In the following I will discuss in detail the approach originally proposed by K.G.Wilson, with special emphasis on the problem of chiral invariance which is of crucial importance in the physics of kaon decays.

2. THE WILSON LATTICE REGULARIZATION FOR FERMIONS

When we introduce a regularization in the process of renormalizing a theory some of the symmetries present in the original (bare) action are in general lost. It is precisely for this reason that anomalies are present in quantum field theories. In most of the versions of lattice QCD we break the Lorentz invariance and, in presence of fermions, we get in troubles with the chiral symmetry of the theory. In this section the problem of the chiral symmetry for a wide class of lattice actions à la Wilson is discussed in detail.

2.1. *Basic definitions and Ward identities*

A quite general form of the fermion action on the lattice is:

$$S_{\psi} = \sum_{x,f} \{ \frac{1}{2a} \sum_{\mu} [\bar{\psi}_f(x)\, (r - \gamma_{\mu})\, U_{\mu}(x)\, \psi(x + \hat{\mu}) + \bar{\psi}_f(x+\mu)\, (r + \gamma_{\mu})\, U^{+}_{\mu}(x)\, \psi(x)] + \bar{\psi}_f(x)\, [M_0 + \tfrac{r}{a}]\, \psi_f(x) \} = \sum_{f} \bar{\psi}_f\, \Delta_f(U)\, \psi_f \tag{2.1}$$

The terms proportional to the Dirac matrices γ^{μ} are the naive lattice transcription of the covariant derivative $\bar{\psi}\gamma^{\mu} D_{\mu}\psi$ (sect.1.3) and an "irrelevant" operator, proportional to the parameter r, has been introduced to avoid the fermion species doubling (Wilson term[3]). The Wilson term acts as a mass term and breaks chiral symmetry explicitly, even in the limit of the vanishing quark mass M_0. As a consequence, the Ward identities for the vector and axial vector currents contain anomalous pieces, and tree-level chiral properties of composite operators are spoiled by the interaction between quarks and gluons. The Ward identities for the vector and axial vector currents can be easily derived by a chiral rotation of the fermion fields in the action of Eq.(2.1):

$$\psi \rightarrow \psi + i(\alpha_V^a + \gamma_5\, \alpha_A^a) \frac{\lambda^a}{2}\, \psi$$
$$\bar{\psi} \rightarrow \bar{\psi} - i\bar{\psi}(\alpha_A^a - \gamma_5\, \alpha_V^a) \frac{\lambda^a}{2} \tag{2.2}$$

One obtains[5]:

$$< \alpha\, |\nabla_\mu\, V_\mu^a(x)|\beta > = - < \alpha\, |\bar{\psi}(x)|\, \left[\frac{\lambda^a}{2}, M_o \right] \psi(x)|\beta > + < \alpha\, |X_V^a(x)|\beta >$$
$$< \alpha\, |\nabla_\mu\, A_\mu^a(x)|\beta > = < \alpha\, |\bar{\psi}(x)|\, \left\{ \frac{\lambda^a}{2}, M_o \right\} \gamma_5\, \psi(x)|\beta > + < \alpha\, |X_A^a(x)|\beta > \tag{2.3}$$

where

$$\nabla_\mu f(x) = f(x) - f(x-\hat{\mu}) \tag{2.4}$$

and

$$V_\mu^a(x) = \frac{1}{2} \left[\bar{\psi}(x)\, U_\mu(x)\, \gamma_\mu \frac{\lambda^a}{2}\, \psi(x + \hat{\mu}) + \text{h.c.}\right]$$
$$A_\mu^a(x) = \frac{1}{2} \left[\bar{\psi}(x)\, U_\mu(x)\, \gamma_\mu \gamma_5 \frac{\lambda^a}{2}\, \psi(x + \hat{\mu}) + \text{h.c.}\right] \tag{2.5}$$

The Ward identities in Eq.(2.3) look very similar to the corresponding identities in the continuum limit but for the last term on the right-hand side. In the free theory, this term is of order 0(a), a being the lattice spacing, and disappears as $a \rightarrow 0$. Unfortunately, in the real case, the interaction promotes these anomalous pieces, so that not only they do not vanish as $a \rightarrow 0$, but become linearly divergent ($\sim 1/a$), as illustrated in Fig.6. For the vector current this is not a problem, since one can show that X_V^a is itself the divergence of a current[5]:

FIGURE 6
The operator X_A^a is of order a (Fig.6a). One-loop corrections, because of the bad ultra-violet behaviour of X_A^a ($\sim p^2$) give a linearly divergent contribution as $a \rightarrow 0$ (Fig.6b)

$$X_V^a = \nabla_\mu^a J_\mu^a \tag{2.6}$$

X_V^a can be moved on the left-hand side of the Ward identity. The current $\hat{V}_\mu^a = V_\mu^a - J_\mu^a$ now obeys the same Ward identity of the continuum theory. This happens because the Wilson term respects vector symmetries (e.g., baryon number), which is not the case for the axial current. In the latter case, however, one can show that the matrix elements of X_A^a between on-shell states α and β can be written as[5,7]:

$$\begin{aligned} < \alpha | X_A^a | \beta > = & - < \alpha | \bar{\psi} \left\{ \frac{\lambda^a}{2}, \bar{M}(M_o) \right\} \gamma_5 \psi | \beta > \\ & - (Z_A - 1) < \alpha | \nabla_\mu A_\mu^a | \beta > + O(a) \end{aligned} \tag{2.7}$$

The relation (2.7) is true at all orders in the coupling constant and is the more general one which is compatible with the symmetries of the action. O(a) indicates matrix elements of operators of dimension larger than four, which vanish as $a \to 0$. One can also show that the coefficient $\bar{M}(M_o)$ is linearly divergent in $1/a$ and that $Z_A = Z_A(g_o)$ is a function only of the bare lattice coupling constant g_o (and r)[7,8]. Let us rewrite the last of equations (2.3) as follows

$$\begin{aligned} Z_A < \alpha | \nabla_\mu A_\mu^a | \beta > = & < \alpha | \left\{ M_o - \bar{M}(M_o), \frac{\lambda^a}{2} \right\} \gamma_5 \psi | \beta > + \\ & < \alpha | \bar{X}_A^a | \beta > \end{aligned} \tag{2.8}$$

where

$$\bar{X}_A^a = X_A^a + \psi \left\{ \frac{\lambda^a}{2}, \bar{M} \right\} \gamma_5 \psi + (Z_A - 1) \nabla_\mu A_\mu^a \tag{2.9}$$

has matrix elements between on-shell states which vanish in the continuum limit [$<\alpha|\bar{X}_A^a|\beta>$ ~ 0(a)]. From Eq.(2.8) we see that we have recovered the usual continuum Ward identity, provided that we identify the good axial current A_μ^a and the bare quark mass m with:

$$\hat{A}_\mu^a = Z_A A_\mu^a \qquad m = M_o - \bar{M}(M_o) \tag{2.10}$$

We also note that Eq.(2.8) is not sufficient alone to fix Z_A and m separately, but only the ratio $\frac{m}{Z_A}$. We need at least another Ward identity to disentangle the two constants. The simplest Ward identity we can think of involves two- and three-point correlation functions of vector and axial vector currents:

$$\begin{aligned} < \nabla_\mu \hat{A}_\mu^a(x) \hat{A}_\nu^b(y) \hat{V}_\rho^c(0) > = & < \bar{\psi}(x) \left\{ \frac{\lambda^a}{2}, m \right\} \gamma_5 \psi(x) \hat{A}_\nu^b(y) \hat{V}_\rho^c(0) > \\ & + i f^{abd} \frac{Z_A}{Z_V} \delta(x-y) < \hat{V}_\nu^d(y) \hat{V}_\rho^c(0) > + i f^{acd} \frac{Z_V}{Z_A} \delta(x) < \hat{A}_\rho^b(y) \hat{A}_\rho^d(0) > + \\ & < \bar{X}_A^a(x) \hat{A}_\nu^b(y) \hat{V}_\rho^c(0) > + \ldots \end{aligned} \tag{2.11}$$

The dots indicate possible Schwinger terms. Z_V is the renormalization constant analogous to Z_A for the vector current V_ρ, [Eq.(2.8)], ($\hat{V}_\rho = Z_V V_\rho$), and m and Z_A have been defined before. The last term on the right-hand side of Eq.(2.11) is non-zero even in the continuum limit. The matrix elements of $\bar{X}^a_A$ vanish between on-shell states. $\bar{X}^a_A$ can, however, still give rise to contact terms [i.e., in Eq.(2.11) terms proportional to $\delta(x-y)$ or $\delta(x)$] in more complicated correlation functions because of extra divergences present when more fields are at the same point. These terms transform the anomalous Ward identity in Eq.(2.11) into the corresponding identity of the continuum:

$$< \nabla_\mu \hat{A}^a_\mu(x)\, \hat{A}^b_\nu(y)\, \hat{V}^c_\rho(0) > = < \bar{\psi}(x) \{ \frac{\lambda^a}{2}, m \} \gamma_5 \psi(x)\, \hat{A}^b_\nu(y)\, \hat{V}^c_\rho(0) >$$

$$+ i f^{abd} \delta(x-y) < \hat{V}^d_\nu(y)\, \hat{V}^c_\rho(0) > + i f^{acd} \delta(x) < \hat{A}^b_\nu(y)\, \hat{A}^d_\rho(0) > + \ldots \qquad (2.12)$$

as it can be proved in perturbation theory[7]. This implies the following relation:

$$\sum_x < [\nabla_\mu A^a_\mu(x) - \bar{\psi}(x) \{ \frac{\lambda^a}{2}, \frac{m}{Z_A} \} \gamma_5 \psi(x)]\, A^b_\nu(y)\, V^c_\rho(0) > =$$

$$i f^{abd} \left(\frac{Z_V}{Z_A^2} \right) < V^d_\nu(y)\, V^c_\rho(0) > + i f^{acd} \frac{1}{Z_V} < A^b_\nu(y)\, A^d_\rho(0) > \qquad (2.13)$$

Equations (2.8) and (2.13) allow us to separate Z_A and m in order to find the axial current and the quark mass of the continuum theory.

The strategy based on continuum Ward identities as a mean of identifying the "good" parameters and renormalized operators (m and $\hat{A}^a_\mu$ this case) is quite general, as it will also be clear from the discussion of the renormalization of four-fermion operators given in the next sub-section. Current algebra relations of the type discussed above have be verified by numerical simulation on the lattice in Ref.[9].

2.2 *Renormalization of the Four-Fermion Operators of the Weak Hamiltonian*

The problem of the construction of renormalized, four-fermion operators is quite intriguing because of the presence of severe power divergences and of the bad chiral behaviour of the bare operators on the lattice, induced by the Wilson term.

It is simple to start by giving a specific example. Let us consider the (8,1) operator:

$$O_\alpha = (\bar{\psi}_L \frac{\lambda_\alpha}{2} \gamma_\mu \psi_L)(\bar{\psi}_R \gamma^\mu \psi_R) \qquad (2.14)$$

where $\psi_{L,R} = \frac{(1 \mp \gamma_5)}{2} \psi$, and λ_α is one of the Gell-Mann matrices acting in flavour space. Even at zero order in the bare strong coupling constant and in the chiral limit, the operator in Eq.(2.14) can mix with an operator of dimension three with a power-divergent coefficient through the diagram in Fig.(7):

$$\delta O_\alpha = \frac{1}{a^3} (\bar{\psi}_L \frac{\lambda_\alpha}{2} \psi_R + \bar{\psi}_R \frac{\lambda_\alpha}{2} \psi_L) \qquad (2.15)$$

δO_α is a $(3,\bar{3})$ operator. The factor $1/a^3$ is there because of dimensional reasons. Usually, for massless quarks, this mixing is not possible because the original operator cannot flip the helicity; on the lattice the mixing is given by the Wilson term which acts as a mass term when inserted into the loop of the diagram in Fig.(7). This example shows that a bare operator, which one would naively expect to be an (8,1) operator, is really a mixture of (8,1) and $(3,\bar{3})$ operators. A definite chiral behaviour and the removal of lattice artifact divergences can be obtained at the same time, using the Ward identities along a line similar to the one followed in the previous subsection for the axial current[7,10].

First, one subtracts from the naive operator a combination of operators with equal or lower dimensions and all possible naive chiralities, in such a way as to obtain an operator with the correct chiral properties with respect to the "good" axial and vector charges. Generally, at this point not all divergences have been eliminated, but one is left with an overall multiplicative renormalization which is determined by requiring the lattice operators to be normalized equally to the continuum ones.

In the chiral limit, this procedure is straightforward. Let us consider a particular operator O_α which transforms, under naive axial transformation, according to some representation R:

$$\frac{\delta O_\alpha}{\delta \varepsilon^a} = i\ R^{a}_{\alpha,\beta}\ O_\beta \tag{2.16}$$

R^a are the axial generators of the particular representation to which O belongs: α and β are flavour labels appropriate to R. The operator which truly transforms according to R is

$$\tilde{O}_\alpha = O_\alpha + \sum_{n,\sigma} d^{(n)}_{\alpha\sigma}\ O^{(n)}_\sigma \tag{2.17}$$

with $O^{(n)}$ transforming according to $R^{(n)} \neq R$. The coefficient $d^{(n)}$ are restricted by the condition that they must respect the conserved vector symmetry.

In the chiral limit, the integrated axial current Ward identity for $\tilde{O}$ reads:

$$\sum_x \nabla_\mu < \alpha\ |T\big(\hat{A}^{a}_{\mu}(x)\ \tilde{O}(0)\big)\beta > =$$
$$\sum_x < \alpha\ |T\big(\bar{X}^{a}(x)\ \tilde{O}(0)\big)|\beta > + i < \alpha\,|\ \frac{\delta \tilde{O}}{\delta \varepsilon^a}\ |\beta > \tag{2.18}$$

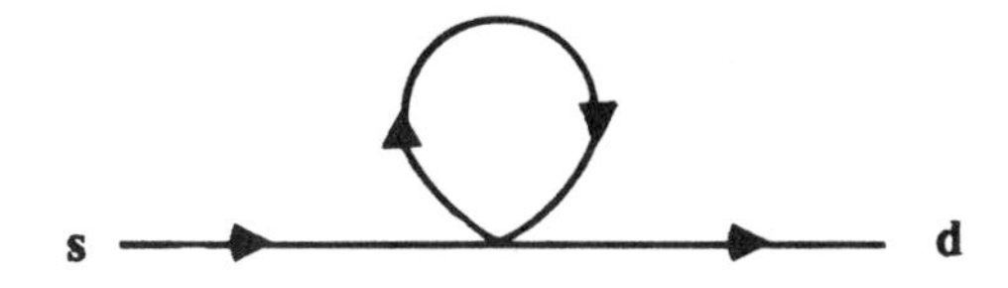

FIGURE 7
The insertion of the Wilson term in the loop of this diagram mixes the four-fermion operator with a two-fermion operator by flipping the helicity of the quarks.

where

$$\frac{\delta\tilde{O}}{\delta\varepsilon} = i\left(RO + \sum_n d^n R^n O^{(n)}\right) \tag{2.19}$$

and we have omitted flavour indices. For the appropriate choice of the coefficients $d^{(n)}$, the contact terms arising in the correlation function of $\bar{X}$ with $\tilde{O}$ will correct the complicated naive chiral variation, Eq.(2.19), in such a way as to reproduce the desired transformation property. The condition for this is:

$$\sum_x < \alpha\ |T(\bar{X}^a(x)\ \tilde{O}_\alpha(0))|\beta > + i < \alpha\,| \frac{\delta\tilde{O}_\alpha}{\delta\varepsilon^a} |\beta > =$$
$$- R^a_{\alpha,\beta} < \alpha\ |\tilde{O}_\beta|\ \beta > \tag{2.20}$$

equations (2.20) are a set of linear, inhomogeneous equation in $d^{(n)}$ which has a unique solution. The $\tilde{O}$ thus constructed is not yet finite, to $a \to 0$. Equations (2.18) and (2.20) show that the only freedom left is to take a linear superposition of operators transforming in the same way under chiral rotations. We thus define

$$\hat{O}(\mu) = Z_{LATT}\ \tilde{O} \tag{2.21}$$

Z_{LATT} being a matrix which mixes equivalent representations only, and it is such that:

$$< \alpha\ |Z_{LATT}\ \tilde{O}|\ \beta > = \text{finite} \tag{2.22}$$

This condition is consistent with the left-hand side of Eq.(2.18). In fact, if both $\hat{A}_\mu$ and $Z_{LATT}\ \tilde{O}$ have finite matrix elements and, possible divergent terms in their product are localized operators whose contribution to the left-hand side of Eq.(2.18) vanishes upon taking the four-divergence and summing over x.

To complete the programme, we must specify more precisely the normalization condition (2.22). In the chiral limit, this requires the introduction of a subtraction point μ, so that:

$$\begin{aligned} Z_{LATT} &= Z_{LATT}(\mu a, g_o) \\ \hat{O}(\mu) &= Z_{LATT}\ \tilde{O} \end{aligned} \tag{2.23}$$

$Z_{LATT}(\mu a, g_o)$ is completely determined, including the finite terms, by the requirement that $\hat{O}(\mu)$, for some value $\mu << a^{-1}$, obeys the same normalization condition as the continuum operator $O(\mu)$.

By construction, the operator $\tilde{O}$ obeys the continuum Ward identity (up to terms of order a)

$$\int dx < \alpha|\nabla_\mu\ \hat{A}^a_\mu(x)\ Z_{LATT}\ \tilde{O}|\beta > = - R^a < \alpha|Z_{LATT}\ \tilde{O}|\beta > \tag{2.24}$$

where all the matrix elements in Eq.(2.24) are finite. Equation (2.24) is equivalent to the statement that the operator $\tilde{O}$ obeys soft pion theorems. In fact, upon integration on x, only the pion pole can give a contribution to the correlation function on the left-hand side of Eq.(2.24), as illustrated in Fig.(8). Thus we obtain

$$f_\pi < \alpha + \pi^a|\tilde{O}|\beta > = R^a < \alpha|\tilde{O}|\beta > = < \alpha|\left[Q_5^a, \tilde{O}\right]|\beta > \qquad (2.25)$$

i.e., the usual soft pion theorem of the continuum theory. This also implies that we can compute the physical $k \to \pi\pi$ amplitude from the $k \to \pi$ matrix element alone. At this point, we may turn the argument around. We can forget the Ward identity in Eq.(2.24) and determine the coefficients $d^{(n)}$ in the expansion (2.17) simply by requiring that the matrix elements of $\tilde{O}$ on the lattice obey the standard soft pion theorems. The latter conditions determine the operator apart from the overall normalization matrix Z_{LATT}.

To complete the analysis, we have to accomplish two further steps (see also Sect.4 and 7):

i) identify in the various cases the operators that we need to subtract from the bare operators which are relevant to the weak Hamiltonian;
ii) give a practical set of rules to determine the coefficients $d^{(n)}$ from the matrix elements on pseudoscalar meson states of the previous operators.

A final remark is necessary. The use of perturbation theory to fix coefficient $d^{(n)}$ which are finite (i.e., the coefficients of operators of dimension six) is completely justified. On the other hand, the use of perturbation theory to fix power-divergent coefficients leads in general to wrong results.

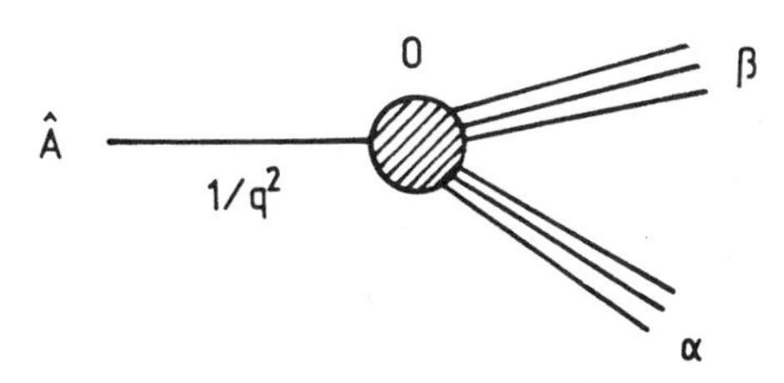

FIGURE 8
Pion pole contribution to the matrix element of the renormalized operator.

This happens because, the coefficients of these lower dimension operators diverges as inverse powers of the lattice spacing (i.e. a^{-N}) for dimensional reasons. Any non-perturbative spourious term given by these operators will then remain even for $g_o \to o$ (in fact $1/a\ e^{-1/2\beta_o g_o^2} \to$ const. as $g_o \to o$).

The mixing problem discussed above is present for all composite operators, as for example the two quark operators $\bar{\psi}\gamma^\mu D^{\nu 1} \ldots D^{\nu 2}\psi$, which are relevant in the computation of the structure functions. The mixing with lower dimensional operators poses the same problems as in the case of the four fermion operators and non perturbative subtractions are needed also in this case[11,12]. Few explicit examples will be discussed in the section on the structure functions.

3. BASIC NOTIONS ON MONTECARLO TECHNIQUES AND ON THE CALCULATION OF HADRONIC MATRIX ELEMENTS

3.1 *Two point functions, hadron masses and meson decay constants*

We start by defining operators carrying the same quantum number of the particles under study. For example, in the case of the pion and of the rho, we may use

$$\pi^+(x) = \bar{u}^A(x)\,\gamma_5 d^A(x)$$

$$x \equiv (\mathbf{x},t) \qquad (3.1)$$

$$\rho^+_\mu(x) = \bar{u}^A(x)\,\gamma_\mu d^A(x)$$

It is straightforward to compute the correlation function for these operators. In the pion case:

$$G(\mathbf{x},t) = \langle\, \pi(\mathbf{x},t)\pi^+(\mathbf{0},0)\,\rangle =$$

$$\frac{\int d[U]\,\Pi_f[\det\Delta_f(U)]\;\mathrm{tr}\,[S^u(x,0)\,\gamma_5 S^d(0,x)\,\gamma_5]\;e^{S_G(U)}}{\int d[U]\,\Pi_f[\det\Delta_f(U)]\;e^{S_G(U)}} \qquad (3.2)$$

$S_G(U)$ is the gluon action; Δ_f has been defined in Eq.(2.1). $S^{u,d}(x,0)$ is the up, down quark propagator between 0 and x:

$$\sum_z \Delta_{u,d}(x,z)\; S_{u,d}(z,o) = \delta_{x,o} \qquad (3.3)$$

The right-hand side of Eq.(3.2) is represented by the diagram in Fig.9. In a Monte Carlo simulation the integral over the gluon fields U_μ is replaced by the sum over gluon field configurations generated by some numerical algorithm (Metropolis, Langevin, ...) on a finite, generally hypercubic lattice. $\Delta(x,0)$ is inverted [see Eq.(3.3)] by some numerical technique, such as the Gauss-Seidel method, for example. All the numerical results discussed in the following have been obtained in the so-called "quenched" approximation, $\det \Delta_f(U) = 1$. The quenched approximation is expected to work reasonably well for several reasons: it is exact for $N_{colour} \to \infty$ at n_f fixed; Zweig rule is exactly satisfied and hadrons are made only by valence quarks and gluons, almost as much in the same way as it happens in real world; all the few known results for the spectroscopy which include the effects of fermion loops seem in good agreement (within ~ 15% 20%) with the results of the quenched approximation. On the other side it is clear that there are quantities that cannot be computed in this approximation, like for example the width of the $\rho \to 2\pi$. Moreover, we expect that fermion loops are very important to establish the existence and the properties of the deconfinement phase transition.

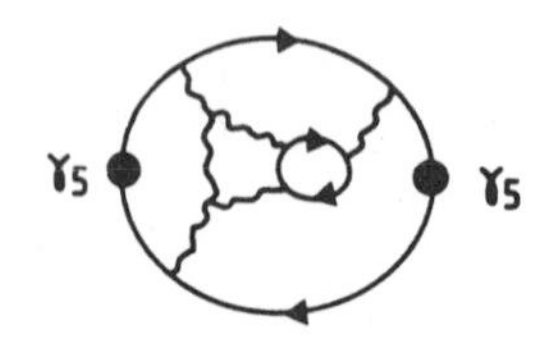

FIGURE 9
Typical diagram for the pion propagator.

We may fix the momentum by summing G(x,t) over the space components:

$$G(\mathbf{q},t) = \sum_{\mathbf{x}} G(\mathbf{x},t)\, e^{i\,\mathbf{q}\,\mathbf{x}} \tag{3.4}$$

$G(\mathbf{q},t)$ will propagate all possible intermediate states of momentum $\mathbf{q}$ carrying the same quantum numbers of the pion.

At zero momentum, considering only one-particle intermediate states, one has:

$$G(t) = \sum_{\mathbf{x}} G(\mathbf{x},t) = \sum_{n} \frac{Z_n}{2m_n}\, e^{-m_n t} \tag{3.5}$$

$m = m_\pi, m_\pi^*, \ldots$ For large time distances the correlation function is dominated by the pole corresponding to the lowest-lying state, the pion, since heavier states are exponentially suppressed as $\exp[-(m_n-m_\pi)t]$:

$$G(t) \underset{t\to\infty}{\to} \frac{Z_\pi}{2m_\pi}\, e^{-m_\pi t} \tag{3.6}$$

On the other side, if one uses eq.(3.6) when t is too small, one has a systematic overestimate of the mass of the lowest-lying state. The logarithm of G(t), as we observe it in a Monte Carlo simulation, is plotted in Fig.10 as a function of t. At large t, $\ell n[G(t)]$ goes like a straight line because only one particle is propagating [Eq.(3.6)]. The slope of the straight line corresponds to the mass of the particle in lattice units ($m_\pi a$ in this case), and the intercept in t=0 allows the determination of Z_π.

Z_π is connected to the matrix element of the operator which we have used as interpolating field:

$$Z_\pi = |\langle 0|\bar{u}^A \gamma_5\, d^A|\pi^+\rangle|^2 \tag{3.7}$$

A check of consistency of the results can be obtained by using some other operator with the same quantum numbers: different operators should lead to the same value of m_π. A possible alternative for the pion at zero momentum, is the fourth component of the axial current

$$A_o^+(x) = \bar{u}^A(x)\, \gamma^o \gamma^5\, d^A(x) \tag{3.8}$$

We may use the operators in Eqs.(3.1) and (3.8) to compute the pseudoscalar meson decay constants, which are of particular interest in the physics of D and B mesons.

Let us consider the following correlation functions:

$$\sum_{\mathbf{x}} \langle 0|A_o(\mathbf{x},t)\pi^+(\mathbf{0},0)|0\rangle \underset{t\to\infty}{\to} \frac{\langle 0|A_o|\pi\rangle\langle\pi|\pi^+|0\rangle}{2m_\pi}\, e^{-m_\pi t} \tag{3.9}$$

$$\sum_{\mathbf{x}} \langle 0|\pi(\mathbf{x},t)\pi^+(\mathbf{0},0)|0\rangle \underset{t\to\infty}{\to} \frac{\langle 0|\pi|\pi\rangle\langle\pi|\pi^+|0\rangle}{2m_\pi}\, e^{-m_\pi t} \tag{3.10}$$

$$\frac{\langle A_o\pi^+\rangle}{\langle\pi\pi^+\rangle} \underset{t\to\infty}{\to} \text{const} = \frac{f_\pi\, m_\pi}{Z_A \langle 0|\pi|\pi\rangle} \tag{3.11}$$

By fitting the propagator in Eq.(3.6), using Eq.(3.11) and taking Z_A computed on the lattice with the Ward identities (see the previous section), we can thus obtain $f_\pi(f_k, ...)$.

In Table 1 the value of the pseudoscalar meson decay constants obtained on the lattice[13,14,15] are compared with predictions from QCD sum rules[16,17] and (when known) with their experimental values.

The systematic error quoted for f_π in the Table originates from the calibration of the lattice spacing with different physical masses (e.g., the ρ or proton mass). A detailed discussion can be found in Ref.13. This error is absent if we compute the other pseudoscalar meson decay constants f_{PS} from the ratio f_{PS}/f_π. This method is equivalent to fixing the lattice spacing using f_π itself. The further advantage is that the poorly-known renormalization constant of the axial current[5,7], Z_A, cancels in the ratio. It is interesting to notice the good agreement of f_k with its experimental value. $f_{D,DS}$ agrees well with the results from QCD sum rules. The B-meson decay constant in the Table has been obtained from f_D by extrapolating with the non-relativistic formula:

$$f\sqrt{M} = \text{const.} \tag{3.12}$$

where M is the heavy meson mass. Equation (3.12) is expected to work well (up to logarithmic) corrections) for really heavy quarks but it is not clear that it can already be used for the charm quark. The value of f_B in the Table is sensibly lower than the result from QCD sum rules.

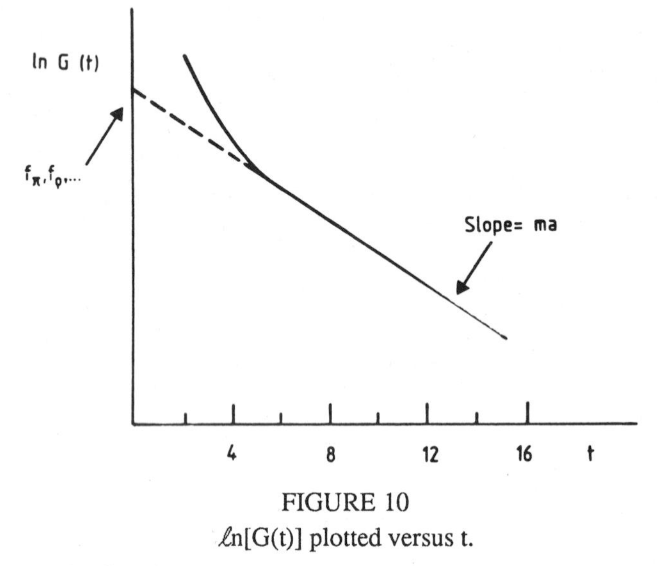

FIGURE 10
$\ell n[G(t)]$ plotted versus t.

3.2 *Three point functions*

We now consider the more complicated case of three-point functions involving two sources of pseudoscalar particles (e.g. $\bar{\psi}\gamma_5\psi$) and a local operator whose matrix elements we want to compute. Three-point correlations are needed for the calculation of the pion and proton structure function and form factors, of the kaon B-parameter, i.e. the matrix element $\langle K^o|(\bar{s}_L \gamma^\mu d_L)\ (\bar{s}_L \gamma_\mu d_L)|\bar{K}^o\rangle$, and of the K-$\pi$ matrix elements of the weak Hamiltonian.

They have the general form:

$$C(t_1, t_2) = \sum_{\mathbf{x}_1, \mathbf{x}_2} < P_5(\mathbf{x}_1, t_1) \;\; O(\mathbf{0},0) \;\; P_5(\mathbf{x}_2, t_2) > e^{i\, q_1 . x_1 + q_2 . x_2} \qquad (3.13)$$

where $P_5 = \bar{\psi}\, \gamma_5\, \psi$ and $O \sim (\bar{\psi}\Gamma\psi)\, (\bar{\psi}\Gamma\psi)$ or $\bar{\psi}\Gamma\psi$, Γ being one of the Dirac matrices.

For $\Delta S = 2$ ($\Delta T = 3/2$) transitions only the "eight" diagram in Fig.11 contributes. As shown in figure all the quark lines stem from the origin and we can use directly the quark propagators $S^{AB}_{\alpha\beta}(x,0)$ (A,B = colour; α,β = spin) used for hadron spectroscopy. The calculation of the three-point correlation for $\Delta T = 1/2$ operators is considerably more difficult because of the presence of the so called "eye" graphs of Fig.12. The same problem is present for quark bilinear operators ($\sim(\bar{\psi}\Gamma\psi)$), which enters in the form factors and structure functions as shown in Fig.13.

TABLE 1

Pseudoscalar decay constants from lattice calculations (second column), QCD sum rules (third column) and their experimental value (fourth column). The results denoted by a * have been obtained from f_{D,D_s} using Eq.(3.12).

Meson	f_{lattice}	$f_{\text{QCD-sum rules}}$	$f_{\text{experiment}}$
π	[140±20±20(syst)] MeV [13]		131 MeV
K	(158±13) MeV [13] (173±83) MeV [14]		164 MeV
D	(180±30) MeV [13] (215±60) MeV [14] (174±52) MeV [15]	(172±15) MeV [16]	< 290 MeV
D_s	(218±30) MeV [13] ~ 251 MeV [14] (234±70) MeV [15]	~ 220 MeV [16]	-
B_d	~ 120 MeV* [13] (105±34) MeV* [15]	(187±24) MeV [16]	-
B_s	~ 150 MeV* [13]		-

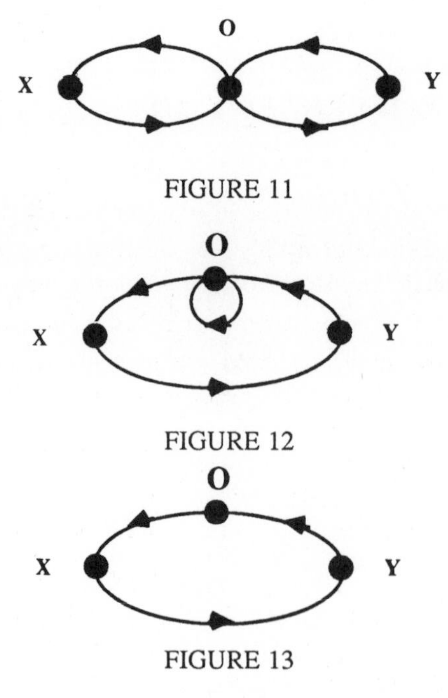

FIGURE 11

FIGURE 12

FIGURE 13

The difficulty arises because the last two cases involve propagators from any point x_1 to any point x_2 on the lattice, besides the propagator from the origin to any point x (typically the number of points is of the order of 10^4 - 10^5). The method usually adopted consists of computing a new set of quark propagators with one of the two pseudoscalar densities inserted at a fixed time (say t_2) and a fixed momentum ($\mathbf{q}_2$)[18,19]:

$$S_1(x,0) = \sum_{\substack{x_2 \\ \text{at } t_2 \text{ fixed}}} S(x, x_2)\, \gamma_5\, S(x_2, 0)\, e^{i\, q_2 . x_2} \tag{3.14}$$

$S_1(x,o)$ satisfies an equation similar to eq.(3.3):

$$\sum_z \Delta(x,z)\, S_1(z,o) = \gamma_5\, S(x,o)\, e^{i\, q.x}\, \delta_{t_x,t_2} \tag{3.15}$$

The computation of S_1 takes about the same computer time as that of S(x,0). When t_2 is large enough, the t_2 dependence is given by the meson propagator and we do not need the correlations at all t_2's. We cannot change either $\mathbf{q}_2$ but we can freely vary $\mathbf{q}_1$ and t_1. In terms of S_1 the diagrams in Fig.12 and Fig.13 reduce to two-point correlation functions as shown in Figs.14 and 15.

All the propagators again stem from the origin where the operator is located. For large time distances ($t_1 = t_k \to -\infty$; $t_2 = t_\pi \to +\infty$) the correlation in eq.(3.13) is dominated by the matrix element of the operator sandwiched between the lightest pseudoscalar mesons:

$$G(t_k, t_\pi) \to \frac{Z_K^{1/2}\; Z_\pi^{1/2}}{(2E_K)\,(2E_\pi)} < \pi(\mathbf{p})|O|K(\mathbf{k})> \exp(-E_K\, |t_k| - E_\pi\, t_\pi) \tag{3.16}$$

In eq.(3.16) we have chosen $\mathbf{q}_1 = \mathbf{k}$ and $\mathbf{q}_2 = \mathbf{p}$ and $Z_{K,\pi}$ and the meson energies $E_{K,\pi}$ can be found by studying the two-point correlation function in eq.(3.6). Eq.(3.16) is then used for the determination of $<\pi|O|K>$.

To compute $K \to 2\pi$ matrix elements we consider the four-point correlation function:

$$G(t_1, t_2, t_3) = \sum_{\mathbf{x}_1\,\mathbf{x}_2\,\mathbf{x}_3} < P_5(\mathbf{x}_1,t_1)\; O(\mathbf{0},0)\; P_5(\mathbf{x}_2,t_2)\; P_5(\mathbf{x}_3,t_3) > \qquad (3.17)$$

For simplicity we have put to zero all spatial momenta. All the formulae will refer to $m_K^2 = m_\pi^2 = m^2$ and $Z_{K,\pi} = Z_5$ in the following. The calculation of the four point correlation in eq.(3.17) requires the use of at least one S_1 (for "eight" diagrams) or two S_1's (for "eye" diagrams) as explained in the two examples reported in Fig.16 and Fig.17 but it does not present any new problem.

At large time distances ($t_1 = t_k \to -\infty$; $t_{2,3} = t_{\pi1,\pi2} \to +\infty$ and $t_{\pi1} \neq t_{\pi2}$) one has:

$$G(t_K, t_{\pi1}, t_{\pi2}) = \frac{Z_5^{3/2}}{(2m)^3} <\pi_1\,\pi_2|O|K> \exp[-m(|t_K| + t_{\pi1} + t_{\pi2})]$$

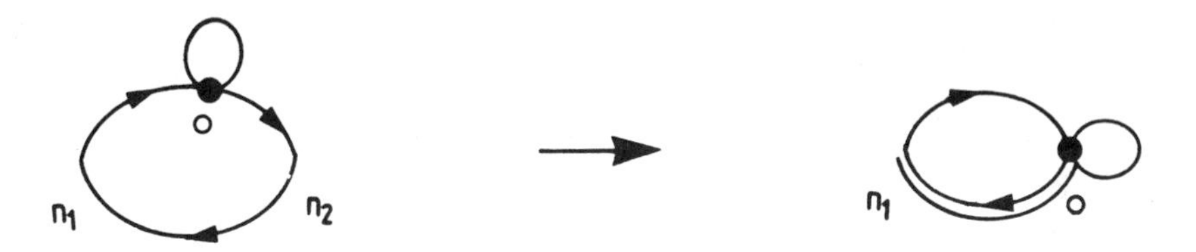

FIGURE 14

FIGURE 15

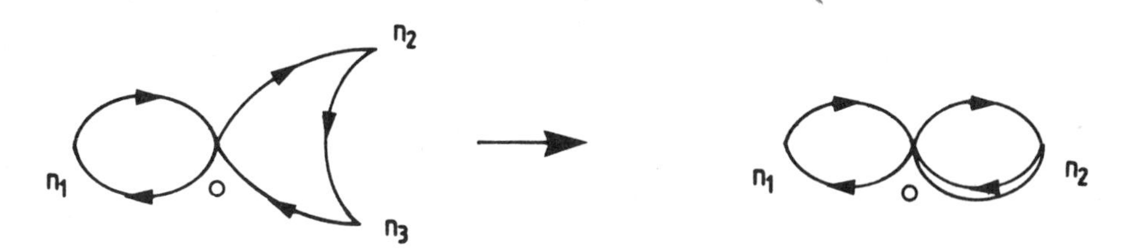

FIGURE 16

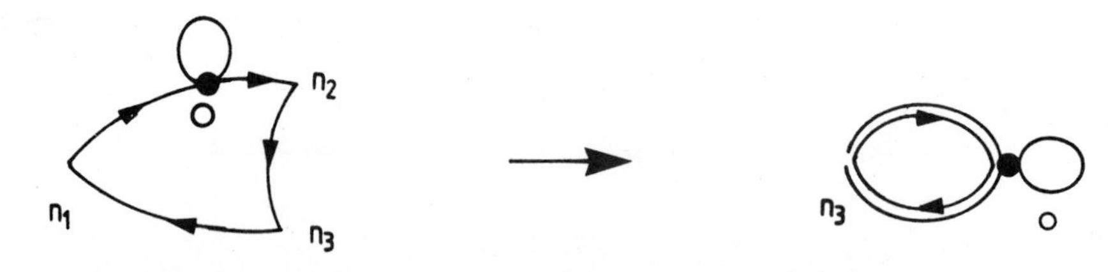

FIGURE 17

Thus from the knowledge of Z_5 and m one is able to compute the $K \to 2\pi$ matrix elements.

It should be noticed that we expect large finite size effects when there are two pions in the final state. In fact, on a relatively small box, like the lattices which are currently used, the two pion distributions tend to overlap so that their final state interaction is larger than in the real world. For $<K|O|\pi>$ matrix elements, electromagnetic form factors and structure functions, finite size effects are extimated (and in some cases were found) always very small in comparison to other more serious systematic effects, such as the "quenched" approximation.

3.3 *Nucleon Correlation Functions and Operator Matrix Elemens*

The nucleon case is slightly more complicated because of the spinor structure. In this case, the matrix elements are obtained by computing the Euclidean three-point correlation function

$$C(t_x, t_y) = \int d^3x\, d^3y <0|T[J_\gamma(0)\ O(y)\ \bar{J}_{\gamma'}(x)]|0> e^{i\,\mathbf{p}\cdot\mathbf{x}}\ e^{i\,\mathbf{q}\cdot\mathbf{y}} \tag{3.18}$$

where J is an interpolating operator for the nucleon (which, for definiteness, for the rest of this section we assume to be a proton), γ and γ' are spinor labels, t_x and t_y are the time components of x and y and we integrate only over the space coordinates. Inserting a complete set of states between each pair of operators in eq.(3.18) and taking $0<t_y<t_x$, with t_y and t_x-t_y sufficiently large so that only the lightest state contributes significantly, we have

$$C(t_x, t_y) = -\frac{e^{-E't_y}\ e^{-E(t_x-t_y)}}{4EE'}$$
$$\sum_{s,s'} <0|J_\gamma(0)|p,s>\ <p,s|O(0)|p',s'>\ <p',s'|\bar{J}_{\gamma'}(0)|0> \tag{3.19}$$

where $|p,s>$ represents a proton states with four-momentum p and a third component of spin s. p and p' are given by

$$p = (E, \mathbf{p}) \qquad\qquad E = \sqrt{\mathbf{p}^2 + m^2}$$

and

$$p' = (E', \mathbf{p} + \mathbf{q}) \qquad\qquad E' = \sqrt{(\mathbf{p} + \mathbf{q})^2 + m^2} \tag{3.20}$$

where m is the mass of the proton. The required matrix element $<p,s|O(0)|p',s'>$ is one of the factors on the right-hand side of eq.(3.19), while the remaining factors can be obtained by computing two-point correlation functions, as will be explained below.

For the proton we may use the following interpolating operator:

$$J_\gamma = \epsilon^{ijk} (u^i C\gamma^5 d^j) u^k_\gamma \tag{3.21}$$

where i,j,k are colour labels and C is the charge conjugation matrix. This isospin-$\frac{1}{2}$ interpolating operator has been shown in several lattice studies of hadron spectroscopy to have a significant overlap with the proton. Using Lorentz- and parity invariance, we can write

$$<0|J_\alpha(0)|p,s> = \sqrt{Z_p}\ \ N^{(s)}_\alpha(p) \tag{3.22}$$

where Z_p is a scalar quantity and $N^{(s)}_\alpha(p)$ is the proton's spinor, which satisfies the Dirac equation. Z_p and m can be obtained from the two-point Euclidean correlation function

$$K(t_x) = \int d^3x <0|T[J_\beta(o)\ \bar{J}_\alpha(x)]|0> e^{i\,\mathbf{p}.\mathbf{x}} \tag{3.23}$$

For $\mathbf{p} = 0$, inserting a complete set of states between the interpolating operators J and taking t_x to be sufficiently large so that the single proton state is the dominant one, we have:

$$K(t_x) = Z_p\ \frac{(1+\gamma^o)}{2}\ {}_{\beta\alpha}\ e^{-\,mt_x} \tag{3.24}$$

We have neglected in (3.24) the contributions from heavier states, in particular that from the opposite parity baryon, $N^*(\frac{1}{2}-)$, which is equal to

$$K(t_x) = Z_{N^*}\ \frac{(1-\gamma^o)}{2}\ {}_{\beta\alpha}\ e^{-\,m't_x}$$

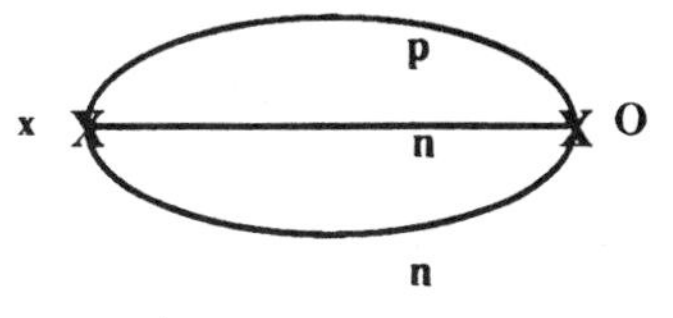

FIGURE 18

One can further suppress its contribution by an appropriate choice of spinor indices (e.g., $\alpha = \beta = 1$ or $\alpha = \beta = 2$). Thus from the behaviour of $K(t_x)$ with t_x, both m and Z_p can be determined and from Eq.(3.19) we can extract the matrix elements <p,s|O|p',s'>.

The evaluation of $K(t_x)$ and $C(t_x, t_y)$ requires the computation of the diagrams in Fig.18 and Fig.19 respectively.

In these figures, the propagators are quark propagators in a gluon background field. The crosses denote the insertion of the interpolating operator (x) and the shaded circles that of the operator whose matrix element we wish to compute*.

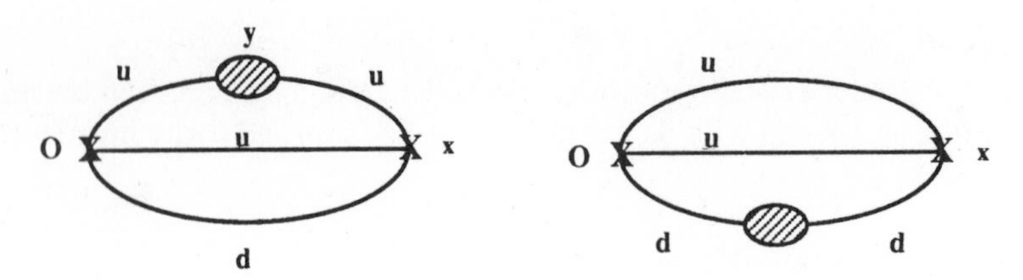

FIGURE 19

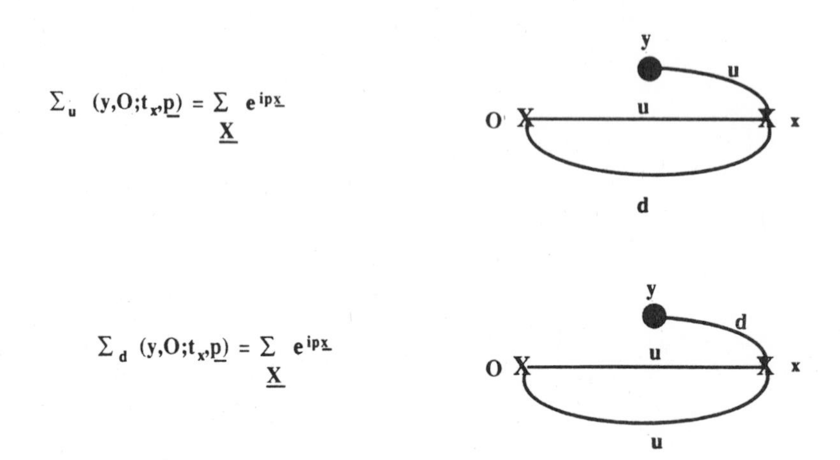

FIGURE 20

For the diagram in Fig.18 it is sufficient to have the set of quark propagators from an arbitrary lattice point to the origin S(x,o). For simplicity we keep the up and down quark masses equal and hence the propagators S(x,o) are identical for both flavours. In order to compute the diagrams in Fig.19 we need the sets of generalized propagators Σ_d and Σ_u (in analogy with the pion case) defined by (see Fig.20)[20]

$$\left[\Sigma_d\,(o,y;\,t_x,\,\mathbf{p})\right]^{XY}_{\rho\sigma} = -\,\epsilon^{ABC}\,\epsilon^{A'XC'}\,(C\gamma_5)_{\alpha\beta}\,(C\gamma_5)_{\alpha'\rho}$$

$$\sum_{x}\left\{ e^{i\,\mathbf{p}.x}\; S^{BY}_{\beta\sigma}(x,y)\,[S^{AA'}_{\alpha\alpha'}(x,o)\,S^{CC'}_{\gamma\gamma'}(x,o) - S^{AC'}_{\alpha\gamma'}(x,o)\,S^{CA'}_{\gamma\alpha'}(x,o)]\right\} \qquad (3.25)$$

---

* I do not consider here the diagrams relevant for hyperon decays but only those which are relevant for form factors and structure functions.

and

$$[\Sigma_u (o,y;\, t_x,\, \mathbf{p})]^{XY}_{\rho\sigma} = - \epsilon^{ABC} (C\gamma^5)_{\alpha\beta} \sum_x e^{i\, \mathbf{p}.\mathbf{x}} \; S^{BB'}_{\beta\beta'}(x,o)$$

$$\{\epsilon^{XB'C'} (C\gamma^5)_{\rho\beta'} [S^{AY}_{\alpha\sigma}(x,y)\, S^{CC'}_{\gamma\gamma'}(x,o) - S^{AC'}_{\alpha\gamma'}(x,o)\, S^{CY'}_{\gamma\sigma'}(x,y)] \tag{3.26}$$

$$+ \epsilon^{A'B'X} (C\gamma^5)_{\alpha'\beta'}\, \delta_{\gamma\rho'} [S^{AA'}_{\alpha\alpha'}(x,o)\, S^{CY}_{\gamma\sigma}(x,y) - S^{AY}_{\alpha\sigma}(x,y)\, S^{CA}_{\gamma\alpha}(x,o)]\}$$

The upper case Latin superscripts and the lower case Greek subscripts (both primed and unprimed) are colour and spinor labels respectively. The interpolating operators for the proton were chosen to be $\bar{J}_{\gamma'}(o)$ and $J_\gamma(x)$ [with J defined in Eq.(3.21)], so that the spinor indices γ and γ' are external labels. There is a different Σ_d and Σ_u for each pair (γ, γ').

To illustrate the sifnificance of the Σ_u's and Σ_d's a, consider the matrix element $<p|\bar{\psi}\Gamma\psi|p>$ where ψ can be an up or down quark and Γ is one of the 16 matrices of Dirac theory. The three-point correlation function eq.(3.18) corresponding to the operator $\bar{\psi}\Gamma\psi$, computed in the quenched approximation, is the weighted average over all configurations of

$$\sum_y \text{tr}\, [\Sigma_\psi\; (o,y;\; t_x,\, \mathbf{p})\, \Gamma\;\, S(y,o)]\; e^{i\, \mathbf{q}.\mathbf{y}} \tag{3.27}$$

where ψ = u or d. Thus the set $\{\Sigma_\psi(o,y;\, t_x,\, \mathbf{p}),\, \psi = u \text{ or } d\}$, together with the set of quark propagators S(y,o), is sufficient to compute the matrix elements of a large class of local operators (including operators containing covariant derivatives).

The Σ_d's and Σ_u's satisfy equations which can be solved using the same numerical techniques as those used to generate the usual quark propagators S(x,o). For example, Σ_d satisfies

$$\sum_y \Sigma_d\, (o,y;\; t_x,\, \mathbf{p})^{XY}_{\rho\sigma}\; \Delta^{YZ}_{\sigma\tau}(y,z) = - e^{i\, \mathbf{p}.\mathbf{z}}\; \epsilon^{AZC}\, \epsilon^{A'XC'}\, (C\gamma_5)_{\alpha\tau}\, (C\gamma_5)_{\alpha'\rho}\;\; \delta_{t_x,t_z}\, \cdot$$

$$[S^{AA'}_{\alpha\alpha'}(z,o)\, S^{CC'}_{\gamma\gamma'}(z,o) - S^{AC'}_{\alpha\gamma'}(z,o)\, S^{CA'}_{\gamma\alpha'}(z,o)] \tag{3.28}$$

where Δ is as before the kernel of the fermionic action.

Each set $\{\Sigma_u(o,y;\, t_x,\, \mathbf{p})\}$ or $\{\Sigma_d(o,y;\, t_x,\, \mathbf{p})\}$ depends on t_x, on $\mathbf{p}$ and on the choice of spinor indices, γ and γ', in the interpolating operators J_γ, $\bar{J}_{\gamma'}$ (this dependence on γ and γ' is implicit throughout the above discussion). Therefore, we have to choose each of these parameters in a way which will optimized the amount of information obtainable.

In particular the choice of the spinor indices γ and γ' depends on the physics one wishes to study. To sum over the two cases $\gamma = \gamma' = 1$ and $\gamma = \gamma' = 2$, enables us to compute the low moments of the unpolarized deep inelastic structure functions, the electric form factor of the proton at several values of momentum transfer and the sigma term. In order to compute other interesting quantities, such as the matrix elements of the axial vector current or the magnetic form factor of the nucleon or the recently-measured lowest moment of the polarized structure function g_1[21], other choices of γ and γ' must be made ($\gamma = \gamma' = 1$ minus $\gamma = \gamma' = 2$).

4. THE PION ELECTROMAGNETIC FORM FACTOR AND STRUCTURE FUNCTION

Structure functions have played an important role in our understanding of the parton model and of quantum chromodynamics. They are measure in deep inelastic experiments by probing the hadron with a photon (W or Z) as shown in Figs.21 and 22. In the parton model, the photon interacts directly with the quarks inside the hadron and measures the probability $q(x,Q^2)$ of finding a parton with a fraction x of the hadron momentum and a momentum transfer $-q^2 = Q^2$ (Fig.21). Scaling violations allow us to measure the gluon distribution $G(x,Q^2)$ as shown in Fig.22.

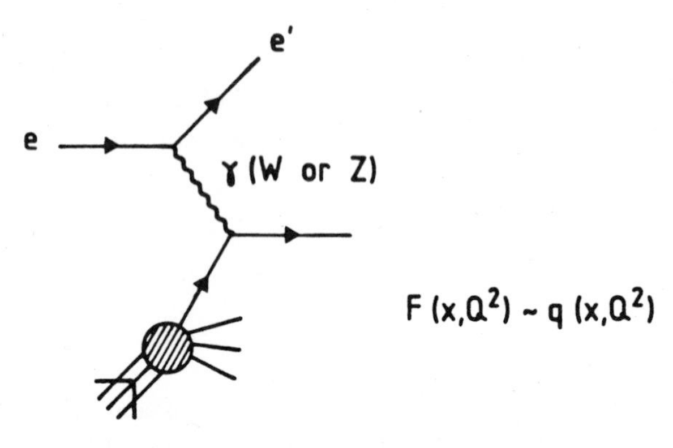

FIGURE 21
The structure function $F(x,Q^2)$ is proportional to the probability of finding a quark with a fraction x of the hadron momentum at $-q^2 = Q^2$, $q(x,Q^2)$.

In principle it is possible to compute directly the structure functions on the lattice through the diagram in Fig.23. This, however, requires that the distance between the two currents satisfies the condition:

$$1/m_h >> z >> a \qquad (4.1)$$

in order to be in the inelastic region ($1/m_h >> z$) and to avoid lattice artefacts (present when $z/a \sim 1$).

To satisfy Eq.(4.1), one needs an inverse lattice spacing of several GeV, and hence a very big lattice in order to contain the hadronic system under study. Such a calculation is not practicable at present. However, using the operator product expansion and the renormalization group one can express the T-product of two currents as a combination of local operator (Fig.24). The problem is thus reduced to the computation of matrix elements of local operators between hadronic states.

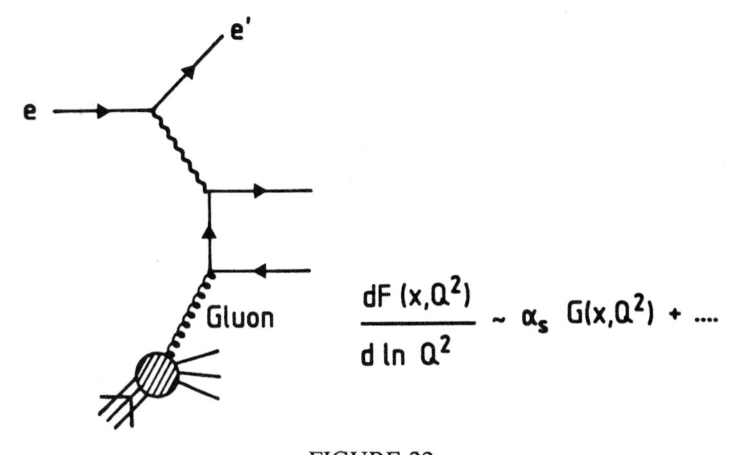

FIGURE 22
Scaling violations allow us to measure the gluon density $G(x,Q^2)$.

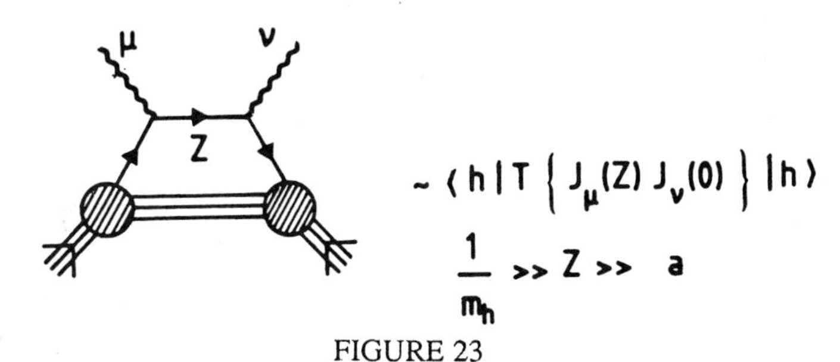

FIGURE 23
Forward Compton scattering which allow us to measure the structure functions.

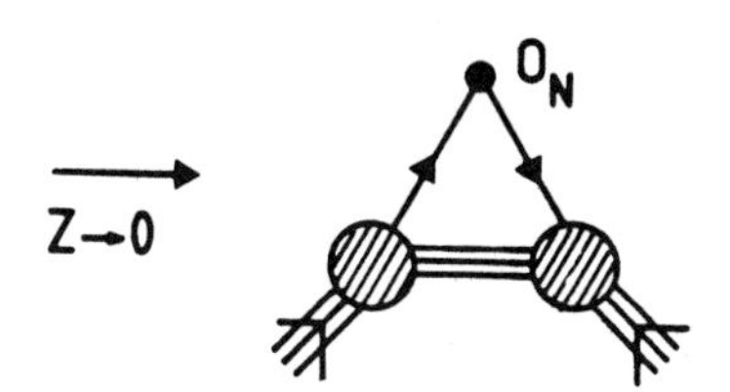

FIGURE 24
Operator product expansion of the diagram in Fig.23.

In the non-singlet case (valence quarks only) and keeping only the contribution of lowest twist operators, the moments of the structure function can be written as:

$$\int_0^1 dx\, x^{N-2}\, F(x,Q^2) = \int_0^1 dx\, x^{N-1}\, q(x,Q^2) = C_N(Q^2/\mu^2, g(\mu))\, A_N(\mu) \qquad (4.2)$$

where $A_N(\mu)$ are obtained from the matrix elements of the operators:

$$O_N^{\mu_1 \ldots \mu_N} = i^N \bar{\psi} \gamma^{\mu_1} D^{\mu_2} \cdots D^{\mu_N} \psi - \text{traces} \tag{4.3}$$

(with suitable flavour indices and symmetrized over Lorentz indices) using:

$$\langle h(p)| O_N^{\mu_1 \ldots \mu_N} |h(p)\rangle = p^{\mu_1} p^{\mu_2} \ldots p^{\mu_N} A_N \tag{4.4}$$

In Eq.(4.2), μ is the renormalization scale of the operator. Since the left-hand side is independent of μ, the relation in Eq.(4.2) gives the renormalization group equation to which the coefficients C_N must obey. Provided $\mu >> \Lambda_{QCD}$, the coefficient functions can be computed in perturbation theory. The non-perturbative physics is contained in the operator matrix elements [Eq.(4.4)]. We can compute them on the lattice.

4.1. *Renormalization of the Relevant Operators*

Following the discussion of Sect.2 we now discuss the renormalization properties of the operators which are relevant for the calculation of the electromagnetic form factor and for the structure functions.

To determine the electromagnetic form factor of the pion, we have to compute the matrix elements of the vector current. We can consider for example the local vector current:

$$V_\mu^{loc}(x) = \bar{q}(x)\, \gamma_\mu q(x) \tag{4.5}$$

As discussed in Section 2. the matrix elements of the conserved current $\hat{V}_\mu$ and V_μ^{loc} between on-shell states are related throught the equation

$$< \alpha |\hat{V}_\mu| \beta > = Z_V^{loc}(g_o) < \alpha |V_V^{loc}| \beta > \tag{4.6}$$

The renormalization constant Z_V^{loc} is a finite function of the bare lattice coupling g_o and can be computed in perturbation theory. On the lattice we can measure these constants non-perturbatively by comparing the matrix elements of V_μ^{loc} with those of $\hat{V}_\mu$.

We can compute for example Z_V^{loc} from the proton matrix elements of the vector currents:

$$Z_V^{loc} = \frac{\langle p|\hat{V}_\mu(o)|p\rangle}{\langle p|V_\mu^{loc}(o)|p\rangle} \tag{4.7}$$

In ref.20 it was found

$$Z_V^{loc} = 0.74 \pm 0.02 \tag{4.8}$$

For sufficiently small lattice spacing Z_V^{loc} is independent of the states between which the conserved and local vector currents are sandwiched and can be calculated in perturbation theory. The result reported in (4.8) and is similar to those obtained with the currents sandwiched between pion states, $Z_V^{loc} \sim 0.71$[12]. They differ, however, from the one-loop perturbative result (0.83)[22,23,24] and from the value obtained with the currents sandwiched between the rho meson and the vacuum (0.57)[9]. These discrepancies are certainly due to the presence of terms of order a (a = lattice spacing) which arise in the renormalization of the local current and provide an indication of some of the systematic uncertainties these calculations at finite lattice spacing.

In order to calculate $\langle x \rangle$, the average momentum carried by the valence quarks in the pion, we need to compute the matrix element

$$<\pi(p)|O_2^{\mu\nu}(0)|\pi(p)> = p^\mu p^\nu A_2(p) \tag{4.9}$$

where $|\pi(p)>$ is a pion state with momentum p and

$$O_2^{\mu\nu}(0) = i^2 \bar{\psi} \gamma^\mu \overleftrightarrow{D}^\nu \psi - \text{traces} \tag{4.10}$$

(with suitable flavour indices and symmetrized over the Lorentz indices). I now discuss the two cases $\mu = \nu$ and $\mu \neq \nu$ in turn.

The three linearly-independent operators with $\mu = \nu$ transform like one of the three-dimensional irreducible representation of the hypercubic group (where the character table for this group can be found in ref.25). The subtraction of the trace removes the singlet component (which is proportional to $(\bar{\psi}\psi/a)$) and hence eliminates the power divergence. Such subtractions have, led to significant numerical cancellations and hence to large statistical errors[11].

The subtraction of power divergences can be avoided by choosing μ to be different from ν. The six operators $O_2^{\mu\nu}$ with $\mu \neq \nu$ transform like the six-dimensional irreducible representation of the hypercubic group, and for this reason there is no mixing with lower-dimensional operators. In addition, there is no numerical subtraction of divergent terms in the calculation itself, and hence the choice $\mu \neq \nu$ is likely to give results with a smaller statistical error. The price to be paid for choosing to have $\mu \neq \nu$ is that the pion must be given a non-zero component of momentum in at least one spatial direction, otherwise the matrix element in eq.(4.9) vanishes. Finally, as was also the case for the vector current, a finite multiplicative renormalization is needed to relate the lattice operator to the corresponding continuum operator defined in a given renormalization scheme. For the $\overline{\text{MS}}$ scheme, this renormalization is given by

$$<\pi(p)|O_2^{\mu\nu}(0)|\pi(p)>_{\overline{MS}} = <\pi(p)|O_2^{\mu\nu}|\pi(p)>_{latt} (1 + \alpha_S C_F \delta/4\pi) \tag{4.11}$$

where $\delta = -2.5$ for $\mu = \nu$ and $\delta = -1.2$ for $\mu \neq \nu$ [26]. At $\beta \sim 6.0$ (g_o=1) the perturbative corrections are negligible and the matrix elements in the two schemes can be taken to be equal.

The average fraction of momentum carried by gluons in a pion can be evaluated from the matrix element

$$<\pi(p)|F^{\mu\rho}(0)\, F^{\nu\rho}(0) - \text{traces}\, |\pi(p)> \tag{4.12}$$

where $F_{\mu\nu}$ is the field-strength tensor. To avoid subtractions of power divergences, we may take $\mu \neq \nu$.

We define the lattice field-strength tensor to be (see also ref.25)

$$F^{\mu\nu}(x) = \frac{1}{4a^2} \sum_{4 \text{ plaquettes}} \frac{1}{2i}(U - U^+) \qquad (4.13)$$

where the sum extends over the four plaquettes in the $\mu\nu$ plane which have the point x at one corner. The matrices U are the products of the four link variables for each plaquette, taken in an anticlockwise sense. With this definition of $F^{\mu\nu}$, no power divergences are encountered when evaluating the matrix element of eq.(4.12). This is no longer true if other definitions of $F^{\mu\nu}$ are used. The essential point is that $F^{\mu\nu}$, as defined in eq.(4.13), transforms like the direct sum of two three-dimensional irreducible representations of the hypercubic group (**E** + **B** and **E** – **B**, each of which transform irreducibly). If one does not symmetrize over the four plaquettes, then this is not the case. Suppose, for example, that we define $F^{\mu\nu}$ using just one of the plaquettes (this is an equally good definition at the tree level). At the quantum level, the matrix element in eq.(4.12) will diverge as inverse powers of the lattice spacing. In fact, $F^{\mu\nu}$ defined by means of a single plaquette transforms like a 24-dimensional reducible representation of the hypercubic group and $\{F^{\mu\rho} F^{\nu\rho}, \mu \neq \nu\}$ like a 192-dimensional reducible representation. This 192-dimensional representation contains the singlet, so that $F^{\mu\rho} F^{\nu\rho}$ can mix with the identity, the mixing coefficients containing power divergences. By using the definition (4.13) we automatically avoid this problem.

The evaluation of the second moment of the momentum distribution of the valence partons requires the calculation of the matrix element

$$<\pi(p)|O_3^{\mu\nu\rho}(0)|\pi(p)> = p^\mu p^\nu p^\rho A_3(p) \qquad (4.14)$$

where

$$O_3^{\mu\nu\rho}(0) = i^3 \bar{\psi} \gamma^\mu \overleftrightarrow{D}^\nu \overleftrightarrow{D}^\rho \psi - \text{traces} \qquad (4.15)$$

(with suitable flavour indices and symmetrized over the Lorentz indices).

The transformation properties of $O_3^{\mu\nu\rho}$ under the hypercubic group have been discussed in detail in refs.[11]. Taking $\mu = \nu = \rho$, we find that $O_3^{\mu\mu\mu}$ transforms like the four-dimensional vector representation ($(\frac{1}{2}, \frac{1}{2})$ in the notation of ref.25), and hence can mix with lower dimensional operators such as $\bar{\psi} \gamma^\mu \psi$. One can therefore not study the matrix elements of $O_3^{\mu\mu\mu}$, but rather of $O^{411} - O^{433}$ (which transforms like the eight-dimensional irreducible representation). To have a non-vanishing matrix element we shall give the pion a component of momentum in the 1 direction. The subtraction in $O^{411} - O^{433}$ involves a subtraction of power divergences, and hence we expect our results for $\langle x^2 \rangle$ to have larger statistical errors than for $\langle x \rangle$.

The operator O^{412}, which transforms like the $\overline{(\frac{1}{2}, \frac{1}{2})}$ four-dimensional irreducible representation (in the notation of ref.25) is also free of power divergences. In this case, however, we must give the pion non-zero components of momentum in the 1 and 2 directions, increasing the systematic errors which grow as $a|\mathbf{p}|$ increases [the granularity of the lattice becomes apparent once $a|\mathbf{p}|$ is O(1); see also below]. Also in this case a renormalization factor is needed to relate the lattice to the continuum result:

$$<\pi(p)|O_3^{\mu\nu\rho}(0)|\pi(p)>_{\overline{MS}} = <\pi(p)|O_3^{\mu\nu\rho}|\pi(p)>_{latt}\,(1+\alpha_S\, C_F\, \delta'/4\pi) \tag{4.16}$$

For the quantities which are obtained from matrix elements of the form $<\pi|O|\pi>$ (e.g., $\langle x\rangle$, $\langle x^2\rangle$ and the electromagnetic form factor), the optimal use of computing resources requires us to have one of the two pion states at the same momentum in all cases. In view of the above discussion, for one of the two pion states we may chose $p_1 \neq 0$ and $p_2 = p_3 = 0$. This means that for the calculation of $\langle x^2\rangle$ we have to use the operator $O_3^{411} - O_3^{433}$, while all the other matrix elements do not require any subtractions.

4.2 The electromagnetic form factor of the pion

For the computation of the electromagnetic form factor of the pion we start with the three-point correlation function $C(t_x, t_y)$ of eq.(3.13) generalized to nonzero momentum transfers

$$C(t_x, t_y) = \sum_{x,y} e^{ip.x}\, e^{iq.y} <0|T[J_\pi^+(0)\, V_\mu(y)\, J_\pi(x)]|0> \tag{4.17}$$

where V_μ is the electromagnetic current (e.g., the conserved or local current defined in Sect.2) and J_π is the pion source (e.g. P_5). For sufficiently large values of t_y and $t_x - t_y$, eq.(4.17) reduces to

$$C(t_x, t_y) = \frac{e^{-E(t_x-t_y)}\, e^{-E't_y}}{4EE'} |<\pi|J_\pi(0)|0>|^2 \cdot Z_V^{-1} \cdot (p+p')_\mu\, F(q^2) \tag{4.18}$$

where $E = \sqrt{m_\pi^2 + \mathbf{p}^2}$, $E' = \sqrt{m_\pi^2 + (\mathbf{p} + \mathbf{q})^2}$ and p' is the four-vector $(E', \mathbf{p} + \mathbf{q})$. $F(q^2)$ is the electromagnetic form factor and Z_V is the renormalization constant for the electromagnetic current being used.

The results for $F(q^2)$ with $q^2 \neq 0$, using the local vector current $\bar{\psi}\, \gamma^\mu\, \psi$ with $\mu = 1$ or 4 are presented in the following.

In Fig.25 the form pion factor calculated is on 15 gauge field configurations on a $20\text{x}10^2\text{x}40$ lattice at $\beta = 6/g_o^2 = 6$ ($a^{-1} \sim 2$ GeV) is reported for a pion with a mass of ~ 1 GeV ($m_\pi\, a \sim 0.52$). For reference, the expression (suggested by the vector dominance model) $1/(1+(-q^2)/m_\rho^2)$, where m_ρ is the mass of the vector meson at this value of the pion mass, is also plotted. The results are encouraging, the form factor falls with q^2 and is "measured" reasonably accurately for at least four of the five momentum transfers.

Of course we are really interested in the form factor of a real pion, and not the one which has a mass of about 1 GeV. We must, therefore, extrapolate to the chiral limit.

In Fig.26 the results, extrapolated to the physical pion mass, are plotted versus $\sqrt{-q^2}$ using $a^{-1} = 2.2$ GeV. We also plot the results from the NA7[27] and Cornell[28] experiments*. The results (in particular for the two most reliable points $\mathbf{p}'=\mathbf{0}$ and $\mathbf{p}'=-\mathbf{p}$) are quite satisfactory.

---

* The Cornell experiments determine the pion's form factor indirectly, by studying the electroproduction processes $ep \to e\pi^+ n$ and $en \to e\pi^- p$.

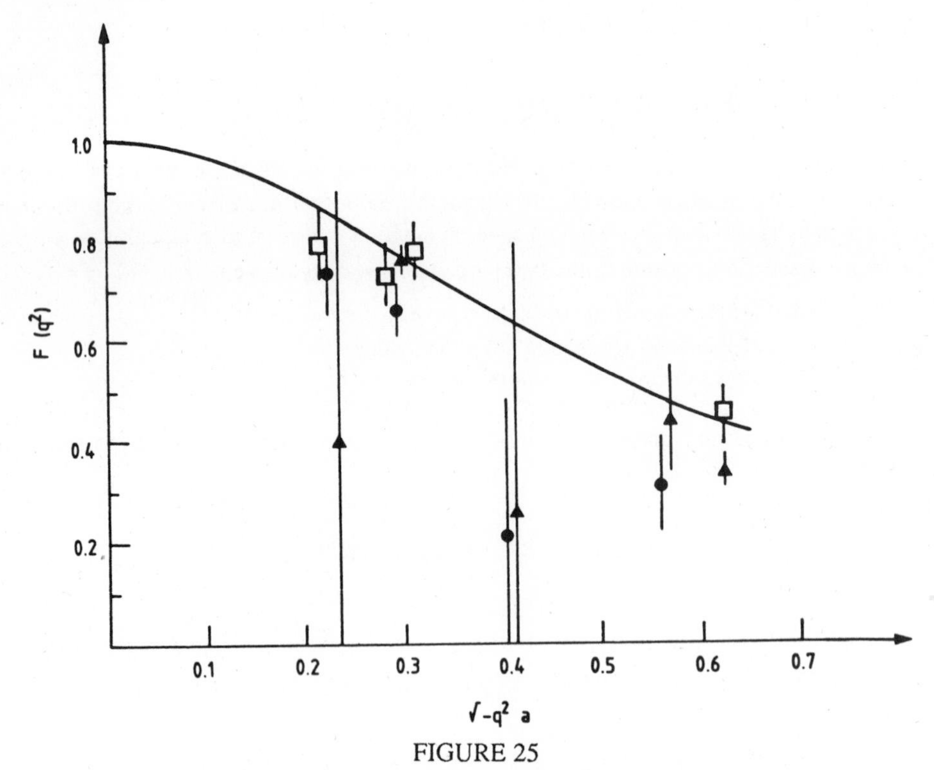

FIGURE 25

The pion's form factor at m_π a=0.52 at β=6 on a $20x10^2x40$ lattice. The curve is the result suggested by the vector dominance model $1/(1+(-q^2)/m_\rho^2)$. The circles and triangles signify that the Lorentz index was chosen to be 1 or 4 respectively. Few points, obtained by using the conserved vector current $\hat{V}_\mu$, are given in the figures as squares.

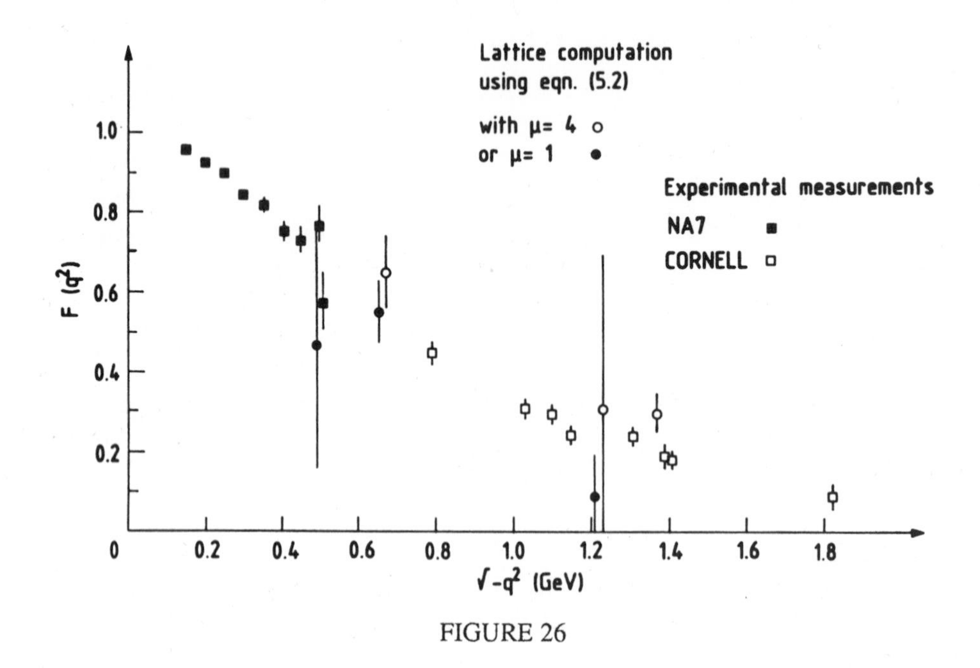

FIGURE 26

4.3. *The deep inelastic structure function of the pion*

We define $\langle x\rangle_{latt}$ and $\langle x^2\rangle_{latt}$ in terms of the three-point correlation functions (3.13) as follows

$$C^{(2)}(t_x, t_y) = \sum_{\mathbf{x},\mathbf{y}} e^{i\mathbf{p}.\mathbf{x}} <0|T[J_\pi^+(\mathbf{0},0)\ O_2^{41}(\mathbf{y},t_y)\ J_\pi(\mathbf{x},t_x)]|0> \rightarrow$$

$$|<\pi|J_\pi(0)|0>|^2\ \frac{p\langle x\rangle_{latt}}{2E}\ e^{-Et_x} \tag{4.19}$$

$$C^{(3)}(t_x, t_y) = \sum_{\mathbf{x},\mathbf{y}} e^{i\mathbf{p}.\mathbf{x}} <0|T[J_\pi^+(\mathbf{0},0)\ \tilde{O}_3(\mathbf{y},t_y)\ J_\pi(\mathbf{x},t_x)]|0> \rightarrow$$

$$|<\pi|J_\pi(0)|0>|^2\ \frac{p^2\langle x^2\rangle_{latt}}{E}\ e^{-Et_x} \tag{4.20}$$

where $\tilde{O}_3 = O_3^{411} - \frac{1}{2}(O_3^{422} + O_3^{433})$, $\mathbf{p} \equiv (p,0,0)$ and $E = \sqrt{m_\pi^2 + p^2}$.

Let $C^{(2)}(t_x, t_y)$, $C^{(3)}(t_x, t_y)$ and $C^{(\hat{V})}(t_x, t_y)$ be the three-point correlation functions corresponding to the operators O_2^{41}, $\tilde{O}_3$ and the conserved vector current $\hat{V}^1$ (with Lorentz index $\mu = 1$), respectively. Then

$$\frac{C^{(2)}(t_x, t_y)}{C^{(\hat{V})}(t_x, t_y)} = 2\,E\langle x\rangle_{latt}, \qquad t_x - t_y \text{ large, } t_y \text{ large,} \tag{4.21}$$

$$\frac{C^{(3)}(t_x, t_y)}{C^{(\hat{V})}(t_x, t_y)} = 4pE\langle x^2\rangle_{latt}, \qquad t_x - t_y \text{ large, } t_y \text{ large,} \tag{4.22}$$

Using (4.21) and (4.22) at $\beta = 6$, on a $20x10^2x40$ lattice it was found[12]:

$$\begin{aligned} \langle x\rangle_{latt} &= 0.60 \pm 0.08 \\ \langle x^2\rangle_{latt} &= 0.23 \pm 0.06 \end{aligned} \tag{4.23}$$

Most of the error in eqs.(4.23) is due to the linear extrapolation to the chiral limit. Using perturbation theory to rewrite these results in continuum schemes for which the coefficient function is close to 1 (such as the $\overline{MS}$ scheme), and at a renormalization scale $\mu \sim 7$ GeV. From eqs.(4.23) one finds:

$$\begin{aligned} \langle x\rangle_{cont} &= 0.46 \pm 0.07 \\ \langle x^2\rangle_{cont} &= 0.18 \pm 0.05 \end{aligned} \tag{4.24}$$

The above result for $\langle x\rangle$ well agrees with the experimental results:

$$\begin{aligned} &\langle x\rangle_{exp} = 0.4 \pm 0.1, && \text{ref.[29],} \\ &\langle x\rangle_{exp} = 0.35 \pm 0.05, && \text{ref.[30],} \\ &\langle x\rangle_{exp} = 0.4 - 0.5, && \text{ref.[31],} \\ &\langle x\rangle_{exp} = \sim 0.34, && \text{ref.[32],} \end{aligned} \tag{4.25}$$

and $\langle x^2\rangle_{cont}$ is well compatible with the experimental observation that the distribution function $V_\pi(x) \sim (1-x)^\alpha$, as $x \to 1$ with the exponent α being close to 1.

5. THE NUCLEON STRUCTURE FUNCTION AND ELECTROMAGNETIC FORM FACTOR

5.1. *Deep inelastic structure functions*

We now come to the computation of the first two moments of the non-singlet structure function of the nucleons (for unpolarized scattering) by using O_2^{41} and $\tilde{O}_3 = O_3^{411} - \frac{1}{2}(O^{422} + O^{433})$ as for the pion case.

The insertion of the operators O_2 and $\tilde{O}_3$ in Eq.(3.18) gives

$$C_{SF}^{(2)}(t_x, t_y) = \sum_{\gamma=\gamma'=1}^{2} \sum_{\mathbf{x},\mathbf{y}} <0|T[J_\gamma(x)\, O_2(y)\, \bar{J}_{\gamma'}(0)]|0> e^{i\,\mathbf{p}.\mathbf{x}} = -\frac{2\langle x\rangle_{latt}\,(E+m)}{E}\, p\, Z_p\, e^{-Et_x} \tag{5.1}$$

and

$$C_{SF}^{(3)}(t_x, t_y) = \sum_{\gamma=\gamma'=1}^{2} \sum_{\mathbf{x},\mathbf{y}} <0|T[\bar{J}_{\gamma'}(0)\, O_3(y)\, J_\gamma(x)]|0> e^{i\,\mathbf{p}.\mathbf{x}} = -\frac{4\langle x^2\rangle_{latt}\,(E+m)}{E}\, p^2\, Z_p\, e^{-Et_x} \tag{5.2}$$

where $\mathbf{p} \equiv (p,0,0)$ $E = \sqrt{m^2+p^2}$ and Z_p is defined in Eq.(3.22). The subscript "latt" reminds us that the quantities are computed with the corresponding operators renormalized in the lattice renormalization scheme, and at a renormalization scale equal to the lattive spacing. Using perturbation theory, we are then able to reduce $\langle x\rangle$ and $\langle x^2\rangle$ in standard continuum renormalization schemes (such as the $\overline{MS}$ scheme).

By using (5.1) and (5.2) one can determine the first two moments of the protons' structure function. In ref.20 they found for the two <u>up quarks</u>:

$$\begin{aligned} \langle x\rangle_{LATT} &\sim 0.45 \\ \langle x^2\rangle_{LATT} &\sim 0.13 \end{aligned} \tag{5.3}$$

and for the <u>down quark</u>:

$$\begin{aligned} \langle x\rangle_{LATT} &\sim 0.22 \\ \langle x^2\rangle_{LATT} &\sim 0.08 \end{aligned} \tag{5.4}$$

The perturbative correction relating the operator $O_2^{\mu\nu}$ renormalized in the lattice renormalization scale to that renormalized in the $\overline{MS}$ scheme is negligibly small. Thus the values given in (5.3) and (5.4) can be taken as being $\langle x\rangle_u$ and $\langle x\rangle_d$, the average fraction of momentum carried by the up and down valence quarks, as defined from the operator $O_2^{\mu\nu}$ in the $\overline{MS}$ scheme at a renormalization scale of about 2 GeV. Within the errors, $\langle x\rangle_u$ is twice $\langle x\rangle_d$,

$$\frac{\langle x\rangle_u}{\langle x\rangle_d} = 2.0 \pm 0.25$$

The values reported above are larger than those measured experimentally,

$$\langle x\rangle_v = \langle x\rangle_u + \langle x\rangle_d \sim 0.4 \text{ at } \mu \sim 2 \text{ GeV} \tag{5.5}$$

Part of the discrepancy (but less than about 0.1 in $\langle x\rangle_v$) can be attributed to the fact that there are no sea quarks in the quenched approximation. The effect of the quenching alone seems however unable to explain the discrepancy of the lattice calculation and the experimental result. One possibility is the presence of effects of order a, as it was the case of the vector current discussed before; another possibility is that for baryons the effects of quenching are much worse than for mesons (see sect.6).
Using the tadpole dominance of lattice perturbation theory one may estimate that the values of $\langle x^2\rangle$ in the $\overline{MS}$ scheme are about 13% smaller than those in the lattice renormalization scheme. Given the large statistical errors in our results for $\langle x^2\rangle$, I do not comment on them further, other than to note that they are reasonable.

5.2. *The Electric Form Factor of the Proton*

For the electric proton form factor the correlation functions which we compute are of the form:

$$C^{\mu}_{FF}(t_x, t_y) = \sum_{\gamma=\gamma'=1}^{2} \sum_{x\, y} \langle 0|T[J_\gamma(x)\ \hat{V}^{\mu}(y)\ \bar{J}_{\gamma'}(0)]|0\rangle\ e^{i\, p.x}\ e^{i\, q.y} \tag{5.6}$$

where $\hat{V}^{\mu}(y)$ is the electromagnetic conserved current. In general, the matrix element depends on the two form-factors F_1 and F_2, defined by (in the notation of ref.33)

$$\langle p,s|\hat{V}^{\mu}(o)|p',s'\rangle = \bar{u}^{(s)}(p)\ [\gamma^{\mu} F_1(q^2) + \frac{i\sigma^{\mu\nu}q_\nu}{2m}\ kF_2(q^2)]u^{(s')}(p') \tag{5.7}$$

with the normalization chosen so that $F_2(0)=1$ and $F_1(0)=1(0)$ for the proton (neutron) and $q \equiv p'-p$. In actual calculations the non-relativistic approximation $E = m+p^2/2m$ holds to very good precision, one sums over $\gamma = \gamma' = 1$ and 2 and the spatial momentum is always in the x-direction (ref.20). Then the correlation function is proportional to the electric form factor, defined by

$$G_E = F_1 + \frac{kq^2}{4m^2} F_2 \tag{5.8}$$

In fact it is easy to show that:

$$C^{4}_{FF}(t_x, t_y) = \frac{Z_p}{2EE'}\, e^{-E't_y}\, e^{-E'(t_x - t_y)}\, 4m^2\, G_E(q^2) \quad (5.9)$$

and

$$C^{1}_{FF}(t_x, t_y) = \frac{-i\, Z_p}{2EE'}\, e^{-E't_y - E'(t_x - t_y)}\, 2m(p_x + p_x')\, G_E(q^2) \quad (5.10)$$

The factor of i in (5.10) comes from the continuation to Euclidean space. G_E, evaluated at the five values of momentum transfer corresponding to the initial state proton using the conserved vector current is reported in Fig.27 against $|q|/m_\rho$.
The solid curve is the prediction obtained using the vector meson dominance model

$$G_E(q^2) = \frac{1}{1 + \frac{(-q^2)}{m_\rho^2}} \quad (5.11)$$

The results have the correct trend of the experimental results[34]. The points lie a little below the vector dominance prediction, and if we replace m_ρ by $\lambda\, m_\rho$ in (5.11), then the best fit gives

$$\lambda = 0.85 \pm 0.05 \quad (5.12)$$

to be compared to $\lambda \sim 0.65$ which corresponds to the best fit to the results of Ref.34.

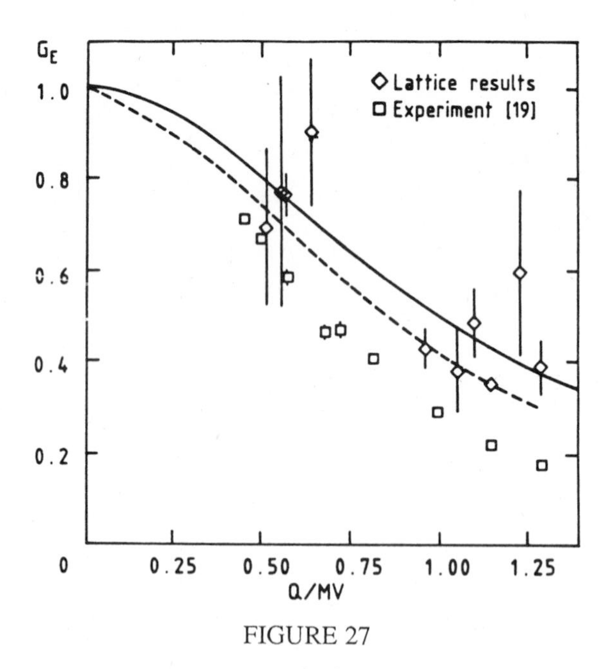

FIGURE 27

6. THE σ-TERM OF THE NUCLEON

The Hamiltonian part which breaks the SU(3) flavour symmetry can be written as:

$$H_{SB} = -\frac{(m_s - m)}{\sqrt{3}} \quad (\bar{q}\, \lambda_8\, q) \tag{6.1}$$

with $H_{TOT} = H_{SYM} + H_{SB}$.
H_{SYM} is the Hamiltonian which is symmetric in the s, u and d quarks and gives a common mass to all the hadrons belonging to the same multiplet, i.e. $\langle B|H_{SYM}|B\rangle = M$, $|B\rangle$ being the state of a given baryon at rest. m_s is the mass of the strange quark and $m = m_u = m_d$ is the common mass of the up and down quarks.

The baryon matrix elements of $(\bar{q}\, \lambda_8\, q)$ can be parametrazed, as usual, in terms of the F and D couplings:

$$\langle B'|\bar{q}\, \lambda_8\, q|B\rangle = F\, tr(B'^{+}[\lambda_8, B]) + D\, tr(B'^{+}\{\lambda_8, B\}) \tag{6.2}$$

B is the matrix which represents the baryon multiplet:

$$B \equiv \begin{pmatrix} \frac{1}{\sqrt{2}}\Sigma^o + \sqrt{\frac{1}{6}}\Lambda^o & \Sigma^+ & p \\ \Sigma^- & -\frac{1}{\sqrt{2}}\Sigma^o + \sqrt{\frac{1}{6}}\Lambda^o & n \\ \Xi^- & \Xi^o & -\sqrt{\frac{2}{3}}\Lambda^o \end{pmatrix} \tag{6.3}$$

$$\lambda_8 = \frac{1}{\sqrt{3}} \begin{pmatrix} 1 & 0 & 0 \\ 0 & 1 & 0 \\ 0 & 0 & -2 \end{pmatrix}$$

From eq.(6.2) and (6.3) one finds:

$$M_\Xi - M_p = (m_s - m)\,(2F) = 0.379 \text{ GeV (exp)}$$

$$M_\Xi + M_p - 2M_\Sigma = (m_s - m)\,(2D) = -0.129 \text{ GeV (exp)} \tag{6.4}$$

$$-\frac{3}{2}\,(M_\Sigma - M_\Lambda) = (m_s - m)\,(2D) = -0.116 \text{ GeV (exp)}$$

and $(D/F) \sim -0.32$.
The above values are related the matrix elements of the scalar densities in the hadrons; in the case of the proton:

$$\langle P|\bar{u}u|P\rangle = A$$

$$\langle P|\bar{d}d|P\rangle = B \tag{6.5}$$

$$\langle P|\bar{s}s|P\rangle = C$$

It is easy to verify that:

$$A - C = 2F$$

$$A - 2B + C = 2D \tag{6.6}$$

We cannot determine A,B and C with only two relations. However, given the ratio D/F, if we assume that the contribution of the strange quarks is negligible, C=0, the above relations fix completely A and B. In particular we can compute the so-called σ-term:

$$\sigma = m <P|\bar{u}u + \bar{d}d|P> =$$

$$\frac{3}{2}\left(\frac{m}{m_s - m}\right)\left[M_\Xi - M_P + \frac{1}{2}(M_\Sigma - M_\Lambda)\right] \tag{6.7}$$

Using the PCAC relation $\frac{m}{m_s - m} = \frac{M_\pi^2}{M_K^2 - M_\pi^2}$ and the experimental meson and baryon masses one then finds:

$$\sigma \sim 26 \text{ MeV} \tag{6.8}$$

On the other hand σ is also related to the I=0 π-Nucleon scattering amplitude through PCAC reduction of the initial and final pions in the soft limit. The value obtained from the scattering amplitude is[35]

$$\sigma \sim 55 - 60 \text{ MeV} \tag{6.9}$$

about a factor of two larger than the value obtained by neglecting the strange quark contribution to the proton mass.
The matrix elements of the scalar density in the proton can be computed on the lattice using the same technique used for the structure function and the electromagnetic form factor[36].
One can also use the Ward identity:

$$m\,(A + B) = m\,\frac{\partial M_P}{\partial m} \qquad M_P = \text{proton mass} \tag{6.10}$$

which is valid in the quenched case.

A priori we expect that the results from the lattice, in the quenched approximation, will be about the same as those obtained by assuming C=0, since we have imposed, by "fiat", that there are no strange quarks in the proton. A collection of results from several groups is reported in Fig.28[20,36,37,38]. All the values have been obtained by the explicit calculation of matrix element of the scalar density or by measuring the dependence of the mass of the proton on the quark mass. Possibly with the exception of ref.37, all the groups find a value of σ which is even lower than the expected value, $\sigma = 26$ MeV.

This is a typical case where the "quenched" approximation does not work well and a full calculation is needed. It should also be noticed that there could be a relationship between the low value of σ found in the quenched case and the observation that, in the same approximation, the proton to ρ mass ratio on the lattice turns out to be systematically larger than its experimental value (see Fig.29). Without excluding other systematic effects (finite size, extrapolation to the chiral limit and so on), this is not surprising, given that σ is related to the slope of the proton mass versus the quark mass. This fact, together with the unsatisfactory results for the proton structure function, and the recent experimental measurement of the proton polarized structure function by EMC, seems to indicate that the quenched approximation is worse of what it was thought before, at least for baryon physics.

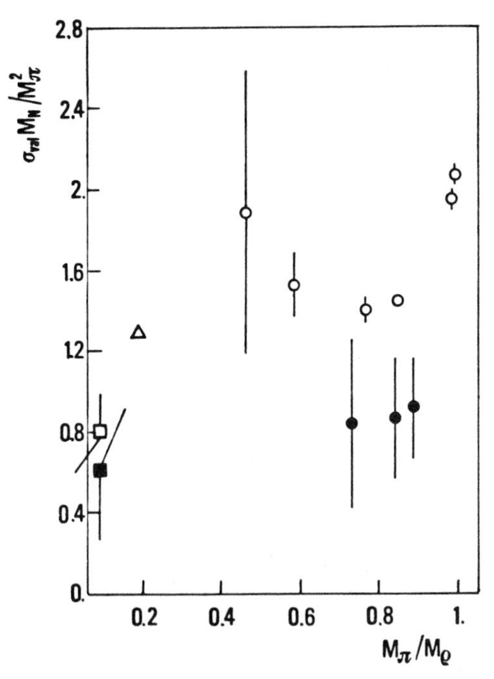

FIGURE 28

The sigma term, defined in eq.(6.7) and multiplied by M_N/M_π^2 (M_N being the nucleon mass) as a function of M_π/M_ρ, as found in ref.36, black circles, and ref.37, white circles, by computing the scalar density matrix element. The black squares (ref.20) and white squares (ref.38) have been obtained by measuring the σ-term from eq.(6.10). The triangle is the "experimental" quenched value, given in eq.(6.8).

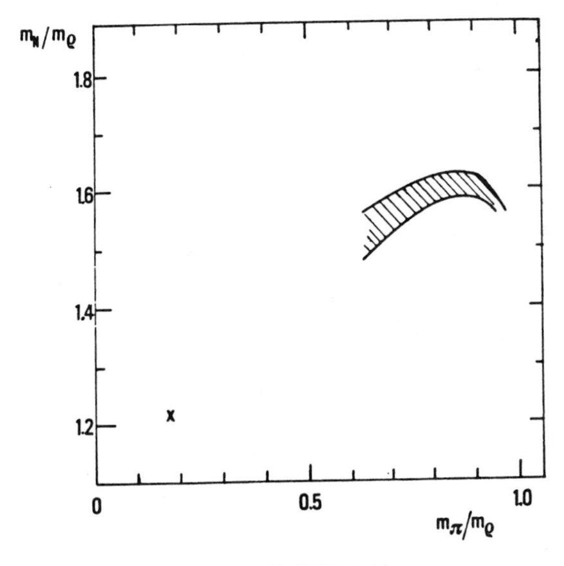

FIGURE 29

M_N/M_ρ as a function of M_π/M_ρ. The shaded area indicates the results of numerical simulations in the quenched approximation. The cross corresponds to the experimental value.

7. WEAK HAMILTONIAN MATRIX ELEMENTS

7.1 *Renormalization of the ΔS=2 and $\Delta I=\frac{1}{2}$ lattice operators*

In the following the general theory of the renormalization of lattice quark composite operators discussed in Sect.2 is applied to the operators which are relevant for kaon systems, i.e., the ΔS=2 operator:

$$O^{\Delta S=2} = (\bar{s}_L \gamma^{\mu} (1 - \gamma_5) d) \ (\bar{s} \gamma_{\mu} (1 - \gamma_5) d) \tag{7.1}$$

and the two (8,1) operators responsible for $\Delta T = \frac{1}{2}$ transitions:

$$O^{-} = [(\bar{s}_L \gamma^{\mu} d_L) \ (\bar{u}_L \gamma_{\mu} u_L) - (\bar{s}_L \gamma^{\mu} u_L) \ (\bar{u}_L \gamma_{\mu} d_L)] - [u \leftrightarrow c]$$

$$O^{+} = \frac{1}{5} \ [(\bar{s}_L \gamma^{\mu} d_L) \ (\bar{u}_L \gamma_{\mu} u_L) + (\bar{s}_L \gamma^{\mu} u_L) \ (\bar{u}_L \gamma_{\mu} d_L) +$$

$$2 (\bar{s}_L \gamma^{\mu} d_L) \ (\bar{d}_L \gamma_{\mu} d_L) + 2 (\bar{s}_L \gamma^{\mu} d_L) \ (\bar{s}_L \gamma_{\mu} s_L)] -$$

$$[(\bar{s}_L \gamma^{\mu} d_L) \ (\bar{c}_L \gamma_{\mu} c_L) + (\bar{s}_L \gamma^{\mu} c_L) \ (\bar{c}_L \gamma_{\mu} d_L)] \tag{7.2}$$

The operator in Eq.(7.1) cannot mix with lower dimension operators since there are no two-quark operators with ΔS = 2(or ΔT=3/2). As a consequence the renormalized operator $O^{\Delta S=2}$ has the form:

$$O(\mu)^{\Delta S=2} = Z_{LATT}^{\Delta S=2} (\mu a, g_o) \ [O^{\Delta S=2} + \sum_i C_i (\bar{s} \Gamma_i d) \ (\bar{s} \Gamma_i d)] \tag{7.3}$$

where the coefficients $C_i = C_i(g_o)$ are finite, depend only on g_o (the lattice bare QCD coupling constant) and can be safely computed in perturbation theory. They are different from zero because of the Wilson term. The corresponding operators $O_i=(\bar{s}\Gamma_i d)$ $(\bar{s}\Gamma_i d)$ (Γ_i are Dirac matrices) respect vector symmetries and CPS, are parity conserving but belong to chiral representations different from the (27,1).

Contrary to the previous case the two operators in Eqs.(7.2) can also mix with operators of dimension five and three or, more precisely, with $(\bar{s} \ \sigma_{\mu\nu} F_{\mu\nu} \ d)$, $(\bar{s} \ \sigma_{\mu\nu} \tilde{F}_{\mu\nu} \ d)$, $(\bar{s}d)$ and $(\bar{s}\gamma_5 \ d)$. On dimensional grounds, the coefficients of the operators of dimension less than six are expected to diverge as powers of a^{-1}. However, it is simple to show that the coefficient of $(\bar{s} \ \sigma_{\mu\nu} \tilde{F}_{\mu\nu} \ d)$ is finite because of CPS symmetry and that the coefficient of $(\bar{s} \ \sigma_{\mu\nu} F_{\mu\nu} \ d)$ can be made finite by the GIM cancellation when we allow the charm quark to propagate in the loop in diagrams like the one shown in Fig.30 (this implies that the condition $m_c a << 1$ must be satisfied). On the other hand the coefficients of the scalar and pseudoscalar densities are always divergent.

FIGURE 30

In summary the renormalized (8,1) operator $O^{\pm}$ have the form:

$$O^{\pm}(\mu) = Z^{\pm}_{LATT}\ (\mu a, g_o)\,[O^{\pm} + \delta_6 O^{\pm} + \delta_5 O^{\pm} +$$

$$\frac{(m_c - m_u)}{a^2}\ C^{\pm}_S(\bar{s}\, d) + \frac{(m_c - m_u)\,(m_s - m_d)}{a}\ C^{\pm}_P(\bar{s}\,\gamma_5\, d)] \tag{7.4}$$

where $\delta_6 O^{\pm}$ and $\delta_5 O^{\pm}$ are the corrections due to the operators of dimension six (as for the $\Delta S = 2$ case) and five whose coefficients are finite and can be computed perturbatively*. The coefficients of the pseudoscalar and scalar densities in Eq.(7.4), being power divergent, cannot be computed in perturbation theory. The reason in that any non-perturbative contribution $\sim e^{-(1/2\beta_o g_o^2)}$ to power divergent coefficients will not disappear even when $g_o \to 0$, i.e., in the continuum limit.

The overall renormalization constants $Z^{\Delta S=2}_{LATT}$ and $Z^{\pm}_{LATT}$ are completely determined, including the finite terms, by the requirement that the renormalized operators $O(\mu)$ obey, for some value of $\mu << a$, the same renormalization conditions on quark states as the continuum operators of the effective Hamiltonian (cfr. Sect.2).

A few comments are necessary at this point.

(i) As shown by the explicit one-loop calculations of Refs.39-41 and by the general theoretical analysis of refs.7,10, the off-diagonal operators in Eq.(7.3) as well as the operators $\delta_6 O^{\pm}$ in Eq.(7.4) are all parity even. This result is true also non-perturbatively and is a consequence of an exact symmetry as discussed in Ref.42.

(ii) The parity violating off-diagonal operators of dimension five and three, i.e., $(\bar{s}\,\sigma_{\mu\nu}\,\tilde{F}_{\mu\nu}\, d)$ and $(\bar{s}\gamma_5\, d)$ have coefficients which vanish for $m_s = m_d$.

(iii) Non-perturbative methods to fix the coefficients of operators of dimension three, $C^{\pm}_{S,P}$ are necessary.

For practical reasons, we have to rely on perturbation theory to fix the coefficients of the operators of dimension six and five. As it.will appear from the discussion of the lattice calculation of the kaon B-parameter, a certain systematic error is entailed in the procedure of constructing the renormalized operators using lowest-order perturbation theory and neglecting terms of order a (as for the vector current).

At lowest order in chiral symmetry breaking, soft pion theorems imply the following relations between matrix elements:

$$\langle 0\,|\,O^{\pm}(\mu)\,|\,K^o\rangle = i(m_k^2 - m_\pi^2)\,\delta^{\pm}$$

$$\langle \pi^+(p)\,|\,O^{\pm}(\mu)\,|\,K^+(k)\rangle = -\frac{m_k^2}{f}\,\delta^{\pm} + \gamma^{\pm}(p\cdot k) \tag{7.5}$$

$$\langle \pi^-\,\pi^+\,|\,O^{\pm}(\mu)\,|\,K^o\rangle = \frac{i}{f}\,(m_k^2 - m_\pi^2)\,\gamma^{\pm}$$

(all the $K \to \pi\pi$ matrix elements being taken for vanishing four-moment transfer of the operator).

On the other hand, for small masses and momenta we also have:

---

* The explicit expressions for $\delta_6 O^{\pm}$ and $\delta_5 O^{\pm}$ as well as the computation of the mixing coefficients can be found in Refs.39, 40, 41.

$$\langle 0|O^{\pm} + \delta_6 O^{\pm} + \delta_5\ O^{\pm}|K^0\rangle = i\ \delta_1^{\pm}$$

$$\langle \pi^+(p)|O^{\pm} + \delta_6 O^{\pm} + \delta_5\ O^{\pm}|K^+(k)\rangle = \delta_2^{\pm} + \gamma_2^{\pm}(p\cdot k) \tag{7.6}$$

where $O^{\pm} + \delta_6 O^{\pm} + \delta_5 O^{\pm}$ are the bare lattice operators in Eqs.(7.2) added with the corrections due to the operators of dimension six ($\delta_6 O^{\pm}$) and five ($\delta_5 O^{\pm}$) whose coefficients are taken from perturbation theory.

Similarly:

$$\langle 0|\bar{s}_L \gamma_5 d\ |K^0\rangle = i\ \delta_p$$

$$\langle \pi^+(p)|\bar{s}\, d\ |K^+(k)\rangle = \delta_s + \gamma_s\,(p\cdot k) \tag{7.7}$$

Then by a suitable linear combination of $O^{\pm} + \delta_6 O^{\pm} + \delta_5 O^{\pm}$ with ($\bar{s}d$) and ($\bar{s}\gamma_5 d$) we can enforce the relations in (7.5). This implies:

$$\gamma^{\pm} = Z^{\pm}_{LATT}\ (\mu a, g_o)\ [\gamma_2^{\pm} - \frac{\delta_2^{\pm}\gamma_s}{\delta_s}] + O(m_k^2) \tag{7.8}$$

In Eq.(7.8) the $O(m_k^2)$ terms depend on $\delta^{\pm}$ but are of higher order in the chiral expansion.

The method explained above requires the computation of the $<\pi|O|K>$ matrix elements at different values of the transferred momentum $q^2 = (p-k)^2$ in order to disentangle δ from γ in Eqs.(7.6) and (7.7). Equation (7.8) thus allows the determination of the $\Delta T = \frac{1}{2}$ $K\rightarrow 2\pi$ amplitude at lowest order in chiral perturbation theory:

$$\frac{A(K_s\rightarrow\pi^+\pi^-)}{m_k} = 2\,G_F \sin\theta_c\,\cos\theta_c\ \frac{m_k^2 - m_\pi^2}{m_k f}\ [C^-(\mu)\,\gamma + C^+(\mu)\,\gamma^+] \tag{7.9}$$

The procedure described above cannot be used for the "penguin" operators relevant for CP violation (i.e., ε'/ε). In fact, short of putting the top quark on the lattice (this would correspond to $m_t a<<1$), the "penguin" operators need two non-perturbative subtractions ($\bar{s}\,\sigma_{\mu\nu} F_{\mu\nu}\, d$ and $\bar{s}d$) for the parity-conserving part. On the other hand, they still require only one subtraction in the parity-violating sector. So in principle one can use the $K\rightarrow 0$ amplitude to fix the coefficient of $\bar{s}\gamma_5 d$ (e.g., by imposing the $<0|O^{\pm}|K^o> = 0$) and compute directly the amplitude $K\rightarrow 2\pi$ of the renormalized operator.

A very nice alternative procedure has been recently proposed in Ref.42. The idea is to evaluate the form factor $K\rightarrow 2\pi$ at the point $p_k^2 = p_\pi^2 = m_k^2 = m_\pi^2 = m^2$. In this limit, because of CPS symmetry, the coefficient of $\bar{s}\gamma_5 d$ (as well as the coefficient of $\bar{s}\,\sigma_{\mu\nu}\,\tilde{F}_{\mu\nu}\,d$) vanishes and the Cabibbo/Gell-Mann theorem is evaded because the operator carries some energy. Moreover all the counter-terms of dimension six, being parity conserving, cannot contribute to the matrix element. Thus we have:

$$\langle \pi^+\pi^-|O^{\pm}(\mu)|K^0\rangle = \langle \pi^+\pi^-|Z^{\pm}_{LATT}\ O^{\pm}|K^0\rangle = i\,\frac{2m^2\,\gamma^{\pm}}{f} \tag{7.10}$$

With this method $\gamma^{\pm}$ is directly obtained from the lattice operator (times the factor $Z^{\pm}_{LATT}$) without any further subtraction.

7.2 The Kaon B-Parameter and the $\Delta I=3/2$ Amplitude

The simplest (and first studied[13,42,43,44,45]) weak transition amplitude concerns the matrix element of the $\Delta S=2$ four-fermion operator:

$$\langle K^o | O^{\Delta S=2}_{(\mu)} | \bar{K}^o \rangle = \frac{8}{3} f_k^2 m_k^2 B_{K^o\bar{K}^o} \tag{7.11}$$

which is usually expressed in terms of the B-parameter appearing on the right-hand side of Eq.(7.11). To compute $B_{K^o\bar{K}^o}$ we extract, at several values of the pseudoscalar meson mass, m, the matrix elements $\langle K^o|O|\bar{K}^o\rangle$ and $\langle 0|O|\bar{K}^o\bar{K}^o\rangle$ from the three-point correlation functions [cf. Eq.(3.13)].

Lowest-order chiral perturbation theory predicts:

$$\langle 0 | O^{\Delta S=2}_{(\mu)} | \bar{K}^o \bar{K}^o \rangle = \alpha + \beta m^2 + \gamma p_{\bar{K}^o} \cdot p_{\bar{K}^o} + O(m^4) = \alpha + (\beta - \gamma) m^2 + O(m^4)$$

$$\langle K^o | O^{\Delta S=2}_{(\mu)} | \bar{K}^o \rangle = \alpha + (\beta + \gamma) m^2 + O(m^4) \tag{7.12}$$

when the mesons are at rest.

For the truly renormalized operator $O^{\Delta S=2}_{REN}$, α and β should vanish. However, since the subtraction has been computed only at lowest order in g_0^2 and terms of order a still play a role, the operator, renormalized in perturbation theory, has α and β different from zero. So we fit the matrix elements of $O^{\Delta S=2}$ to the expressions in Eq.(7.14) and take as our best estimate of the matrix element the coefficient γ. In so doing we are certainly left with a systematic error. In fact the residual unknown subtraction will not only cancel α and β but also modify γ ($\gamma_O \to \gamma_{OREN}$ only as $a \to 0$). As discussed before the measurement of the renormalization constant of the local vector on the lattice, which can be evaluated quite precisely in a non-perturbative way, strongly suggests that the main systematic effects in the renormalization of the operators are not due to an inadequate accuracy of perturbation theory but to effects of order $\Lambda_{QCD}a$, which are certainly present at finite lattice spacing[20].

Actually, to evaluate the B-parameter, it is convenient to fit the matrix elements of the operator as a function of the squared matrix element of the axial current. The coefficient of the fit is the B-parameter (up to an overall renormalization constant $8/3Z_A^2$). In Fig.31, the ratio $\langle 0|O^{\Delta S=2}|\bar{K}^o\bar{K}^o\rangle/Z_5$ ($Z_5 = |\langle 0|\bar{s}\gamma_5 d|K^o\rangle|^2$) and $\langle K^o|O^{\Delta S=2}|\bar{K}^o\rangle/Z_5$ are reported as a function of $8/3(\langle 0|A_0|\bar{K}^o\rangle)^2/Z_5 = 8/3\, f_K^2\, m_K^2 /(Z_A^2\, Z_5)$ at $\beta = 6.2$ on a 16^3x48 lattice (15 configurations). These combinations of matrix elements are particularly convenient for reducing the errors since they are directly computed by taking ratios of correlation functions.

The value of the B-parameter obtained from a linear and a quadratic fit to the points in Fig.31 and the analogue at $\beta = 6$ and at $\beta = 5.7$ are reported in Table 2. We take as our best estimate for the B-parameter the results of the quadratic fits, given the fact that terms of order m^4 are still important in the range of masses considered. To combine the results at $\beta = 5.7$, $\beta = 6$ and $\beta = 6.2$ we must take into account that the B-parameter is a function of the scale, and use the renormalization group evolution of the corresponding operator.

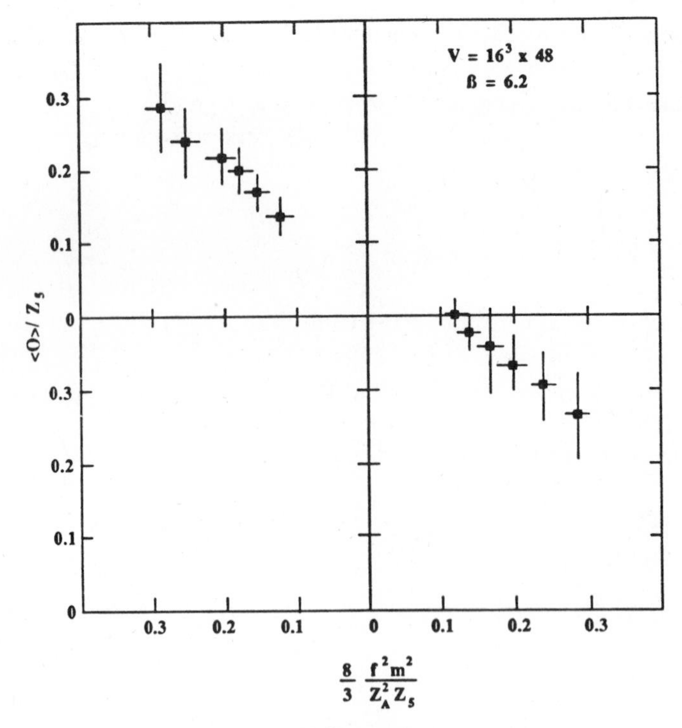

FIGURE 31

$<O>/Z_5$ as a function of $8/3[(<0|A_0|\bar{K}^o>)^2/Z_5 = 8/3\ f_K^2\ m_K^2\ /(Z_A^2\ Z_5)$ at $\beta = 6.2$ on the 16^3x48 lattice. The points correspond to the matrix elements $<0|O^{\Delta S=2}|\bar{K}^o\bar{K}^o>$, i.e., $p_K.p_K = -\ m_K^2$ on the left-hand side, and $<K^o|O^{\Delta S=2}|K^o>$, i.e., $p_K.p_K = +\ m_K^2$ on the right-hand side. Six different combinations of quark masses have been used.

TABLE 2

Results for the B-parameter obtained from a linear and a quadratic fit to the matrix elements of $O^{\Delta S=2}$.

B-parameter		
Linear fit	Quadratic fit	
0.88±0.10	0.60±0.20	$\beta = 6.2$ V=16^3x48 15 CONFS
0.81±0.16	0.75±0.20	$\beta = 6.0$ V=20x10^2x40 30 CONFS
0.91±0.11	0.65±0.10	$\beta = 5.7$ V=16x12x10x32 110 CONFS

Combining the results of the Table from the quadratic fit we obtain:

$$B_{K^o\bar{K}^o}\ (\mu) = 0.60 \pm 0.08 \qquad \text{at}\ \mu \sim \frac{1}{a} \sim 3\ \text{GeV} \tag{7.13}$$

which in terms of the renormalization group invariant B-parameter corresponds to:

$$B_{K^o\bar{K}^o} = [\alpha_s(\mu)]^{-6/25}\ B_{K^o\bar{K}^o}\ (\mu) = 0.81 \pm 0.10 \tag{7.14}$$

for Λ_{QCD} = 200 MeV. The difference between the linear and the quadratic fit in Table 2 gives an idea of the systematic error entailed in the evaluation of the linear coefficient from the matrix element at various m^2. On the other hand, with staggered fermions, the chiral behaviour of $\langle K^o | O^{\Delta S=2} | \bar{K}^o \rangle$ is automatically enforced and it is possible to compute the matrix element directly at the physical value of the meson mass (including quadratic effects). The Sharpe group founds[46]:

$$B^{STAGGERED}_{K^o\bar{K}^o}\ (\mu \sim 2\ \text{GeV}) = 0.70 \pm 0.02$$

in good agreement with the results of Table 2 but with a much smaller statistical error.

At this point the major source of uncertainty comes from the coefficient to be used to relate the lattice result to the continuum B-parameter and from the quenched approximation. To reduce the first source of systematic error a two loop calculation on the lattice is required.

Using SU(3)xSU(3) chiral perturbation theory[47] the value of $B(\mu)$ can be translated into the weak non-leptonic $\Delta I = 3/2$ amplitude $K^+ \rightarrow \pi^+\pi^o$:

$$\frac{\langle \pi^+ \pi^o | H_W | K^+ \rangle}{m_K} = G_F \frac{\sin\theta_c \cos\theta_c}{2} \frac{3}{4} \frac{m_K^2 - m_\pi^2}{m_K}\ C_4(\mu)\ x \left[\frac{\langle K^o | O^{\Delta S=2}(\mu) | \bar{K}^o \rangle}{f_K\, m_K^2} \right]$$

One finds[13]:

$$A_{3/2} = \frac{\langle \pi^+ \pi^o | H_W | K^+ \rangle}{m_K} = [7 \pm 2 \pm 1\ (\text{syst})]\ 10^{-8} \tag{7.15}$$

to be compared with the experimental value $A^{exp}_{3/2} = 3.7 \text{x} 10^{-8}$. Almost all the theoretical extimates tends to give a value for $A_{3/2}$ which is larger than the experimental one.

These extimate however do not include other contributions such as isospin-breaking and electromagnetic corrections which are very difficult to evaluate quantitatively. It could well be that these effects reduce the ΔI=3/2 amplitude as predicted on the basis of eq.(7.15) only. This would reconcile the extimate of the B parameter from ref.47 with the lattice and $1/N_c$ calculations and it would imply that a smaller relative enhancement is needed for $A_{1/2}$ to explain the large ratio $\frac{A_{1/2}}{A_{3/2}} \sim 20$ observed experimentally.

With the techniques exposed in Section 3, using four-point correlation functions, it has also been possible to compute directly the ΔI = 3/2 matrix element $<\pi^+\pi^o|H_W|K^+>$. Chiral perturbation theory gives the following relations when all mesons are at rest and $m_K^2 = m_\pi^2 = m^2$:

$$\langle \pi^+ | H_W | K^+ \rangle = g_{3/2}\, m^2$$

$$\langle \pi^+ \pi^- | H_W | K^o \rangle = \frac{2i}{f}\, g_{3/2}\, m^2$$

$$\langle \pi^- | H_W | \pi^- K^o \rangle = -\frac{6i}{f}\, g_{3/2}\, m^2 \qquad (7.16)$$

An explicit lattice calculation[13] of the last two matrix elements has shown that the relation:

$$\frac{\langle \pi^- | H_W | K^o \pi^- \rangle}{\langle \pi^+ \pi^- | H_W | K^o \rangle} \simeq -3$$

is satisfied with a good approximation as expected from CPS arguments[42]. The values of the coupling $g_{3/2}\, a^2$, a being the lattice spacing, from the K-π and K-$\pi^-\pi$ transitions of eqs.(7.16) at $\beta = 6$ and 6.2 are reported in Table 3. From the Table we see that soft-pion theorems and lowest-order chiral relations are satisfied at the 30% level.

The results for $-1/3 \langle K^o\pi^- | O_4 | \pi^- \rangle$ as a function of m^2 at $\beta = 6$ and $\beta = 6.2$ are reported in Figs.32a and 32b respectively. Only a linear fit in m^2 has been done. With the present accuracy, we cannot exclude corrections of order m^4 or effects of order a such that the matrix element does not really vanish as $m \to 0$. However, the expected chiral behaviour seems to be well reproduced by the numerical results.

Given the off-diagonal subtractions are only needed for K-π amplitudes, and that meson masses are never very small, the K-π and K-$\pi\pi$ comparison from Table 3 is an excellent test of our computation and of the approach to the chiral limit. Soft pion relations are indeed a very powerful way to control the renormalization of the relevant operators on the lattice. A similar test should also be performed in the more complicated case of $\Delta I = \frac{1}{2}$ transition amplitudes.

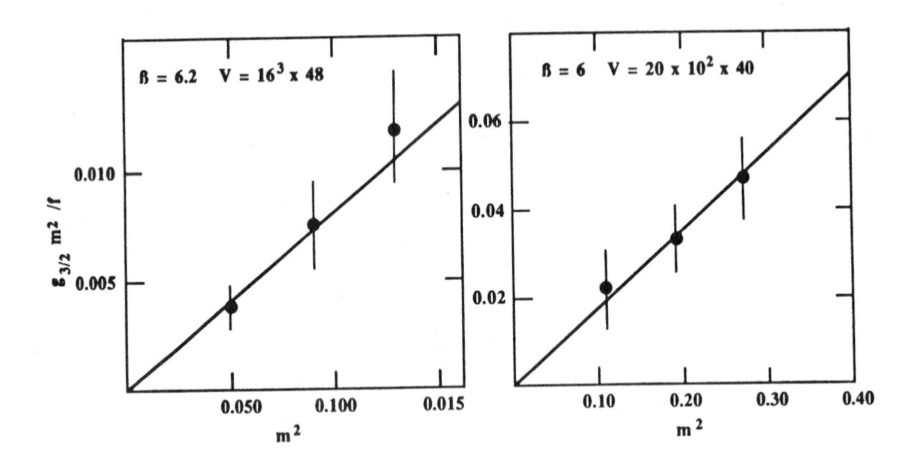

FIGURE 32a and b

The quantity $g_{3/2}m^2/f$, defined in Eqs. (7.16) and derived from the matrix element $-1/3\langle K^o\pi^- | O_4 | \pi^- \rangle$ is reported at $\beta = 6.2$ and $\beta = 6$ as a function of the pseudoscalar meson mass $m_K^2 = m_\pi^2\ m^2$.

TABLE 3

The coupling $g_{3/2}$ defined in Eqs.(7.16) as computed from the K-π and K-$\pi\pi$ matrix elements. Notice the good agreement between the two values both at $\beta = 6$ and $\beta = 6.2$. Similar results have been obtained at $\beta = 5.7$ with a much larger statistics.

$g_{3/2}\, a^2$		
K^+-π^+	K^o-$\pi^+\pi^-$	
$(2.0\pm0.8)10^{-2}$	$(1.3\pm0.4)10^{-2}$	$\beta = 6$ 15 configurations
$(5.9\pm1.2)10^{-3}$	$(4.4\pm1.2)10^{-3}$	$\beta = 6.2$ 15 configurations

7.3 *The* $\Delta I = \frac{1}{2}$ *amplitude*

To lowest order in masses and momenta, chiral perturbation theory gives the expressions for the $K_S \rightarrow \pi^+\pi^-$ amplitude

$$\frac{A(K_S \rightarrow \pi^+ \pi^-)}{m_K} = 2\, G_F \sin\theta_c \cos\theta_c \, \frac{m_K^2 - m_\pi^2}{m_K f_\pi} \, [C^-(\mu)\, \gamma^- + C^+(\mu)\, \gamma^+] \tag{7.17}$$

where $\gamma^\pm$ are defined according to:

$$\langle \pi^+ \pi^- | O^\pm(\mu) | K^o \rangle = \frac{i}{f_\pi} (m_K^2 - m_\pi^2)\, \gamma^\pm \tag{7.18}$$

In the same approximation, $\gamma^\pm$ appear in the following K-π and K-$\pi\pi$ matrix elements:

$$\langle \pi^+ (o) | O^\pm(\mu) | K^+(\mathbf{q}) \rangle = \delta^\pm m^2 + \gamma^\pm E(\mathbf{q}) m \tag{7.19}$$

$$\langle 0 | O^\pm(\mu) | K^+(o)\, \pi^-(o) \rangle = \delta^\pm m^2 - \gamma^\pm m^2 \tag{7.20}$$

$$\langle \pi^+ (o) \pi^-(o) | O^\pm(\mu) | K^o(o) \rangle = \frac{2i}{f_\pi} m^2 \gamma^\pm \tag{7.21}$$

$$\langle \pi^+ (o) | O^\pm(\mu) | K^o(o) \pi^+ (o) \rangle = -\frac{2i}{f_\pi} m^2 \gamma^\pm \tag{7.22}$$

where we have taken K and π degenerate in mass ($m_K^2 = m_\pi^2 = m^2$), $\mathbf{q}$ is the spatial momentum, ($E = \sqrt{m^2 + \mathbf{q}^2}$, and $f_\pi \simeq 131$ MeV.

The calculation on the lattice of the matrix elements on the left-hand side of Eqs.(7.19) to (7.22) allows a determination of $\gamma^{\pm}$, i.e., of the physical amplitude, to lowest order in chiral perturbation theory.

As discussed in Refs.7 and 10 (see also Section 7.1), in the SU(3) symmetric limit, the renormalized operators $O^{\pm}(\mu)$ are related to the lattice operator $O^{\pm}$ as follows:

$$O^{\pm}(\mu) = Z^{\pm}_{LATT}(\mu a, g_0)\,[O^{\pm}_{pert} + c_S(\bar{s}\,d) + c_P(\bar{s}\,\gamma_5\,d)] \qquad (7.23)$$

where $O^{\pm}_{pert} = O^{\pm} + \delta_6\, O^{\pm} + \delta_5\, O^{\pm}$ is the naive lattice operator plus the appropriate operators of dimensions six and five.

To fix $C^{\pm}_S$ one can impose:

$$\langle \pi^+(o) | O^{\pm}(\mu) | K^+(o) \rangle = 0 \qquad (7.24)$$

which corresponds to $\delta^{\pm} = -\gamma^{\pm}$* (eqs. 7.19 - 7.20).

The smoothness assumption on the behaviour of the matrix elements as a function of m^2 made in Eqs.(7.20) - (7.22) may fail in the presence of octet scalar particles comparatively light with respect to the pseudoscalar mass m. Exchanges of scalar particles as shown for example in the diagram of Figs.33 gives non-smooth contributions to the matrix elements of Eqs.(7.20) - (7.22), enhanced by the scalar propagator when $M^2_S = 4m^2$. The matrix element in Eq.(7.19) is relatively unaffected, because the momentum transfer is space-like. For quark masses at which current numerical studies are done, the scalar octet mass M_S is close to 2m**.

In such a situation the only theoretically safe way to extract $\gamma^{\pm}$ is from the matrix element of Eq.(7.19). By imposing the subtraction condition Eq.(7.24), one has:

$$\gamma^{\pm} = \frac{\langle \pi^+(o) | O^{\pm}(\mu) | K^+(\mathbf{q}) \rangle}{m[E(\mathbf{q}) - m]} \qquad (7.25)$$

Because of the subtraction invoved, the signal in the numerator of Eq.(7.25) is small and affected by large statistical fluctuations. The direct calculation of the K-ππ amplitude needs no subtraction, is subject to smaller statistical fluctuations, but may be contaminated by spurious octet scalar contributions unless $M^2_S >> 4m^2$.

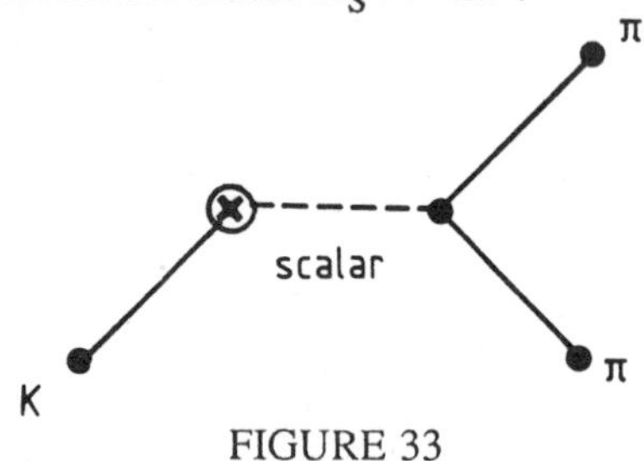

FIGURE 33

Feynman diagram relative to the octet scalar contribution K-ππ transitions. The symbol ⊗ indicates the insertion of the weak Hamiltonian.

---

* As discussed in Ref.10, $\delta^{\pm}$ do not appear in the physical decay amplitude and in fact are arbitrary. Different value of $\delta^{\pm}$ correspond to different and equivalent subtraction prescriptions needed to define $O^{\pm}(\mu)$.

** A scalar mass even smaller than 2m is found by high statistics results from the APE collaboration.

K-π and K-$\pi\pi$ matrix elements are computed from three- and four-point correlations of the operators and pseudoscalar sources. In the following we will denote the contribution to $\gamma^{\pm}$ from the eye and from the other (eight-shaped) diagrams, by $\gamma^{\pm}_{eye}$ and $\gamma^{\pm}_{eight}$ respectively.

We discuss first K-π matrix elements[48].

The values of $\gamma^{\pm}_{eye}$ obtained using Eq.(7.25) and extrapolated to the chiral limit are reported in Table 4.

Using Eq.(7.17) with the experimental values for m_K, m_π and f_π and using a^{-1} = 1.8 GeV we find the amplitude for $K_S \to \pi^+\pi^-$ reported in Table 6 where we have used $c^{(-)}(\mu)$ = -1.7 and $c^{(+)}(\mu)$ = 0.75 at $\mu \sim$ 1.8 GeV. In the Table, R is the following ratio:

$$R = \frac{\langle \pi^+ \pi^- | H_W | K_S \rangle}{\langle \pi^+ \pi^o | H_W | K^+ \rangle} \qquad \exp \sim 21.2 \qquad (7.26)$$

is also reported.

The results for $A(K_S \to \pi^+\pi^-)$ are still compatible with zero, given that the statistical error is ~100%. The result for γ^{+}_{eye} is definitely worse than the one for γ^{-}_{eye}. The trend of the data is, however, encouraging. In particular they favour a positive sign for R, in agreement with the fact that $B(K_S \to \pi^+\pi^-) > 2B(K_S \to \pi^o\pi^o)$. More statistics are clearly needed.

A scalar pole may affect also the K-$\pi\pi$ calculation, where subtractions are not needed because of CPS symmetry[42]. If this is the case this method is reliable only at low pseudoscalar masses. The values of $2m^2\gamma^{\pm}_{eye}/f(m)$, reported in Table 5, have been obtained from the average of the matrix elements, Eqs.(7.21) and (7.22). f(m) is the pseudoscalar meson axial coupling as a function of the quark mass. In the Table, also $2m^2\gamma_4/f(m)$, which refers to the $\Delta I = 3/2$ operator is given:

$$O_4 = [(\bar{s}_L \gamma_\mu d_L)\ (\bar{u}_L \gamma^\mu u_L) + (\bar{s}_L \gamma_\mu u_L)\ (\bar{u}_L \gamma^\mu d_L)] - (\bar{s}_L \gamma_\mu d_L)\ (\bar{d}_L \gamma^\mu d_L)] \qquad (7.27)$$

Assuming a constant behaviour in m^2 of all γs, we obtain the physical amplitudes reported in Table 6.

Within errors the K$\to\pi\pi$ results are quite compatible with those obtained from the $q \neq 0$ K$\to\pi$ matrix elements, and do not show any non-smooth variation with the pseudoscalar meson mass m. Within the present data, we are unable to decide whether any contribution of the scalar particle is hidden in the errors, or is really small. A more recent study at $\beta = 5.7$ with a very large statistics (110 gauge field configurations) has shown that there is no signal for K$\to\pi\pi$ $\Delta I = \frac{1}{2}$ transition. Probably the effects of the scalar particle propagator change rapidly when one varies β, and the only conclusion than can be drawn at the moment is that the K-$\pi\pi$ method is unreliable at the accessible values of β and of the quark masses. The analysis of the K-π matrix elements at $\beta = 5.7$ has not been completed yet.

In conclusion, the method based on the calculation of K-π matrix elements with a non-vanishing space-momentum, **q**, supports the observed enhancement of the $\Delta I = \frac{1}{2}$ amplitude, although within very large statistical fluctuations. The eye-diagrams are the source of the enhancement, an indirect confirmation of the scheme of Ref.49 since penguin diagrams are indeed generated by eye-diagrams. The vacuum saturation approximation gives $\gamma^{-}_{eye}/\gamma^{+}_{eye} = -\frac{1}{2}$, while we find a large ratio which seems to be just the continuation of the short-distance enhancement.

TABLE 4

$\gamma^{+}_{eye\ eight}$ extrapolated to the chiral limit from K-π matrix elements. I and II refer to two different extrapolation method as discussed in detail in the first of refs.48.

	$\gamma^{-}_{eye}(q\neq0)\times10^3 \times a^2$	$\gamma^{+}_{eye}(q\neq0)\times10^3 \times a^2$	$\gamma^{-}_{eight}\times10^3 \times a^2$	$\gamma^{+}_{eight}\times10^3 \times a^2$
chiral limit	(-8±9)(I) (-20±12)(II)	(+2±12)(I) (+5±19)(II)	-2.9±1.5	3.9±1.5

TABLE 5

$\gamma^{+}_{eye\ eight}$ from K-ππ matrix elements at two different values of the pseudoscalar meson mass at β=6.2.

$\frac{2m^2}{f(m)}\gamma^{-}_{eye}$ $\times10^4 \times a^3$	$\frac{2m^2}{f(m)}\gamma^{+}_{eye}$ $\times10^4 \times a^3$	$\frac{2m^2}{f(m)}\gamma^{-}_{eight}$ $\times10^4 \times a^3$	$\frac{2m^2}{f(m)}\gamma^{+}_{eight}$ $\times10^4 \times a^3$	$\frac{2m^2}{f(m)}\gamma_4$ $\times10^4 \times a^3$	K
-8.0±5.4 -5.5±3.7	5.4±2.3 3.7±1.7	-0.33±0.28 -0.14±0.24	0.51±0.07 0.38±0.06	0.59±0.13 0.37±0.11	0.1500 0.1510

TABLE 6

Results from Ref.48 obtained at two different values of the strong coupling constant $\beta = 6/g_0^2$, β = 6 and β = 6.2.

β	SOURCE	$A(K_S\to\pi^+\pi^-)/m_K$	$A(K^+\to\pi^+\pi^0)/m_K$	R
6.0	K-π(q≠0)	$(1.2\pm1.3)10^{-6}$ (I) $(2.5\pm1.8)10^{-6}$ (II)	$(16.0\pm6.0)10^{-8}$	8±24(I) 16±40(II)
6.2	K-ππ	$(2.1\pm1.3)10^{-6}$	$(5.9\pm1.4)10^{-8}$	35±31
Experiment		$0.78\ 10^{-6}$	$3.8\ 10^{-8}$	21.2

The K-ππ method gives unstable results and its validity is very doubtfull.

7.4 *Electropenguin contribution to ε'/ε*

The electropenguin contribution to the CP-violating parameter ε' induced by the electropenguin diagrams has been widely discussed in recent literature[50]. After integration of t and c quark virtual loops, the electropenguin contribution to the I=2, K→ππ amplitude, A_2, is determined by the matrix elements of the following four-fermion operators:

$$O_{LR} = \frac{1}{3} \{(\bar{s}_L \gamma_\mu d_L) [\bar{u}_R \gamma^\mu u_R - \bar{d}_R \gamma^\mu d_R] + (\bar{s}_L \gamma_\mu u_L)(\bar{u}_R \gamma^\mu d_R) \quad (7.28)$$

$$(O_c)_{LR} = \frac{1}{3} \{(\bar{s}_L \gamma_\mu t^A u_L) [\bar{u}_R \gamma^\mu t^A u_R - \bar{d}_R \gamma^\mu t^A d_R] + (\bar{s}_L \gamma_\mu t^A u_L)(\bar{u}_R \gamma^\mu t^A d_R)\} \quad (7.29)$$

The operators in Eqs.(7.28) and (7.29) have pure ΔI=3/2. From the point of view of flavour SU(3), they are a superposition of 27-plet and decuplet, unlike the fully left-handed operator O_4 given in eq.(7.27) which is a pure 27.

In a recent paper[51], the K-π and K-ππ matrix elements of the operators in Eqs.(7.28) and (7.29) have been computed on the lattice. The results of Ref.49 can be summarized as follows.

(i) In the chiral limit, the K-π matrix elements of O_{LR} and $(O_c)_{LR}$ are both very close to their vacuum saturation value, the latter being also computed on the lattice. More precisely they find:

$$B_{LR} (a^{-1} \simeq 2.2 \text{ GeV}) = 1.0 \pm 0.1$$

$$(B_c)_{LR} (a^{-1} \simeq 2.2 \text{ GeV}) = 0.95 \pm 0.1 \quad (7.30)$$

(ii) Chiral relations are obeyed within about 50% accuracy, similarly to what is found in Ref.13 for O_4.

(iii) For both O_{LR} and $(O_c)_{LR}$, the matrix elements $<\pi^- \bar{K}^o | O | \pi^->$, which allow for a scalar octet intermediate state, are between 5 to 10 times larger than the $<\pi^+\pi^- | O | K^o>$ and $<\pi^- | O | \pi^- K^o>$ matrix elements, which do not. This is strongly suggestive of an important role of the scalar octet also in the K→ππ, $\Delta I = \frac{1}{2}$ amplitude. It suggests that a reliable calculation of $\Delta I = \frac{1}{2}$ CP-conserving and CP-violating K→ππ amplitudes requires quark masses much closer to the chiral limit.

The results on the electropenguin contribution to ε'/ε can be stated in two different ways:

a) from a full-lattice calculation of $(\text{Im } A_2/A_2)_{EMP}$ one finds:

$$\left| \frac{\epsilon'}{\epsilon} \right|_{EMP} = (0.74 \pm 0.27)\, 10^{-3} \left(\frac{A}{1.05} \right)^2 \left(\frac{\lambda}{0.221} \right)^4 \left(\frac{\rho}{0.6} \right) \sin\varphi \quad (7.31)$$

where the error is purely statistical; A, λ, ρ and φ are the parameters of the C-K-M matrix in the Wolfenstein parametrization[52]; alternatively:

$$\epsilon' = -i\, e^{i(\delta_2 - \delta_o)} \frac{\omega}{\sqrt{2}} \frac{\text{Im } A_o}{A_o} (1 + \Omega_{EMP} - \Omega_{\eta,\eta'}) \quad (7.32)$$

with $\Omega_{\eta,\eta'}$ arising from π-η and π-η' mixing. One gets[51]:

$$\Omega_{EMP} \simeq 0.17/B_6 \tag{7.33}$$

where B_6 is the B-parameter of the penguin operator O_6. The above value for Ω_{EMP} is of the same order of the estimated value of $\Omega_{\eta,\eta'}$ but with the opposite sign.

This result is quite different from what was obtained using the $1/N_c$ expansion. It agrees with previous lattice determinations of the B-parameter of left-right operators[42,53]. In Ref.53 they also estimated the ordinary penguin B-parameter with the result $B_6 = 0.5\pm0.2$. According to Eq.(7.33), and assuming $B_6 = 0.5$, this would imply that the electropenguin effects can be as large as 35%.

8. HEAVY FLAVOUR PHYSICS FROM LATTICE QCD

Two groups[54,55] have recently studied the matrix element $\langle K^-|J_\mu|D^o\rangle$ in lattice QCD. In ref.55 the form factor at zero momentum transfer has been found $f^+(0) = 0.74 \pm 0.17$, in excellent agreement with the experimental measurement. From a computation of the matrix element $\langle \pi^-|J_\mu|D^o\rangle$ it has been also found that the corresponding form factor at zero momentum transfer is given b $f^+(0) = 0.70 \pm 0.20$ and that, within statistical errors, Vector Dominance gives a good description of the form factor at the values of the momentum transfer accessible on the lattice which was used.

The matrix elements $\langle K^-|J_\mu|D^o\rangle$ ($\langle \pi^-|J_\mu|D^o\rangle$), where $J_\mu = \bar{s}\,\gamma_\mu\, c$ ($\bar{d}\,\gamma_\mu\, c$), contain all the information about the hadronic physics present in semileptonic decays of D^o into light pseudoscalar mesons. Precise knowledge of these matrix elements would allow for a direct determination of the element V_{cs} of the Cabibbo-Kobayashi-Maskawa (C-K-M) matrix[56], from an experimental measurement of semileptonic decay rates. Below I will compare the results of ref.55 with the data from the Tagged Photon Spectrometer Collaboration[57] and the MARK III collaboration[58].

On the other hand the determination of the as yet unknown C-K-M matrix element $|V_{ub}|$, the effects of CP-violation and attempts to determine the possible existence of new interactions beyond the Standard Model[59] all rely up to now on phenomenological models for the decay of B-mesons. This is due to the fact that B-physics is at the interface of the perturbative and non-perturbative QCD regimes. The light quark inside the meson is in fact subject to QCD non-perturbative effects because its typical momentum is of order $\mu \ll m_Q$. Lattice QCD offers the possibility of computing the spectrum and the decay matrix elements of B-mesons from first principles and without any free parameter as it is the case for D-mesons. In practice however we cannot put a b-quark directly on the lattices which are currently used. This is because the typical inverse lattice spacing $a^{-1} \sim$ 2-3 GeV and the condition that the ultra-violet cut-off is much larger than the characteristic scale (i.e. $m_Q a \ll 1$) cannot be satisfied. Fortunately when m_Q is very large we do not need to put a dynamical heavy quark on the lattice. It has been shown by E.Eichten[60] that, in the limit where the heavy quark Q has a very large mass the long distance physics (i.e. physics at distances which are large compared with $1/m_Q$) can be described by an effective field theory with a static colour source. All the relevant physical quantities (masses, decay constants, weak matrix elements) can be then systematically expanded in powers of $1/m_Q$[60]. The first attempts to compute on the lattice the B-meson lattice binding energy and pseudo-scalar decay constant at the lowest order in $1/m_Q$ encountered several difficulties and I will not discuss then further.

8.1 *Semi-leptonic Decays of D-mesons*

Using Lorentz and parity invariance the $D \to K$ matrix element can be parametrized in terms of two form factors, which (using the helicity basis) we choose to be $f^+(q^2)$ and $f^o(q^2)$, defined by

$$<K^-|J_\mu|D^o> = (p_D + p_K - \frac{m_D^2 - m_K^2}{q^2} q)_\mu f^+(q^2) + \frac{m_D^2 - m_K^2}{q^2} q_\mu f^o(q^2) \qquad (8.1)$$

where q is the momentum transfer, $q = p_D - p_K$. We only need to compute matrix elements of the vector current, since between two pseudoscalar states is zero. At zero momentum transfer $f^+(0) = f^o(0)$.

Lattice computations of electromagnetic form factors of hadrons containing quarks of equal mass have shown that different choices for the lattice definition of the vector current can lead to results which differ by up to 25% or so. These differences are due to corrections of O(a), where a is the lattice spacing. The optimal choice of vector current is the one which is conserved in the lattice theory itself:

$$J_\mu = 1/2\, [\bar{s}(x)(\gamma_\mu-1)U_\mu(x)c(x+\hat{\mu})+\bar{s}(x+\hat{\mu})(\gamma_\mu+1)\, U^+_\mu(x)c(x)] \qquad (8.2)$$

J is the current which, in the case of degenerate quark masses is conserved in the lattice theory defined by the Wilson action which was used in ref.55. The matrix elements are calculated by computing the three-point correlation functions

$$C_\mu(t_x,t_y) \equiv \sum_{\underline{x},\underline{y}} \exp(i\underline{q} \cdot \underline{y}) \exp(i\underline{p} \cdot \underline{x}) <0|T[M_K(x)J_\mu(y)M^+_D(0)]|\,0> \qquad (8.3)$$

where M_D and M_K are interpolating operators for the D and K mesons respectively, (throughout this work we use the pseudoscalar densities $(\bar{c}\,\gamma_5\,u)$ and $(\bar{u}\,\gamma_5\,s)$ for these operators).

The matrix element is then extracted from the correlation function (8.3) by using the methods described in Section 3.

By assuming the validity of the vector dominance model:

$$\frac{f^+(q^2)}{f^+(0)} = \frac{m^2_{D*}}{q^2 - m^2_{D*}}, \qquad (8.4)$$

and using the masses of the vector meson found at the correspondinh values of the light quark masses the authors of ref.55 found:

$$f^+_\pi(0) = 0.70 \pm 0.20$$

$$(8.5)$$

$$f^+_K(0) = 0.74 \pm 0.17$$

These results are extremely satisfying. The statistical errors are reasonably small and the $f_K^+(0)$ is in excellent agreement with recent experimental data from the Tagged Photon Spectrometer Collaboration[57]. This collaboration measures the relevant branching ratio to be $B(D^o \rightarrow K^-e^+\nu_e) = (3.8 \pm 0.5 \pm 0.6)\%$. From this value and the D^o lifetime, and assuming that the q^2 behaviour of the form factor is given by (7), they obtain the result, $|V_{cs}|^2|f^+(0)|^2 = 0.50 \pm 0.07$. Assuming three genrations and imposing unitarity on the C-K-M matrix, (i.e. taking $|V_{cs}| = 0.975$) this result corresponds $f^+(0) = 0.73 \pm 0.05 \pm 0.07$. The MARK III[58] collaboration obtain an almost identical result for the branching ratio, $B(D^o \rightarrow K^-e^+\nu_e) = (3.9 \pm 0.6 \pm 0.6)\%$, and hence the deduced value of $f^+(0)$ is also very similar. Both experiments have studied the behaviour of $f^+(q^2)$ with q^2 and found it compatible with vector meson dominance.

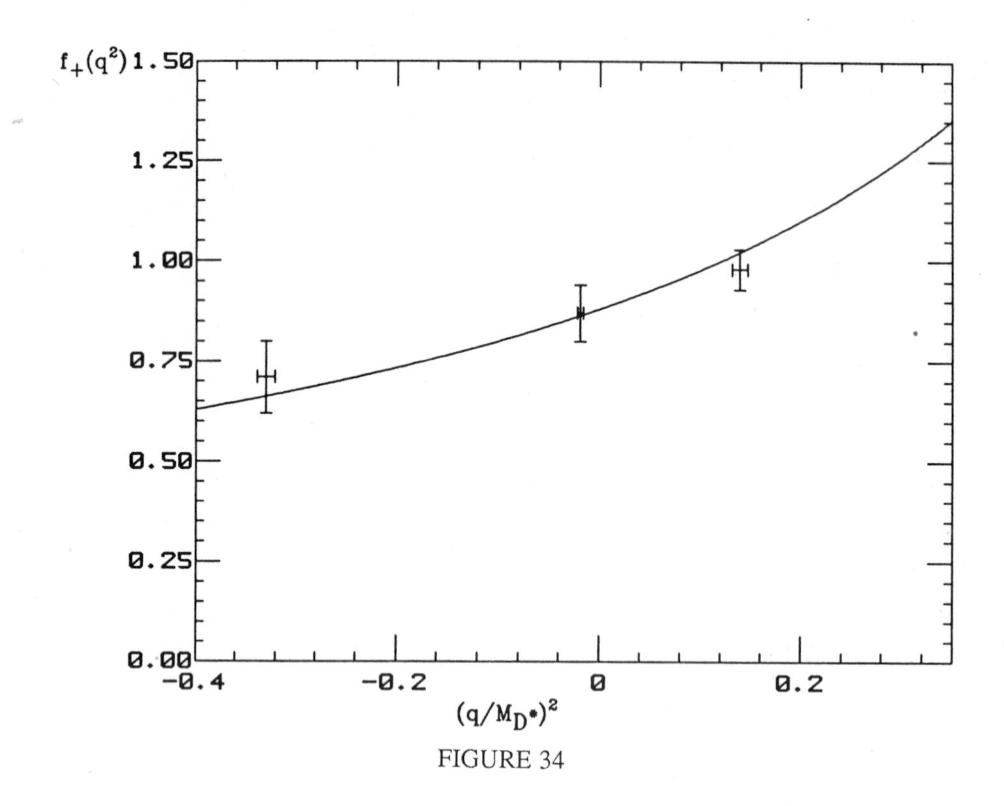

FIGURE 34

In Figure 34 the results for the form factor f^+ as a function of q^2/m^2_{D*} is reported. For comparison with the vector dominance model the curve obtained by using eq.(8.4), with $(m_{D*}a) = 0.92$ and $f^+(0)_{K=0.1515}=0.88$ is also plotted.

Many interesting phenomenological models which have been proposed to study the semi-leptonic D form factors (see for example refs.(60-62) and references therein). These have a variety of model assumptions and parameters and consequently a range of predictions, (including $f^+(0) = 0.6 \pm 0.1$[61]; 0.76[62]; 0.58[63]).

9. CONCLUSION

The quality of the results in the computation of hadronic matrix elements on the lattice is already comparable to other theoretical approaches. The same method is used in all the cases (structure functions, form factors, σ-term, weak decays, ...) and relies only on calculations from first principles which use the renormalization group and the Wilson operator expansion, as basic tools.

The error on the evaluation of the matrix elements ranges between 20 and 100%. The main problem is to reduce the systematic errors in the final results. Two facts seem to be at the moment the major sources of uncertainties: the effects of order a and the quenched approximation (finite size effects seem reasonably small for all the cases). For the effects of order a, one could think to improve the situation by using the Symanzik proposal, reduce them to order g^2a [64]. On the contrary it is rather difficult to do at the moment a complete, unquenched calculation on large lattices and at realistic quark masses. It is already rather encouraging that the lattice approach has started to give its first predictions, such as the kaon B-parameter, the decay constants of the D and D_s mesons, f_{D,D_s}, and the electro-penguin contribution to ϵ'/ϵ. A new field of activity which have important phenomenological implications is just starting with the study of B-meson semileptonic decays and B-$\bar{B}$ mixing, the route to solve the $\Delta I = \frac{1}{2}$, althought difficult, it not hopeless and will be done.

REFERENCES

1. C.N.Yang, R.L.Mills, Phys.Rev. 96 (1954) 191.
2. H.D.Politzer, Phys.Rev.Lett. 30 (1973) 1346; D.J.Gross and F.Wilczek, Phys.Rev.Lett. 30 (1973) 1343; Phys.Rev. D8 (1973) 3633.
   For a detailed introduction to QCD see E.S.Abers, B.W.Lee, Phys.Rep. 9 (1973) 1.
3. K.G.Wilson, Phys.Rev. D10 (1974) 2445; "New Phenomena in Subnuclear Physics", A.Zichichi ed, Plenum Press, New York (1977).
4. G.C.Wick, Phys.Rev. 96 (1954) 1124.
5. L.H.Karsten and J.Smit, Nucl.Phys. B183 (1981) 103.
6. H.B.Nielsen and N.Ninomiya, Nucl.Phys. B185 (1981) 20; B193 (1981) 173.
7. M.Bochicchio, L.Maiani, G.Martinelli, G.C.Rossi and M.Testa, Nucl.Phys. B262 (1985) 331.
8. G.Curci, Phys.Lett. B167 (1986) 425.
9. L.Maiani and G.Martinelli, Phys.Lett. 168B (1986) 265.
10. L.Maiani, G.Martinelli, G.C.Rossi and M.Testa, Phys.Lett. 176B (1986) 445; Nucl.Phys. B289 (1987) 505.
11. G.Martinelli and C.T.Sachrajda, Phys.Lett. B190 (1987) 151.
12. G.Martinelli and C.T.Sachrajda, Phys.Lett. B196 (1987) 184; Nucl.Phys. B306 (1988) 865.
13. M.B.Gavela, L.Maiani, G.Martinelli, S.Petrarca, F.Rapuano and C.T.Sachrajda, Nucl.Phys. B306 (1988) 677; M.B.Gavela, L.Maiani, O.Pène and S.Petrarca, Phys.Lett. B206 (1988) 113.
14. T.A.De Grand and R.D.Loft, Preprint COLO-HEP-165 (nov. 1987) to appear in Phys.Rev.D38 (1988) 954.
15. C.Bernard et al., Nucl.Phys. (Proc.Suppl.) B9 (1989) 155 and refs. therein.
16. S.Narison, Phys.Lett. B198 (1987) 104.
17. C.A.Dominguez and N.Paver, Phys.Lett. B197 (1987) 423.

18. For the exponentiation method see:
F.Fucito, G.Parisi, S.Petrarca, Phys.Lett. 115B (1982) 148; G.Martinelli, G.Parisi, R.Petronzio, F.Rapuano 116B (1982) 432; C.Bernard et al., Phys.Rev.Lett. 49 (1982) 1076; S.Gottlieb, P.B.Mackenzie, H.B.Thacker, D.Weintgarten, Nucl.Phys. B263 (1986) 704.
19. For the source method see:
C.Bernard in "Gauge Theories on a lattice", 1984 eds. Zachos et al., Nat.Tech. Information Service, Springfield, VA (1984); R.Gupta et al., Phys.Lett. 164B (1985) 347.
20. G.Martinelli and C.T.Sachrajda, Nuch.Phys. B316 (1989) 305.
21. J.Ashman et al. (EMC Collaboration), Phys.Lett. B206 (1988) 364.
22. B.Meyer and C.Smith, Phys.Lett. 123B (1983) 62.
23. G.Martinelli and Y.C.Zhang, Phys.Lett. 123B (1983) 77.
24. R.Groot, J.Hoek and J.Smit, Nucl.Phys. B237 (1984) 111.
25. J.Mandula, G.Zweig and J.Govaerts, Nucl.Phys. B228 (1983) 9.
26. G.Martinelli and Zhang Yi Cheng, Phys.Lett. B125 (1983) 7.
27. S.R.Amendolia et al., Nucl.Phys. B277 (1986) 168.
28. C.J.Bebek et al., Phys.Rev. D17 (1987) 1693 and refs. therein.
29. C.Newman et al., Phys.Rev.Lett. 42 (1979) 951.
30. J.Badier et al., Proc. EPS Conference on HEP, Geneva (1979) p.751.
31. R.Barate et al., Phys.Rev.Lett. 43 (1979) 1541.
32. B.Betev et al., Z.Phys. C28 (1985) 15.
33. J.D.Bjorken and S.D.Drell, "Relativistic Quantum Mechanics" (Mc Graw Hill, 1965)
34. G.Höler et al., Nucl.Phys. B114 (1976) 505 and references therein.
35. G.Höler, F.Kaiser, R.Koch and E.Pietarinen, "Handbook of pion-nucleon scattering" (Fachs-information-Zeutrum, Karlsruhe, 1979) p.427 and references therein.
36. L.Maiani, G.Martinelli, M.L.Paciello, B.Taglienti, Nucl.Phys. B293 (1987) 420.
37. S.Güsken et al., CERN-TH 5027/88 (1988).
38. Bacilieri et al., (Ape Collaboration), presented by F.Rapuano at the XXIV HEP Conference, Münich, August 1988, R.Kothaus and J.H.Kühn eds, Springer-Verlag (1989) 804.
39. G.Martinelli, Phys.Lett. 141B (1984) 395.
40. C.Bernard, A.Soni and T.Draper, Phys.Rev. D36 (1987) 363.
41. G.Curci, E.Franco, L.Maiani and G.Martinelli, Phys.Lett. 202B (1988) 363.
42. C.Bernard, T.Draper, G.Hockney and A.Soni, Talk given by C.Bernard at the "Int. Symposium: Field Theory on the Lattice", Seillac (France), Nucl.Phys. B (Proc. Sup.) 4 (1988) 483.
43. N.Cabibbo, G.Martinelli and R.Petronzio, Nucl.Phys. B244 (1984) 381.
44. R.C.Brower, M.B.Gavela, R.Gupta and G.Maturana, Phys.Rev.Lett. 53 (1984) 1318.
45. L.Maiani and G.Martinelli, Phys.Lett. B181 (1986) 344.
46. G.Kilcup, Proceedings of the Rencontre de Moriond, March 1989.
47. J.F.Donoghue, E.Golowich and B.R.Holstein, Phys.Lett. B199 (1986) 442.
48. M.B.Gavela, L.Maiani, G.Martinelli, O.Pène and S.Petrarca, Phys.Lett. 211B (1988) 139.
49. M.A.Shifman, a.I.Vainshtein and V.J.Zakharov, Nucl.Phys. B120 (1977) 346; Sov.Phys. Jept. 45 (1977) 670.
50. J.Bijens and M.B.Wise, Phys.Lett. B137 (1984) 245; J.F.Donoghue, E.Golowich, B.R.Holstein and J.Trampetic, Phys.Lett. B179 (1986) 361 and B188 (1987) 511; A.Buras and J.M.Gèrard, Phys.Lett B192 (1987) 156; S.Sharpe, Phys.Lett. B194 (1987) 551.
51. E.Franco, L.Maiani, G.Martinelli and A.Morelli, Nucl.Phys. B317 (1989) 63.
52. L.Wolfenstein, Phys.Rev.Lett. 51 (1983) 1945.

53. R.Gupta, G.Guralnik, W.Kilcup, A.Patel and S.R.Sharpe, Phys.Lett. B192 (1987) 149.
54. C.Bernard, A.El-Khadra and A.Soni, Proceedings of the Conference on Lattice Field Theory, Fermilab 22-28 September 88, Nucl.Phys.B (Proc.Suppl.) 9 (1989) 186.
55. M.Crisafulli et al., Phys.Lett. B233 (1989) 90.
56. N.Cabibbo, Phys.Rev.Lett. 10 (1963) 531; M.Kobayashi and K.Maskawa, Prog.Theor.Phys. 49 (1972) 282.
57. The Tagged Photon Spectrometer Collaboration, J.C.Anjos et al., Fermilab Preprint FNAL 88/141-E (1988).
58. MARK 3 Collaboration, D.Hitlin, Proceedings of the 1987 International Symposium on Lepton and Photon Interactions at High Energies, ed. W.Bartel and R.Ruckl, North-Holland, Amsterdam (1987); R.Schindler, Proceedings of the XXXIII International Conference on High Energy Physics, vol. 1, p.745, ed.S.C.Loken, World Scientific (1987); D.Coffman, Ph.D. Thesis, California Institute of Technology, CALT-68-1415 (1987).
59. For a review see for example P.J.Franzini, Phys.Rep. C173 (1989) 1.
60. E.Eichten, Nucl.Phys.B (Proc.Suppl.) 4 (1988) 170.
61. T.M.Aliev et al., Sov.J.Nucl.Phys. 40 (1984) 527.
62. M.Wirbel, B.Stech and M.Bauer, Z.Phys. C29 (1985) 637.
63. B.Grinstein et al., University of Toronto Preprint, UTPT-88-12 (1988).
64. G.Heatlie, G.Martinelli, C.Pittore, G.C.Rossi, C.T.Sachrajda, in preparation.

EXPERIMENTAL STATUS OF CP VIOLATION

J. Steinberger

CERN, Geneva, Switzerland
and
Scuola Normale Superiore, Pisa, Italy

ABSTRACT

The experiments on CP violation as well as the phenomenology of CP violation in K decay are reviewed.

The two most recent experiments searching for direct CP violation—the University of Chicago-Saclay experiment at FNAL, as well as the CERN experiment which has given a positive result—are discussed in some detail.

1. EXPERIMENTS ON CP VIOLATION

In the unfolding of the story of the elementary particles and their interactions, underlying symmetries have often played an important role. An excellent example is the discovery of parity violation in 1957 [1]. This stimulated a number of vital experiments, and within a year the V − A current structure of the weak interaction was known [2, 3].

The discovery of CP violation has not had a similar impact, and the few manifestations of CP violation subsequently discovered had no immediate effect on other areas of particle physics; instead, the way in which CP violation enters the general framework of particle theory has remained obscure.

The advent of the electroweak theory, and especially the realization [4] that CP violation could quite naturally occur in it—through the quark mixing matrix—has generated a new experimental incentive to test this hypothesis, in particular the experiments searching for 'direct' CP violation that are reviewed in Sections 3 and 4.

1.1 Discovery of CP violation

The experiment of Christenson et al. [5] was probably motivated by the perplexing result (later understood to be erroneous) of 'anomalous regeneration' [6]. The experiment consisted of a double-armed spectrometer looking at a He-filled region for decays taking place in a neutral beam sufficiently far from its target of origin so that the K_S component had decayed (see Fig. 1). The K_{e3} decays were rejected by means of threshold Cherenkov counters. The signal was dominated by the $\pi^+\pi^-\pi^0$ and $\mu^\pm\pi^\mp\nu$ decays of the long-lived kaon. The invariant mass distribution, assuming that both dectected particles are pions, shows no clear peak at the K^0 mass above the three-body background, but the mass slice at the K^0 has a peak in the forward direction (in contrast to neighbouring mass bins) (Fig. 2), which demonstrated the

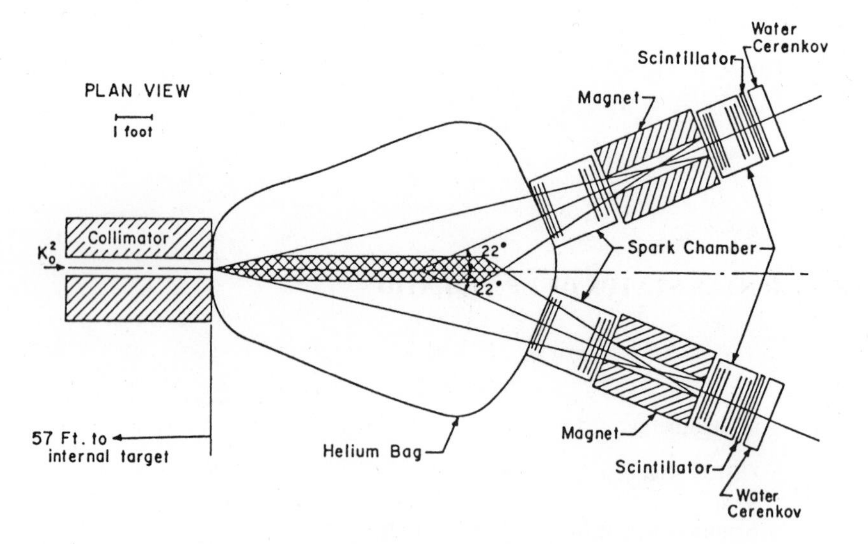

Fig. 1 Apparatus of the experiment in which CP violation was discovered [5].

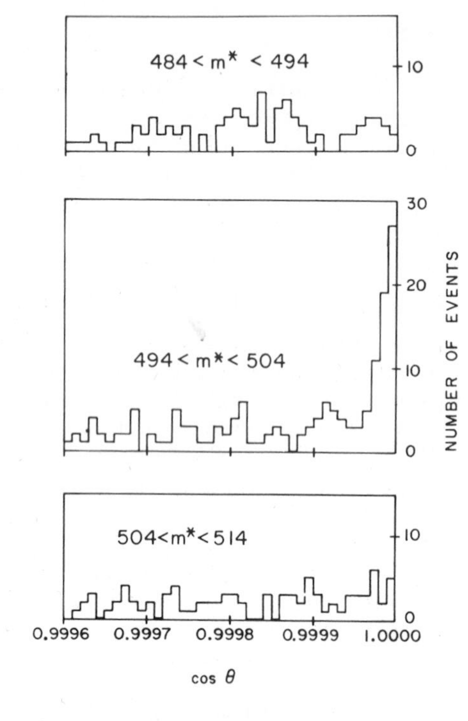

Fig. 2 Angular distributions of the observed events near the forward direction in three mass bins. The forward peak at the K^0 mass demonstrated CP violation [5].

$\pi^+\pi^-$ decay of the long-lived kaon, and therefore CP violation. The branching ratio was found to be small

$$\frac{\Gamma_{L,+-}}{\Gamma_{L,all}} = (2 \pm 0.4) \times 10^{-3} ,$$

corresponding to

$$|\eta_{+-}| \equiv \frac{\langle + - |T|L\rangle}{\langle + - |T|S\rangle} \simeq 2.3 \times 10^{-3} .$$

1.2 $K_L^0 \to 2\pi^0$

In addition to the decay $K_L \to \pi^+\pi^-$, CP violation has been observed in only two channels: the decay $K_L \to \pi^0\pi^0$ and the charge asymmetry in K_L semileptonic decay.

The $2\pi^0$ decay presented considerable difficulty at the time, and the first observation [7, 8] yielded results for the partial width that were in disagreement with the values accepted at present. The first results in agreement are due to the bubble chamber experiment of Budagov et al. [9] and the counter experiment of Banner et al. [10]. The apparatus of the counter experiment is shown in Fig. 3. A neutral beam of very low momentum (200–300 MeV/c) produced at the Princeton–Penn 1.3 GeV proton accelerator passes through a helium bag at a sufficient distance from the target to eliminate the K_S component. The photons from neutral K_L decays were observed in a lead-plate/spark-chamber sandwich arrangement surrounding three sides of the bag, and a magnetic pair spectrometer on the fourth side. The $2\pi^0$ (4γ) decay is overconstrained by the observation of the directions of the four photons as well as the energy of one of them. This permits an adequate separation from the $3\pi^0$ decay, as can be seen in the invariant mass plot of Fig. 4. The result for $\eta_{00} \equiv \langle 00|T|L\rangle/\langle 00|T|S\rangle$, $|\eta_{00}| = (2.3 \pm 0.3) \times 10^{-3}$, is in good agreement with the bubble chamber result of Budagov et al. [9], $|\eta_{00}| = (2.2 \pm 0.4) \times 10^{-3}$.

Until recently, the most precise determination was that of Holder et al. [11]. Figure 5 shows the detector. The four photons are converted in thin lead sheets, each followed by spark chambers to measure their positions and directions. The innovation was the use of a segmented calorimeter, in this instance a lead-glass array, to measure the photon energies. All more recent experiments follow this calorimetric technique. The experimental problem is the isolation of the $2\pi^0$ decay from the hundreds of times more frequent $3\pi^0$ decay. If two photons are missed in this decay, the invariant mass is smaller, whereas true $2\pi^0$ decays must reconstruct to the K^0 mass. The results (Fig. 6) showed a very clean peak corresponding to the CP-violating $K_L \to 2\pi^0$ decay, and the value for the relative rates of CP violation in the $\pi^0\pi^0$ and $\pi^+\pi^-$ decays, $|\eta_{00}/\eta_{+-}|^2 = 1.0 \pm 0.12$.

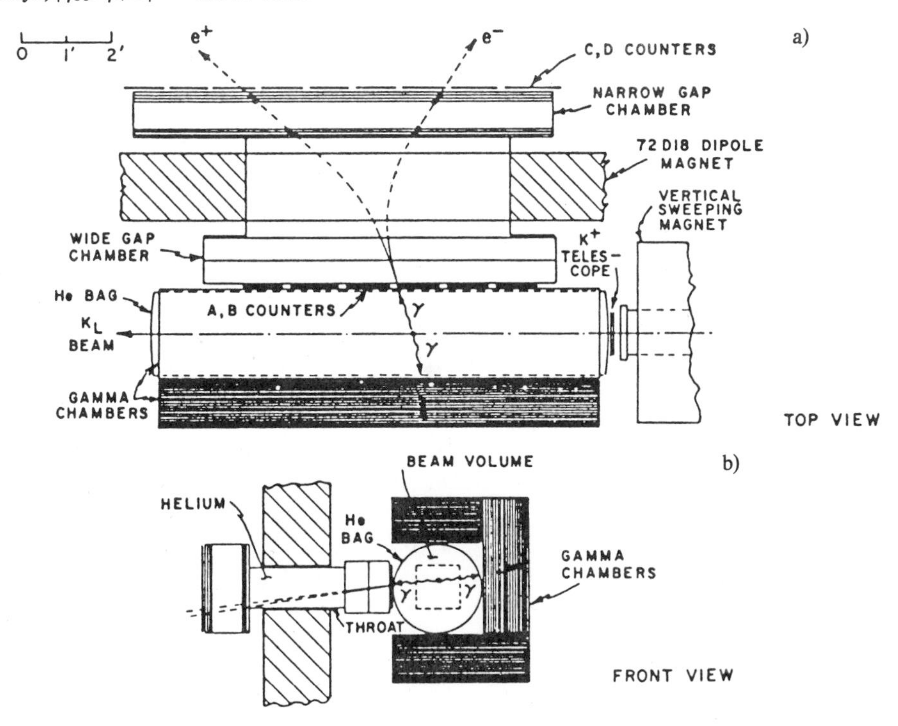

Fig. 3 Apparatus of Banner et al. for the measurement of $\Gamma_{L,00}$. At least one of the four photons is required to be detected in the pair spectrometer and its momentum vector is therefore well measured. The other photons are converted and observed in the lead/spark-chamber ('gamma chamber') stacks on the other three sides.

Fig. 4 Invariant mass distributions for events identified as $2\pi^0$ decays and for the $3\pi^0$ background. Experiment of Banner et al. [10].

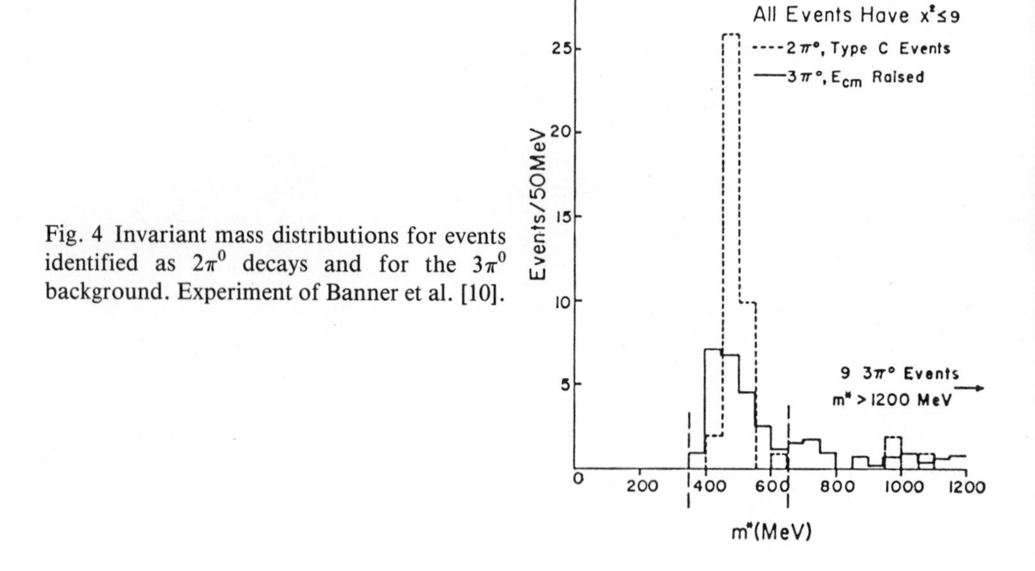

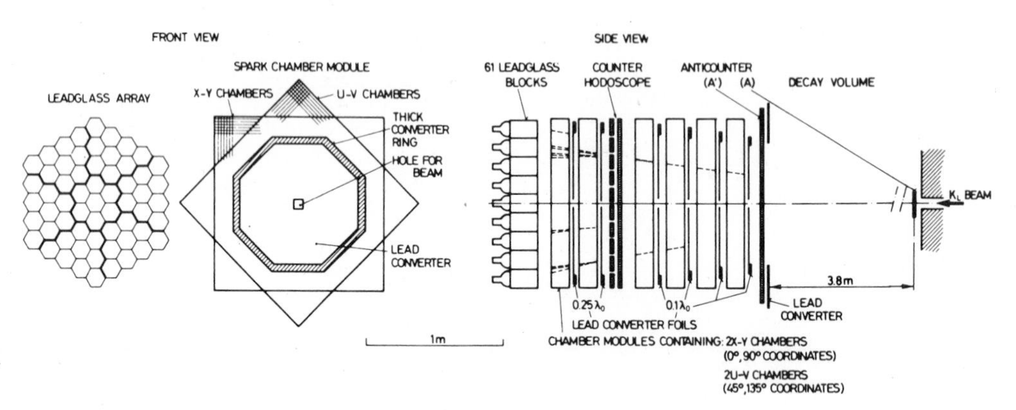

Fig. 5 Apparatus of Holder et al. [11] to measure $|\eta_{00}/\eta_{+-}|^2$.

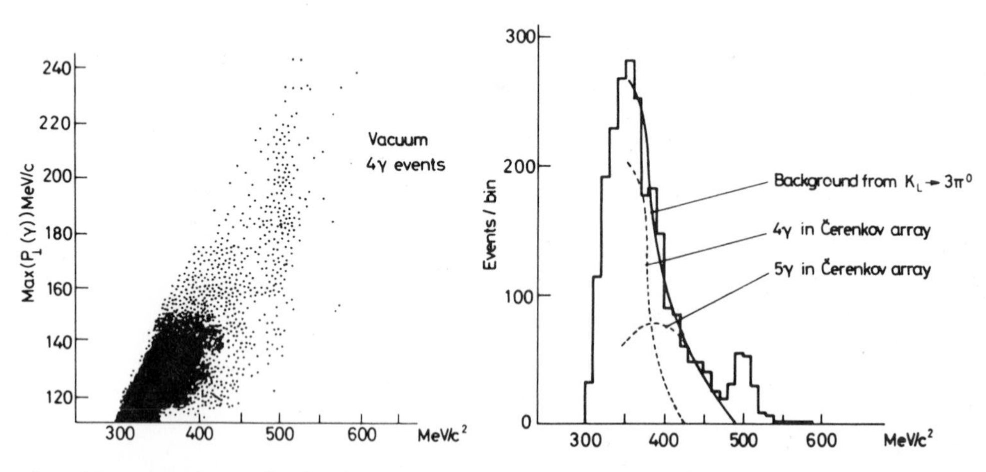

Fig. 6 Invariant mass distributions for the decays of K_S and K_L into four photons, from Ref. [11]. In K_S decay the $2\pi^0$ mode is completely dominant. The K_L decay shows a small $2\pi^0$ peak at the K^0 mass, containing 167 events, above the dominant background of $3\pi^0$ decays.

1.3 The charge asymmetry in K_L^0 semileptonic decay

The charge asymmetry in the decays $K_L^0 \to \ell^\pm + \pi^\mp + \nu$,

$$\delta_e = \frac{N_{e^+} - N_{e^-}}{N_{e^+} + N_{e^-}} ,$$

violates CP symmetry. It was first observed in two simultaneous experiments: one at Brookhaven in the electron decay (Bennett et al. [12]), the other at SLAC in the muon decay (Dorfan et al. [13]). The problem here is not in the identification of the decay mode—both are common decay channels and easy to identify—but rather in obtaining adequate statistics as well as low enough systematic biases to measure the small asymmetry of $\sim$ 0.003. The BNL apparatus is shown in Fig. 7. The electrons are identified in a gas threshold Cherenkov counter. Their sign is selected in a magnet, and the asymmetry is determined by reversing the magnetic field.

At present, the most precise results for both decays are those of Geweniger et al. [14]. In this experiment the asymmetry is measured sufficiently close to the K-producing target so that the K_S-K_L interference (see Section 2.8) can be observed as well and therefore the $\Delta S = \Delta Q$ rule also checked. The layout of the apparatus is shown in Fig. 8. The observed time-dependent asymmetries are shown in Fig. 9. The results for δ_ℓ are

$$\delta_e = (3.41 \pm 0.18) \times 10^{-3} \quad \text{and} \quad \delta_\mu = (3.13 \pm 0.29) \times 10^{-3} .$$

The biggest systematic correction in these asymmetry measurements is due to the difference in the nuclear interaction cross-sections of the π^+ and π^- mesons, which must pass through a small amount of matter before detection.

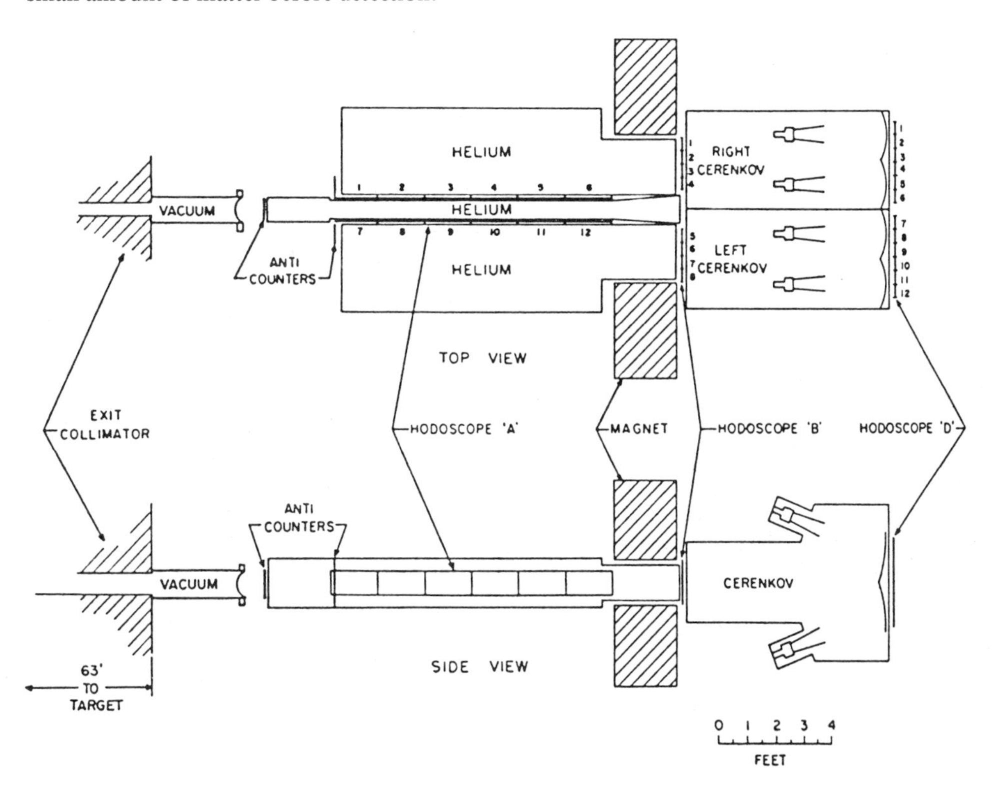

Fig. 7 Apparatus of Bennett et al. [12] for determining the charge asymmetry in K_{e3} decay. The electrons and positrons are charge selected by the magnet and identified in the gas (ethylene) threshold Cherenkov counter.

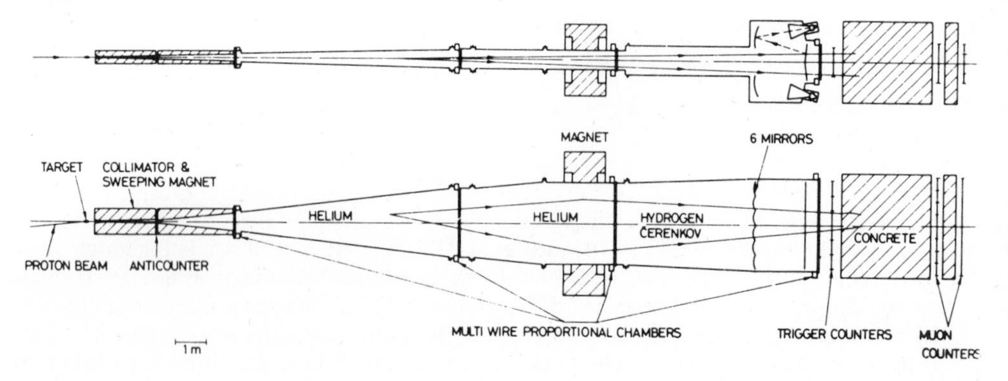

Fig. 8 Apparatus of Geweniger et al. [14] for the determination of K_S–K_L interference in the charge symmetry of the decays $K^0 \to e^{\pm} + \pi^{\mp} + \nu$ and $K^0 \to \mu^{\pm} + \pi^{\mp} + \nu$.

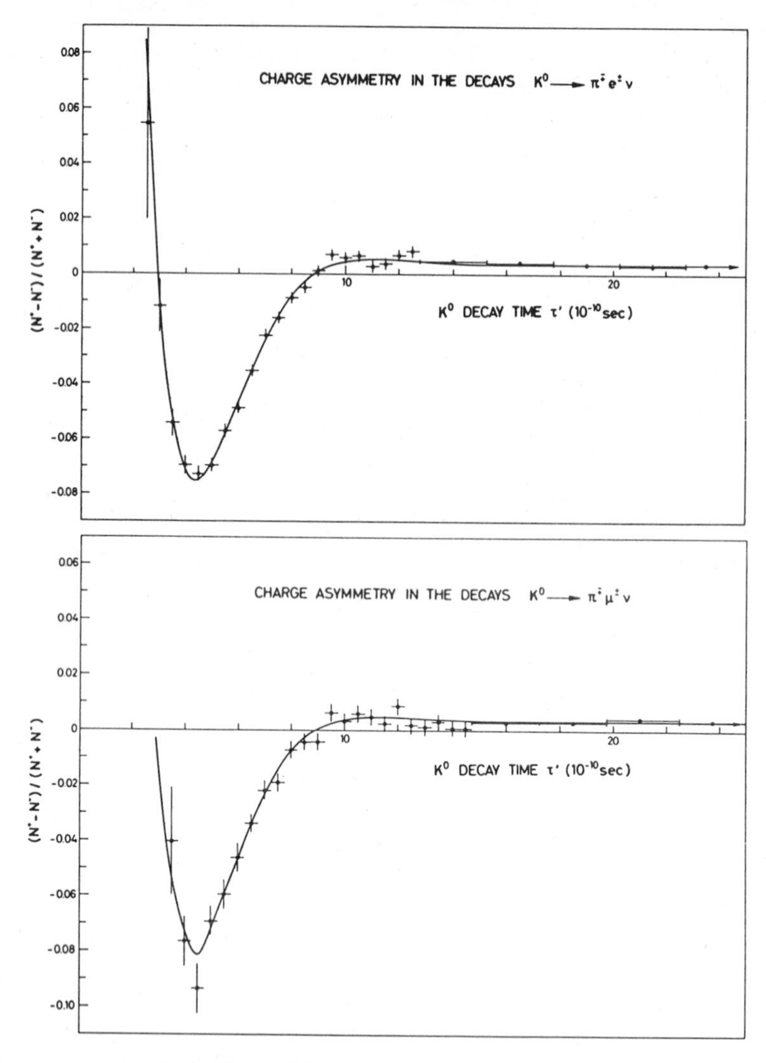

Fig. 9 Charge asymmetries in K_{e3} and $K_{\mu 3}$ decays as function of τ, the time in the kaon rest frame since its production. The asymmetry at long times is the CP-violating asymmetry δ_ℓ. The oscillating asymmetry at short times is due to K_S–K_L interference and does not violate CP (Geweniger et al. [14]).

1.4 CP violation search in the three-pion decay

The most probable 3π decay is in the state of zero orbital angular momenta, as is known since the τ–θ puzzle days of 1956 from the τ^+ Dalitz plot analysis. This is CP-allowed for K_L and CP-forbidden for K_S. The $3\pi^0$ decay is strictly CP –. The $\pi^+\pi^-\pi^0$ decay can be CP + as well, but for odd orbital angular momenta, which have smaller amplitude. The search for CP violation in these decay modes is difficult, because pure K_S beams do not exist; just after the production target, the K_L intensity is equal to that of the K_S, and, relative to the K_S, increases rapidly with distance. It is therefore necessary to look for a K_L–K_S interference immediately following the production target. No experiment with sensitivity close to the expected level has so far been achieved. There is some hope that this sensitivity will be accessible to the experiment just beginning at CERN, which uses stopped antiprotons in hydrogen [15].

1.5 Interference experiments in the two-pion decay

A coherent mixture of K_S and K_L will, in its time evolution, show a CP-violating interference, which can be used to measure the phases of η_{+-} and η_{00}. The mixed state can be produced by passing a K_L beam through a 'regenerator' or by observing the decay sufficiently close to the neutral-beam production target so that the K_S amplitude is still large enough. The expected time distribution has the form

$$\frac{dN_{\pi\pi}}{d\tau} \propto |\varrho|^2 e^{-\Gamma_S\tau} + |\eta|^2 e^{-\Gamma_L\tau} + 2|\varrho\eta|\, e^{-(\Gamma_S+\Gamma_L)\tau/2} \cos(\Delta m\tau + \phi_\varrho - \phi_\eta) \ .$$

Here τ is the time as measured in the K rest frame and Δm is the $K_L - K_S$ mass difference, which, fortunately for these experiments, is of the order of Γ_S. In the case of a regenerator in a K_L beam, ϱ is the regeneration amplitude and ϕ_ϱ its phase. In the case of a K ($\bar{K}$) beam, $\varrho = 1\,(-1)$.

The first observation of this interference was made by Fitch et al. [16]. In this experiment, the integrated $\pi^+\pi^-$ decay rate in a relatively short space region (see the apparatus in Fig. 10) was compared for a) air only in the region, b) low-density Be as regenerator, and c) solid Be as regenerator. The observed rate for the low-density regenerator was not just the sum of the regenerated and K_L-free decay rates; it also reflected the interference term. The magnitude of the interference depends on Δm, as can be seen from the result of the experiment in Fig. 11. At that time, Δm was hardly known.

The experiments of Alff-Steinberger et al. [17] and of Bott-Bodenhausen et al. [18] observed the time dependence of the decay rate following a regenerator. This rate is dominated at short times by the K_S decay and at long times by the CP-violating decay; at intermediate times it shows a large interference effect. The apparatus of Alff et al. is shown in Fig. 12, and the result in Fig. 13a. The interference can be isolated (Fig. 13b) and provides a good measurement of the mass difference, $\Delta m/\Gamma_S = 0.455 \pm 0.034$, as well as of the difference between the CP-violating phase and the regeneration phase, $\phi_{+-} - \phi_\varrho = 1.41 \pm 0.18$.

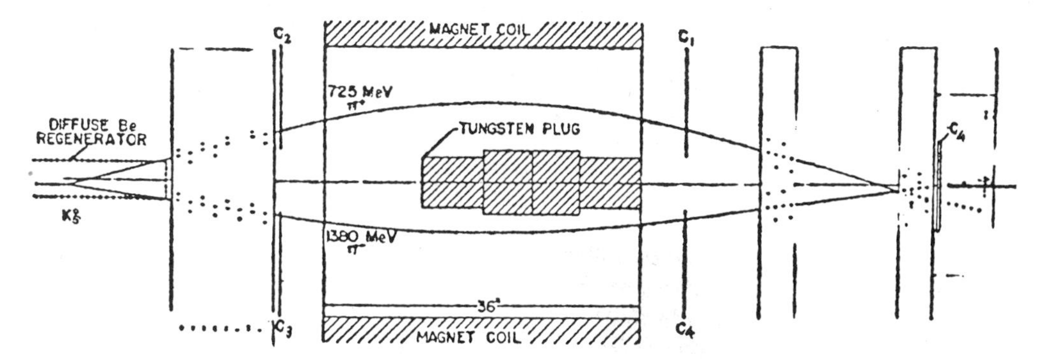

Fig. 10 Spark-chamber spectrometer of Fitch et al. for the demonstration of K_S–K_L interference in $\pi^+\pi^-$ decay [16].

Fig. 11 Result of Fitch et al. [16] for the phase $\alpha = \phi_\varrho - \phi_\eta$ as a function of the mass difference $\Delta m/\Gamma_S$, which at the time was poorly known.

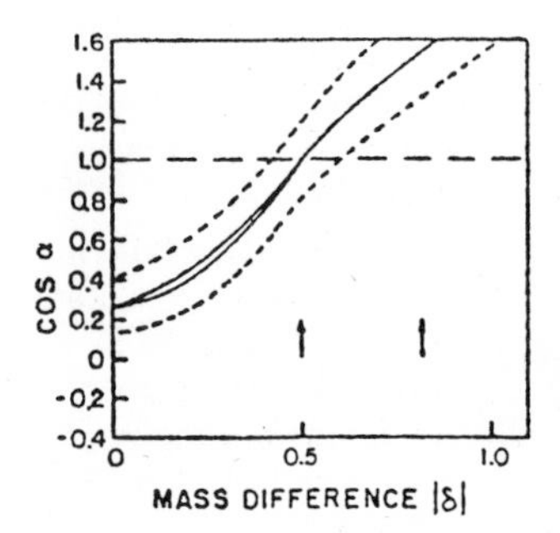

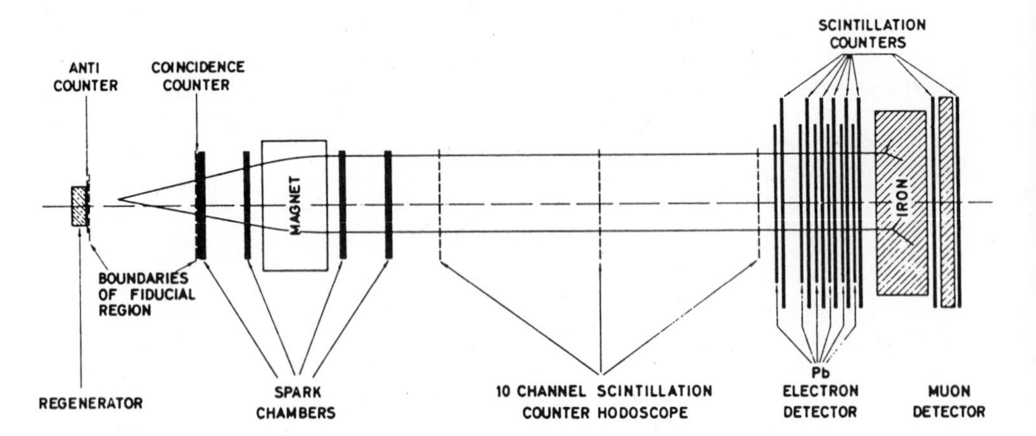

Fig. 12 Spark-chamber spectrometer of Alff-Steinberger et al. [17] used in the observation of the K_S-K_L interference in $\pi^+\pi^-$ decay, and its time dependence.

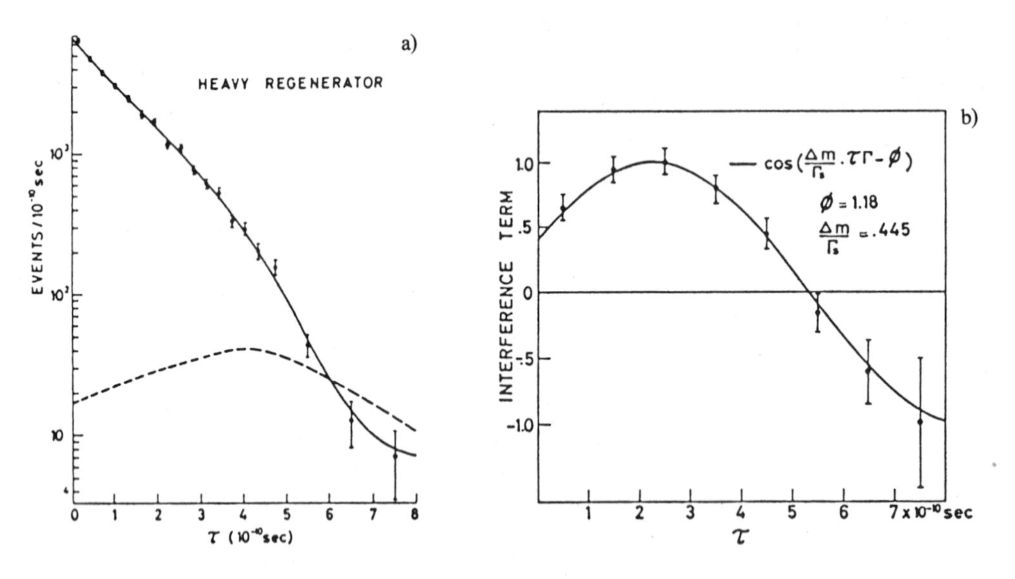

Fig. 13 a) Proper time distribution following a 12 cm thick copper regenerator. The dotted line is the acceptance of the apparatus. At short times the K_S decay dominates; at longer times the interference is very clear. The observed times are too short to reach the region ($\tau > 12 \times 10^{-10}$ s) dominated by K_L decay. b) The isolated interference term permits a measure of $\phi_\eta - \phi_\varrho$ and $\Delta m/\Gamma_S$.

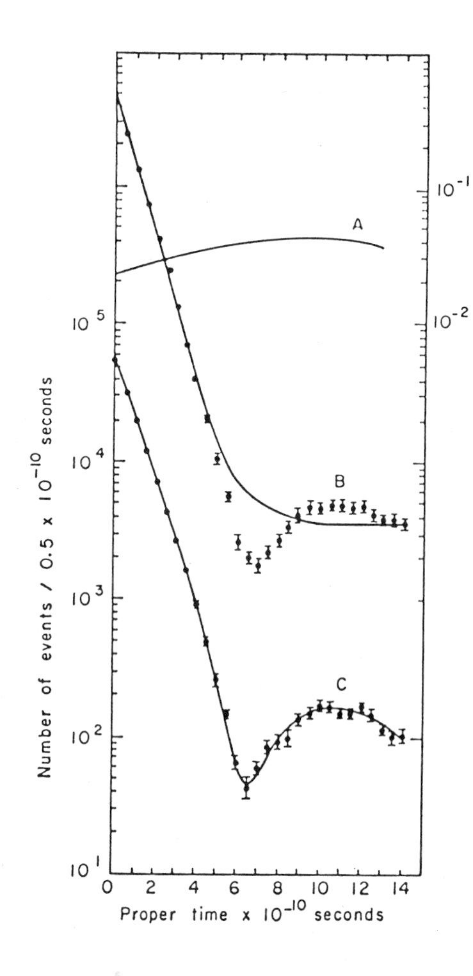

Fig. 14 K_S-K_L interference in the $\pi^+\pi^-$ decay following a carbon regenerator (Carithers et al. [19, 20]).

Figure 14 shows the result of the more recent and much more precise experiment of Carithers et al. [19, 20]. The regeneration phase was determined independently by measuring the $K_{\ell 3}$ charge asymmetry following the same regenerator. The combined result is $\phi_{+-} = (45.5 \pm 2.8)°$.

The experiment of Böhm et al. [21] in 1969 represented a substantial step forward in the experimental technique. It was the first to use a *short* neutral beam following the production target, so that the surviving K_S amplitude in the decay acceptance region was sufficient to permit observation of the $K_S - K_L$ interference and measurement of ϕ_{+-} directly, without regenerator. Figure 15 shows the layout. This method was subsequently used by Geweniger et

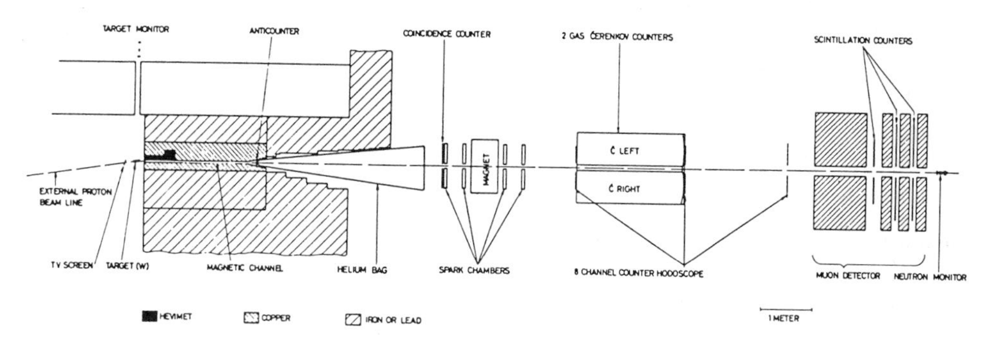

Fig. 15 Apparatus of Böhm et al. [21] for the measurement of the time-dependent K_S-K_L interference in the $\pi^+\pi^-$ decay.

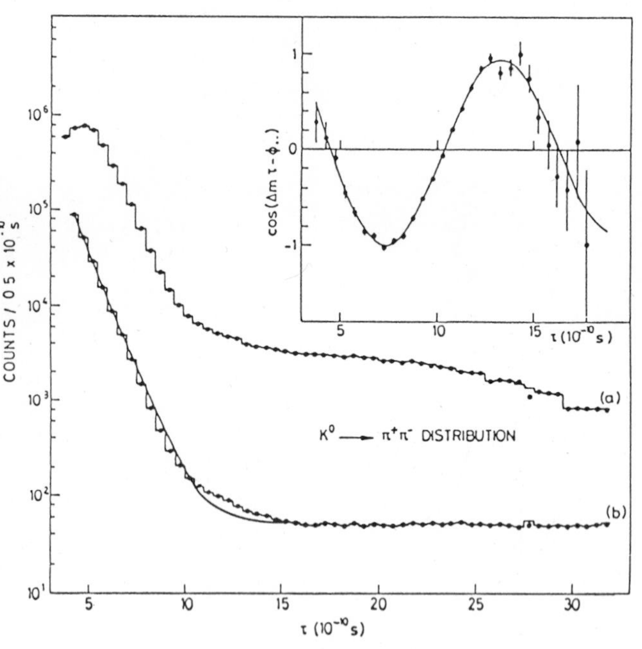

Fig. 16 Proper time distribution of $\pi^+\pi^-$ decay following a target in which mostly K (rather than $\bar{K}$) are produced (Geweniger et al. [22]). The upper histogram is raw data. In the lower histogram the data are corrected for acceptance, and the solid curve is a best fit neglecting interference. The inset shows the isolated interference term which gives a precise value of ϕ_{+-}.

al. [22, 23] in the most precise measurements available up to now. The experimental arrangement was basically that of Böhm et al. [21]; however, the experiment exploited the multiwire proportional chambers (MWPCs) invented in the mean time by Charpak [24]. It was the first application of this powerful new technique, and large chambers and associated circuitry had first to be developed [25]. The MWPCs permit rates that are hundreds of times higher than those possible with spark chambers, and they can be self-triggered, thus avoiding trigger scintillators, so that there can be less material in the detector. The apparatus is the same (Fig. 8) as that used also in the $K_{\ell 3}$ charge asymmetry measurements already cited [14, 16]. The $K_S - K_L$ interference in the $\pi^+\pi^-$ decay is shown in Fig. 16. The results were:

$$\phi_{+-} = (45.9 \pm 1.6)^\circ \ ,$$
$$|\eta_{+-}| = (2.30 \pm 0.035) \times 10^{-3} \ .$$

In addition, the experiment yielded a more precise measurement of Γ_S:

$$\Gamma_S = (1.119 \pm 0.006) \times 10^{10}\,s^{-1} \ .$$

Similar experiments were performed to measure ϕ_{00}, but this proved more difficult. The first results were those of Chollet et al. [26] and of Wolff et al. [27]. The four photons from a decay region following a copper regenerator are detected in the aluminium-plate/spark-chamber arrangement shown in Fig. 17, which permits the measurement of position and direction of each of the four γ-rays, as well as a crude measurement of each energy. From the proper time distribution of Fig. 18, ϕ_{00} was found to be: $\phi_{00} = (38 + 25)^\circ$.

Using the apparatus (Fig. 5) of Holder et al. [11], Barbiellini et al. [28] compared the $K \to 2\pi^0$ intensities following dilute and dense copper regenerators; they found $\phi_{00} - \phi_\varrho = (99.6 \pm 17)^\circ$. Making a comparison with previous results on $\pi^+\pi^-$ decay, $\phi_{+-} - \phi_\varrho = (92 \pm 5.5)^\circ$, the phase difference is $\phi_{00} - \phi_{+-} = (7.6 \pm 18)^\circ$.

Until very recently, the most precise experiment on ϕ_{00} and the phase difference was that of Christenson et al. [29], using a short neutral beam, a proportional wire spectrometer for the charged mode, and lead-glass Cherenkov detector blocks for the neutral mode, as shown in Fig. 19. The efficiency-corrected proper time distribution and the isolated interference term are shown in Fig. 20. The results are the following:

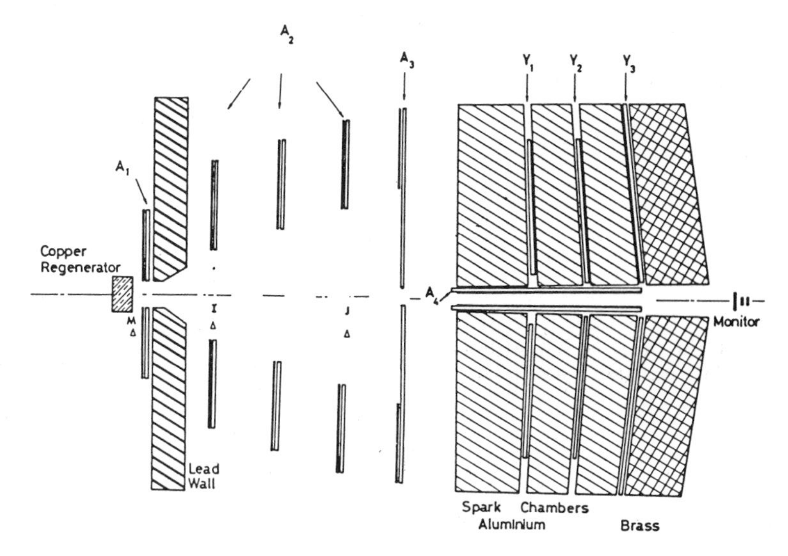

Fig. 17 Layout of the experiment of Chollet et al. [26] and of Wolff et al. [27] for the measurement of ϕ_{00}. The four photons from the decay region following a copper regenerator are detected in an aluminium-plate/spark-chamber sandwich arrangement which permits measurements of position, angle, and energy of the converted photons.

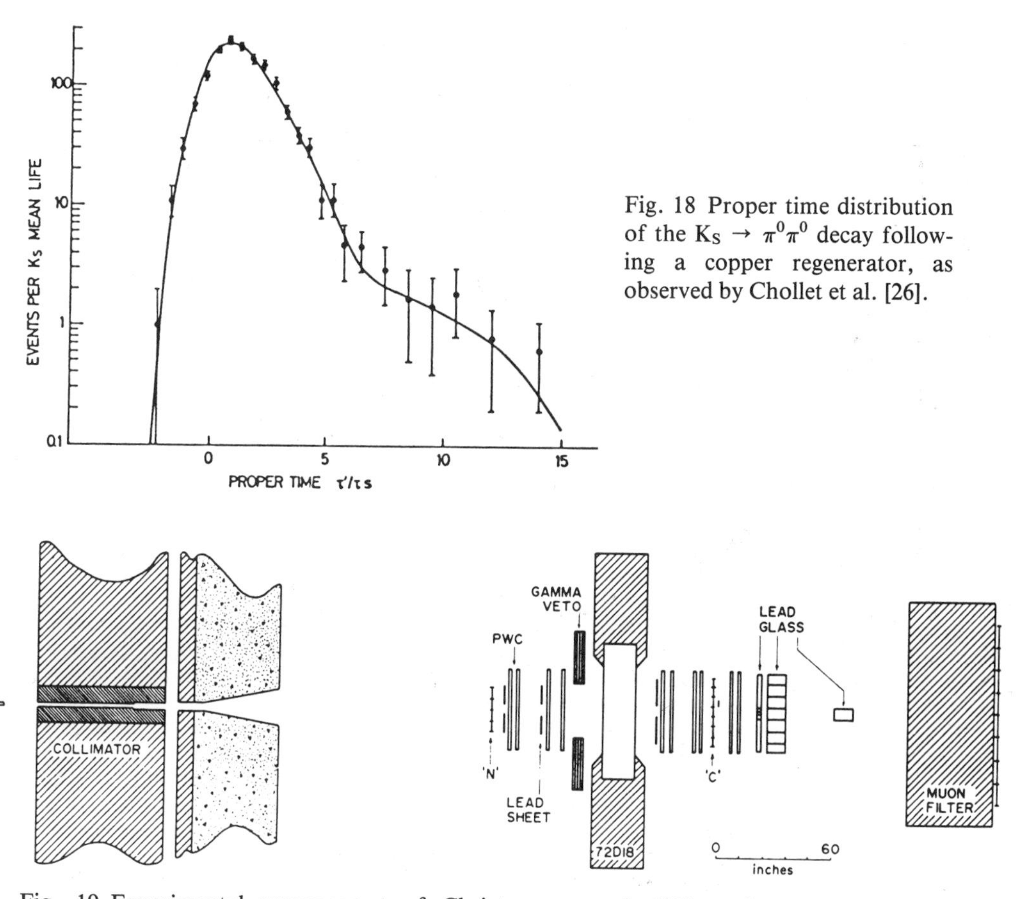

Fig. 18 Proper time distribution of the $K_S \rightarrow \pi^0\pi^0$ decay following a copper regenerator, as observed by Chollet et al. [26].

Fig. 19 Experimental arrangement of Christenson et al. [29] used to measure K_S–K_L interference in the neutral and charged two-pion decay modes.

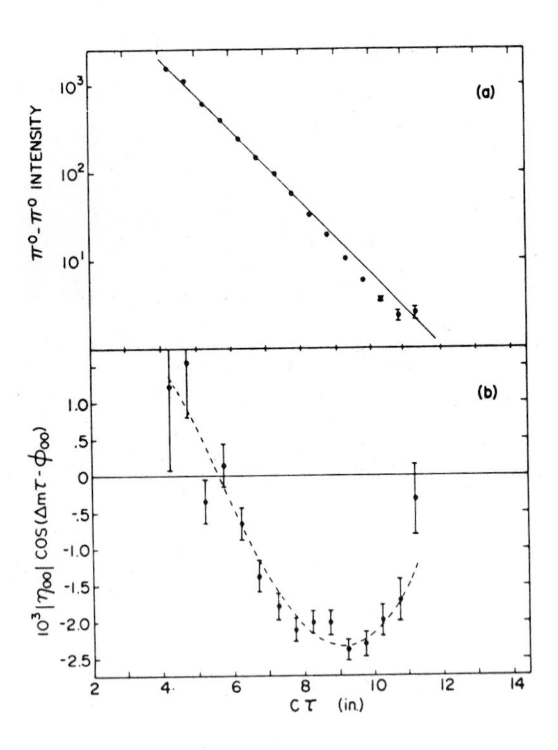

Fig. 20 Result of Christenson et al. 1979 [29]: a) time dependence of decay rate. The solid line is the expected rate without interference; b) isolated interference term.

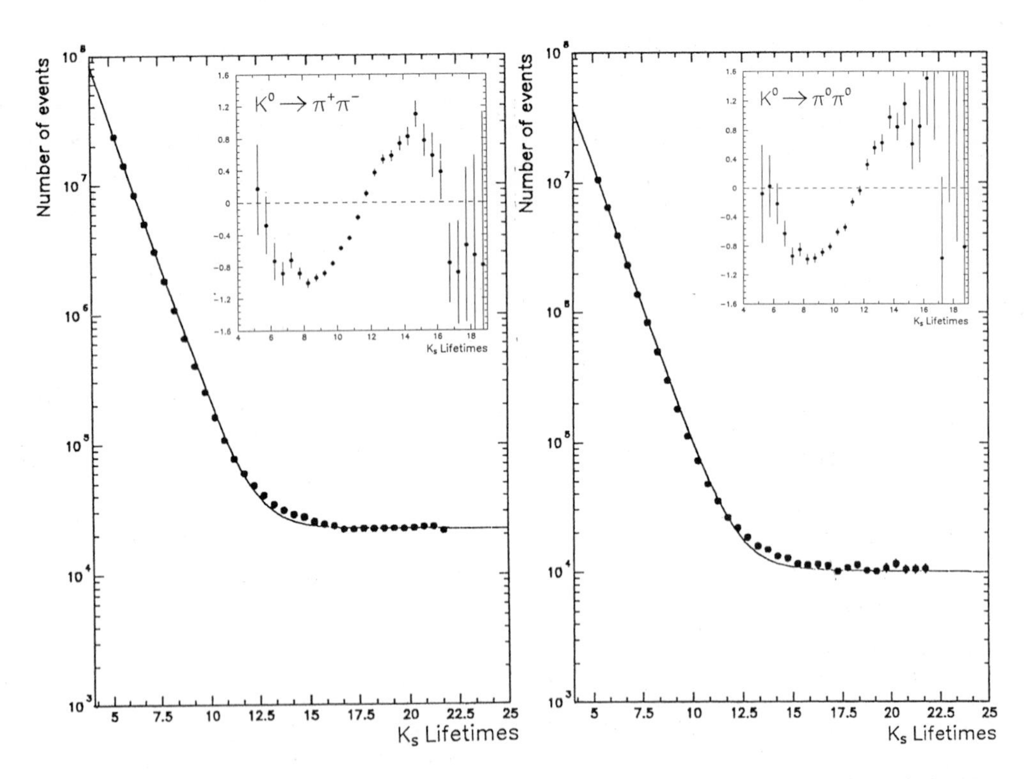

Fig. 21 The rate of decay of kaons to charged and neutral pions as a function of the K_S lifetime. Superimposed are the fitted lifetime distributions without interference. The insets show the interference terms extracted from the data. (From Ref. [30].)

$$\phi_{00} = (55.7 \pm 5.8)° \ , |\eta_{00}| = (2.33 \pm 0.18) \times 10^{-3} \ ,$$

$$\phi_{+-} = (41.7 \pm 3.5)° \ , |\eta_{+-}| = (2.27 \pm 0.12) \times 10^{-3} \ ,$$

$$\phi_{00} - \phi_{+-} = (12.6 \pm 6.3)° \ , |\eta_{00}/\eta_{+-}| = 1.00 \pm 0.09 \ .$$

This result for the phase difference is two standard deviations from a null result. A non-zero result of this order of magnitude would require CPT violation (see Section 2). Since all other experimental results are compatible with CPT conservation, this result was of great interest.

Happily, a new measurement was reported very recently by the CERN group [30]. The detector is described in Section 4, in connection with the result of this experiment for ϵ'/ϵ. The two decay modes are detected simultaneously. The phase is determined from the ratios of the decay rates for two beams whose target positions differ by 14 m, corresponding to a difference in proper time of $\sim 2\tau_s$ averaged over the spectrum. The acceptance of the apparatus drops out in these ratios. Figure 21 shows the observed decay time distributions for the two modes. The results for the phases are as follows:

$$\phi_{+-} = (46.8 \pm 1.5)° \ ,$$

$$\phi_{00} = (47.1 \pm 2.3)° \ ,$$

$$\phi_{00} - \phi_{+-} = (0.2 \pm 2.9)° \ .$$

The uncertainty for the phase difference is considerably reduced, and the value itself is consistent with zero.

1.6 Summary of experimental results on CP violation

$|\eta_{+-}|$

				Ref.
$(2.3 \pm 0.2) \times 10^{-3}$	Christenson et al.	BNL	'64	[5]
$(2.5 \pm 0.3) \times 10^{-3}$	Abashian et al.	BNL	'64	[31]
$(3.0 \pm 0.7) \times 10^{-3}$	de Bouard et al.	CERN	'65	[32]
$(2.3 \pm 0.2) \times 10^{-3}$	Galbraith et al.	Nimrod	'65	[33]
$(2.23 \pm 0.05) \times 10^{-3}$	Messner et al.	SLAC	'73	[34]
$(2.30 \pm 0.035) \times 10^{-3}$	Geweniger et al.	CERN	'74	[22]
$(2.25 \pm 0.05) \times 10^{-3}$	De Voe et al.	Chicago	'77	[35]
$(2.27 \pm 0.12) \times 10^{-3}$	Christenson et al.	NYU	'79	[29]

Best value $(2.27 \pm 0.025) \times 10^{-3}$

ϕ_{+-}

$(51 \pm 11)°$	Bennett et al.	Columbia-BNL	'68	[36]
$(41 \pm 15)°$	Böhm et al.	CERN	'69	[21]
$(37.1 \pm 7.1)°$	Bennett et al.	Columbia-BNL	'69	[37]
$(44.7 \pm 5)°$	Jensen et al.	Chicago	'69	[38]
$(49.3 \pm 6.8)°$	Faissner et al.	CERN	'69	[39]
$(36.2 \pm 6.1)°$	Carnegie et al.	Princeton-BNL	'72	[40]
$(45.9 \pm 1.6)°$	Gjesdal et al.	CERN	'74	[23]
$(45.5 \pm 2.8)°$	Carithers et al.	Columbia-BM	'75	[20]
$(41.7 \pm 3.5)°$	Christenson et al.	NYU	'79	[29]

Best value $(45.2 \pm 1.4)°$

$|\eta_{00}|$

$(2.2 \pm 0.4) \times 10^{-3}$	Budagov et al.	CERN	'68	[9]
$(2.3 \pm 0.3) \times 10^{-3}$	Banner et al.	Princeton	'68	[10]
$(3.6 \pm 0.6) \times 10^{-3}$	Gaillard et al.	CERN	'69	[41]
$(2.0 \pm 0.23) \times 10^{-3}$	Barmin et al.	Serpukhov	'70	[42]

$|\eta_{00}/\eta_{+-}|$

(1.03 ± 0.07)	Banner et al.	Princeton	'72	[43]
(1.00 ± 0.06)	Holder et al.	CERN	'72	[11]
(1.00 ± 0.09)	Christenson et al.	NYU	'79	[29]

Best value (1.01 ± 0.04)

ϕ_{00}

$(51 \pm 30)°$	Chollet et al.	CERN	'70	[26]
$(38 \pm 25)°$	Wolf et al.	CERN	'71	[27]

$\phi_{00} - \phi_{+-}$

$(7.6 \pm 18)°$	Barbiellini et al.	CERN	'73	[28]
$(12.6° \pm 6.2)°$	Christenson et al.	NYU	'79	[29]
$(0.2 \pm 2.9)°$	Barr et al.	CERN	'89	[30]

Best value $(2.4 \pm 2.6)°$

δ_ℓ

$(2.24 \pm 0.36) \times 10^{-3}$	Bennett et al.	Columbia–BNL	'67	[12]
$(4.05 \pm 1.35) \times 10^{-3}$	Dorfan et al.	SLAC	'67	[13]
$(3.46 \pm 0.33) \times 10^{-3}$	Marx et al.	Columbia–BNL	'70	[44]
$(2.8 \pm 0.5) \times 10^{-3}$	Piccioni et al.	SLAC	'72	[45]
$(3.33 \pm 0.5) \times 10^{-3}$	Williams et al.	Yale–BNL	'73	[46]
$(3.18 \pm 0.38) \times 10^{-3}$	Fitch et al.	Princeton	'73	[47]
$(3.41 \pm 0.18) \times 10^{-3}$	Geweniger et al.	CERN	'74	[14]
$(3.13 \pm 0.29) \times 10^{-3}$	Geweniger et al.	CERN	'74	[14]

Best value $(3.33 \pm 0.15) \times 10^{-3}$
Search in 3π decay

η_{000}

$0.08 \pm 0.18 + i(0.05 \pm 0.27)$	Barmin et al.	Serpukhov	'83	[48]

η_{+-0}

$0.13 \pm 0.2 + i(0.17 \pm 27)$	Metcalf et al.	CERN	'72	[49]
$0.05 \pm 0.17 + i(0.15 \pm 0.33)$	Barmin et al.	Serpukhov	'85	[50]
$0.04 \pm 0.03 + i(0.015 \pm 0.07)$	Border et al.	Serpukhov	'86	[51]

2. PHENOMENOLOGY OF CP VIOLATION IN K DECAY

2.1 Early history

In the beginning [52] there were 'V^0's'—new neutral particles decaying to two charged. The K^0 and its decay into two pions emerged clearly in 1952 [53]. Following the discovery of the isotopic spin–electric charge relationships of the strange particles in 1953 by Gell-Mann [54], Gell-Mann and Pais [55] were the first to notice that there must be two K^0's, with opposite assignment of isospin projection and strangeness, K^0 and $\bar{K}^0$. This was beautifully confirmed by Landé et al. in 1956 [56].

It follows that if CP were conserved in the decay interaction, the two states $|K_1\rangle = |K\rangle + |\bar{K}\rangle$ and $|K_2\rangle = |K\rangle - |\bar{K}\rangle$ would be eigenstates of the weak interaction, with well-defined but different lifetimes, slightly different (second order in the weak interaction) masses, and with the K_1 decaying to CP-even and the K_2 to CP-odd final states. Both K_1 and K_2 evolve exponentially with time:

$$\frac{d|K_1\rangle}{d\tau} = -iM_S|K_1\rangle\,, \qquad \frac{d|K_2\rangle}{d\tau} = -iM_L|K_2\rangle\,,$$

with

$$\begin{aligned} M_S &= m_S - i\Gamma_S/2 & M_L &= m_L - i\Gamma_L/2 \\ m_S &= 497.7\ \text{MeV} & m_L &= 497.7\ \text{MeV} \\ \Gamma_S &= 1.121 \pm 10^{10}\ \text{s}^{-1} & \Gamma_L &= 1.93 \pm 10^{7}\ \text{s}^{-1}\,. \end{aligned}$$

The dominant branching ratios are

$\pi^+\pi^-$	69%	$\pi^+\pi^-\pi^0$	12.4%
$\pi^0\pi^0$	31%	$\pi^0\pi^0\pi^0$	21.5%
		$\pi^\pm\mu^\mp\nu$	27.1%
		$\pi^\pm e^\mp\nu$	38.7%

The mass difference is

$$\Delta m \equiv m_L - m_S = 0.477\Gamma_S \simeq 10^{-16}\,m_K\,.$$

The K_1 and K_2 are orthogonal, and there can be no interference between K_1 and K_2 in decays to any CP eigenstate.

2.2 K_1–K_2 mixing

CP violation changes all this. The phenomenology of CP violation in K^0 decay was first discussed by Wu and Yang [57]. There are still two exponentially decaying eigenstates of the weak interaction, but these are now mixtures of K_1 and K_2:

$$|K_S\rangle = |K_1\rangle + (\tilde{\epsilon} + \delta)\,|K_2\rangle \qquad \text{and} \qquad |K_L\rangle = |K_2\rangle + (\tilde{\epsilon} - \delta)\,|K_1\rangle\,.$$

The observed CP violation is small, so we anticipate that $\tilde{\epsilon}$ and δ are small and neglect higher-order terms.

K_S and K_L are no longer orthogonal:

$$\langle L|S\rangle = 2\,\text{Re}\,\tilde{\epsilon} + 2i\,\text{Im}\,\delta\,.$$

2.4 Mass matrix

The time evolution of a general neutral-kaon state $\alpha|K\rangle + \beta|\bar{K}\rangle$ can also be expressed by the mass matrix:

$$\frac{d}{d\tau}\begin{pmatrix}\alpha\\ \beta\end{pmatrix} = -i\begin{pmatrix}M & A\\ B & \bar{M}\end{pmatrix}\begin{pmatrix}\alpha\\ \beta\end{pmatrix}.$$

The elements of this matrix, in terms of M_S, M_L, and the mixing parameters $\tilde{\epsilon}$ and δ are then:

$$M = \frac{M_S + M_L}{2} + \delta(M_S - M_L),$$

$$\bar{M} = \frac{M_S + M_L}{2} - \delta(M_S - M_L),$$

$$A = \frac{M_S - M_L}{2}(1 + 2\tilde{\epsilon}),$$

$$B = \frac{M_S - M_L}{2}(1 - 2\tilde{\epsilon}).$$

2.4 CPT conservation

If CPT is conserved—and therefore T is violated—then

$$\langle K|T|K\rangle = \langle \bar{K}|T|\bar{K}\rangle .$$

Therefore

$$M = \bar{M}$$

and

$$\delta = 0 .$$

2.5 T conservation

If, instead, T were conserved—and therefore CPT violated—then

$$\langle K|T|\bar{K}\rangle = \langle \bar{K}|T|K\rangle .$$

Therefore

$$A = B$$

and

$$\tilde{\epsilon} = 0 .$$

As we will see, the experiments show clear T violation and are consistent with CPT conservation.

2.6 Unitarity [58]

Unitarity here is the statement of probability conservation: the decay of the kaon is compensated by the appearance of the decay product,

$$-\frac{d}{d\tau}|\langle\psi^*|\psi\rangle|^2 = \sum_F |\langle F|T|\psi\rangle|^2 .$$

Here $|\psi\rangle$ is a general K^0 state: $|\psi\rangle = \alpha|S\rangle + \beta|L\rangle$, and F is the final state. Since α and β are arbitrary, this results in the two intuitively immediate relations

i) $\Gamma_S = \Sigma_F |\langle F|T|S\rangle|^2$

and

ii) $\Gamma_L = \Sigma_F |\langle F|T|L\rangle|^2$,

but also in the less obvious result from the cross-term:

$$\text{iii)} \left(-i\Delta m + \frac{\Gamma_S + \Gamma_L}{2}\right)\langle L|S\rangle = \sum_F \langle F|T|L\rangle^* \langle F|T|S\rangle . \tag{2}$$

Both sides of Eq. (2) are CP-violating: the left, because $\langle L|S\rangle = 0$ if CP were conserved; the right, because the final states of S and L could not be the same.

Let η_F be the CP-violating amplitude ratio:

$$\eta_F = \frac{\langle F|T|L\rangle}{\langle F|T|S\rangle} \qquad \text{if F is CP +},$$

and

$$\eta_F = \frac{\langle F|T|S\rangle^*}{\langle F|T|L\rangle^*} \qquad \text{if F is CP} - .$$

Then Eq. (2) becomes

$$\left(-i\Delta m + \frac{\Gamma_S + \Gamma_L}{2}\right) \langle L|S\rangle = \sum_F \eta_F^* \Gamma_F . \tag{2'}$$

The right-hand side is dominated by the two-pion decay. Present upper limits on CP violation in other channels ensure that these contribute at most several per cent. It is therefore an adequate approximation for the following argument to write Eq. (2′):

$$\left(-i\Delta m + \frac{\Gamma_S + \Gamma_L}{2}\right) \langle L|S\rangle = \eta_{+-}^* \Gamma_{S,+-} + \eta_{00}^* \Gamma_{S,00} . \tag{2''}$$

Experimentally, $\eta_{+-} = (2.27 \pm 0.025) \times 10^{-3}$, $\phi_{+-} = (45.2 \pm 1.3)°$, $(\eta_{00}/\eta_{+-}) = 1.01 \pm 0.04$, and $\phi_{00} - \phi_{+-} = (2 \pm 2.5)°$, so that Eq. (2″) becomes:

$$\left(-i\frac{\Delta m}{\Gamma_S} + \frac{1}{2}\right) \langle L|S\rangle = \eta_{\pi\pi}^* ,$$

where $|\eta_{\pi\pi}| = (2.27 \pm 0.03) \times 10^{-3}$ and $\phi_{\pi\pi} = (45.8 \pm 1.6)°$.

2.6.1 Unitarity, CPT conservation, and T violation

Since CP is violated, either CPT is violated, or T is violated, or both

If CPT is good	**If T is good**
$\delta = 0$	$\tilde{\epsilon} = 0$
$\langle L\|S\rangle = 2\,\mathrm{Re}\,\tilde{\epsilon}$	$\langle L\|S\rangle = 2i\,\mathrm{Im}\,\delta$
$2\left(\frac{i\Delta m}{\Gamma_S} + \frac{1}{2}\right) \mathrm{Re}\,\tilde{\epsilon} = \eta_{\pi\pi}$,	$2\left(\frac{\Delta m}{\Gamma_S} - \frac{i}{2}\right) \mathrm{Im}\,\delta = \eta_{\pi\pi}$,
$\therefore \phi_{\pi\pi} = \tan^{-1}\frac{2\Delta m}{\Gamma_S} = (43.7 \pm 0.3)°$	$\therefore \phi_{\pi\pi} = \tan^{-1}\frac{-\Gamma_S}{2\Delta m} = (138.7 \pm 0.3)°$

The experiment is in good agreement with CPT conservation. *T-violation is clearly demonstrated.* From now on we therefore take $\delta = 0$.

2.7 Isospin analysis of the two-pion decay

The states of well-defined charge can be resolved into states of well-defined isospin:

$$|+-\rangle = \sqrt{\frac{2}{3}}\,|I = 0\rangle + \sqrt{\frac{1}{3}}\,|I = 2\rangle ,$$

$$|00\rangle = -\sqrt{\frac{1}{3}}\,|I = 0\rangle + \sqrt{\frac{2}{3}}\,|I = 2\rangle .$$

Let $A\,e^{i\delta_0} = \langle I = 0|T|K\rangle$, where δ_0 is the $\pi\pi$ scattering phase; then

$$A_0^*\,e^{i\delta_0} = \langle I = 0|T|\bar{K}\rangle .$$

This is a consequence of CPT conservation and unitarity (Watson's theorem).

Similarly, let

$$A_2\, e^{i\delta_2} = \langle I = 2|T|K\rangle\,,$$

then

$$A_2^*\, e^{i\delta_2} = \langle I = 2|T|\bar{K}\rangle\,;$$

A_0 and A_2 are real if CP is conserved. Let

$$\epsilon = \frac{\langle I = 0|T|L\rangle}{\langle I = 0|T|S\rangle} = i\,\frac{\text{Im}\,A_0}{\text{Re}\,A_0} + \tilde{\epsilon}\,,$$

and

$$\epsilon' = \frac{1}{\sqrt{2}}\left[\frac{\langle I = 2|T|L\rangle}{\langle I = 0|T|S\rangle} - \tilde{\epsilon}\,\frac{\langle I = 2|T|S\rangle}{\langle I = 0|T|S\rangle}\right]$$

$$= \frac{i}{\sqrt{2}}\,e^{i(\delta_2 - \delta_0)}\left(\frac{\text{Im}\,A_2}{\text{Re}\,A_0} - \frac{\text{Im}\,A_0}{\text{Re}\,A_0}\,\frac{\text{Re}\,A_2}{\text{Re}\,A_0}\right).$$

Then

$$\eta_{+-} = \epsilon + \epsilon' \qquad \text{and} \qquad \eta_{00} = \epsilon - 2\epsilon'\,,$$

where we have ignored terms that are smaller by $\text{Re}\,A_2/\text{Re}\,A_0$. This ratio measures the admixture of $I = 2$ in the decay amplitude, and is obtained from the K_S branching ratios:

$$\text{Re}\,A_2/\text{Re}\,A_0 \simeq \left[\frac{\Gamma_{S,+-}}{\Gamma_{S,00}}\,\frac{1 - (2m_{\pi^0}/m_K)^2}{1 - (2m_{\pi^+}/m_K)^2} - 2\right] \div [\cos(\delta_2 - \delta_0)\,6\sqrt{2}]\,.$$

$$\simeq 0.049\,.$$

The measurements of the $\pi\pi$ scattering phase $\delta_2 - \delta_0$,

$\delta_2 - \delta_0$	$= (41 \pm 8)^\circ$	Devlin and Dickey 1979 [59]
	$= (29 \pm 3)^\circ$	Biswas et al. 1981 [60]
	$= (56 \pm 3)^\circ$	Kamal 1987 [61]

are not in agreement within their errors. Here it will be adequate to use $\delta_2 - \delta_0 = (45 \pm 15)^\circ$ as proposed in a recent review by Wahl [62].

The quantity $\eta_{+-}\Gamma_{+-} + \eta_{00}\Gamma_{00}$ occurring in the unitarity relation, is equal to $\epsilon\Gamma_S$ with high precision. The phase of ϵ is therefore predicted by unitarity. Since the phase of ϵ' is also known, the phenomenology doubly constrains the two complex amplitudes η_{+-} and η_{00}; ϵ and ϵ' are expected to be nearly collinear, as illustrated in Fig. 22.

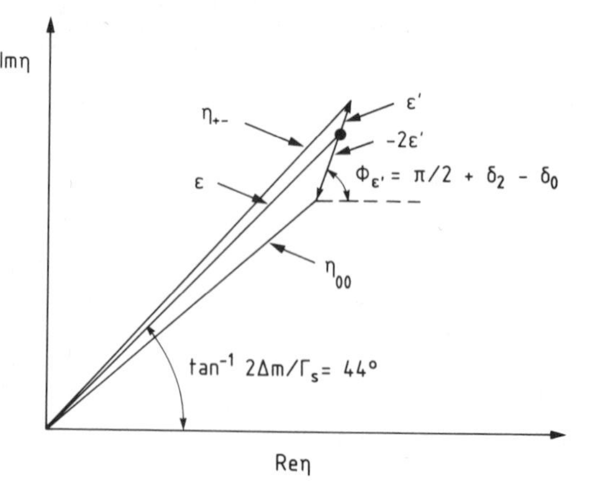

Fig. 22 Relationship between the amplitude ratios η_{+-}, η_{00}, ϵ and ϵ'.

2.8 Charge asymmetry in semileptonic decay

If x is the ratio of the $\Delta S = \Delta Q$ violating amplitude to the $\Delta S = \Delta Q$ conserving amplitude in the K^0 semileptonic decay:

$$x = \frac{\langle \ell^- |T|K\rangle}{\langle \ell^+ |T|K\rangle},$$

then the time-dependent charge asymmetry for a state that is initially $|K\rangle$ is

$$\delta_\ell(\tau) = \frac{2(1-|x|^2)\,[e^{-\Gamma_S\tau} + \text{Re}\,\epsilon\, e^{-\Gamma_L\tau} + e^{(\Gamma_S+\Gamma_L)\tau/2}\cos\Delta m\tau]}{|1+x|^2 e^{-\Gamma_S\tau} + |1-x|^2 e^{-\Gamma_L\tau} + 4\,\text{Im}\,x\, e^{-(\Gamma_S+\Gamma_L)\tau/2}\sin\Delta m\tau}.$$

For a state that is initially $|\bar{K}\rangle$ the sign of the interference in the numerator changes. In the quark model, $x = 0$, and the asymmetry simplifies:

$$\delta_\ell(\tau) = \frac{2\,[e^{-\Gamma_S\tau} + \text{Re}\,\epsilon\, e^{-\Gamma_L\tau} + e^{-(\Gamma_S+\Gamma_L)\tau/2}\cos\Delta m\tau]}{e^{-\Gamma_S\tau} + e^{-\Gamma_L\tau}}.$$

The oscillating term here is CP-conserving. The CP-violating asymmetry is the asymmetry left over at long times:

$$\delta_\ell = 2\,\text{Re}\,\epsilon\,.$$

The interference (not CP-violating) at shorter times has had an important application in the determination of the regeneration phase for those experiments measuring ϕ_{+-} and ϕ_{00} following a regenerator [19, 37].

2.9 Mixing only. The superweak model [63]

If CP violation in K decay is manifested by the mixing parameter $\tilde{\epsilon}$ only, then $\eta_F = \tilde{\epsilon}$ for all decay channels F of well-defined CP. In particular, $\epsilon' = 0$. There is then only one parameter, the strength of the mixing, since the phase is fixed by unitarity. There are five measured quantities and therefore four checks on this model:

	Experiment	**Mixing only**
$\lvert\eta_{+-}\rvert$	$(2.27 \pm 0.025) \times 10^{-3}$	-
$\lvert\eta_{00}/\eta_{+-}\rvert$	1.01 ± 0.04	1
ϕ_{+-}	$(45.8 \pm 1.4)°$	$(43.7 \pm 0.3)°$
$\phi_{00} - \phi_{+-}$	$(2.4 \pm 2.6)°$	0
δ_ℓ	$(3.33 \pm 0.15) \times 10^{-3}$	$(3.30 \pm 0.04) \times 10^{-3}$

The four checks, each at a level of a few per cent, support the model.

3. CP VIOLATION IN ELECTROWEAK THEORY

In the electroweak theory, the charged weak quark currents are flavour-mixed. The mixing matrix is unitary. In the case of two families [64, 65],

$$\begin{pmatrix} u \\ d \end{pmatrix} \text{ and } \begin{pmatrix} c \\ s \end{pmatrix},$$

if $j^\mu_{\alpha\beta} = V_{\alpha\beta}\bar{\psi}_a\gamma_\mu(1-\gamma_5)\psi_\beta$,

$$V_2 = \begin{pmatrix} V_{ud} & V_{us} \\ V_{cd} & V_{cs} \end{pmatrix}$$

With the proper choice of the arbitrary phases of the quark states,

$$V_2 = \begin{pmatrix} \cos\theta_C & \sin\theta_C \\ -\sin\theta_C & \cos\theta_C \end{pmatrix}$$

There is a single mixing angle and the currents are CP-conserving.

In the case of three families, the unitary mixing matrix has three mixing angles and one complex phase, the latter giving rise to CP violation within the model. This was first noted by Kobayashi and Maskawa (KM) in 1973, long before the third family had been discovered [4]. For three families, the KM mixing matrix is

$$V_3 = \begin{pmatrix} V_{ud} & V_{us} & V_{ub} \\ V_{cd} & V_{cs} & V_{cb} \\ V_{td} & V_{ts} & V_{tb} \end{pmatrix} = \begin{pmatrix} c_1 & c_3 s_1 & s_1 s_3 \\ -c_2 s_1 & c_1 c_2 c_3 - s_2 s_3 e^{i\delta} & c_1 c_2 s_3 - c_3 s_2 e^{i\delta} \\ s_1 s_2 & -c_1 c_3 s_2 - c_2 s_3 e^{i\delta} & -c_1 s_2 s_3 + c_2 c_3 e^{i\delta} \end{pmatrix}$$

where $s_1 = \sin\theta_1$, $c_1 = \cos\theta_1$, etc.; s_1 is essentially the Cabibbo angle, determined from hyperon and kaon semileptonic decay: $s_1 = 0.22$; s_3 is known with modest precision from the B lifetime, $s_3 = 0.05 \pm 0.01$; and an upper limit on s_2 can be put from the experimental upper limit on the $b \to u$ branching ratio, $s_2 < 0.01$.

The phase δ gives rise to CP-violation effects. It has been shown by Jarlskog [66], Greenberg [67], and Wu [68] that these effects are proportional to $J = \mathrm{Im}\,(V_{us}V_{cb}V_{ub}^*V_{cs}^*) = \mathrm{Im}\,(V_{ud}V_{cs}V_{us}^*V_{cd}^*) = c_1c_2c_3s_1^2s_2s_3 \sin\delta$.

The smallness of the observed CP-violation effects is immediately understood in terms of the smallness of J. For instance, $\tilde{\epsilon}$ would be expected to be of the order of $J/s_1^2 \lesssim 0.0025$; the agreement is much too good.

Attempts at a real calculation of $\tilde{\epsilon}$ on the basis of the box diagram

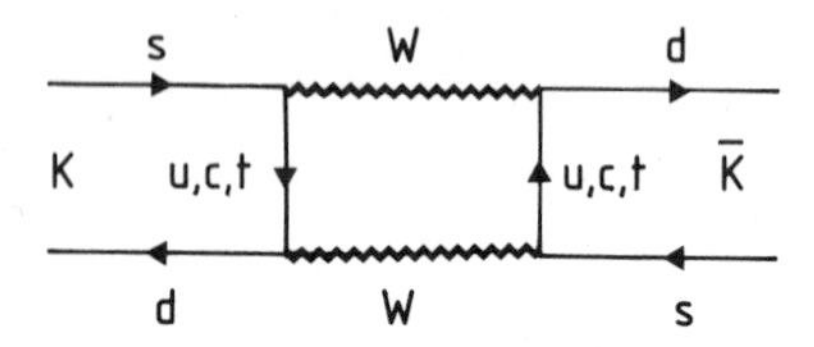

meet with difficulties associated with the problems of dealing with mesons rather than quarks, the long-distance or dispersive contributions, and the unknown top-quark mass (see, for example, the review of Wolfenstein [69]). The model definitively predicts non-zero ϵ', but again the calculations, based on the 'penguin' diagram

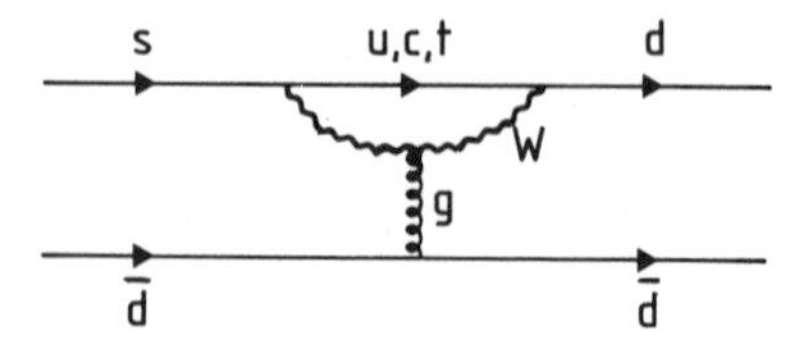

suffer from the same uncertainties. A recent summary of these calculations (Donoghue et al. [70]) gives the result

$$1 \times 10^{-3} < |\epsilon'/\epsilon| < 7 \times 10^{-3}.$$

Such small values of ϵ'/ϵ were not accessible to the older experiments.

Since the phase of ϵ' is within $\sim$ 15° of the phase of ϵ (Section 2), the magnitude of ϵ' is best seen as a difference in the magnitudes of η_{+-} and η_{00} rather than of their phases. With adequate accuracy $|\eta_{+-}| = |\epsilon| + |\epsilon'|$ and $|\eta_{00}| = |\epsilon| - 2|\epsilon'|$, so that $\epsilon'/\epsilon = {}^{1}/_{6}(1 - |\eta_{00}|/|\eta_{+-}|^2)$.

The summary result $|\eta_{+-}/\eta_{00}| = 1.01 \pm 0.04$ corresponds to $\epsilon'/\epsilon = (3 \pm 12) \times 10^{-3}$, and is not sensitive to the levels expected theoretically. This has encouraged new experimental efforts to reach the required sensitivity.

4. ϵ'/ϵ. DIRECT CP VIOLATION

Since $|\eta_{+-}|^2 = \Gamma_{L,+-}/\Gamma_{S,+-}$ and $|\eta_{00}|^2 = \Gamma_{L,00}/\Gamma_{S,00}$, the measurement of ϵ'/ϵ requires the measurement of the double ratio $(\Gamma_{L,00}/\Gamma_{S,00})/(\Gamma_{L,+-}/\Gamma_{S,+-})$ with adequate systematic and statistical accuracy. To achieve the systematic accuracy, it is desirable to observe all four reactions simultaneously, or at least two at a time, so that beam intensities and detection efficiencies cancel. This was not the case for the earlier experiments. The rates should be such that at least 10^5 events are observed in each channel. The most difficult of the four channels is $K_L \rightarrow 2\pi^0$, where the earlier experiments were limited to $\sim$ 100 events. The required statistical factor of $\sim$1000 represented a great challenge, not so much for the kaon beam as for the detector.

The first of the 'new wave' experiments were published in 1985. The sensitivity was considerably improved, but not sufficiently. The Brookhaven–Yale experiment of Black et al. [71] found $\epsilon'/\epsilon = (2 \pm 8) \times 10^{-3}$, and the Chicago–Saclay experiment of Bernstein et al. [72] at Fermilab found $\epsilon'/\epsilon = (-5 \pm 5) \times 10^{-3}$. For the rest of this review we give a brief account of the two most recently published experiments: the second round of the Fermilab experiment [73], and the CERN experiment of Burkhardt et al. [74].

4.1 Second Fermilab experiment

Some of the details of this description are taken from the thesis of Jarry [75]; the result is that of Woods et al. [73].

In this experiment the K_L and K_S decays are measured simultaneously. There are two parallel beams of K_L, and in one of them a regenerator is inserted ahead of the sensitive decay volume of the detector. The 2π decays from this beam are essentially K_S. The regenerator is alternated from one beam to the other between bursts. The K_L and K_S decays are observed simultaneously. However, the detector is not exactly the same for neutral and charged detection, so the two decay modes are observed alternately.

4.1.1 The detector

The experimental arrangement is shown in Fig. 23. The 17 m long decay volume is terminated by a scintillator used in the trigger. For neutral-mode observation, 0.5 mm of lead is inserted in front of this scintillator, and one of the four γ's is required to convert in it. The materialized pair serves to assign the event to one of the two beams. This piece of lead is the essential difference in the detector for the two decay modes.

The evacuated decay region is extended by 20 m to permit the decay particles to separate; at this point it is terminated by a thin Kevlar–Mylar window.

This is followed by an 18 m long spectrometer to measure the charged tracks. The spectrometer consists of a pair of drift chambers and a magnet, followed by another pair of drift chambers. Two scintillator hodoscopes, one before the magnet and one after the last drift chamber, serve to trigger charged decays. In the detection of the neutral mode, a compensating magnet just following the lead converter is activated. It is adjusted in such a way that the pair, after traversing the two magnets, is reunited in the photon detector, so that to this detector it looks like a photon.

The photon detector is an arrangement of 804 lead-glass blocks (see Fig. 24), each 5.7 cm $\times$ 5.7 cm laterally and 20 radiation lengths long. Calibration is effected by means of electrons and of π^0's whose origin is known. The spatial resolution obtained is $\approx$ 3.5 mm in x and y. The energy resolution is found to be

$$(\Delta E/E)^2 = \sqrt{(6\%/\sqrt{E})^2 + (3\%)^2}\ .$$

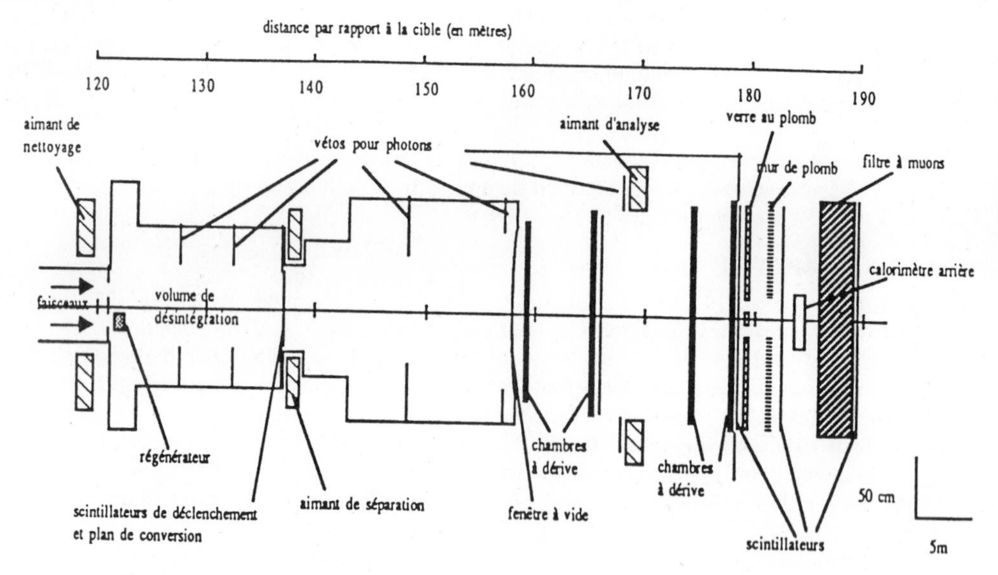

Fig. 23 Experimental arrangement for the experiment of Woods et al. [73].

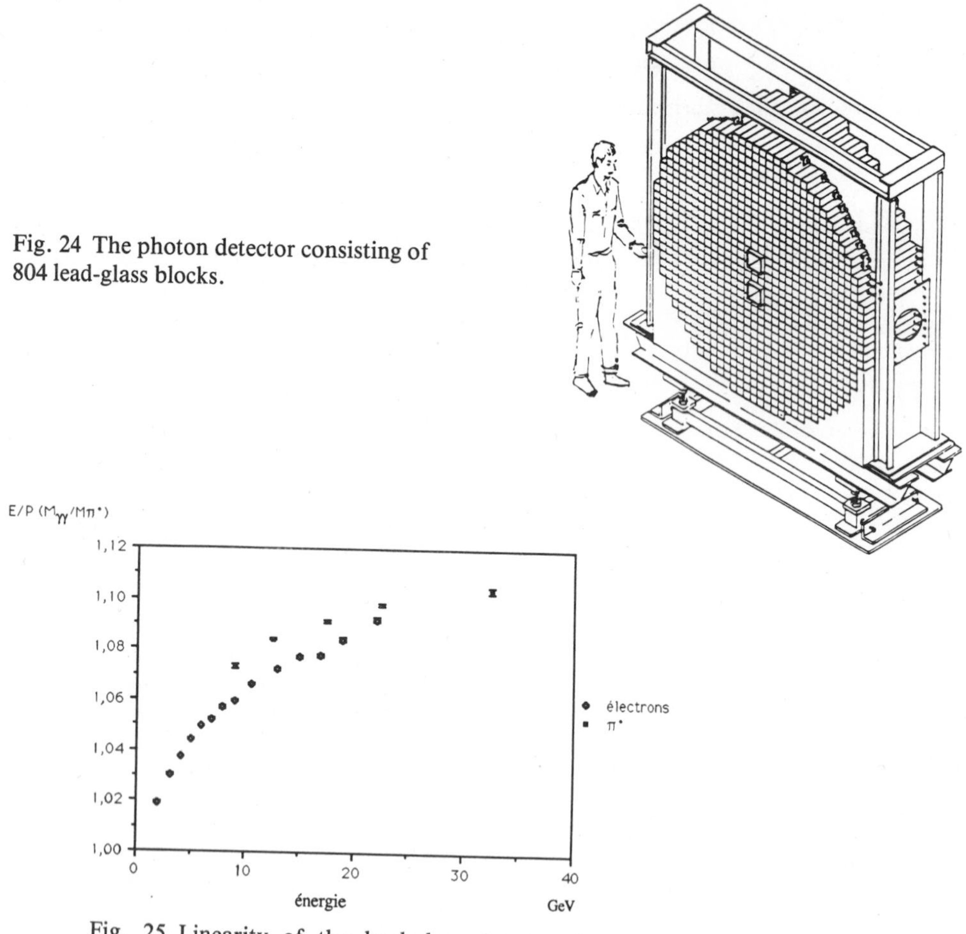

Fig. 24 The photon detector consisting of 804 lead-glass blocks.

Fig. 25 Linearity of the lead-glass detector for electrons and photons.

There is a substantial non-linearity as shown in Fig. 25.

The lead-glass array is followed by 12 cm of lead and a scintillator hodoscope which serves to signal hadron contamination in neutral events.

Finally, following a 3 m thick iron wall, there is a scintillator hodoscope to detect muons, and to filter the large $K_{\mu 3}$ background.

Positioned along the decay and detection path are seven annular lead/scintillator sandwiches which serve to veto events in which charged particles or photons are emitted outside the acceptance of the detector. This limits, in particular, the backgrounds due to $3\pi^0$ and $\pi^+\pi^-\pi^0$ decays.

4.1.2 Data and background subtraction

In Figs. 26a–c, the invariant mass distribution and the z distributions for K_S and K_L (K_S = regenerated beam, K_L = free beam) are shown for the charged decays after cuts, together with results of the Monte Carlo simulation. The same distributions are presented in Figs. 27a–c for the neutral decay. The energy distribution for charged events in the K_S beam is shown in Fig. 28. The K_L beam is very similar.

The subtracted background is of two sorts: for the K_L beam it is what is left of three-body decays after the cuts; for the K_S beam it is the decays from diffraction-regenerated and inelastic-regenerated K_S^0. In the K_S case, the subtractions are performed on the basis of the

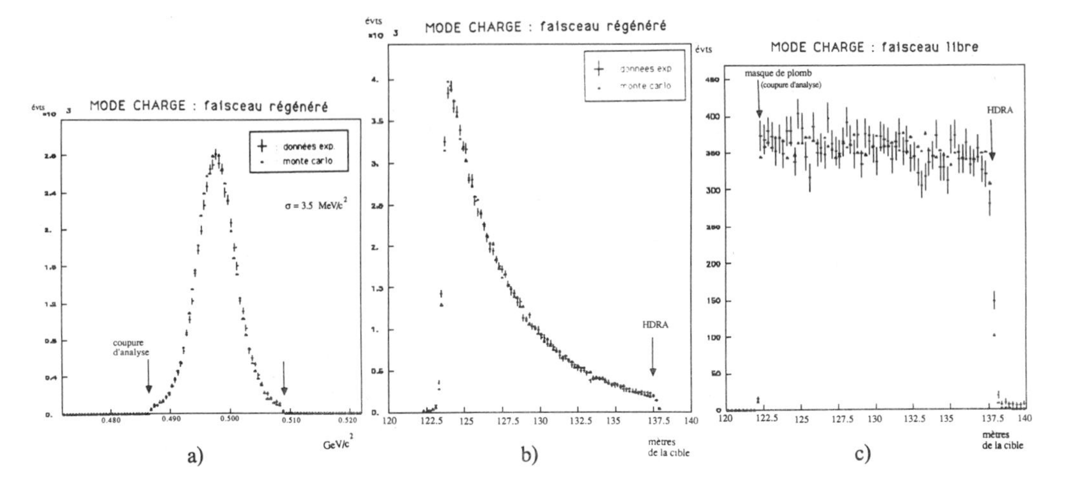

Fig. 26 Charged decay distributions after cuts (error bars) as well as Monte Carlo simulation (no error bars) for a) the invariant mass distribution, b) K_S z-distribution, and c) K_L z-distribution.

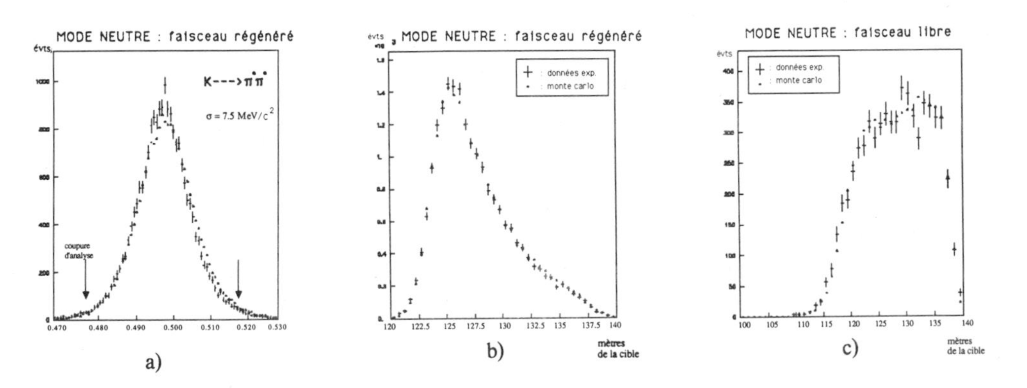

Fig. 27 Same as Fig. 26, but for the neutral decay.

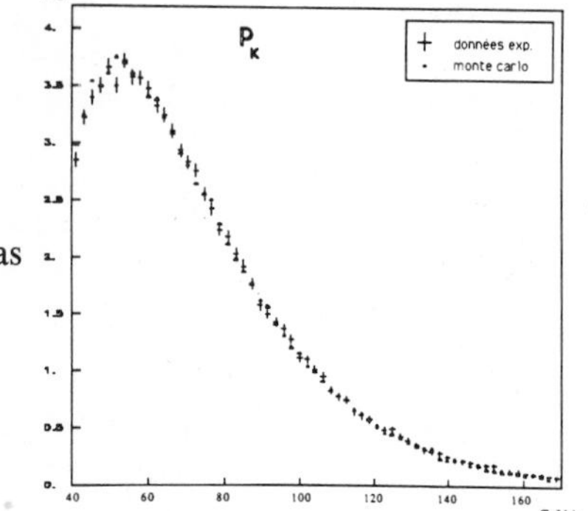

Fig. 28 Energy spectrum of K_S charged events, as well as Monte Carlo simulation.

distributions in the square of the transverse momentum. The unwanted decays give tails in this distribution, and these form the basis for extrapolation to the region of small p_T^2 that is accepted. In the K_L case, the observed mass distributions are used to estimate the backgrounds, although in the $K_L \rightarrow 2\pi^0$ channel the p_T^2 distribution is used as well.

For the K_S subtractions, the p_T^2 dependence of the diffraction regeneration is calculated according to a model; the p_T^2 dependence of the inelastic regeneration is assumed to be of the form $\alpha\, e^{-\beta p_T^2}$, where α and β are fitted to the data. These are shown in Fig. 29 for the neutral decays and in Fig. 30 for the charged decays. The neutral background is evaluated at (3 $\pm$ 0.3)%, the charged at (0.56 $\pm$ 0.1)%. The difference between the two is due to the fact that the charged p_T^2 cut for acceptance is 400 (MeV/c)^2 whilst the neutral cut is ten times higher.

For the $K_L \rightarrow 2\pi^0$ channel, the background is the $3\pi^0$ mode, where one or both extra photons overlay the other four. On the basis of the mass distribution of Fig. 31, the background is estimated to be (2.9 $\pm$ 1)%. Essentially the same result is obtained if one extrapolates in p_T^2. For the charged K_L channel, the background is calculated on the basis of the wings of the mass distribution, interpolating to the accepted mass interval with a Monte Carlo calculation, assuming the background to be K_{e3} and $K_{\mu 3}$ (see Fig. 32). A background of (1.9 $\pm$ 0.3)% is found.

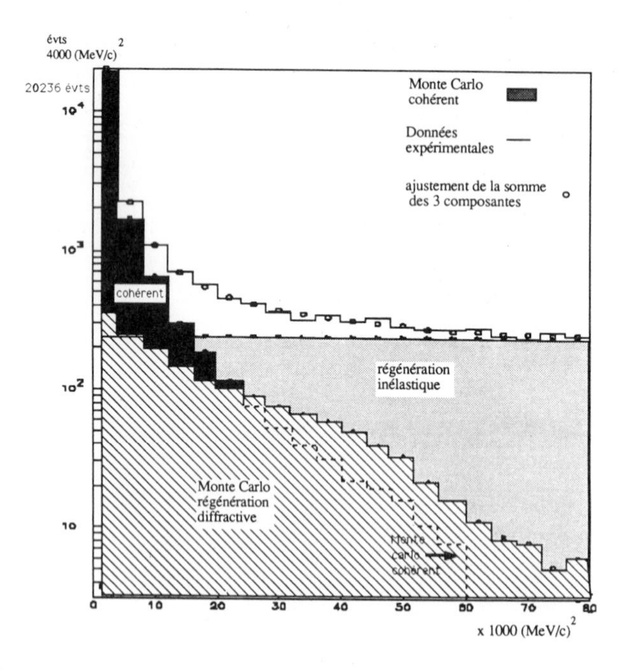

Fig. 29 The p_T^2 distribution for K_S neutral decay. The fitted diffractive and inelastic backgrounds are shown. The p_T^2 cut for good events is 4000 (MeV/c)^2.

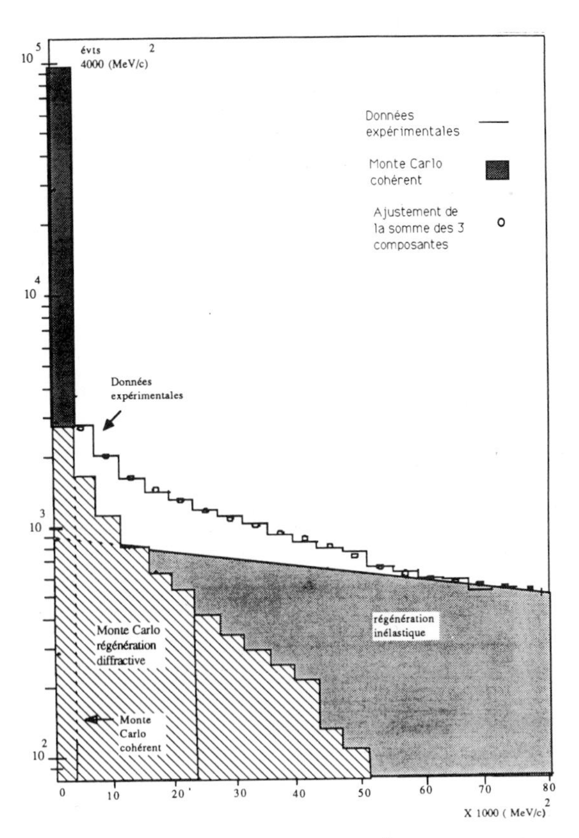

Fig. 30 Same as Fig. 29 but for K_S charged decay. The p_T^2 cut for good events is 400 $(MeV/c)^2$.

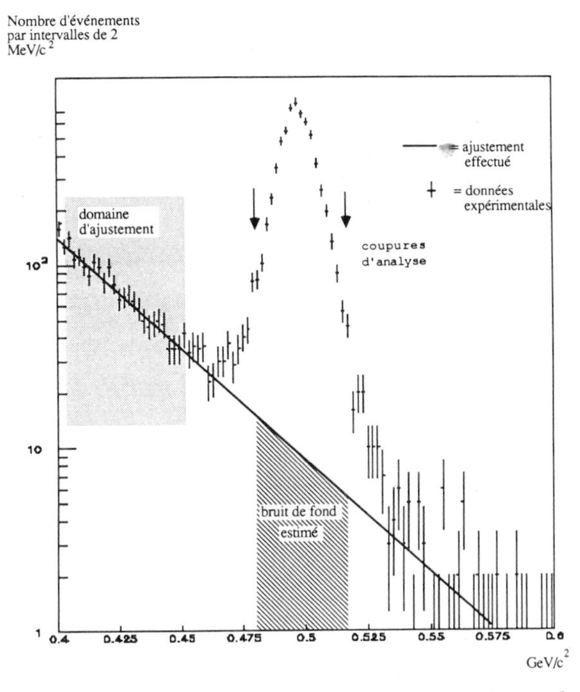

Fig. 31 Invariant mass distribution in K_L neutral decay, and the method of background subtraction.

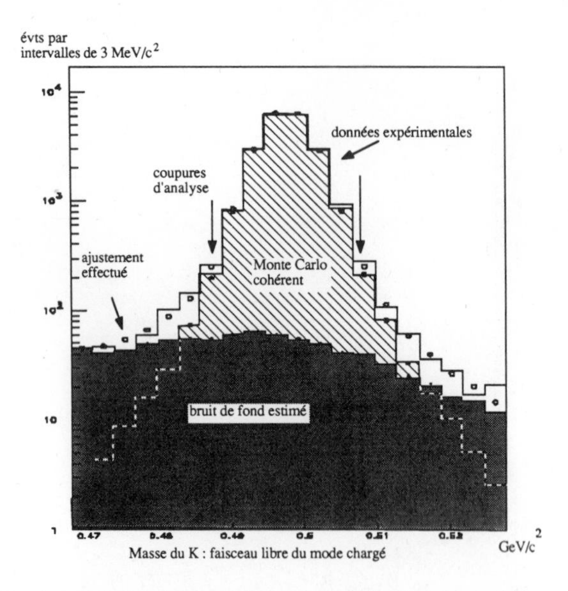

Fig. 32 Mass distribution in K_L charged decay, with indication of how the background is estimated.

4.1.3 Results

The event numbers, after background subtraction, are

$$K_S \to \pi^+\pi^-: \quad 88{,}696$$

$$K_S \to \pi^0\pi^0: \quad 19{,}629$$

$$K_L \to \pi^+\pi^-: \quad 20{,}242$$

$$K_L \to \pi^0\pi^0: \quad 6{,}230$$

The statistics are limited by the $K_L \to 2\pi^0$ channel, as is always the case.

The double ratio is obtained in energy bins but not in z bins, because the z region is inadequately scanned by the K_S decays, as can be seen from Fig. 26. Instead, the z-distributions are integrated. This necessitates a correction for the fact that neutral and charged acceptances have different z-dependences. This correction is performed on the basis of a Monte Carlo calculation, which can be seen in Fig. 33.

The result is $\epsilon'/\epsilon = 0.0032 \pm 0.0028$ stat. $\pm$ 0.0012 syst.

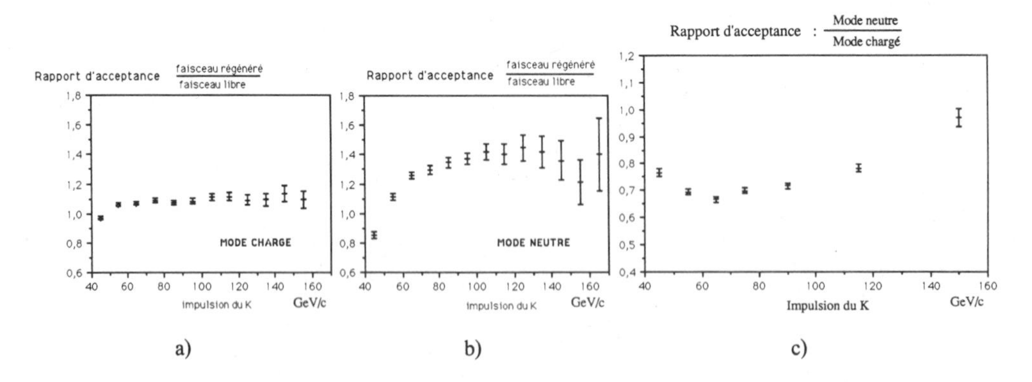

Fig. 33 Monte Carlo calculations of acceptance ratios as a function of kaon momentum: a) charged mode, K_S/K_L; b) neutral mode, K_S/K_L; c) double ratio neutral/charged, K_S/K_L.

4.2 The CERN experiment [74]

4.2.1 The beams and the detector

Details of the detector and its performance are given in Burkhardt et al. [76]. Figure 34 shows the experimental layout. In this experiment, K_S and K_L data are taken alternately, and the two decay modes simultaneously. The K_S beam (i.e. the target, the sweeping magnet, the collimator, and the anticounter) (Fig. 35) is on wheels so that the entire 48 m decay region can be uniformly explored despite the rather short mean-free path of K_S decay. The decay region,

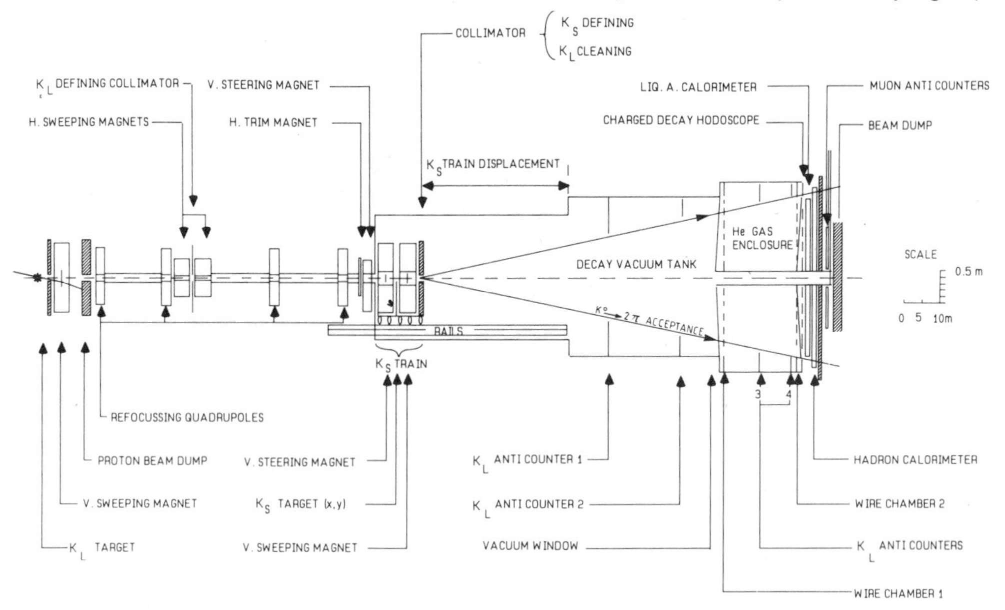

Fig. 34 Layout of CERN experiment.

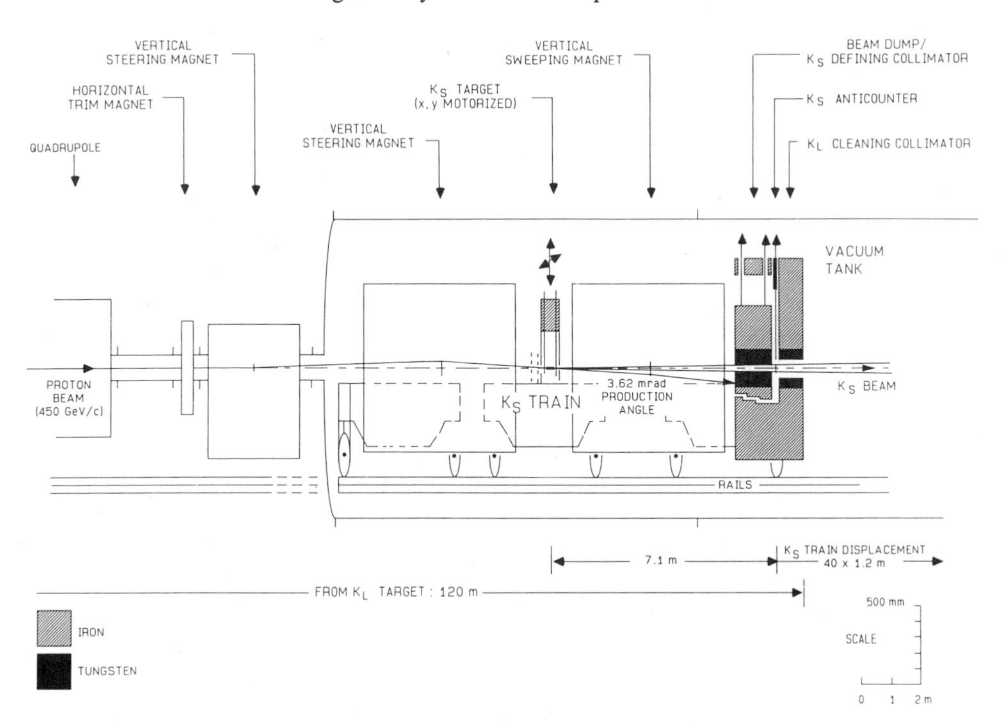

Fig. 35 The K_S beam train.

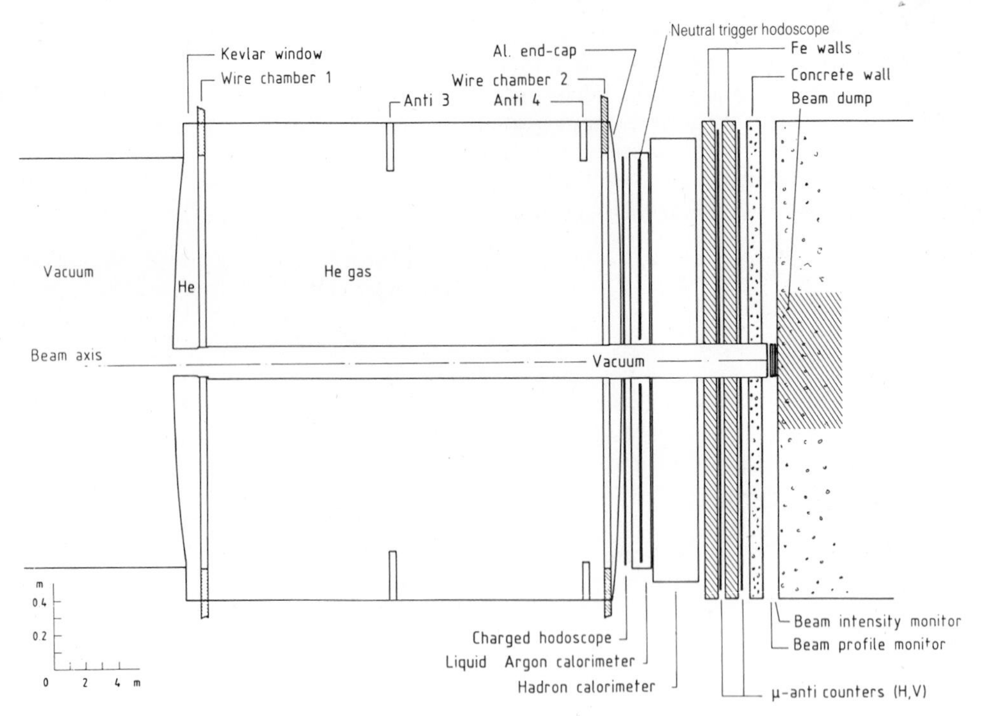

Fig. 36 Detector layout.

and the following 50 m region that serves to let the decay products spread sufficiently to enter the sensitive region of the detector, are evacuated. The neutral beam then continues through an evacuated pipe, and the decay products leave the vacuum region by a 1 mm hemispherical Kevlar window and then enter the helium-filled detector region.

The detector consists of the following elements (see Fig. 36).

1) Minidrift chambers. These are two proportional wire chambers, each with four wire planes, horizontal, vertical, 53°, and 143°. The wire spacing is 6 mm; the position is interpolated by means of drift-time measurement. The precision is $\approx$ 300 μm in each projection. These chambers serve to measure the tracks of the charged decay products. Reconstruction precision of the decay vertex along the decay axis is $\approx$ 35 cm.

2) A scintillator hodoscope that serves as the trigger for charged decays.

3) A lead/liquid-argon calorimeter. There are 80 layers, each consisting of a 1.5 mm lead sheet clad on both sides with 0.2 mm aluminium, a liquid-argon gap of 2 mm, and a printed circuit board covered on both sides with copper-strip anodes with a pitch of 12.5 mm, and again 2 mm of liquid argon. In succeeding gaps the strips alternate from horizontal to vertical. The total thickness corresponds to 25 radiation lengths.

The energy resolution for photons is $7.5\%/\sqrt{E(\mathrm{GeV})}$. The space resolution is $\approx$ 0.5 mm in x and y.

The stack is read out independently in quadrants and in two groups of 40 layers in depth each. In the central plane of the calorimeter there is a scintillator hodoscope that serves to trigger the neutral decays.

4) A hadron calorimeter consisting of 48 sandwiches of 25 mm iron plate and 5 mm scintillator. The scintillators are arranged in strips, 12 cm wide, alternately horizontal and vertical. The stack is read out in quadrants and in two depth layers. The energy resolution for pions of the two calorimeters is $0.65/\sqrt{E}$.

5) A muon anticoincidence arrangement, consisting of two layers, each with a 70 cm thick iron absorber and a scintillator hodoscope.

6) Four photon anticoincidence annular rings along the decay and detector length, to eliminate some of the $3\pi^0$ and $\pi^+\pi^-\pi^0$ background. They consist of lead sheets and scintillation counters.

7) In the case of the K_S beam only, there is an anticoincidence counter preceded by 1.5 radiation lengths of lead, just following the collimator. This serves a twofold purpose: a) to define the beginning of the fiducial region for neutral and charged decays equally, thus avoiding corrections for differences in z resolution; b) to calibrate and monitor the absolute energy scale of the photon calorimeter with the required one tenth of one per cent precision.

One of the main design considerations was to have the beam entirely in vacuum, so that the neutrons in the beam would not contribute to dead-time or background. This was achieved. A second important objective was to have a large acceptance. The acceptance, averaged over the 50 m fiducial length and the accepted energy interval, $70 < E_K < 170$ GeV, was 22% for the neutral decay and 43% for the charged decay.

4.2.2 Operation

The rates were in general limited by the background trigger rates, which were kept at a level of $\approx 10^5$ per 2 s burst (at 12 s intervals). This resulted in dead-times of the order of 12% in the K_L beam and 4% in the K_S beam. Approximately 10^3 events per burst were written on tape in the K_L beam, and about 300 in the K_S beam. The rates per burst, for good events after cuts, were as follows: 1 $K_L \to 2\pi^0$, 3 $K_L \to \pi^+\pi^-$, 30 $K_S \to 2\pi^0$, and 70 $K_S \to \pi^+\pi^-$.

In order to permit an understanding of backgrounds, looser trigger conditions were recorded at scaled rates, as well as randomly triggered events to permit analysis of losses due to accidentals.

The liquid-argon calorimeter was tested with known energy electrons for resolution, uniformity, and linearity, and the combined calorimetric system was tested with pions.

For the purpose of positioning the K_S target, the decay region was divided into 40 'train positions' at 1.2 m intervals. The running was organized into 'miniperiods', of which there were 36, all in 1986. Each miniperiod consisted of one 24 h exposure in the K_L beam, and ten exposures of $\sim$ 3/4 h each in the K_S beam. These K_S exposures were spaced by four train positions, so that one miniperiod had exposures at, say, positions 2, 6, 10, etc.; the next would use positions 4, 8, 12, etc. In four miniperiods, each position in the decay region was covered.

The K_L and K_S beams were produced by protons striking the target at the same angle, 3.6 mrad. Nevertheless the spectra were appreciably different, owing to the preferential decay of lower-energy kaons in the 8 m distance between K_S target and anticounter. The spectra of the observed decays are shown in Fig. 37 for the two beams.

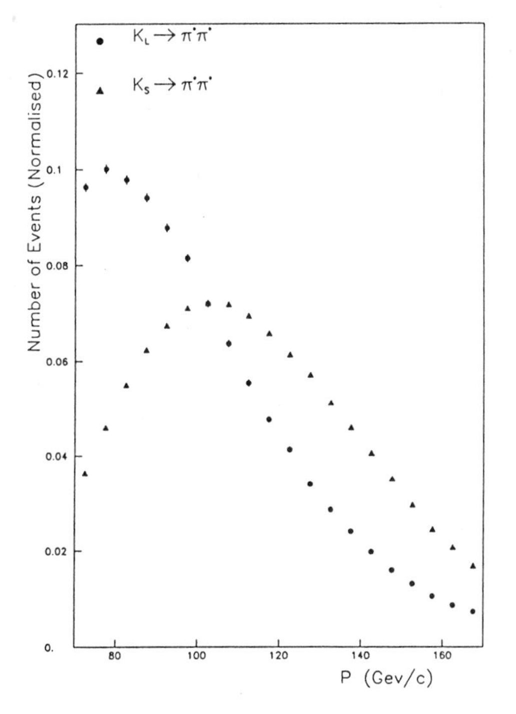

Fig. 37 Observed energy spectra in the K_S and K_L beams.

4.2.3 Data reduction and background subtraction

The on-line trigger and off-line analysis were arranged so that for the two decay modes the dead-times were the same, and accidental effects were as nearly the same as possible. For instance, an extra space point in the front chamber invalidated charged events; it therefore did the same for neutrals. Or, an extra photon (the fifth for neutral decays, and the first for charged decays) disqualified both types of events. However, an extra photon is not quite the same when there are already four present, or there are not, because of overlap problems; nor is an extra space point quite the same if two space points are already there, or if they are not. It is therefore necessary to make a correction for the differences in the effects of accidentals on the four event classes.

4.2.3.1 Neutral events

For each event, four photons of more than 5 GeV are required. They must be separated by 5 cm in at least one projection. The centre of energy in the liquid-argon calorimeter must be in the beam; therefore,

$$r = \left[\left(\sum E_i x_i\right)^2 + \left(\sum E_i y_i\right)^2\right]^{1/2} / \sum E_i$$

is required to be less than 10 cm. Figure 38 shows the distribution in r for the K_S beam.

The decay vertex is reconstructed, using the K mass, according to

$$z_v = 1/m_K \left\{\sum_{i<j}^{4} E_i E_j[(x_i - x_j)^2 + (y_i - y_j)^2]\right\}^{1/2} .$$

The resolution in z_v is $\approx$ 1 m.

Given the decay vertex z_v, the invariant masses of photon pairs may be calculated:

$$m_{ij} = 1/z_v \sqrt{E_i E_j[(x_i - x_j)^2 + (y_i - y_j)^2]} .$$

Those of the three combinations for which the two masses are nearest to the π^0 mass are chosen. For $2\pi^0$ decays, these masses will be close to the π^0 mass, but not, in general, for $3\pi^0$ decays for which only four of the six photons have been detected. The distribution in these masses is therefore used in the background subtraction.

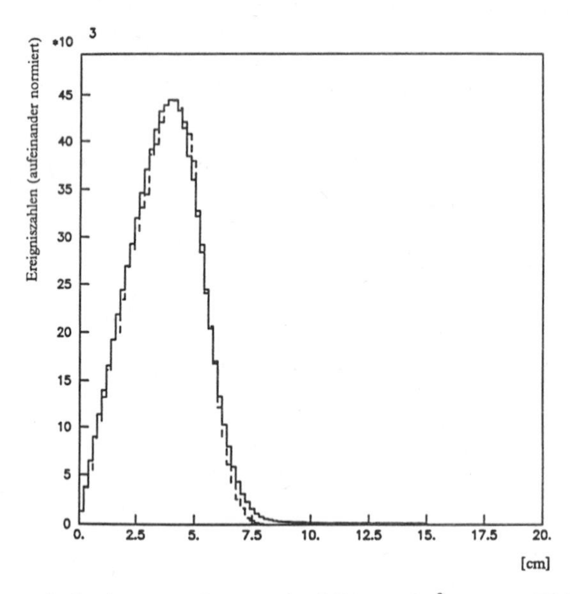

Fig. 38 Distribution of the 'centre of energy' of $K_S \rightarrow 2\pi^0$ events. This shows the radial distribution of the K_S beam in the plane of the liquid-argon calorimeter. The dashed histogram is the Monte Carlo simulation.

It is possible to proceed inversely: to reconstruct the decay vertex using the photon pairs, assuming that they have the π^0 mass, and then to use this vertex, $z_v(\pi^0)$, to find the invariant mass of the event:

$$m^2 = \sum_{i<j}^{4} E_i E_j [(x_i - x_j)^2 + (y_i - y_j)^2]/z_v^2(\pi^0) .$$

This mass is not used in the event selection, but illustrates the resolution of the detector. It is shown in Fig. 39 for K_S events.

Event selection is based on the two-dimensional $\gamma_i\gamma_j$ mass distribution. It is shown in Fig. 40 for K_S events as a Lego plot and in Fig. 41 for K_L events as a scatter plot. The K_S events have essentially zero background, whilst the $3\pi^0$ background in the K_L decay is uniformly distributed. In order to select good events and to subtract the K_L background, constant χ^2 ellipses of linearly increasing area are constructed and the event numbers in the annular regions are plotted (see Fig. 42). Of the good events, 98.5% are in the inner ellipse, as seen in the background-free K_S case. In the K_L case, extrapolation to the inner ellipse gives the $3\pi^0$ background of $(4.0 \pm 0.2)\%$.

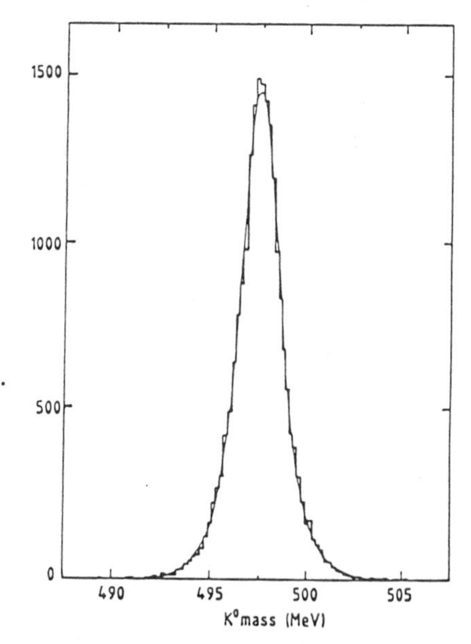

Fig. 39 Mass resolution for $2\pi^0$ decay in the K_S beam.

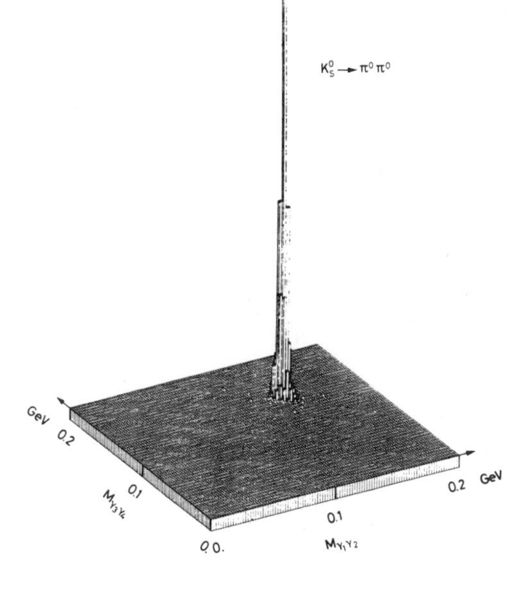

Fig. 40 Two-dimensional histogram of K_S neutral events in the plane $m_{\gamma_1\gamma_2}$ versus $m_{\gamma_3\gamma_4}$.

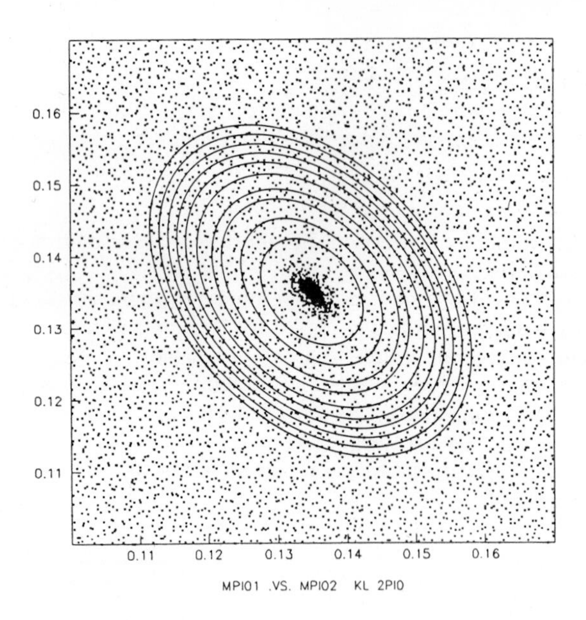

Fig. 41 Scatter plot in the plane $m_{\gamma_1\gamma_2}$ versus $m_{\gamma_3\gamma_4}$ for K_L neutral events. The ellipses correspond to $\chi^2 = 9, 18, 27, \ldots$.

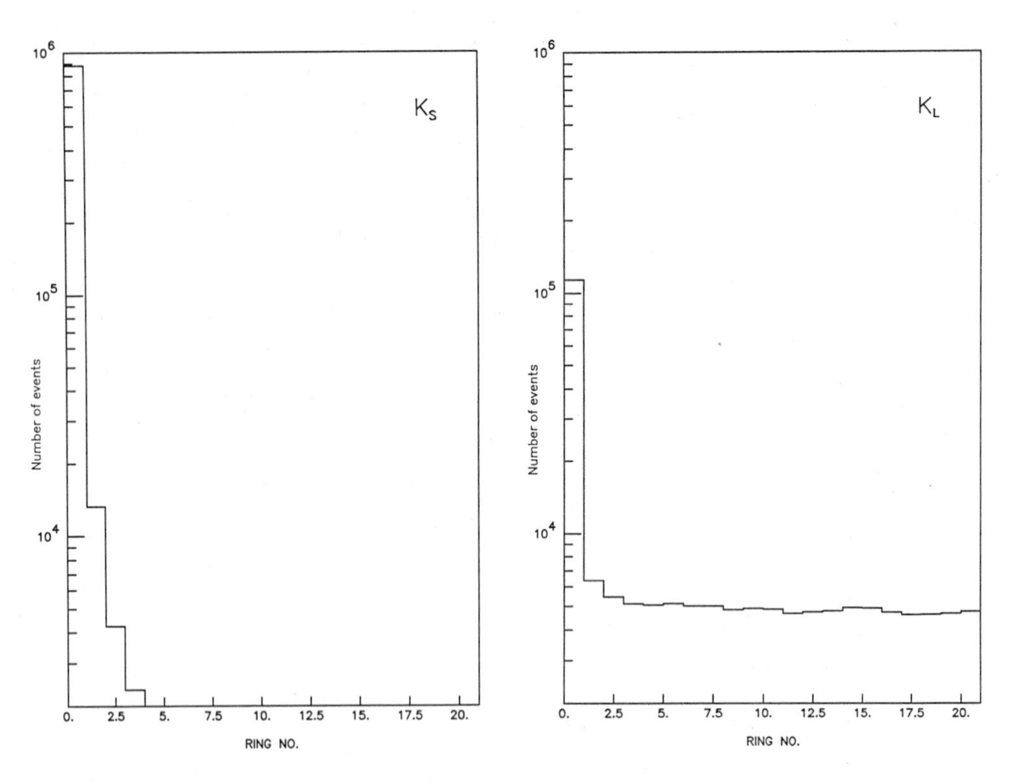

Fig. 42 Event numbers in the annulus between successive ellipses for K_S and K_L events; 98.5% of events are contained in the inner ellipse. Events in rings numbers 5 to 10 are used to extrapolate to zero for background subtraction in the K_L case.

4.2.3.2 Charged events

The energy of an event is not determined calorimetrically, but geometrically from the opening angle θ and the *ratio* R of the calorimetric energies of the two pions. This is more accurate, and is less dependent on calorimetry calibration:

$$E^2 = \frac{2 + R + (1/R)}{\theta^2} \{m_K^2 - m_\pi^2[2 + R + (1/R)]\} .$$

The resolution deteriorates with R, so the events are restricted to $0.4 < R < 2.5$. This cut is also effective against the $\Lambda \to \pi^- p$ background. The average energy resolution is $\approx$ 1%.

A very important cut requires that at least 20% of the energy of *each* pion be deposited in the hadron calorimeter. This cut loses 50% of the signal, but reduces the K_{e3} background to an acceptable level.

A further cut, again to reduce the three-body decay background, requires that the distance from the beam centre to the line joining the two points in the back chamber be less than 10 cm.

The invariant mass of each event is calculated from the calorimetric pion energies and the opening angle. The resolution depends on the kaon energy, and a 2.1σ cut is applied.

The K_S sample has no significant background. The K_L background is determined on the basis of the distribution of the events in the distance between the event plane and the target where the kaons are produced. This is shown in Fig. 43 (the K_S distribution is scaled geometrically so that it can be compared directly). The signal region is taken to be less than 5 cm, and the control region for determining the background is between 7 and 12 cm. This background is mainly due to K_{e3} events and, to a lesser extent, to $K_{\mu3}$ and $K_{\pi3}$ events, and their relative contributions can be determined from the many auxiliary events in which the decay is identified. It is extrapolated to the signal region by means of a Monte Carlo calculation, as shown. The background subtracted is (0.65 ± 0.2)%, where the systematic error is included.

It has already been pointed out that the energy scales of the two types of events must be the same to $\sim$ 0.1%. We have also seen that the energy in charged decays is determined geometrically. In order to achieve the equality of the energy scales, the overall calibration of

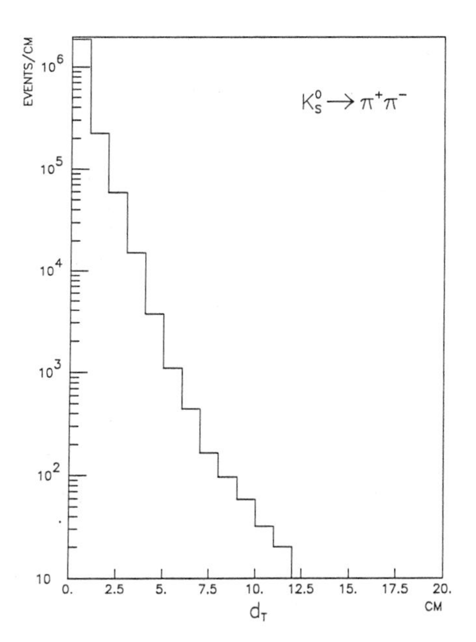

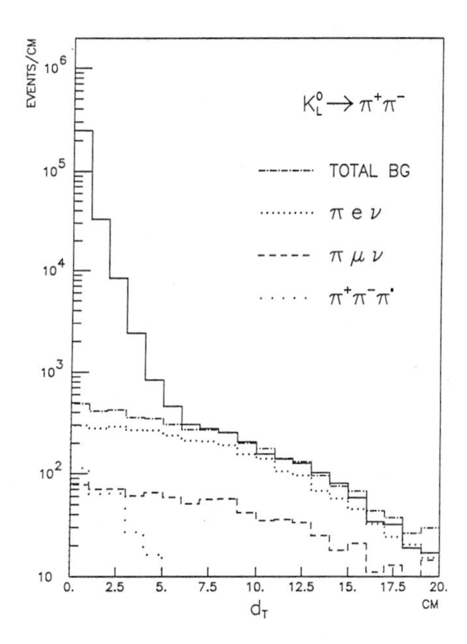

Fig. 43 Distributions in the distance between the production target and the decay plane for charged events. The Monte Carlo extrapolation of the three-body background from the tail to the signal region is indicated.

the liquid-argon calorimeter is also determined geometrically. It is chosen by requiring that the reconstructed anticounter position in the K_S beam is the same for both neutral and charged decay. Since, in the neutral decay, only photon energies and positions in the liquid argon are measured, a 1% change in overall energy scale will translate to approximately 1 m in anticounter position. The 0.1% equality in energy scales is easily achieved.

4.2.4 Results

In the final tabulation the energy interval $70 < E < 170$ GeV and the decay region $10.5 < z < 48.9$ m were retained. The z interval at the beginning of the decay region, $z < 10.5$ m, was dropped to avoid complication caused by a low level of regeneration in the final K_L collimator. The results are summarized in Table 1 and the contributions to the systematic error are given in Table 2.

The statistics are limited essentially by the 109,000 events in the $K_L \to 2\pi^0$ channel. The data were analysed in bins of energy and of position of the decay along the beam axis, and for each miniperiod. They were corrected by a Monte Carlo calculation for binning effects, for the slight difference in the K_S and K_L beam shapes, and for the scattering of K_S in the anticounter and collimator. The net Monte Carlo correction is 0.3%.

Table 1
Event statistics and corrections

	Signal events (× 1000)	Back-ground (%)	Scat-tering (%)	Inefficiencies Pretrigger (%)	Inefficiencies Trigger (%)	Accidental losses (%)
$K_L \to 2\pi^0$	109	4.0	< 0.1	0.06 ± 0.06	0.20 ± 0.10	2.6 ± 0.07
$K_L \to \pi^+\pi^-$	295	0.6		0.37 ± 0.07	0.05 ± 0.06	2.6 ± 0.05
$K_S \to 2\pi^0$	932	< 0.1	0.3	0.04 ± 0.02	0.12 ± 0.03	2.5 ± 0.05
$K_S \to \pi^+\pi^-$	2300	< 0.1		0.48 ± 0.03	0.01 ± 0.01	2.8 ± 0.05
Effect on R			0.3	−0.12 ± 0.10	−0.03 ± 0.12	−0.34 ± 0.10

Table 2
Systematic uncertainties on the double ratio R
(in percent)

Background subtraction for $K_L \to 2\pi^0$	0.2
Background subtraction for $K_L \to \pi^+\pi^-$	0.2
$2\pi^0/\pi^+\pi^-$ difference in energy scale	0.3
Regeneration in the K_L beam	< 0.1
Scattering in the K_S beam	0.1
K_S anticounter inefficiency	< 0.1
Difference in K_S/K_L beam divergence	0.1
Calorimeter instability	< 0.1
Monte Carlo acceptance	0.1
Gains and losses by accidentals	0.2
Pretrigger and trigger inefficiency	0.1
Total systematic uncertainty	±0.5%

The result for R is

$$R = 0.980 \pm 0.004\,(\text{stat.}) \pm 0.005\,(\text{syst.})\,.$$

This translates to

$$\epsilon'/\epsilon = (3.3 \pm 1.1) \times 10^{-3}\ .$$

This is the first statistically significant evidence for 'direct' CP violation; that is, for an effect other than that due to K_1-K_2 mixing. It clearly is in disagreement with the superweak model [63], but agrees in sign and order of magnitude with the predictions of the electroweak theory [69, 70], if it is assumed that the electroweak currents are responsible for CP violation [4].

4.3 Present status of ϵ'/ϵ experiments

The positive result of the CERN experiment for ϵ'/ϵ is interesting but, with three standard deviations, near the limit of its sensitivity. The Fermilab and CERN experiments are therefore continuing. At the recent conference on Lepton and Photon Interactions at High Energies (SLAC, Aug. 89), Winstein reported preliminary results based on a small fraction of data obtained at Fermilab during the autumn and winter of 1987. These data were obtained without the thin lead sheet converter for photons, so that neutral and charged modes could be observed together. It is the first experiment in which the four rates are measured simultaneously. On the basis of 52,000 neutral K_L, 43,000 charged K_L events, and approximately four times as many K_S events, the result is $\epsilon'/\epsilon = (-0.5 \pm 1.5) \times 10^{-3}$ (preliminary), a null result differing by two standard deviations from the CERN result.

The CERN experiment has continued with small modifications designed to increase the rate modestly and to reduce the systematic uncertainty. The runs in '88 and '89 should quadruple the statistics.

REFERENCES

[1] C.S. Wu et al., Phys. Rev. **105** (1957) 1413.
[2] R.P. Feynman and M. Gell-Mann, Phys. Rev. **109** (1958) 193.
[3] E.C.G. Sudarshan and R. Marshak, Phys. Rev. **109** (1958) 1860.
[4] M. Kobayashi and K. Maskawa, Prog. Theor. Phys. **49** (1973) 652.
[5] J.H. Christenson, J.W. Cronin, V.L. Fitch and R. Turlay, Phys. Rev. Lett. **13** (1964) 138.
[6] R. Adair et al., Phys. Rev. **132** (1963) 2285.
[7] J.M. Gaillard et al., Phys. Rev. Lett. **18** (1967) 20.
[8] J.W. Cronin et al., Phys. Rev. Lett. **18** (1967) 25.
[9] I.A. Budagov et al., Phys. Lett. **B28** (1968) 215.
[10] M. Banner et al., Phys. Rev. Lett. **21** (1968) 1107.
[11] M. Holder et al., Phys. Lett. **B40** (1972) 141.
[12] S. Bennett et al., Phys. Rev. Lett. **19** (1967) 993.
[13] D. Dorfan et al., Phys. Rev. Lett. **19** (1967) 987.
[14] C. Geweniger et al., Phys. Lett. **B48** (1974) 483.
S. Gjesdal et al., Phys. Lett. **52B** (1974) 113.
[15] L. Adiels et al., CP violation studies at LEAR, presented by P. Bloch at the 25th Anniversary of the Discovery of CP Violation, Blois (France), 1989.
[16] V.L. Fitch et al., Phys. Rev. Lett. **15** (1965) 73.
[17] C. Alff-Steinberger et al., Phys. Lett. **20** (1966) 207 and **21** (1966) 595.
[18] M. Bott-Bodenhausen et al., Phys. Lett. **20** (1966) 212 and **23** (1966) 277.
[19] W.C. Carithers et al., Phys. Rev. Lett. **34** (1975) 1240.
[20] W.C. Carithers et al., Phys. Rev. Lett. **34** (1975) 1244.

[21] A. Böhm et al., Nucl. Phys. **B9** (1969) 605.
[22] C. Geweniger et al., Phys. Lett. **B48** (1974) 487.
[23] S. Gjesdal et al., Phys. Lett. **B52** (1974) 119.
[24] G. Charpak et al., Nucl. Instrum. Methods **62** (1968) 262.
[25] P. Schilly et al., Nucl. Instrum. Methods **91** (1971) 221.
[26] J.C. Chollet et al., Phys. Lett. **B31** (1970) 658.
[27] B. Wolff et al., Phys. Lett. **B36** (1971) 517.
[28] G. Barbiellini et al., Phys. Lett. **B43** (1973) 529.
[29] J.H. Christenson et al., Phys. Rev. Lett. **43** (1979) 1209 and 1212.
[30] G.D. Barr et al., A measurement of the phases of the CP-violating amplitudes in $K^0 \to 2\pi$ decays and a test of CPT invariance, to be submitted to Phys. Lett. B.
[31] A. Abashian et al., Phys. Rev. Lett. **13** (1964) 245.
[32] X. de Bouard et al., Phys. Lett. **15** (1965) 58.
[33] W. Galbraith et al., Phys. Rev. Lett. **14** (1965) 383.
[34] R. Messner et al., Phys. Rev. Lett. **30** (1973) 876.
[35] R. DeVoe et al., Phys. Rev. **D16** (1977) 565.
[36] S. Bennett et al., Phys. Lett. **B27** (1968) 248.
[37] S. Bennett et al., Phys. Lett. **B29** (1969) 317.
[38] D.A. Jensen et al., Phys. Rev. Lett. **23** (1969) 615.
[39] H. Faissner et al., Phys. Lett. **B30** (1969) 204.
[40] R.K. Carnegie et al., Phys. Rev. **D6** (1972) 2335.
[41] J.M. Gaillard et al., Nuovo Cimento **A59** (1969) 453.
[42] V.V. Barmin et al., Phys. Lett. **B33** (1970) 377.
[43] M. Banner et al., Phys. Rev. Lett. **28** (1972) 1597.
[44] J. Marx et al., Phys. Lett. **B32** (1970) 219.
[45] R. Piccioni et al., Phys. Rev. Lett. **29** (1972) 1412.
[46] H.H. Williams et al., Phys. Rev. Lett. **31** (1973) 1521.
[47] V.L. Fitch et al., Phys. Rev. Lett. **31** (1973) 1524.
[48] V.V. Barmin et al., Phys. Lett. **B128** (1983) 129.
[49] M. Metcalf et al., Phys. Lett. **40B** (1972) 703.
[50] V.V. Barmin et al., Nuovo Cimento **A85** (1985) 67.
[51] P.M. Border et al., Proc. 23rd Int. Conf. on High-Energy Physics, Berkeley, 1986 (World Scientific, Singapore, 1987), Vol. II, p. 845.
[52] C.C. Rochester and G.D. Butler, Nature **160** (1947) 855.
[53] R.W. Thompson et al., Third Rochester Conference, 1952, and Phys. Rev. **90** (1953) 329. R. Armenteros et al., Phil. Mag. **43** (1951) 113.
[54] M. Gell-Mann, Phys. Rev. **92** (1953) 1833.
[55] M. Gell-Mann and A. Pais, Phys. Rev. **97** (1955) 1387.
[56] K. Landé et al., Phys. Rev. **103** (1956) 1901.
[57] T.T. Wu and C.N. Yang, Phys. Rev. Lett. **13** (1964) 380.
[58] The argument presented here is due to J.S. Bell, Proc. Conf. on Elementary Particles, Oxford, 1965 (Rutherford Lab., Chilton, Didcot, 1966), p. 195.
[59] T.T. Devlin and J.O. Dickey, Rev. Mod. Phys. **51** (1979) 237.
[60] N.N. Biswas et al., Phys. Rev. Lett. **47** (1981) 1378.
[61] A.N. Kamal, J. Phys. **G12** (1986) L43.
[62] H. Wahl, preprint CERN–EP/89–86, talk given at the Rare Decay Symposium, Vancouver, 1988.
[63] L. Wolfenstein, Phys. Rev. Lett. **13** (1964) 562.
[64] N. Cabibbo, Phys. Rev. Lett. **10** (1963) 531.
[65] S.L. Glashow, J. Iliopoulos and L. Maiani, Phys. Rev. **D2** (1970) 1285.
[66] C. Jarlskog, Phys. Rev. Lett. **55** (1985) 1039.
[67] D.W. Greenberg, Phys. Rev. **D32** (1985) 1841.
[68] D.D. Wu, Phys. Rev. **D33** (1986) 860.
[69] L. Wolfenstein, Annu. Rev. Nucl. Sci. **36** (1986) 137.
[70] J.F. Donoghue, J.F. Golowich and B.R. Holstein, Phys. Rep. **131** (1986) 320.
[71] J.K. Black et al., Phys. Rev. Lett. **54** (1985) 1628.
[72] R.H. Bernstein et al., Phys. Rev. Lett. **54** (1985) 1631.

[73] M. Woods et al., Phys. Rev. Lett. **60** (1988) 1695.
[74] H. Burkhardt et al., Phys. Lett. **B206** (1988) 163.
[75] P. Jarry, Thèse, Orsay, 1987.
[76] H. Burkhardt et al., Nucl. Instrum. Methods **A268** (1988) 116.

LEE-WICK MODEL and SOLITON STARS

R. VINH MAU

Division de Physique Théorique*, I.P.N. 91406 Orsay Cedex
and LPTPE*, Université P. et M. Curie - 4, Place Jussieu
75252 Paris Cedex 05.

INTRODUCTION

Massive stars after exhausting their thermonuclear fuel die in supernovae explosions. Some residues of these massive stars become neutron stars with the following properties : mass ~ 1.4 $m_\odot$, radius ~ 10 km, large density.

Recently, new types of cold stellar configurations called soliton stars are contemplated by Lee and collaborators[1]. They consist of non topological solitons of the Lee-Wick model[2] filled with a large number of particles (> 10^{60}). Lee et al. show that very large soliton stars with mass ~ 10^{12} $m_\odot$ and radius ~ 10^{-1} L.y. can exist.

In the following, the properties of soliton stars, more specifically fermions soliton stars, are reviewed[3]. In particular, it will be explained how these properties are strongly dependent on the choice of parameters in the Lee-Wick model. For example, it will be shown that the features of the soliton stars described in reference[1] were obtained with a very special choice of parameters in the Lee-Wick model, and a different choice can lead to soliton stars with very different properties, more like those of neutron stars. Then the possibility of a phase transition between these two types of soliton stars when a temperature dependence is introduced in the Lee-Wick model, is discussed. This gives some indications on a possible mechanism of evolution of Lee-Wick soliton stars from the early Universe.

---

* Laboratoire Associé au C.N.R.S.

Particle Physics: Cargèse 1989
Edited by M. Lévy *et al.*
Plenum Press, New York, 1990

The Lee-Wick model is defined by the following Lagrangian

$$\pounds(x) = \frac{1}{2}\partial_\mu\sigma\partial^\mu\sigma - U(\sigma) + \bar{\psi}\not{\partial}\psi - m\bar{\psi}\left(1 - \frac{\sigma}{\sigma_o}\right)\psi \qquad (1)$$

where the fermions $\psi(x)$ interact via the scalar field $\sigma(x)$.

The self interaction of the σ field usually given in the form $U(\sigma) = \frac{a\sigma^2}{2} + \frac{b}{3!}\sigma^3 + \frac{c}{4!}\sigma^4$ can be rewritten as

$$U(\sigma) = \frac{1}{2} m_\sigma^2 \sigma^2 \left(1 - \frac{\sigma}{\sigma_o}\right)^2 + B\left[4\left(\frac{\sigma}{\sigma_o}\right)^3 - 3\left(\frac{\sigma}{\sigma_o}\right)^4\right] \qquad (2)$$

The shape of $U(\sigma)$ is shown in figure 1 where the minimum at σ=0 is assigned to the vacuum.

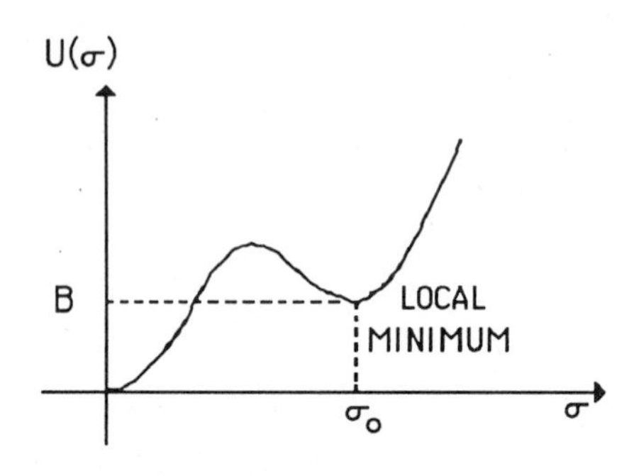

Fig. 1

Note that in eq. (1) one can consider that fermions have an effective mass $M(r) = m\left[1 - \frac{\sigma(r)}{\sigma_o}\right]$. Variations of the energy with respect to the σ and ψ fields lead to the Euler-Lagrange equations for ψ and σ.

The Lee-Wick model has been used to describe various physical phenomena. For example, abnormal nuclear matter[2)] where the fermions are nucleons with zero effective mass M(r), or hadron spectroscopy[4),5)]. In the latter case, the fermions are quarks with an effective mass which is vanishing inside the soliton (where $\sigma = \sigma_o$) and very large outside (m : large and σ = 0) . The field σ is assumed to account for all non perturbative effects of the gluon fields in the QCD Lagrangian, and thus the Lagrangian of eq. (1) is an effective Lagrangian for quarks confined by the σ field. Supplemented by the one gluon exchange interaction which is assumed to account for the residual interaction between quarks, the model is used for the calculation of the hadron mass spectra and static properties. The results obtained are in fairly good agreement with experiment.

SOLITON STARS

In reference 1), the following questions were raised.

i) what happens if the number N of fermions inside the soliton is very large $(N > 10^{60})$?

ii) can one still have a stable configuration for the soliton ?

iii) what are the effects of gravity and in which conditions a soliton star collapses into a black hole ?

In the absence of gravitation, the soliton total energy is given by

$$E = E(f) + E(\sigma) \tag{3}$$

where E(f) is the energy of the fermions and $E(\sigma)$ the energy of the σ field.

1) The fermion energy E(f)

N being very large, one can adopt the Thomas-Fermi approximation. In this case, the fermion number density is given by

$$\nu = \frac{2}{8\pi^3} \int d^3k \; n_k \tag{4}$$

$$\text{with} \quad n_k = \theta\,(k - k_F) = \begin{cases} 0 & \text{if } k > k_F \\ 1 & \text{if } k < k_F \end{cases}$$

the fermion energy density by

$$w = \frac{2}{8\pi^3} \int d^3k \; E_k \, n_k \tag{5}$$

$$\text{with} \quad E_k = \left[k^2 + m^2 \left(1 - \frac{\sigma}{\sigma_o} \right)^2 \right]^{1/2}$$

If k_F is assumed to be a smooth function of r, and taken as constant the total fermion number is

$$N = 4\pi \int r^2\, dr\, v = 4\pi \int_0^R r^2 dr \frac{1}{\pi^2} \int_0^{k_F} k^2 dk = \frac{4}{\pi} \frac{R^3}{3} \frac{k_F^3}{3} \tag{6}$$

where R is the soliton radius.

The total fermion energy is

$$E(f) = 4\pi \int r^2 dr\, w = \frac{1}{\pi} \frac{R^3}{3} k_F^4 \tag{7}$$

since the fermion effective mass is zero inside the soliton.

Using eq. (6) to eliminate k_F, one gets

$$E(f) = \frac{1}{2} \left(\frac{3}{2}\right)^{5/3} \pi^{1/3} \frac{N^{4/3}}{R} \tag{8}$$

2) <u>The σ field energy $E(\sigma)$</u>

$$E(\sigma) = \int d^3x \left[\frac{1}{2} (\vec{\nabla}\sigma)^2 + U(\sigma)\right] \tag{9}$$

where $U(\sigma)$ is given by eq. (2).

In the mean field approximation it is assumed

$$\sigma = \sigma_0 \quad \text{for} \quad r < R$$
$$\text{and } \sigma = 0 \quad \text{for} \quad r > R \tag{10}$$

Actually, the change of the σ field from the value σ_0 to 0 can be made to occur in a shell of width d as represented in figure 2.

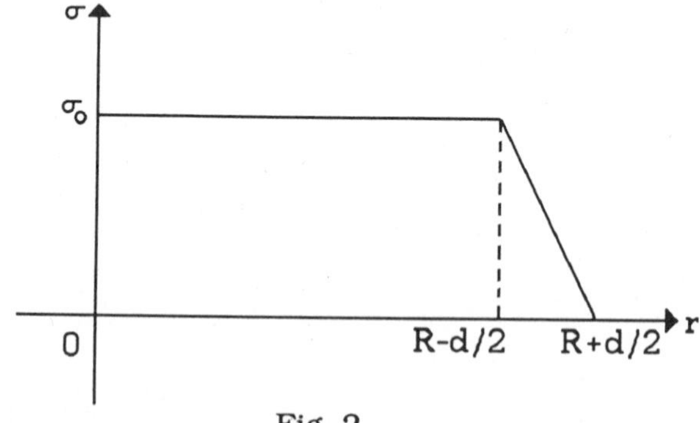

Fig. 2

In this case, the contributions from the kinetic term $(\vec{\nabla}\sigma)^2$ and from the first term of U(r) in eq. (2) to the energy $E(\sigma)$ can be shown to be of the surface-type, with the value $4\pi\, m_\sigma\, \sigma_0^2\, R^2/\sqrt{30}$. In contrast, the contribution due to the second term of $U(\sigma)$ in eq. (2) is of the volume-type with the value $\frac{4\pi}{3} B\, R^3$.

This gives for the soliton total energy

$$E = a\frac{N^{4/3}}{R} + 4\pi s R^2 + \frac{4\pi}{3} B\, R^3 \tag{11}$$

with $\quad a = \frac{1}{2}\left(\frac{3}{2}\right)^{5/3} \pi^{1/3}$ and $s = \frac{m_\sigma\, \sigma_0^2}{\sqrt{30}} \simeq \frac{1}{6} m_\sigma\, \sigma_0^2$

The soliton is stabilized by the surface tension s and the volume pressure B.

3) The Soliton Mass and Radius

In reference 1), the constant B is taken to be zero in the self-interaction $U(\sigma)$. With B=0 the vacuum is degenerate as shown in figure 3. It is this property along with the assumption of massless fermions that leads to very large soliton stars, as shown below.

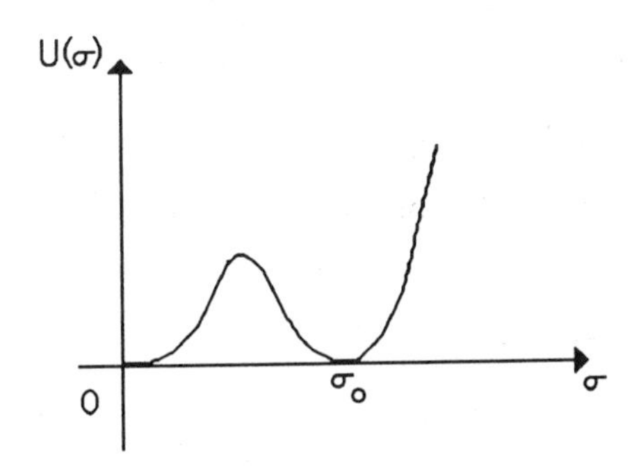

Fig. 3

Effectively, minimizing the soliton energy (eq. (11) with B=0) with respect to R leads to a soliton mass of

$$M = \frac{3}{2} a^{2/3} (8\pi s)^{1/3} N^{8/9} \tag{12}$$

with a radius

$$R = \left(\frac{a}{8\pi s}\right)^{1/3} N^{4/9} \qquad (13)$$

and a number density

$$\rho = \frac{6s}{a} N^{-1/3} \qquad (14)$$

As pointed out in reference 1), the exponent of N in eq. (12) is <1. This ensures, for large N, the stability of the soliton against decay into N free particles.

The effects of gravitation were calculated in reference 1) via general relativity. The detailed analyzis can be found therein. However, if one is interested only in the stablity conditions of the soliton against gravitational collapse, it is sufficient to use the Schwarzschild limit.

Taking $s = \frac{1}{6} m_\sigma \sigma_0^2$ with $m_\sigma = \sigma_0 = 30$ GeV as in reference 1), the Schwarzschild limit $R_c \sim 2GM_c$ together with eqs. (12)-(13) leads to the critical fermion number $N_c \sim 10^{76}$ which, in turn, gives a critical mass $M_c \sim 10^{13}\, M_\odot$ and $R_c \sim$ L.y . Thus, when B = 0, a soliton star can reach the mass of a galaxy before becoming gravitationally unstable to black hole formation. The density ρ is nevertheless very low for large N.

If, however, the condition for a degenerate vacuum is relaxed and one takes B > 0, then for large N, actually for $N > (4\pi s)^3 \left(\frac{3\pi}{4B}\right)^{9/4} a^{3/4}$, the volume term in eq. (11) dominates the surface term and one gets a soliton mass of

$$M = \frac{4}{3} a^{3/4} (4\pi B)^{1/4} N \qquad (15)$$

with a radius

$$R = \left(\frac{a}{4\pi B}\right)^{1/4} N^{1/3} \qquad (16)$$

Since the mass is now proportional to N, the soliton is no longer stable against dispersion. Moreover, taking $B^{1/4} = 100$ MeV (the value commonly used in hadron spectroscopy) one gets, from the Schwarzschild limit and eqs. (9) and (10), $N_c \sim 10^{58}$ which yields $M_c \sim 3M_\odot$, and $R_c \sim 30$ km. These characteristics are more like those of neutron stars. The density in this case is much larger than in the previous case.

Thus, the properties of soliton stars depend dramatically on the choice of the self-interaction $U(\sigma)$ in the Lee-Wick model. The soliton stars can have a galatic mass in one case, and a solar mass in the other case.

The above analyzis has been made at zero temperature. We now explore the possibility of a phase transition between the two types of solutions when a temperature dependence is introduced in the Lee-Wick model.

SOLITON STARS AT FINITE TEMPERATURE

Let us consider the case where at zero temperature the volume term in eq.(11) dominates the surface term. As shown above, a "small" soliton star can be formed. Its total energy is

$$E = a\frac{N^{4/3}}{R} + \frac{4\pi}{3} BR^3 \tag{17}$$

When the methods for introducing temperature dependence in field theories are applied[6] to the Lee-Wick Lagrangian (eq.(1)), and in the high temperature approximation, the free energy of the system at a temperature $k_B T = 1/\beta$ is given by

$$F = \frac{4\pi}{3} R^3 \left\{ B - \frac{1}{\beta^4}\left[\frac{7\pi^2}{180} - \frac{3}{2}(\beta^3 \rho)^2 + O\left((\beta^3\rho)^4\right)\right]\right\} \tag{18}$$

Minimizing F gives the minima for the free energy and the density

$$Fmin = \sqrt{6}\left(B - \frac{7\pi^2}{180\beta^4}\right)^{1/2} \beta N \tag{19}$$

$$\rho_{min} = \sqrt{\frac{2}{3}}\left(B - \frac{7\pi^2}{180\beta^4}\right)^{1/2} \frac{1}{\beta} \tag{20}$$

These quantities vanish at the critical temperature $k_B T_c = \frac{1}{\beta_c} = \left(\frac{180B}{7\pi^2}\right)^{1/4}$. For $B^{1/4} \sim 100$ MeV, $T_c \sim 100$ MeV. Thus, near the critical temperature, one must include the surface energy to stabilize the soliton, and from eq. (18)

$$F = \frac{9}{8\pi} \beta_c^2 \frac{N^2}{R^3} + 4\pi s R^2 \tag{21}$$

The minimum now becomes

$$F_{min} = 5\left(\frac{3\pi}{4}\right)^{1/5} \beta_c^{4/5} s^{3/5} N^{4/5} \tag{22}$$

$$\rho_{min} = \frac{4}{3}\left(\frac{3\pi}{4}\right)^{1/5} (\beta_c)^{-6/5}\, s^{3/5}\, N^{-1/5} \tag{23}$$

$$\text{for} \quad R = \left(\frac{3}{4}\right)^{3/5} \left(\frac{1}{\pi}\right)^{2/5} \left(\frac{\beta_c^2}{s}\right)^{1/5} N^{2/5} \tag{24}$$

The exponent of N in eq. (22) is <1, ensuring the stability of the star against dispersion, and from eq. (23) the density decreases with N. The soliton star now possesses the same characteristics as those considered in reference 1) but at a temperature T_c.

Such a soliton star could have been formed in the early Universe at a time $\sim 10^{-5}$ sec after the big bang corresponding to the critical temperature $T_c \sim 100$ MeV.

I wish to thank Dr. S. Bahcall for having shown me his work on Q-stars[7] during the course of this School.

REFERENCES

1) T.D. Lee, Phys. Rev. D35, 3637 (1987); R. Friedberg, T.D. Lee and Y. Pang, Phys. Rev. D35, 3640 (1987); Phys. Rev. D35, 3658 (1987); T.D. Lee and Y. Pang, Phys. Rev. D35, 3678 (1987).

2) T.D. Lee and G.C. Wick, Phys. Rev. D9, 2291 (1974).

3) Boson soliton stars have been studied by R. Friedberg, T.D. Lee and Y. Pang of ref. 1) and by R. Ruffini and S. Bonazzola, Phys. Rev. 187, 1767 (1969); W. Thirring, Phys. Lett. B127, 27 (1983); E. Takasugi and M. Yoshimura, Z. Phys. C26, 241 (1984); J.D. Breit, S. Gupta and A. Zaks, Phys. Lett. B140, 329 (1984); M. Colpi, S.L. Shapiro and I. Wasserman, Phys. Rev. Lett. 57, 2485 (1986); J. van der Bij and M. Gleiser, Phys. Lett. B194, 482 (1987); M. Gleiser, Phys. Rev. D38, 2376 (1988); S. Selipsky, D.C. Kennedy and B.W. Lynn, Nuclear Physics B321, 430 (1989); B.W. Lynn, Nuclear Physics B321, 465 (1989).

4) R. Friedberg and T.D. Lee, Phys. Rev. D15, 1694 (1977); D16, 1096 (1977).

5) See, for example, L. Wilets in Chiral Solitons edited by K.F. Liu, World Scientific, p. 362 (1988) and references cited therein.

6) For the detailed calculations see W.N. Cottingham and R. Vinh Mau. In preparation.

7) S. Bahcall, B.W. Lynn and S.B. Selipsky, Stanford University Preprints SU-ITP-854 and 856 (1989).

GLUON CONFINEMENT IN CHROMOELECTRIC VACUUM

Rahul Basu

The Institute of Mathematical Sciences
Taramani, Madras 600 113 INDIA

While the study of perturbative QCD has yielded important insights into the nature of the strong force at least in the high energy/momentum region, there are still outstanding problems in the low energy sector. For example, the Bloch-Nordsieck mechanism[1] for cancellation of infra-red(IR) divergences doesn't in general work in QCD (unlike QED)[2] unless one is looking at 'suitably' inclusive cross sections. The main reason for this is the presence of interacting massless particles which typically give rise to infra-red or collinear singularities. It has been suggested that if we could do perturbation theory about a different vacuum which has no interacting massless particles then such problems could be ameliorated.

The work of Shifman et. al.[3] has also demonstrated (from a study of charmonium decay analysis) that the gluon condensate in QCD has a non-zero expectation value. This result cannot be obtained from perturbative QCD. Various attempts have been made to construct a QCD vacuum that would give a non-zero value for the gluon condensate. One of the earlier attempts in this direction was that of Saviddy[4] who constructed a chromomagnetic vacuum for an SU(2) theory with a non-zero value for the gluon condensate. The Saviddy vacuum was lower than the perturbative vacuum and at a non-zero value of the chromomagnetic field. However it was soon demonstrated by Nielsen and Oleson[5] that the Saviddy vacuum was unstable due to the presence of tachyons. A number of attempts have

Particle Physics: Cargèse 1989
Edited by M. Lévy *et al.*
Plenum Press, New York, 1990

since been made[6] to construct a stable vacuum for QCD with the above desirable properties but usually these have had the usual feature of many massless interacting modes leading to the usual IR problems. However a common qualitative understanding that has arisen from all these studies is that any attempt to define the QCD vacuum should include the effects from the cubic and particularly the quartic interaction terms (which are usually dropped in the Gaussian approximation) in the calculation of the one-loop correction to the perturbative vacuum.

Keeping in view all the wisdom gleaned from these earlier studies, we have developed a procedure for studying non-Abelian gauge theories in general using Minkowski functional integral technique. This corresponds to a first order formalism for gauge theories. It has the virtue of incorporating the effects of the cubic and quartic gluon vertices in defining the perturbative vacuum. In an earlier study[7] we examined the chromomagnetic vacuum of the Saviddy type and demonstrated that one can get a stable vacuum for a range of the coupling constant. However this vacuum is not so useful since it breaks Lorentz invariance.

In this paper we would like to present a perturbation theory for pure QCD which by construction is manifestly Lorentz covariant. In gauge theories this is not a necessary constraint for the gauge non-invariant sector (gauge generators and Lorentz generators do not commute with each other). But this choice proves to be useful in performing explicit calculations. This perturbation theory preserves both (spatial) rotational and global colour invariance manifestly. We look at the chromoelectric condensate vacua and find that in general, a few colour degrees of freedom are absent as physical excitations. This is a new mechanism for confinement of colour degrees of freedom. Next we address the IR problem in the theory and find that if we resum the naive perturbation theory and consider the Dyson or 1PI perturbation theory, then there is considerable improvement in the IR region.

Before we explain our method a remark is in order about Euclidean functional integral (EFI). The gluon condensate in

EFI is positive and any analytical continuation to Minkowski space keeping this fixed can at most yield only chromomagnetic vacua. In Minkowski space the gluon condensate can be of either sign, therefore only Minkowski functional integral can be used to study chromoelectric vacuum.

We begin with a brief review of the Minkowski functional integral for pure non-Abelian gauge theory[8]:

$$Z = \int [dA_\mu] \exp\left\{\frac{i}{4g^2}\int d^4x\, F^a_{\mu\nu} F^{\mu\nu a}\right\} \tag{1}$$

where

$$F^a_{\mu\nu} = \partial_\mu A^a_\nu - \partial_\nu A^a_\mu + f^{abc}A^b_\mu A^c_\nu \tag{2}$$

and use the metric $\eta^{\mu\nu}=(+,-,-,-)$ in 4 dimensions. This functional integral can be rewritten upto an overall g^2 dependent term as

$$Z = \int [dA_\mu][dG_{\mu\nu}]\exp\left[\frac{-i}{4}\int d^4x\left\{g^2 G^a_{\mu\nu}G^{\mu\nu a} - 2G^a_{\mu\nu}F^{\mu\nu a}\right\}\right] \tag{3}$$

where we have introduced an auxiliary field $G^a_{\mu\nu}$ and used

$$1 = \int [dG_{\mu\nu}]\exp\left[\frac{-i}{4}\int d^4x\left\{gG^a_{\mu\nu} - \frac{1}{g}F^a_{\mu\nu}\right\}^2\right] .$$

We already notice from the form of this functional integral that treating $G^a_{\mu\nu}$ as a fundamental field, the only interaction term is $G^a_{\mu\nu}F^{\mu\nu a}$ which is cubic in the interaction. There is no quartic term. Its effect has been absorbed in the $G^a_{\mu\nu}G^{\mu\nu a}$ term which can be seen from the the classical equations of motion for the $G^a_{\mu\nu}$ field $G^a_{\mu\nu} = F^a_{\mu\nu}/g^2$.

Next we perform the usual background field calculation wherein we expand A^a_μ and $G^a_{\mu\nu}$ about some background $\bar{A}^a_\mu$ and $\bar{G}^a_{\mu\nu}$ and calculate the one-loop effective potential by retaining terms quadratic in the quantum fields. We thus have

$$\begin{aligned} A^a_\mu &= \bar{A}^a_\mu + a^a_\mu \\ G^a_{\mu\nu} &= \bar{G}^a_{\mu\nu} + \mathcal{G}^a_{\mu\nu} \end{aligned} \tag{4}$$

We choose $\bar{A}^a_\mu$ and $\bar{G}^a_{\mu\nu}$ to be space-time independent.

Specifically $\bar{A}^a_\mu$ and $\bar{G}^a_{\mu\nu}$ can be considered to be the zero momentum components of the Fourier expansion of the A^a_μ and $G^a_{\mu\nu}$

fields and a^a_μ and $\mathcal{G}^a_{\mu\nu}$ the non-zero momentum Fourier component. The most general space-time independent $\bar{G}^a_{\mu\nu}$ in four dimensions is

$$\bar{G}^a_{\mu\nu} = e(n_\mu E^a_\nu - n_\nu E^a_\mu) + h\,\varepsilon_{\mu\nu\lambda\sigma} n^\lambda H^{\sigma a} \tag{5}$$

where n_μ is a time-like vector normalised to unity, $n^2 = 1$, e and h are scalar constants and E^a_μ ,H^a_μ are group vectors with the property

$$n.H^a = 0 = n.E^a$$

In general one can proceed with an ansatz such as in (5) but we find that typically if $h \neq 0$ there are tachyons present thereby invalidating the expansion. We will therefore restrict ourselves to $e \neq 0$ and $h = 0$. Note that we make a unique choice of E^a_μ upto (colour) group rotations and space rotations.

This functional integral needs to be gauge fixed. We find it convenient to work in the background gauge and its associated ghost fields. Substituting (4) in (3) and including the gauge fixing and ghost terms, the partition function has linear, quadratic and cubic (interaction) terms only besides the tree level contribution. As we are interested in constructing a perturbation theory having only 1PI graphs, the linear terms are dropped[9] . Even otherwise here, the only linear term is of the form $\int \bar{\mathcal{G}}_{\mu\nu} G_{\mu\nu}$ which (by the orthogonality condition between zero and non-zero Fourier components of the $G_{\mu\nu}$ field) is zero. The appropriate partition function up to one-loop is thus given by

$$Z=\int d\bar{G}_{\mu\nu}\, d\bar{A}_\mu \exp\left[\frac{-i}{4}\int d^4x g^2 \bar{G}^a_{\mu\nu}\bar{G}^{\mu\nu a} - 2\bar{G}^a_{\mu\nu}\bar{F}^{\mu\nu a}\right]$$

$$\int [d\bar{c}\, dc][da_\mu][d\mathcal{G}_{\mu\nu}] \exp\left[\frac{-i}{4}\int d^4x\left\{g^2\mathcal{G}^a_{\mu\nu}\mathcal{G}^{\mu\nu a} - 2f^{abc}\bar{G}^a_{\mu\nu}a^{\mu b}a^{\nu c}\right.\right.$$

$$\left.\left. -2\mathcal{G}^a_{\mu\nu}(\bar{D}^{\mu ab}a^{\nu b} - \bar{D}^{\nu ab}a^{\mu b}) - 2(\bar{D}^{ab}_\mu a^{\mu b})^2 - 4\bar{c}^a\,\partial_\mu\,\bar{D}^{\mu ab}c^b\right\}\right] \tag{6}$$

where $\bar{D}^{ab}_\mu = \partial_\mu \delta^{ab} + f^{acb}\bar{A}^c_\mu$.

and the interaction term in this perturbation theory is

$$\exp\left[\, i \int d^4x \left\{ \frac{1}{2} f^{abc} \mathcal{G}^a_{\mu\nu}\, a^{\mu b} a^{\nu c} + f^{abc} \bar{c}^a\, a^b_\mu \bar{D}^{\mu cd} c^d \right\} \right] \tag{7}$$

We have explicitly retained the ordinary integral over $\bar{G}^a_{\mu\nu}$ and $\bar{A}^a_\mu$ for clarity. We make the ansatz $\bar{A}^a_\mu = 0$ to maintain manifest translation invariance. Then the $d\bar{A}_\mu$ integration is trivial and the $d\bar{G}^a_{\mu\nu}$ integral is performed by the stationary phase approximation, viz. minimising the effective potential with respect to e. It is evident that the remaining $d\bar{G}_{\mu\nu}$ integrations are over the compact space rotation group and colour rotation group which enforce rotational and global colour invariance on all Green's functions. Concentrating first on the functional integral we make the following definition

$$Z_0 \equiv \int [d\mathcal{G}_{\mu\nu}][da_\mu][d\bar{c}dc] \exp\left[\frac{-i}{4}\int d^4x \left\{ g^2 \mathcal{G}^a_{\mu\nu}\mathcal{G}^{\mu\nu a} - 2f^{abc}\bar{G}^a_{\mu\nu} a^{\mu b} a^{\nu c} \right.\right.$$
$$\left.\left. - 2\mathcal{G}^a_{\mu\nu}(\partial^\mu a^{\nu a} - \partial^\nu a^{\mu a}) - 2(\partial_\mu a^{\mu a})^2 - 4\bar{c}^a \partial_\mu \partial^\mu c^a \right\}\right] \tag{8}$$

We first perform the $[d\mathcal{G}_{\mu\nu}]$ integral and then the $[da_\mu]$ integral. From this partition function (8) the various propagators in momentum space are found to be

$$\langle a^a_\mu a^b_\nu \rangle_0 = -\frac{ig^2}{p^2+i\varepsilon}\left[\eta_{\mu\nu}\delta^{ab} - n_\mu n_\nu \delta^{ab} - (E_\mu \frac{1}{E^2} E_\nu)^{ab}\right]$$
$$-i\left(\frac{g^2}{p^4+g^4e^2E^2}\right)^{ac}\left[n_\mu n_\nu\, p^2\, \delta^{cb} + (E_\mu \frac{p^2}{E^2} E_\nu)^{cb} + eg^2(n_\mu E^{cb}_\nu - n_\nu E^{cb}_\mu)\right] \tag{9}$$

$$\langle \mathcal{G}^a_{\mu\nu} \mathcal{G}^b_{\alpha\beta} \rangle_0 = \frac{-i}{g^2}(\eta_{\mu\alpha}\eta_{\nu\beta} - \eta_{\nu\alpha}\eta_{\mu\beta})\delta^{ab} - \frac{1}{g^4}\left[p_\mu p_\alpha \langle a^a_\nu a^b_\beta\rangle_0 \right.$$
$$\left. - p_\mu p_\beta \langle a^a_\nu a^b_\alpha \rangle_0 - p_\nu p_\alpha \langle a^a_\mu a^b_\beta \rangle_0 + p_\nu p_\beta \langle a^a_\mu a^b_\alpha \rangle_0 \right] \tag{10}$$

$$\langle \mathcal{G}^a_{\mu\nu} a^b_\lambda \rangle_0 = \frac{i}{g^2}\left[p_\mu \langle a^a_\nu a^b_\lambda \rangle_0 - p_\nu \langle a^a_\mu a^b_\lambda \rangle_0 \right] \tag{11}$$

$$< \bar{c}^a \, c^b >_0 = \frac{i}{p^2+i\varepsilon} \, \delta^{ab} \tag{12}$$

We have the following definition for the matrix E_μ.

$$(E_\mu)^{ab} = f^{acb} E^c_\mu \tag{13}$$

and noting that only E_i are non-vanishing, we see that there is always a choice for E_i satisfying an SU(2) algebra

$$[E_i, E_j] = \varepsilon_{ijk} E_k \; . \tag{14}$$

For definiteness we have picked our explicit ansatz for E_i to be an SU(2) subgroup of SU(N), viz. $(E_1)^{ab} = 2f^{2ab}$, $(E_2)^{ab} = 2f^{4ab}$, and $(E_3)^{ab} = 2f^{6ab}$. Note that since $E_0 = 0$ from (5)

$$[E^2, E_\mu] = 0 \tag{15}$$

where $E^2 \equiv E_\mu E^\mu$ is a positive matrix with our choice of metric. For SU(3) we can choose E_i such that $E^2 > 0$.

We now make a few remarks about the propagators. The unit vector n_μ in (9) which originated in our ansatz (5) appears to destroy Lorentz invariance. However this is illusory since the propagator (9) can be completely expressed in terms of $G^a_{\mu\nu}$ and $(G^a_{\mu\nu})^2$. The $<a^a_\mu a^b_\nu>_0$ propagator shows that all the poles give manifestly Lorentz invariant dispersion relations. For SU(3) there are 8(d-2) massless poles which are regulated by the 'Feynman $i\varepsilon$' prescription. The remaining are all 'unstable' poles, viz. in the complex energy plane they lie off the real axis if $e \neq 0$ and hence in position space this propagation decays with time. Thus these propagate only for intermediate times and do not correspond to asymptotic states. We will have more to say on this matter further on.

We next analyse (9) at short distances or large p^2 behaviour. We find that

$$<a^a_\mu a^b_\nu >_0 \xrightarrow[p^2 \to \infty]{} - \frac{ig^2}{p^2} \eta_{\mu\nu} \, \delta^{ab} \tag{16}$$

the standard Feynman gauge propagator. In other words, at short distances all the degrees of freedom are manifested. The other propagators in this theory eq.(10), (11) and (12) have similar properties.

In this perturbation theory we have only a momentum independent interaction given by eq.(7) (apart from the standard ghost term). As to the interpretation of the propagating modes in (9), they cancel against the ghost modes. For example, in computing the one-loop effective action the contribution from propagating modes exactly cancels against the ghost modes (without any UV regularisation scheme). Note that the propagators (9) to (12) along with (7) for the case $e = 0$ generate the usual perturbation theory for QCD.

We now compute the 1PI effective action using dimensional regularisation scheme in Minkowski space. This is given by the following expression.

$$\Gamma^{ren}_{eff} = \frac{g^2 e^2 \mathrm{Tr} E^2}{2N} + \frac{g^4}{16\pi^2} \mathrm{Tr} \left[e^2 E^2 \ell n \left(\frac{e^2 E^2}{\mu^4} \right) \right] \tag{17}$$

Minimising with respect to e^2, and solving for e^2, we get

$$e^2_{min} = \mu^4 \exp \left[- 1 - \frac{8\pi^2}{Ng^2} - \frac{\mathrm{Tr}\ E^2 \ell n\ E^2}{\mathrm{Tr}\ E^2} \right] \tag{18}$$

which gives us the non-zero value of e about which we have expanded in our perturbation series.

We now make a few remarks about our technique. Our construction of a new perturbative vacuum was motivated by the IR problems plaguing standard perturbation theory. In the process of constructing a stable vacuum with gluon condensate (chromoelectric) we find a new mechanism for confinement namely some degrees of freedom of the gluons propagate only for short distances or time scales. This is manifested by the poles of the propagator being off the real axis in the complex energy plane. This is a natural consequence of the chromoelectric condensate. Basically the energy momentum dispersion relation for these gluons are of the type

$$p_0^2 - p^2 = \pm ie$$

In space-time configuration the propagators of these modes go as

$$\exp(-|et|)\ \exp(\pm i|p|t - i\, p.x)$$

so that the amplitude vanishes when $t \rightarrow \pm \infty$. Thus these modes cease to exist as asymptotic states and are therefore irrelevant to the standard S matrix. Interestingly enough for large p^2 (UV region) there are off-shell effects (16) suggesting that the parton model picture can still be at work. The net picture of these excitations of the vacuum is that certain degrees of freedom are confined and do not exist as asymptotic or long-lived states but all the basic degrees of freedom are present at short distances. This picture of confinement although new can be reconciled with the old understanding of the ultraviolet behaviour of QCD.

Let us then summarise what we have shown in this paper. We have constructed a stable vacuum for pure QCD with a chromoelectric gluon condensate and in the process found a new mechanism for confinement *viz.* some degrees of freedom of the gluons propagate only over short distances or time scales. We have also discovered considerable IR softening of the theory at least to one-loop order since to this order all the massless poles cancel against the ghost modes. This is because as we had mentioned earlier we have only one cubic interaction term in (7) (apart from ghosts) involving the massless modes in contrast to usual perturbation theory which has quartic terms. Whether the above cancellation persists to all orders needs to be checked with the full Slavnov-Taylor identities. Note that if the bare propagators (9), (10) and (11) were such that all modes propagate over short distances then one could conclude that in the Gaussian perturbation theory gluons and gluonia do not propagate. However in our case this is not so. We have here a situation wherein colour singlet gluonia can still propagate with confinement of colour degrees of freedom.

A couple of caveats should be mentioned here. The results we have obtained are relevant only to non-Abelian gauge theories but not to a U(1) theory where the background electric field can couple only to some charged particle like the electron. Such a coupling destabilises the vacuum due to the production of e^+e^- pairs as was shown by Schwinger[10]. However, in non-Abelian theories we have envisaged a background electric field coupling only to the dipole moment of the gluon. This, as shown, is stable.

The second caveat is that the 'first order' formalism that we have used here for the construction of the chromoelectric vacuum does not work for the Euclidean functional integral. This is easily seen by harking back at our technique of introducing the auxiliary field (eq(1)-(3)).

We conclude by drawing attention to some of the more obvious open problems that have not been addressed here. The presence of complex poles necessitates rechecking the UV renormalisability and unitarity of the theory though we do not envisage any problem in this regard. An interesting fact that should be mentioned here is that the coupling constant g is only in the propagator (i.e. wave function). The bare vertex is independent of momenta and coupling constant. Whether this feature survives the loop expansion needs to be checked with the Slavnov-Taylor identities. Finally we have to look at the full theory with fermions to see if a similar mechanism for confinement works in the fermionic sector. Work is presently in progress on this last mentioned aspect of the theory.

Acknowledgements

The work reported here was done in collaboration with R. Anishetty and R. Parthasarathy. I would like to thank the organisers of the 1989 Summer School in Particle Physics at Cargèse (Corse) France for inviting me to the school and giving me the opportunity to present my work. In particular, I would like to express my gratitude to Professor Raymonds Gastmans and Ms. Marie-France Hanseler for their excellent organisation and hospitality.

References

1. F. Bloch and A. Nordsieck, Phys. Rev. **52**(1937)54.
2. R. M. Doria, J. Frenkel and J. C. Taylor, Nucl. Phys. **B168** (1980)93.
3. M. A. Shifman, A. I. Vainshtein and V. I. Zakharov, Nucl. Phys. **B147** (1979) 385.
4. G. K. Saviddy, Phys. Lett. **B71** (1977)133.

5. N. K. Nielsen and P. Oleson, Nucl. Phys. **B144**(1978) 376.
6. H. Leutwyler, Nucl. Phys. **B179**(1981)129; S. L. Adler, Phys. Rev. **D23**(1981)2905; Phys. Lett. **B110**(1982)302.
7. R. Anishetty, R. Basu and R. Parthasarathy, IMSc. preprint # IMSc/88/24 - to appear in Jour. Phys. G.
8. R. Anishetty, R. Basu and R. Parthasarathy, IMSc. preprint # IMSc/89/5 - submitted for publication.
9. R. Jackiw, Phys. Rev. **D9**(1974)1686.
10. J. Schwinger, Phys. Rev. **82**(1951) 664.

MEAN FIELD THEORY AND BEYOND FOR GAUGE-HIGGS-FERMION THEORIES

Stephen W. de Souza

Physics Department
University of Edinburgh
Mayfield Road
Edinburgh EH9 3JZ
Scotland

1 INTRODUCTION

The reasons for giving this paper here are as follows. I hope to complement Martinelli's lectures in two ways. This calculation will show that it is possible to do analytic work on the lattice. Also I will describe the other common method of tackling the fermion doubling problem. One of the main equations which I will derive is the gap equation, which has already been discussed by Bjorken at this school. Now for the physical motivations behind this work.

It is well known that gauge theories, with or without fermions, confine in the strong coupling limit. For $U(N)$ and $SU(N)$ this confinement is associated with the dynamical generation of mass. For a theory with a chirally symmetric lagrangian the fact that the composite fermions have a dynamical mass is equivalent to saying that the chiral symmetry is broken. This symmetry breaking is required for QCD to explain such features of the hadron spectrum as the large ratio of m_ρ/m_π. There is no evidence to suggest that this picture is qualitatively changed in the weak coupling limit, which is the regime of physical interest. In the electroweak sector of the standard model the gauge symmetry is spontaneously broken by the addition of scalar fields which couple to both the fermion and gauge fields. This symmetry breaking gives the observed gauge quanta a mass which can be regarded as a measure of the scale of the symmetry breaking. The Yukawa coupling of these scalars to the known fermions is meant to give the fermions a mass which is small relative to this scale. If the theory still confines, with the addition of the scalars, then the mass of the observed fermions might be expected to be not smaller than, but of the same order as, the scale set by the scalars. It is therefore of considerable importance to investigate chiral symmetry breaking for a gauge-Higgs-fermion model. Chiral symmetry breaking is usually a non-perturbative phenomenon and a suitable calculational scheme is to use the lattice regularisation.

2 THE LATTICE ACTION

We choose to study the theory on a d-dimensional hypercubic lattice, of side L, at zero temper-

Particle Physics: Cargèse 1989
Edited by M. Lévy *et al.*
Plenum Press, New York, 1990

ature. The lattice action (with colour indices suppressed) is,

$$S = S_F + S_H + S_G + S_m \tag{1}$$

with,

$$S_F = \frac{1}{2}\sum_x \sum_{l=1}^{d} [\eta_{x,l}\overline{\chi}(x)U(x,l)\chi(x+l) + h.c.] \tag{2}$$

$$S_H = -\beta_h \sum_x \sum_{l=1}^{d} [\phi^\dagger(x)U(x,l)\phi(x+l) + h.c.] \tag{3}$$

$$S_G = \frac{\beta_g}{4N} \sum_{plaquettes} Tr[UUU^\dagger U^\dagger] \tag{4}$$

$$S_m = m\sum_x [\overline{\chi}(x)\chi(x)] \tag{5}$$

The letter x labels the site; the total number of sites is n; l is a unit vector along the lth direction of the lattice. The lattice spacing has been set equal to one. The $\chi(x)$ and $\overline{\chi}(x)$ are N-component, anti-commuting, staggered fermion fields assigned to each site, they lie in the fundamental N-dimensional representation of the gauge group; the $\eta_{x,l}$ are the staggered fermion equivalent of Dirac gamma matrices. As Martinelli has explained each species of fermion on the lattice gives rise to 16 species (in 4 dimensions) of continuum Dirac fermions. This is the infamous fermion doubling problem. Staggered fermions are a partial resolution of this problem: they have only one spinor component, the number is made up to the four required to make a continuum Dirac fermion by using the doubles. This results in 4 species of continuum fermions. Staggered fermions are used in this study because they retain a continuous chiral symmetry and it is the dynamical breaking of this symmetry which we wish to study. (Wilson fermions explicitly break this symmetry on the lattice to beat the doubling problem and require fine-tuning to get the symmetry back in the continuum). The lattice Higgs fields, $\phi(x)$ and $\phi^\dagger(x)$, also lie in the fundamental representation and have fixed length: this does not imply that the continuum fields are also of fixed length. The $N \times N$ dimensional matrices $U(x,l)$ transform in the adjoint representation of some unitary gauge group. The action is chirally symmetric when the fermion mass term m is taken to zero. The mass term is merely inserted as a source for the order parameter $\langle\overline{\chi}\chi\rangle$ of chiral symmetry breaking.

3 INTEGRATION OVER GAUGE AND HIGGS FIELDS.

The partition function is,

$$Z(m) = \int \prod_x d\chi(x)d\overline{\chi}(x)exp[-S_m]\hat{Z} \tag{6}$$

where $\hat{Z}$ is the effective partition function for the fermions once the gauge and Higgs fields have been integrated over. Firstly we make a gauge transformation on each site to fix to unitary gauge, $\phi^\dagger\phi = 1$. The action no longer depends on the Higgs fields and the integration over them becomes trivial. The next step is to take the strong coupling limit, $\beta_g = 0$. This makes the effective partition function $\hat{Z}$ factor into a product of single link integrals. These integrals must give a result that is a gauge invariant object and this makes it simple to do them for any gauge group. The result is an effective action for the fermions containing higher than quadratic interactions between them. In fact due to the anticommuting nature of the fermion fields for the gauge group $U(N)$ or $SU(N)$ and f flavours of staggered fermions the highest interaction will contain $4fN$ fields. For simplicity we now specialise to the gauge group $U(1)$ and use only one flavour of staggered fermions (this corresponds to 4 flavours of Dirac fermions in the continuum limit). After doing the integration and re-exponentiating the result one arrives at an effective action for the fermions[1].

$$S_{eff} = -\frac{1}{4}(1-r^2)\overline{\chi}(x)\chi(x)\overline{\chi}(x+l)\chi(x+l) + \frac{1}{2}r\eta_{x,l}(\overline{\chi}(x)\chi(x+l) - \overline{\chi}(x+l)\chi(x)) \tag{7}$$

The dependence on the Higgs coupling enters through $r = I_1(2\beta_h)/I_0(2\beta_h)$. To proceed further it is necessary to approximate the effect of the four-fermi term. The mean field approximation [1] involves the replacement

$$\sum_{x,l} \overline{\chi}(x)\chi(x)\overline{\chi}(x+l)\chi(x+l) = 2d\langle\overline{\chi}\chi\rangle \sum_{x} \overline{\chi}\chi. \tag{8}$$

The next section will develop a more systematic approach which reproduces the results of the mean field approximation at lowest order.

4 INTEGRATION OVER FERMIONS: THE $\frac{1}{d}$ EXPANSION

In this section we follow the approach of ref. [3] and integrate out the fermions exactly by introducing an auxiliary field and then do the functional integral over this field by the loop expansion. Introduce the variables $M(x) = \sqrt{\frac{d}{2}}\overline{\chi}(x)\chi(x)$ and $\overline{m} = \frac{m}{\sqrt{2d}}$, and the operator $V(x,y) = \frac{1}{2d}\sum_{l=1}^{d}[\delta_{x,y+l} - \delta_{x,y-l}]$. We can now linearise the M dependence of the effective action by introducing an auxiliary field λ. This is achieved by the following identity,

$$exp[\frac{1}{2}(1-r^2)\sum_{x,y} M(x)V(x,y)M(y)] =$$

$$\int \prod_{x} d\lambda(x) exp[\frac{-1}{2(1-r^2)}\sum_{x,y}[\lambda(x)V^{-1}(x,y)\lambda(y)] - \sum_{x}[\lambda(x)M(x)]] \tag{9}$$

If we now shift variables, $\lambda \to \lambda - 2\overline{m}$ then the fermions can be integrated out exactly to give,

$$Z = \int \prod_{x} d\lambda(x) exp[\frac{-1}{2(1-r^2)}\sum_{x,y}[(\lambda(x) - 2\overline{m})V^{-1}(x,y)(\lambda(y) - 2\overline{m})] + trln[D]] \tag{10}$$

where,

$$D(x,y) = \sqrt{\frac{d}{2}}\lambda(x)\delta_{x,y} - \frac{r}{2}\sum_{l}[\eta_{x,l}(\delta_{x,y+l} - \delta_{x,y-l})] \tag{11}$$

The functional integral over the auxiliary field can now be done approximately by the loop expansion. The leading order term is given by finding the stationary point of the action. This gives an implicit equation for λ_o the mean-field. Differentiating the partition function with respect to m gives $\lambda_o - 2\overline{m} = (1-r^2)\sqrt{\frac{d}{2}}\langle\overline{\chi}\chi\rangle$. This enables us to get an equation for $\langle\overline{\chi}\chi\rangle$ itself, which is the gap equation.

$$\langle\overline{\chi}\chi\rangle = \frac{G}{n}\sum_{p_l}[G^2 + r^2 S_2(p)]^{-1} \tag{12}$$

where $G = \frac{d}{2}(1-r^2)\langle\overline{\chi}\chi\rangle + m$, $S_2(p) = \sum_l sin^2\frac{\pi p_l}{L}$ and p_l runs over odd integers from $-(L-1)$ to $(L-1)$. This result agrees with the naive mean field calculation[1]. If the mass is set equal to zero chiral symmetry is broken for sufficiently small r and restored above $r_c = 0.553465$ (on an infinite lattice). Numerical simulations necessarily work with non-zero mass on a finite lattice. This equation predicts how $\langle\overline{\chi}\chi\rangle$ scales with both the mass and the lattice size and may therefore be used to extrapolate[2] the results of simulations to the physical limit.

The critical exponents may be extracted from this equation by expanding in the critical region (i.e. $\langle\overline{\chi}\chi\rangle$, $r^2 - r_c^2$ and m all small). The scaling equation is

$$A\frac{m}{\langle\overline{\chi}\chi\rangle^3} = B\frac{(1-r^2)^2}{r^2} + C\frac{r^2 - r_c^2}{\langle\overline{\chi}\chi\rangle^2}, \tag{13}$$

where A, B and C are r independent constants. This equation has Landau exponents viz. $\beta = \frac{1}{2}$ and $\delta = 3$.

Fluctuations around the mean-field may be included by using Laplace's method on the integral involved in the partition function. The first correction comes from the second derivative of the action $Z_{L=1} = [\det(-S'')]^{-\frac{1}{2}} = -\frac{1}{2}\mathrm{tr}[\ln(1 + VD^{-1}D^{-1})]$. This logarithm may now be expanded as a power series in $\frac{1}{d}$. Odd terms vanish and it is easy to see that successive non-vanishing (even) terms are down by powers of $\frac{1}{d}$. This gives the lowest order correction to the free energy per degree of freedom $W = \frac{1}{n}\ln[Z]$. The result agrees with previous calculation done for the limit $r = 0$[3]. Minimising the corrected free energy gives a new additive correction to the consistency equation.

The correction is the leading term in the $\frac{1}{d}$ expansion of the one loop correction. We now explain why the loop expansion is itself an expansion in $\frac{1}{d}$. As stated above the propagator arising from eq. 10 can be expanded in $\frac{1}{d}$; this expansion in fact starts at $\frac{1}{d}$. The vertices, which arise from the $\mathrm{trln}D$ term, depend on d through $\frac{r}{\sqrt{d}}$ so the L-contribution starts at most at order $\frac{1}{d^L}$. In the next section we discuss the results of solving the corrected consistency equation.

5 PRELIMINARY RESULTS AND CONCLUSIONS

The lowest order mean-field theory compares favourably with the results of the simulations. Together they show that the addition of scalars to a theory of interacting fermions at strong coupling leads to the existence of a chiral symmetry restoring transition. Chiral symmetry is restored at high values of the coupling. The fact that chiral symmetry is broken at low values of the coupling makes a confining version of the electroweak theory unlikely. The aim of this study was to improve the agreement between analytic and numerical work by going beyond leading order. Preliminary reults indicate that the lowest order corrections do not make a significant difference. Further details and a discussion of the infra red divergences of the correction will be presented elsewhere.

I would like to acknowledge Brian Pendleton, David Wallace and other members of the Edinburgh theory group for useful discussions. This work is supported by the SERC.

References

[1] I-H. Lee and R.Shrock, Chiral-Symmetry-Breaking Phase Transition in Lattice Gauge-Higgs Theories with Fermions, *Phys.Rev.Lett.* **59** (1987) 14

[2] A.Horowitz, Critical Exponents and Continuum Limit of the Chiral Transition in the U(1) Higgs Model, *Phys.Lett.* **219B** (1989) 329

[3] H.Kluberg-Stern, A.Morel, and B.Petersson, Spectrum of Lattice Gauge Theories with Fermions From a $\frac{1}{d}$ Expansion at Strong Coupling, *Nucl.Phys.* **B215[FS7]** (1983) 527

INDEX

Altarelli-Parisi equation, 229, 233
Aplanarity, 38
Asymmetry
  forward-backward, 19

Bjorken energy density, 83
Borel transform, 225
Born approximation, 112-115
b-quark
  b-$\bar{b}$ tagging, 12
  fragmentation, 10
  production, 59
Breit-Wigner shape, 34, 132
BRST charge, 168, 189

Cabibbo angle, 316
Cabibbo-Kobayashi-Maskawa matrix, 104, 290, 316
Calabi-Yau manifold, 207
Central extension, 160
Chan-Paton factor, 194
Chiral limit, 244
Chiral symmetry, 355-356
Compactification
  toroïdal, 191
Composite model, 2
Continuum limit, 244
Correlation length, 244-246
c-quark, 59

Debye color screening, 82, 90
Dedekind eta-function, 165, 188, 201
Dehn twist, 197
Di-lepton events, 18
Drell-Yan pair, 55, 89
Drift chamber, 27

Faddeev-Popov trick, 166
Fourth generation, 56

Gauss-Seidel method, 254
Ghost
  no- theorem, 161
  sector, 166
GIM cancellation, 278
Gluino, 66
Gluon, 79
  quark-gluon phase, 93
  structure function, 231
Goldstone boson, 246
Gravitational collapse, 340

Haar measure, 243
Hadron gas, 81, 93
Higgs boson, 48, 136
  charged, 42
  mass, 118
Higgs sector, 105ff
Hydrodynamic model, 90
Hypercubic group, 267

IR softening, 350

Jacobi theta-functions, 187, 202

Lagrange multiplier, 155
Lepton cross-sections, 59

Majorana spinor, 176, 180, 186
Majorana-Weyl spinor, 192, 202
Mass formula, 149, 152

Nambu-Goto action, 144-146, 154
Neutrinos
  number of light -, 36
Neveu-Schwarz sector, 178, 188, 204
Nonplanar events, 42-43

Penguin operator, 278, 289
Photino, 66
Plasma phase, 82
Polyakov action, 144
Pseudorapidity, 84

Quark-gluon phase, 93
Quark masses, 64
Quenched approximation, 273, 276

Radiative corrections, 50
Ramond sector, 181, 188, 204
Regge poles, 228, 235
Riemann identity, 188

Saviddy vacuum, 343
Scaling violation, 264
Schwinger term, 160
Sigma-field energy, 338
Squark, 66
SU(5) symmetry, 129
Supersymmetry, 111ff

Teichmüller space, 198, 200

Unitarity
  one-loop, 161
Uranium ions, 93

Virasoro generators, 149, 152, 156, 160, 172

Ward identities, 247
Weinberg angle, 36, 102, 117ff
Weyl invariance, 145, 201
Wilson term, 247, 249
W mass, 49, 57
Wolfenstein parametrization, 289
World-sheet gauge invariance, 145

Zeta-function regularization, 164, 223
Z-particle
  mass, 49, 57
  width, 2
Zweig rule, 254